An Indispensable Truth

Francis F. Chen

An Indispensable Truth

How Fusion Power Can Save the Planet

Francis F. Chen
Department of Electrical Engineering
University of California at Los Angeles
Los Angeles, CA
USA
ffchen@ee.ucla.edu

ISBN 978-1-4419-7819-6 e-ISBN 978-1-4419-7820-2
DOI 10.1007/978-1-4419-7820-2
Springer New York Dordrecht Heidelberg London

Library of Congress Control Number: 2011922489

Printed on acid-free paper

Springer is part of Springer Science+Business Media (www.springer.com)

Preface

Al Gore's book and video, *An Inconvenient Truth*, has raised the public consciousness about the dangers of global warming and climate change. This book is intended to convey the message that there is a solution. A solution not only to global warming caused by anthropogenic emissions of carbon dioxide, but also to the depletion of fossil fuels and to the wars in the Middle East related to our dependence on their supply of oil.

The solution is the rapid development of hydrogen fusion energy. This energy source is inexhaustible (it is seawater); no greenhouse gases are emitted; and the dangers of nuclear power are avoided.

Most legislators and journalists have regarded fusion as a pipe dream with very little chance of success. They are misinformed, because times have changed. Achieving fusion energy is difficult, but the progress made in the past two decades has been remarkable. Mother Nature has actually been kind to us, giving us beneficial effects that were totally unexpected. The physics issues are now understood well enough that serious engineering can begin. An Apollo 11-type program can bring fusion online in time to stabilize climate change before it is too late.

Seven nations have joined together to form and share the cost of ITER, a large machine which is an important step in achieving fusion. These nations contain more than half the world's population. A community of international workers, as well as schools for their children, has been set up at the ITER site in Cadarache, France. More on ITER will come later. There is a plan and a timetable to pursue the ultimate solution to civilization's most pressing problems. There is no downside to fusion.

So much has been written about climate change and alternate energy sources that almost every magazine has an article on these topics. By repeating the data given by Al Gore, journalists have found an easy way to meet their deadlines. Readers are hard pressed to distinguish fact from conjecture and sensationalism. We therefore start with a summary of climate change and energy sources, trying to give a concise, impartial picture of the facts. Here, I am out of my depth; I am not an expert on these topics. I get my information from the same newspapers, magazines, and websites that you do. But I think it is important to put fusion in the proper context within the general scheme of the world's future.

However, that is not what this book is about; it is about controlled fusion. The physics of fusion is highly technical, but the difficult problems and ingenious solutions can be explained so that everyone can appreciate what has been done. This is a difficult task, and I ask you to be patient. Although our explanations are longer and gentler than the succinct language of scientific journals, you cannot flip through the pages as with an ordinary book. This book is written for a variety of readers, from "green" enthusiasts with no science background to Scientific American magazine subscribers. There is a lot of information contained in many new concepts, but they can be understood by anyone with a college, or even high school, education. If you get stuck, do not give up. Your can skip ahead to more practical and less scientific material. The bottom line – what has yet to be done, how long it will take, and how much it will cost – may surprise you.

Los Angeles, CA, USA Francis F. Chen

Prologue: Toward a Sustainable World

Several hundred million years ago, light from the sun produced trees on the earth, and these were eventually converted into fossil fuels in the earth's crust. This legacy of easy energy allowed mankind to develop the advanced civilization that we enjoy today. But it is fast running out. The sun is the ultimate source of 90% of the energy we use, but it is mostly in fossil form. The everyday influx of solar power is too dilute to supply all energy that we use. We depend on the fossil fuels stored away from forests grown by the sun eons ago. Controlled nuclear fusion, or "fusion" for short, is about making an artificial sun on earth. It is not easy; but we hope to show that it is not only possible, but necessary (Fig. 1).

Fig. 1 The sun, the source of our energy

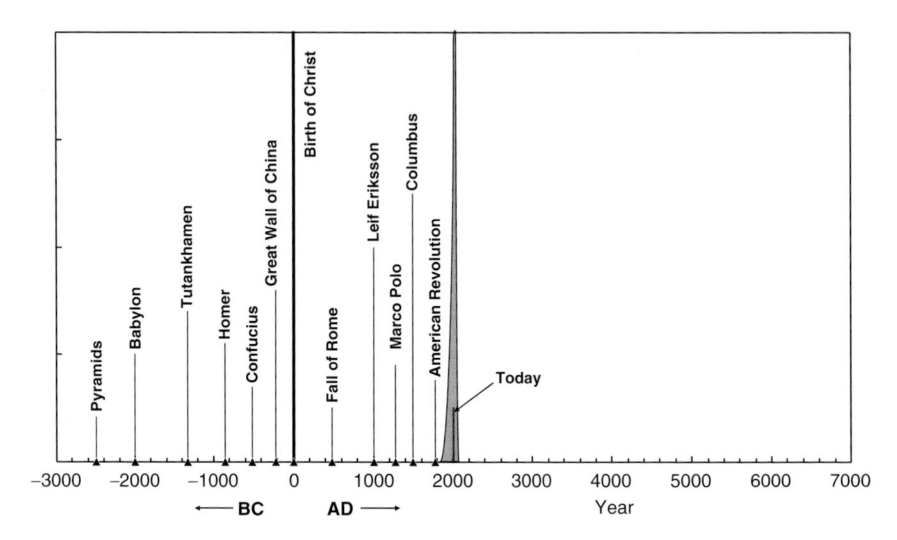

Fig. 2 A timeline of our civilization extending 5,000 years in the past to 5,000 years in the future, should we survive that long. Dates of a few historical markers are shown. The *shaded peak* is actually a plot of the annual usage of fossil fuels and shows the narrow segment of human history that it occupies

Let us take a look at how fossil fuels fit into the scheme of human history. Figure 2 shows a timeline from the beginning of recorded history to several thousand years in the future, showing several significant events along the way. The large, narrow peak in the center, known as Hubbert's Peak, represents the rate of mining and use of fossil fuels. It begins with industrialization in the 1800s and will end less than 100 years from now with the depletion of readily accessible deposits. This will happen within the lifetimes of our children and grandchildren. We are extremely lucky to be here during this very brief slice of time in the history of mankind. If our civilization is to continue as far into the future as it has existed in the past, it is clear that fossil fuels will have to be replaced by other energy sources. Energy conservation and known renewable energy sources will not be enough to sustain our civilization.

In considering either climate change or energy sources, it is important to separate three very different time scales that are involved. The first time scale is a short one, a few months to a few years, the time it takes to implement immediate but temporary solutions. For climate change that might be making an agreement like the Kyoto Protocol or issuing carbon credits which can be traded on the market. For oil or gas shortage, that might be limiting the speed limit to 55 miles per hour, offering tax credits for renewable energy installations, or starting a war in the Middle East. The second time scale is longer, 10–50 years, the time it takes to develop new sources of energy which will not burn fossil fuels and generate CO_2. The third time scale is far into the future, 100–5,000 years, perhaps the life of human civilization on this planet as we know it today. The band-aid solutions of the near term are mostly political. The problems of the far future cannot be solved now, since we do not know what they will be. However, the problems of the second (intermediate) period are upon us now, and there is barely time for effective action. Global warming and sea level rise will accelerate in the next ten years. Fuel prices will rise as fossil fuels become scarce and hard to burn cleanly. It is time to complement the efforts spent on temporary solutions with a serious program to solve the bigger problem.

Fusion power is a solution which will take time and money to bring to reality, but no more so than putting a man on the moon. We live in a glorious age when we can afford to send satellites to explore the solar system and to build huge particle accelerators to probe the structure of matter on the smallest scales. But we are not taking care of our future. The outlook is not quite that bad, however. As will be described in future chapters, the International Thermonuclear Experimental Reactor, ITER, is being supported by seven nations representing more than half the world's population. Costing some $21 billion and located in France, it will test sustainability of a fusion reaction – a continuous "burn." It is to be completed in 2019 and operated for ten years or more. Another large machine will be needed simultaneously to solve engineering problems not included in the ITER project. After that, the first power-producing fusion reactor, DEMO, is planned, but not before the year 2050. The path is clear, but the rate of progress is limited by financial resources. In the USA, fusion has been ignored by both the public and Congress, mainly because of the lack of information about this highly technical subject. People just do not understand what fusion is and how important it is. Books have been written light-heartedly dismissing fusion as pure fantasy.[1] The fact is that progress on fusion reactors has been steady and spectacular. The 50-year time scale presently planned for the development of fusion power can be shortened by a concerted international effort at a level justified by the magnitude of the problem. It is time to stop spinning our wheels with temporary solutions.

The following chapters will tell the fascinating story of how the tricky problems of creating a miniature sun on the earth are being solved, as well as give a realistic account of what is left to be done and the likelihood of success. *Controlled fusion energy is not a pipedream*. It can replace fossil fuels and curb global warming. The world will benefit from a concerted effort to bring fusion reactors into the power grid sooner rather than later.

[1] For instance, C. Seife, *Sun in a Bottle, The Strange History of Fusion and the Science of Wishful Thinking* (Viking Books, 2008).

Contents

Part I
Why Fusion Is Indispensable

Read This First!

I know most people do not bother with the introductions, but please read this guide to the book. Part I shows why fusion power is necessary. Chapter 1 summarizes climate change: what is known, what is predicted, and what can be done about it. This chapter is necessary because 40% of people still doubt that climate change is real. The facts and statements given here are backed up by references given in the footnotes and References section. Chapter 2 shows in detail the situation on fossil fuels. Chapter 3 explains how each renewable energy source works and what new developments are on the horizon.

Many readers get bogged down by the density of information in Part I and would rather just take my word for the conclusions. In that case, you can start with Part II, which gets on with fusion. Part II starts afresh and does not depend on Part I. In fact, Part II was written first. *If you like, start reading this book at Part II.* Taming the fusion reaction has been called the greatest scientific challenge of our time. Its achievement would be comparable to the invention of fire. Part II tells this fascinating story.

Again, notes are indicated by superscript numbers and references are indicated by bracketed numbers like [5]. The notes are organized by chapter.

Chapter 1
The Evidence for Climate Change*

Is Global Warming Real?

The following two graphs have served as icons to raise public consciousness of climate change caused by man's activities. The first (Fig. 1.1) shows the meticulous measurements of carbon dioxide in the atmosphere, taken on Mauna Loa in Hawaii, by Charles D. Keeling over 47 years from 1958 to his death in 2005. A continuous increase can be seen from 315 ppm (parts per million) to 380 ppm. The data are precise enough to show the very regular seasonal variations occurring every year.

The second graph (Fig. 1.2) is the "hockey stick" curve, popularized by Michael Mann in 1998, showing the surface temperature in the northern hemisphere over the past 1,000 years. The curve was relatively flat, on average, for the first 900 years. Then, around 1900, it took a sharp turn upwards and has continued to rise at a steep rate. The shape of the curve reflects the bend in a hockey stick. Though the historic data had to be gathered from tree rings and ice cores, the current rise is measured with thermometers and is much more accurate.

Are these graphs related? Is the increase in CO_2 levels causing the rise in temperature? Is man responsible for the rise in CO_2 levels? The answer is now quite certain, though there have been and still are many skeptics. It is YES to all three questions. We will first discuss the doubts; then we will show why most scientists think that global warming is real and, furthermore, is anthropogenic; that is, caused by man.

Two doctors at the Oregon Institute of Science and Medicine have published papers [1] giving data from various sources showing that warming and cooling have occurred in the past due to natural causes such as solar variability, and that shortening of glaciers started well before the industrial age. They enlisted the support of Frederick Seitz, formerly a well-known physicist, who later in life engaged in activities like consulting for the R.J. Reynolds Tobacco Company. The most outspoken critic of the global warming hypothesis has been Senator James Inhofe (R-Okla), Chairman of the Senate Committee on Environment and Public Works. His "Skeptic's Guide to Debunking Global Warming Alarmism" was delivered to

*Numbers in superscripts indicate Notes and square brackets [] indicate References at the end of this chapter.

F.F. Chen, *An Indispensable Truth: How Fusion Power Can Save the Planet*,
DOI 10.1007/978-1-4419-7820-2_1, © Springer Science+Business Media, LLC 2011

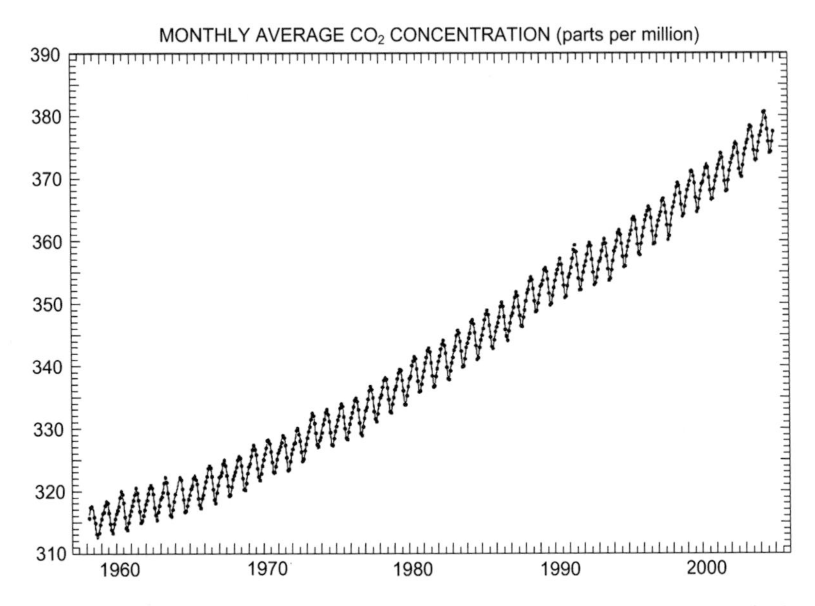

Fig. 1.1 The Keeling curve of CO_2 concentration in the atmosphere (Scripps Institution of Oceanography reports)

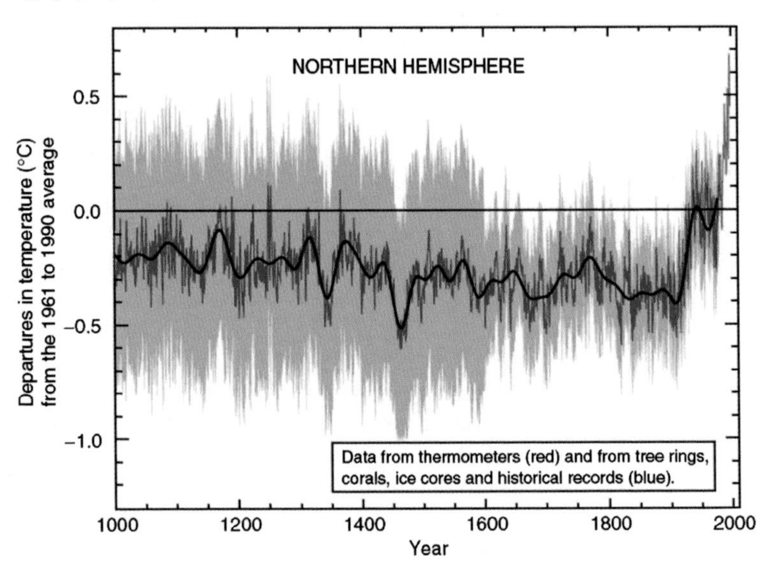

Fig. 1.2 The rise in earth's surface temperature over the past 1,000 years (reprinted with permission from Intergovernmental Panel on Climate Change [3])

the US Senate in 2006, and his 233-page December 2008 updated report [2] claimed that 650 scientists supported his position.

These critics relied on a graph of historical temperatures showing a Medieval Warm Period in the eleventh and twelfth centuries, followed by a Little Ice Age in the fifteenth and sixteenth centuries. This graph showed that temperature fluctuations of the magnitude that we have now have occurred naturally in the past. However, it turns out that these data were taken locally in the Sargasso Sea and do not represent global averages. The 2001 IPCC report [3] specifically refutes the significance of those data and instead presents the more accurate data of Fig. 1.2, in which these periods are not noticeable. Inhofe correctly cautions, however, that one cannot trust what one reads in the press. He cites articles in the media in the 1920s and 1960s warning about global warming, intertwined with articles in the 1950s and 1970s warning against a coming ice age. These critics of anthropogenic climate change are not scientists, and they clearly have their own agenda. Nonetheless, there are physicists who have studied past variations in solar radiation and believe that these could have caused global warming [4].[1] Regardless of the past, however, the best estimates by climate experts, as we shall see below, show that greenhouse gases (GHGs) generated by man will definitely raise the earth's temperature.

The Intergovernmental Panel on Climate Change (IPCC), formed in 1988, issues a detailed report every six years or so. The Third Assessment Report (AR3), issued in 2001, already gave ample evidence of man-made influence on the earth's climate. The Fourth Assessment Report (AR4) of 2007 incorporated tremendous advances in climate science in the intervening years. Many more ice cores, satellite observations, ocean and ice measurements, for instance, had been made to expand the database. In six years, the speed of computer chips has increased dramatically, as we all know. More importantly, the programs used for computer modeling of climate change have become much more trustworthy. The result is that we can predict with more accuracy what our future holds.

The IPCC-AR4 is divided into a Synthesis Report of about 100 pages, followed by the reports of three Working Groups (WGs). Each of these is just short of 1,000 pages and five pounds in weight. The data shown here come mostly from the WG1 report, *The Physical Science Basis*, the work of 152 authors summarizing the work of 650 scientific experts. There were disagreements, of course, and these have been resolved in over 30,000 arguments; this is a true consensus. In a way, science at the forefront is self-monitoring. If there are several researchers working on the same problem, you can be sure that each will examine the methods and results of the others with great care. The IPCC report is impressively careful about statistical errors. Each fact or prediction has a probability of being correct, and this certainty level is stated in words backed up by numbers. The WG2 report deals with the impacts of climate change, and the WG3 report with the methods of mitigation. For popular consumption, each WG report and the synthesis starts with a summary for policymakers. The entire report can be downloaded free of charge from the IPCC website.[2]

The massive compilation of data by the IPCC would not have made an impact on the media and the public if not for the efforts of former Vice-President Al Gore. By reducing the problem to its basics in his video and book *An Inconvenient Truth*, Mr. Gore has made us all aware, logically and emotionally, of the CO_2 problem. His antics may have been over-dramatized, and his predictions of disasters may be unproven, but he has done the hard part that scientists cannot do: get the public interested. What he started was a media frenzy, with an article on global warming appearing in almost every issue of every magazine, most of them simply repeating the material that he had already given.

Many books have been written and new journals started since warming became a household word. After the first wave, articles began appearing on the economics or politics of climate change, rather than the science. But the world runs on money, and platitudes will not lead to action. Al Gore's efforts have galvanized the public on all levels to take action on the climate problem. The USA did not sign the Kyoto Protocol primarily because it would have cost too much to enforce. Being a country with considerable fossil reserves, the USA was not desperate to find alternate energy sources. Fortunately, green energy is now becoming profitable, partly due to government subsidies, and companies in solar and wind power are growing fast. Large companies have installed alternate energy sources in their own buildings. It has become not only fashionable, but also profitable to go green. This is a healthy development, but these energy sources cannot serve mankind in the long run. We aim to show that fusion power is the ultimate solution both to global warming and to fossil depletion, and we should not wait to develop it.

Physics of Temperature Change

How CO_2 raises the earth's temperature is not as simple as people are led to believe. The popular notion is that the sun's rays go through the atmosphere and are absorbed by land and water, which radiate the energy back up at a longer wavelength. GHGs prevent this radiation from getting back through the atmosphere, thus trapping the energy and heating the earth. This notion is not wrong, but it is oversimplified. Indeed, the gases in the atmosphere are quite transparent to sunlight, which has wavelengths near those we can see. When land and water absorb this light, they radiate part of the energy back to the sky at infrared wavelengths, which we cannot see. The main constituents of the air, N_2 (nitrogen) and O_2 (oxygen), allow the infrared to get out, but "greenhouse" gases such as CO_2, CH_4 (methane), and N_2O (nitrous oxide) absorb the infrared and are heated up. They then re-radiate the energy both upwards and downwards. Only the downwards part is the energy "trapped" by the greenhouse effect. Actually, the energy radiated to the earth's surface by the atmosphere is larger than the energy coming directly from the sun [5]. If it were not for GHGs, the average temperature on the earth's surface would be −19°C (0°F) rather than 16°C (60°F) as it is now. Already we can see what a large effect CO_2 has on the earth's temperature, and why even a small change in its abundance would be worrisome.

The situation is complicated by the fact that water vapor is also a strong GHG, and its amount in the atmosphere changes constantly as water evaporates, forms clouds, and then is removed by rain and snow. But H_2O is a *short-lived* GHG, going in and out of the atmosphere every two weeks or so, while CO_2 is a *long-lived* GHG with an average residence time of four years.[3] Furthermore, water forms clouds, which reflect sunlight strongly, and rain and cloud cover vary greatly depending on where you are. It would be impossible to predict the details of changing cloud cover, so the H_2O effect has to be treated as an average. This is not as bad as it sounds because the saturation humidity level, as we all know, increases or decreases with temperature in a predictable way.

Because the water content in the atmosphere changes constantly, climate scientists cannot treat H_2O as a long-lived GHG like CO_2 but only as a *modifier* of the effects caused by those gases. One can calculate that doubling the CO_2 concentration will cause a 1.1°C (2.1°F) rise in temperature, but the presence of H_2O will cause a larger change by *positive feedback*. Positive feedback is a self-enhancing effect like a stock market crash. As stock prices plunge, more people will try to sell their stocks, causing the prices to fall faster. Here, as the temperature rises, more water is evaporated into the atmosphere, where it radiates energy back to earth, further increasing the temperature. It finally settles down at a high value 29°C (85°F). It is the convection of warm air upwards that brings this down to the observed value of 16°C (60°F). It is actually the stoppage of air currents that makes greenhouses work, not the trapping of radiation [5].

Without such mitigating factors, there can be runaway feedback, in which an increase in temperature (caused by CO_2) evaporates more water, which "traps" more solar energy, raising the temperature further, until all the water on the planet has been evaporated. This is apparently what happened to Venus, where the surface temperature is about 460°C, enough to melt lead. The runaway can also go in the other direction if the planet gets so cold that it snows everywhere, reflecting sunlight away so that it gets colder, causing more snow and ice to form. The planet can turn into an ice ball. In geologic times, the earth has had numerous ice ages and interim warm periods but has always escaped from catastrophic runaway feedback. We do not know why, though there are many theories. This is one of the lucky breaks that allowed life, even sentient human life, to arise in an interglacial period.

Quantifying Global Warming

Predicting how the earth's climate will change is a huge job, even with the help of the largest, most advanced computers. Here, we wish to give some idea of how the problem is being tackled. Each factor that can change the earth's average temperature (T) is evaluated for its ability to change T. This ability, called a "forcing," is expressed in watts per square meter (W/m^2), as if sunshine intensity were increased by that many W/m^2, all else staying the same. Forcings have to be computed using a model. For instance, to compute the forcing due to CO_2, one has to take into account the amount of CO_2 in the atmosphere and how long it stays, its rate of absorption and emission of radiation, and feedback effects such as the rise in T due to the increase in water vapor caused by the temperature rise that the CO_2 caused initially. Obviously, the result is only as good as the computer model used to calculate it, but these models are carefully checked, and the uncertainties are clearly stated. More on this will come later. For the GHGs, the forcing is known within ±10% with 90% confidence, but other effects (the small ones) can have errors of

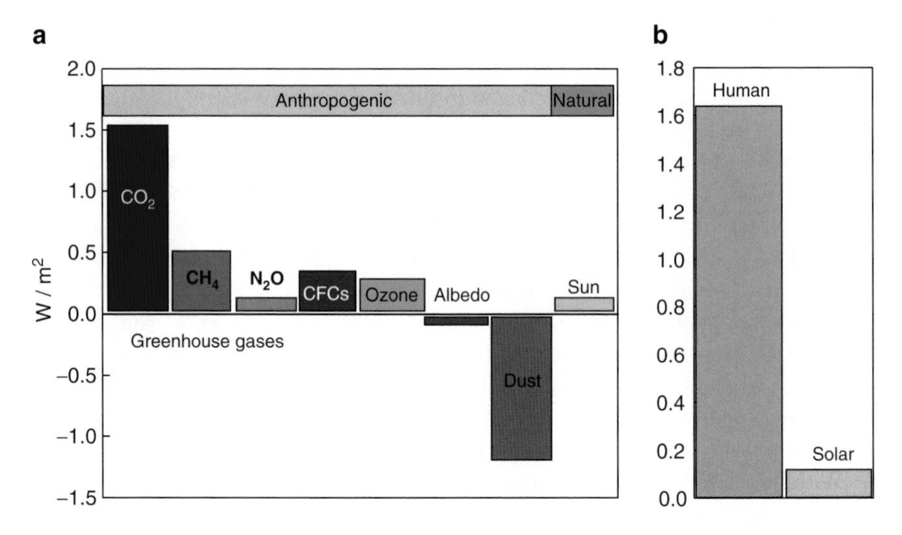

Fig. 1.3 (**a**) Major radiative forcings; (**b**) total anthropogenic vs. natural forcings. Data from Intergovernmental Panel on Climate Change [6]

±100% or so. Figure 1.3a compares the major *radiative* forcings; that is, the effectiveness of the main agents that can change *T* by altering the absorption of solar radiation.

These forcing numbers seem very small, less than 2 W/m^2, compared with the peak solar irradiance of about 1,300 W/m^2, or even the 342 W/m^2 averaged over a hemisphere or the 240 W/m^2 that reaches the earth's surface. But a small change in *T* can have catastrophic effects, as we shall see. The man-made forcings have both positive (warming) and negative (cooling) values. Let us see where these figures come from. The three main GHGs dominate the warming effects. CH_4 has 26 times the warming potential of CO_2 and N_2O, 216 times; but their concentrations are much lower than CO_2's, and CO_2 is dominant. The ozone-depleting chlorine-containing gases which were banned by the Montreal protocol are lumped under the rubric *CFCs*. That value comes from 60 different gases which were evaluated one by one in the IPCC report of 2007 [6]. The value for *ozone* does not depend on the state of the ozone hole, because high-altitude ozone has a small role here. The ozone that contributes to warming is in the lower atmosphere and is generated on the ground by natural processes such as rotting of biological matter. What we have called "dust" is the sum of all *aerosols* emitted by factories and volcanoes. Industrial aerosols are mainly sulfate and carbon particles of varying sizes and reflectivities. You would think that black carbon would absorb well, but remember that black not only absorbs well but also emits well. More importantly, particulate matter can seed cloud formation, and clouds reflect sunlight efficiently. The net result is that aerosols have a large *negative* forcing and give a cooling effect. *Albedo* is the change in the reflectivity of the earth's surface, and this small effect comes from the balance between two effects. Black dust on snow will reduce the albedo of the snow and cause warming.

Deforestation and other land modifications by man will replace trees with farms or buildings, thus increasing the albedo. In this case, land use wins, and changes in albedo are a negative forcing. The result is very uncertain, but it is small in any case.

The natural forcings come from volcanoes and solar variability. Volcano dust stays in the atmosphere only a few years, and eruptions are rare and unpredictable. On the other hand, *solar variability* follows the 11-year sunspot cycle closely, and this 8% effect is accurately predictable. However, what concerns us is not the 11-year cycle but the long-term trend. Changes in the earth's orbit or the tilt of its axis occur over tens of thousands of years, so only a very small part of these changes could have occurred in modern times. Recently obtained data on solar irradiance from 1,750 to the present yield a forcing of +0.12 W/m^2, with a 90% chance of the exact value's being within 50% of this. Figure 1.3b compares the net anthropogenic forcing with the natural forcing caused by solar variability. The man-made part is 13 times larger. Skeptics[1] who say that present global warming is a natural phenomenon would imply that climate scientists are wrong by over an order of magnitude. Even if that were true, it is irrelevant. The present rate of CO_2 emissions by man is not conjectural, and their effect on temperature can be calculated with ± 10% accuracy.

Evidence for Climate Change

Paleoclimate

What the earth's temperature and CO_2 levels were can be determined, surprisingly enough, as far back as 650,000 years ago. For the last millennium, accurate records of temperatures recorded with thermometers can be found. Before that, there are ancient documents telling of extreme weather events, the dates of spring planting, or occurrence of plagues from which some idea of the weather can be gleaned. For prehistoric eras, there were no direct observations, but data can be found indirectly from what are called *proxies*. Tree rings, ice cores, and cores of layered sediments in soil or sea bottoms give annual records that can be counted ring by ring. Trapped air bubbles in ice cores give the CO_2 concentration hundreds of millennia ago. The fractional abundance of oxygen or hydrogen isotopes in ice cores and coral yields the temperature, as do other ratios, such as Mg to Ca. These proxies can be correlated with one another to give higher accuracy in recent times for which there are more data. The result from Antarctica ice is shown in Fig. 1.4. As the earth undergoes long glacial ages and short interglacial warm periods, the CO_2 and CH_4 abundances follow the temperature quite closely. Of course, we cannot tell which is the cause and which is the effect here. The present warm period, which allows life to exist, looks no different from previous interglacial periods, except for the spike seen at the far right. For that, we now know that CO_2 is the cause, and the temperature rise the effect.

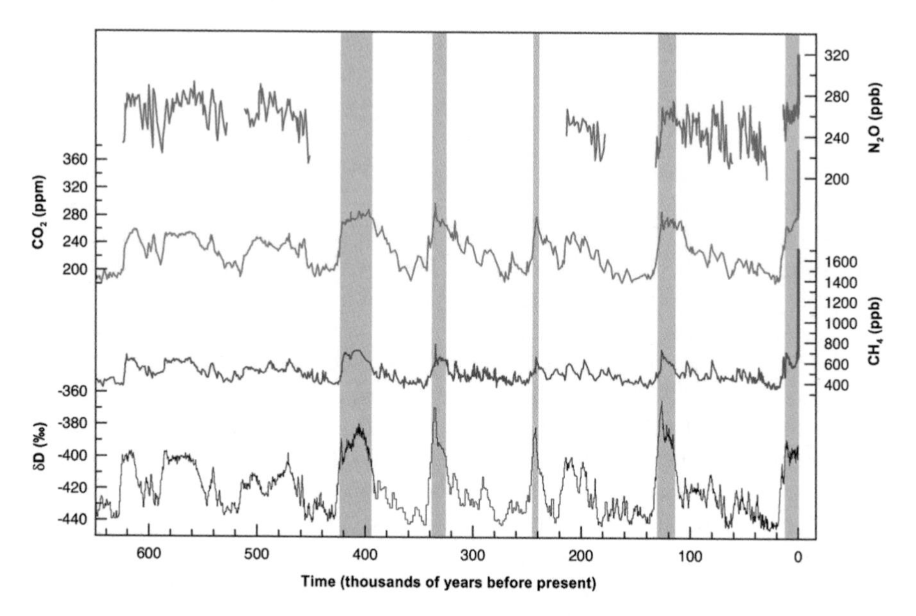

Fig. 1.4 Paleoclimatic data on the variation of temperature and CO_2, CH_4, and N_2O abundances from Antarctic ice cores [6]. The temperature is represented by the deuterium abundance proxy (*bottom curve*). The shadings indicate interglacial warm periods

When considering the climate tens of thousands of years back, we have to take into account changes in the earth's orbit. The earth's spin axis is not perpendicular to the plane of its orbit but is tilted at 23.5°, thus causing winter in the northern hemisphere while it is summer in the south. This tilt can change from 22° to 24.5° over a period of 20,000 years or so. This does not change the total sunlight on the earth, but it distributes differently between the northern and southern hemispheres. Since there is more land in the north and more water in the south, this re-distribution of sunlight can affect the climate. A bigger effect comes from the precession of the equinoxes, when the earth's axis spins around like a gyroscope. The effects come from an interaction with the ellipticity of the earth's orbit, which means that solar radiation is stronger when the earth is near the sun (perihelion) than when it is far away (aphelion). Thus, in one orientation, the northern hemisphere has summer during perihelion; and, 10,000 years later, the southern hemisphere gets the hotter summers. The shape of the earth's orbit can also change between more circular and more elliptical due to the pull of other planets, mainly Jupiter. This happens every 100,000 years or more. The ice ages may have started at a coincidence of these orbital forcings, triggering runaway feedback, as we described before. The recovery into warm periods is equally remarkable. There is an intriguing theory that the most recent recovery (the last shaded bar in Fig. 1.4) may have been caused by humans when they started farming about 11,000 years ago [7]. Methane is produced by decaying vegetative and animal matter produced in agriculture, and deforestation decreases CO_2 absorption by trees.

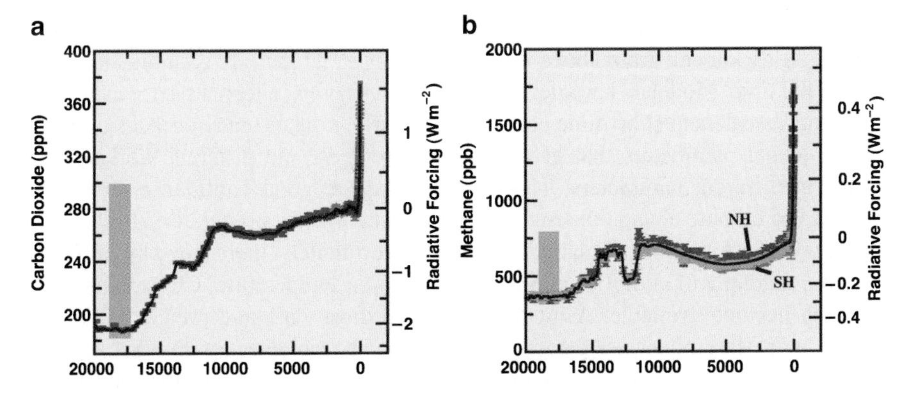

Fig. 1.5 CO_2 levels in parts per million (ppm) and CH_4 levels in parts per billion (ppb) vs. year before 2005, as measured from different sources [6]. NH and SH stand for northern and southern hemisphere, respectively

The paleoclimate data for the last 20,000 years on how CO_2 and CH_4 abundances changed with time are so good that observations from different proxies agree amazingly well. This is shown in Fig. 1.5. The CO_2 level increased slowly from 190 ppm to the preindustrial level of 280 ppm, followed by the recent rapid increase to 379 ppm in 2005. The present level is much higher than any level that existed over the past 650,000 years (indicated by the gray bar at the left). The current spike is also seen in the methane data.

Computer Modeling

This science has improved greatly since the 2001 ICPP report, and predictions are therefore more reliable. It is a very complicated problem [8]. There are standard physics equations that tell you how air and water move and how heat is transferred from one medium to another, but weather varies with location and changes by the hour. To predict climate, one has to divide the space into a finite number of cells, few enough for computers to handle. These cells are about 200 km laterally and 1 km vertically (in the atmosphere), decreasing vertically to maybe 100 m near the ground. To divide up time, 30-min averages are taken for climate, and shorter time steps for weather forecasting. The computer program then takes the average conditions in one cell and predicts what the conditions will be in the next time step. The conditions include, for instance, temperature, wind speed, water vapor, snow cover, and all the effects mentioned earlier in this chapter. We did not mention the history of CO_2. About 45% of the CO_2 that man generates goes into the atmosphere, 30% into the oceans, and the rest into plants. The CO_2 absorbed by oceans diffuses downward over many years. The CO_2 in the atmosphere has a mix of different lifetimes; roughly speaking, half goes away in 30 years and half stays for centuries. All such effects have to be accounted for in the models.

The key word is "average." How does one find the average conditions in a $100 \times 100 \times 1$ km cell 1 km above Paris, for instance? Clouds are forming and moving all the time. Modelers have developed *parametrization*, a technique for averaging over small-scale and short-time conditions. Clearly, it takes many decades of experience to get parameters that give the right averages, and different workers will arrive at different parameters. This does not inspire great confidence, and most skeptics of climate change distrust modeling and correctly point out that this is the weak point in forecasts of impending disaster. Fortunately, there is a way to check. Starting a couple of centuries ago, accurate data on temperature, CO_2 content, and so forth became available. A modeler can take those data and predict what happened later using his or her parameters. Then he/she can check with what actually happened and adjust his parameters to give the correct result. The only uncertainly is then whether or not the parameters of a century ago are the same today. We will show that different workers have varying success in their predictions, but all show that the current global warming is man-made.

Modern Data

Before showing the modeling results for the modern era, it is instructive to show the amount of data now available for analysis, as opposed to what is used for paleoclimates. When one computes the global average temperature, isn't that just a weighted average over a finite number of places on the earth, say, a few hundred? Now that we have satellites, the coverage is much better. Here are three examples. Figure 1.6 shows the tremendous increase in the number of measurements of ocean temperature between the 1950s and the 1990s. Figure 1.7 shows the fine detail that satellite coverage gives on the altitude change in each part of Greenland and Antarctica. The loss of ice thickness can be seen clearly where glaciers and ice sheets have slid into the sea. Figure 1.8 shows the distribution of aerosols over the globe as obtained by opacity measurements by satellites. This is supplemented by a finite number of ground-based observations which can also determine the size and material of the particulate matter.

Global Temperature Rise

Here, we show in detail the present-day peak seen in Figs. 1.4 and 1.5, followed by projections of the temperature rise in the future as computed by climate modeling using the extensive observational database illustrated in the previous section. Figure 1.9 shows the temperature variation over the past 1,000 years as deduced by various methods (proxies). There is considerable disagreement up to about 1850, but with better data since then, all the proxies agree on the most recent temperature rise. The agreement is quite amazing since the range of the entire graph is only 1.8°C.

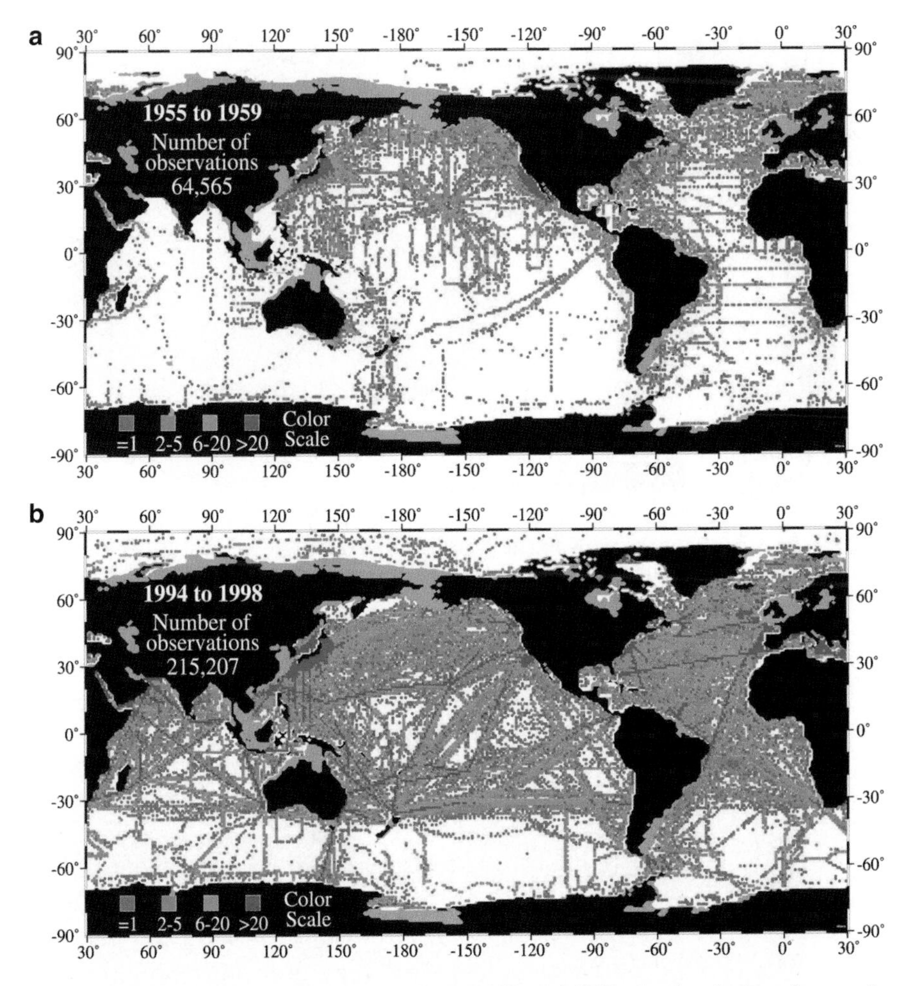

Fig. 1.6 Ocean temperature measurements in the 1950s and 1990s showing the large increase in the database [6]. The *color scale* shows how many measurements are represented by each *dot*

The decreasing uncertainly is seen more clearly in Fig. 1.10, where the weighted global average temperature deviations are shown with error bars for the calculated standard deviation.

Figure 1.11 shows the data from 1850 to the present for the northern and southern hemispheres and their average. The North has higher recent temperatures, probably because there is more industry; but other than that, the histories are similar, showing that the trend is truly global. The error bars on each point are significant: they indicate that there is only a 5% chance that the true value lies outside those ranges. It is quite clear from this that the earth's temperature has risen from about –0.3 to +0.6°C (relative to 1980) since the preindustrial period.

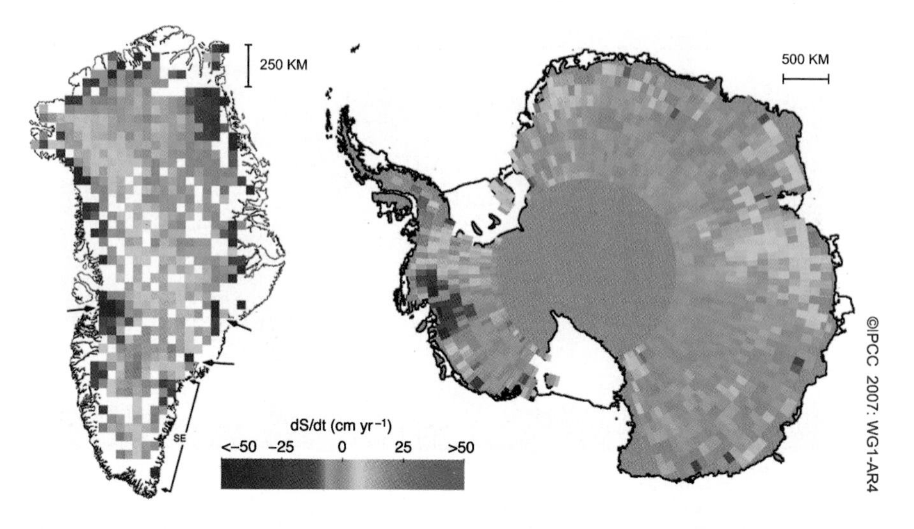

Fig. 1.7 Satellite measurements of the rate of change of elevation in Greenland and the Antarctic, showing the loss of glaciers and ice sheets (*blue*) and accumulation of snow (*red*) [6]

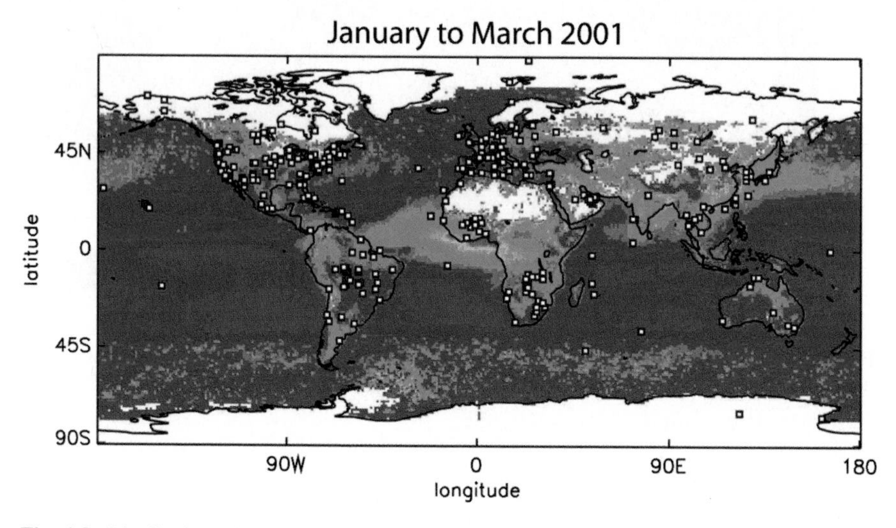

Fig. 1.8 Distribution of aerosols from satellite observations (*color*) and from surface stations (*dots*) [6]. *Red* indicates a lot; *blue*, little; and *white*, no data

The question is now whether the temperature increase is anthropogenic or not. Climate modelers have calculated the natural forcings and those caused by man, as shown in Fig. 1.3. Remember that these forcings depended on the "parameters" that the modelers chose to find the average over fine-scale variations in space or time.

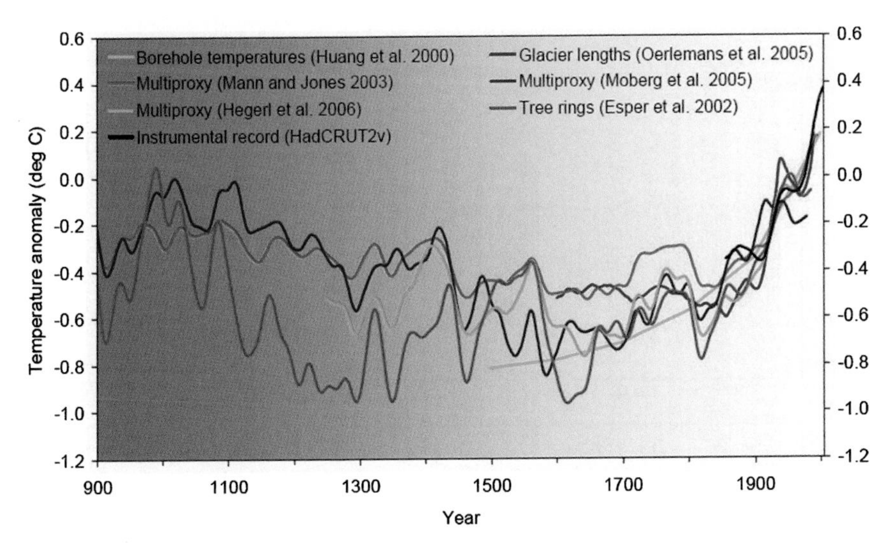

Fig. 1.9 Temperature variations from the peak in year 1000, as measured in different ways (reprinted with permission from National Research Council [9])

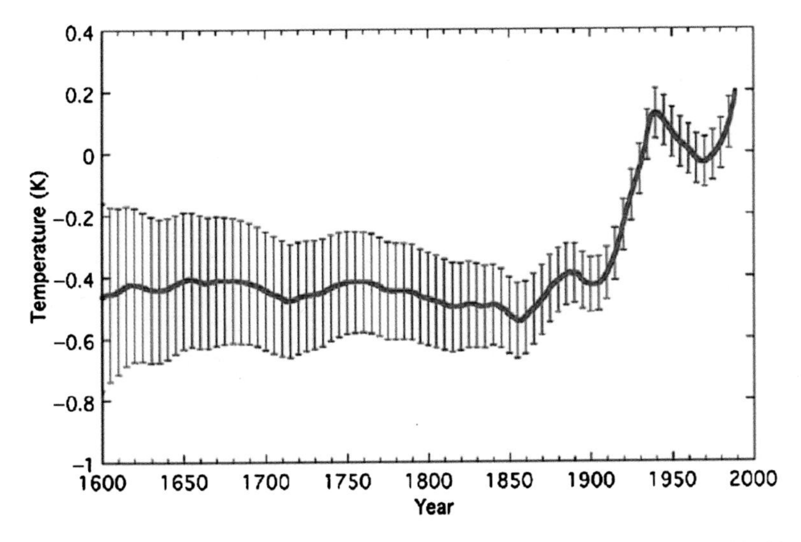

Fig. 1.10 Temperature variations in degree Celsius (=K) with error bars for 1600–1990 [8]

Their projections, shown in Fig. 1.12, all agree up to year 2000 by design. The parameters had been chosen so that the twentieth century data were correctly predicted by the models from the data from the century before that data. This is how the models are calibrated. The models can predict the future as long as the parameters do not change. Nonetheless, different models give different results for the future, and there

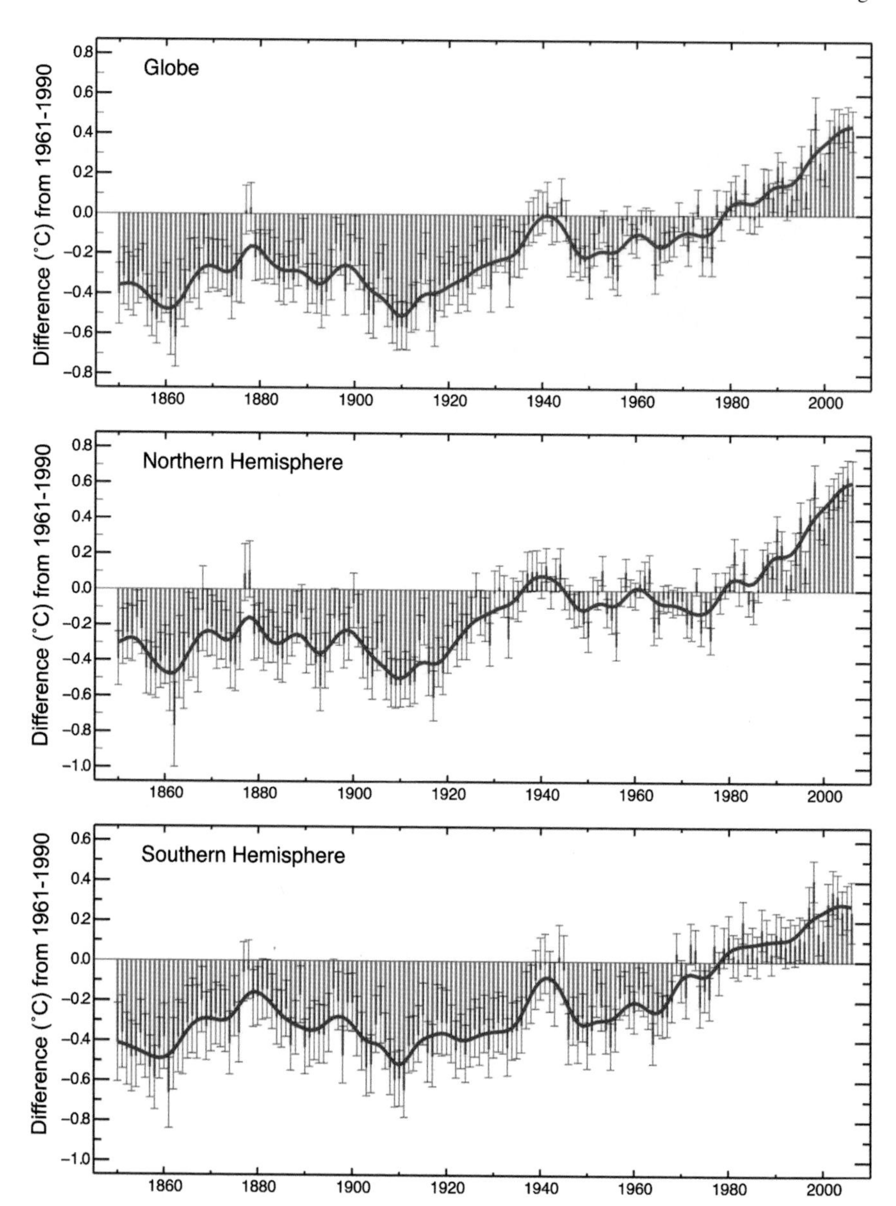

Fig. 1.11 Temperature variations from 1850 to the present as averaged over the northern and southern hemispheres and over the whole globe [9]

is a large range of uncertainty. The lowest curve in Fig. 1.12 is what would happen if the GHG level were held constant at the 2,000 levels with no further emissions. The temperature will not go down because the CO_2 in the atmosphere stays there for hundreds of years. The three models shown predict a temperature rise of 1.8–3.6°C

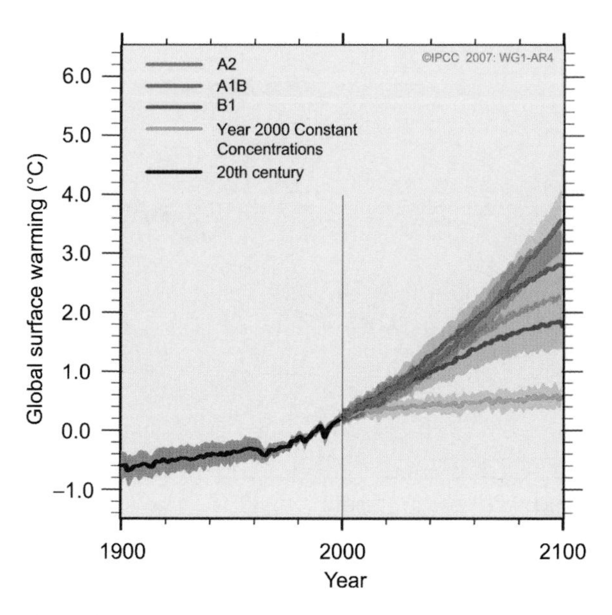

Fig. 1.12 Predictions of temperature increase by various climate models [6]

by the year 2100. The 2007 IPCC report [6] gives the results of six scenarios ranging from optimistic to pessimistic. The most optimistic scenario predicts a temperature rise of 1.1–2.9°C in the next 100 years, and the most pessimistic one is a rise of 2.4–6.4°C. The range given for each model represents the 66% probability level.

I have chosen graphs which give an idea of the uncertainties in both the data and the models because the IPCC report has been challenged by individual scientists who have arrived at different conclusions.[4] Though the ICPP's Working Group 1 had input from over 600 scientists, only a fraction were involved with any one problem, and arguments are bound to arise. Nonetheless, it seems clear that GHG emissions will be harmful to some extent in the future, and these can be suppressed by replacing fossil fuels with other energy sources. There is no need to argue.

Disasters and Catastrophes

Consequences of a global temperature increase have provided fodder for journalists always looking for a new angle. We have all read about recent hurricanes, floods, droughts, and heat waves, as well as the dangers posed to coral, birds, and other species of wildlife. The connection to *global* warming is circumstantial and conjectural at best, but the connection with *local* warming can be established with more certainty. Some phenomena can be modeled quite successfully; the most certain of these is sea-level rise.

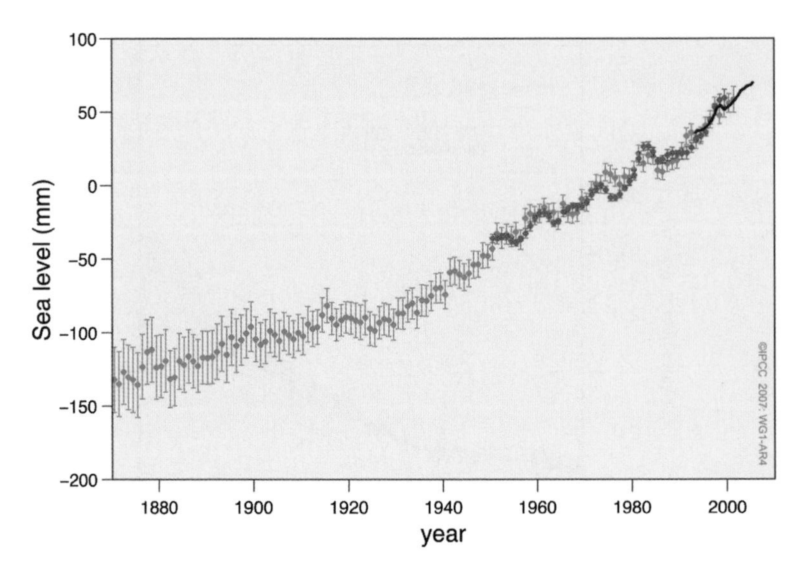

Fig. 1.13 Sea level relative to 1975, with the latest data taken by satellites. The error *bars* show the 90% confidence level [6]

Sea level has been rising at the rate of 3 mm (1/8 in.) per year, which would amount to an inch in eight years, or about foot a century even if the rate does not increase. Low-lying places like the Netherlands, Indonesia, and Bangladesh would be the first to feel the loss of hundreds of square miles of land area. There is some evidence that the rise seen in Fig. 1.13 has accelerated since the onset of industrialization. Most of this can be attributed to global warming.

The three main causes are thermal expansion of water as it is heated, the melting of glaciers that have slid into the sea, and the melting of ice sheets on land. The contributions from each of these sources are shown in Fig. 1.14. The bottom part of each column is the sea-rise rate averaged over the past 42 years, while the recent average is given by the total height of the column. The rate-of-rise scale is in millimeters per year (mm/year). The four columns add up to the 3 mm (1/8 in.) figure quoted above. In each case, it is clear that the rate of rise has accelerated. The breakdown into the four effects required computer modeling, since the water from melting glaciers, for instance, cannot be measured directly. However, the sum of the calculated effects can be shown to agree quite closely with the sea-level rise actually measured. This gives us confidence in the accuracy of modeling procedures.

Icebergs that are already floating will not change sea level as they melt because the part that is underwater (85–90% of the iceberg, depending on the temperature and salinity of the seawater) occupies exactly the volume that the iceberg will fill when it melts.[5] Glaciers, ice caps, and ice sheets that are on land, however, are a different story. As land ice melts, it not only adds water to the oceans, but it also wets the ground under glaciers, making them slide into the ocean faster. Glaciers are melting at the rate of two cubic miles per week,[6] and the shrinking of glaciers

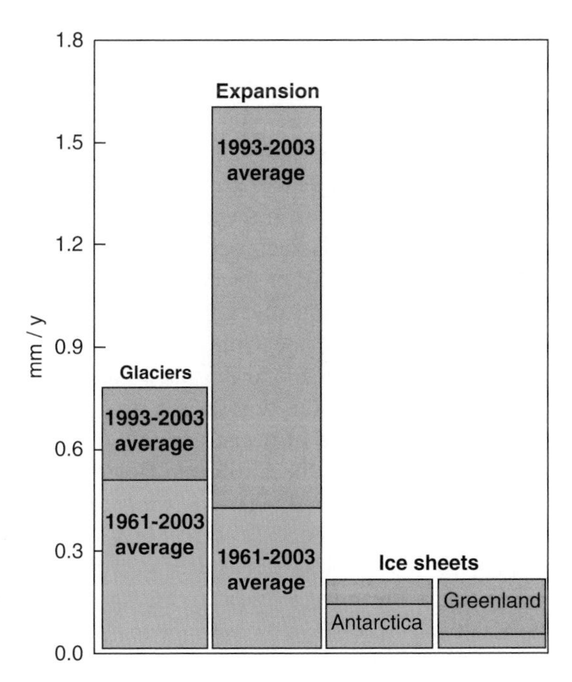

Fig. 1.14 Contributions to sea-level rise by glaciers, thermal expansion, and ice sheets in Antarctica and Greenland. The *lower* part of each column is the 42-year average rate; the most recent 10-year average is the height of the entire column. Data from Intergovernmental Panel on Climate Change [6]

over the past decade can be seen in many photographs. This is direct evidence of rising temperatures, but the unseen feedback effect is more treacherous. Ice has a high albedo, reflecting sunlight efficiently. As it melts, ground is uncovered, and this absorbs more sunlight, causing higher temperatures. As permafrost in Greenland is defrosted, exposed vegetation can rot, giving off CO_2 and methane. Although the total forcing from albedo change is negative, as seen in Fig. 1.3, it is the *local* heating where ice cover is disappearing that causes the runaway effect.

Permanent ice covers only 10% of land surface and 7% of oceans, which is why the catastrophic changes in glaciers that we can see is not the main cause of sea-level rise. As seen in Fig. 1.14, the main effect is simply the expansion of water when it is warmed. Not all consequences of ice melt are negative. Ice over the North Pole is definitely getting thinner, as directly measured by submarines there.[7] The long sought-after Northwest Passage is becoming a reality. Trees growing on newly exposed ground can absorb CO_2. The negative aspects, however, are dominant. If all the snow and ice on Greenland and Antarctica were to melt, the sea level would rise by 7 and 57 m, respectively [6]. This has happened in geologic eras, and the earth has undergone hot and cold periods before, even in human history; but what is new here is that it is happening extremely fast, before mankind can slowly adapt to the changes as it did previously.

The Gulf Stream

The melting of arctic ice injects fresh water into the north Atlantic, possibly disrupting the warm ocean currents that make Europe comfortably habitable. Although this is unlikely, the consequences are so unsettling that this subject has drawn undeserved attention. London is at the same latitude as Calgary, Canada; and Rome is in line with Boston, Massachusetts. Tromsø, Norway, is 250 miles north of the Arctic Circle; yet its harbor never freezes over. That is why most polar expeditions start there. Technically known as the Atlantic Meridional Overturning Circulation (or MOC), the Gulf Stream picks up heat from the Atlantic Subtropical Gyre and carries it to the Subpolar Gyre. These gyres, or circulating currents thousands of miles across, are driven by winds above the water. Figure 1.15 shows the system of ocean currents over the whole earth. In the north Atlantic, water warmed in the Caribbean flows along the shore of the USA up to Cape Hatteras, and then breaks off eastward toward Iceland and England.[8] When it reaches high latitudes, the seawater cools, becomes denser, and sinks to lower depths. The cooled, salty water then flows back to the south underneath the warm water. Fresh water from ice melting from Greenland, however, is lighter than saltwater and stays on top, opposing the northward flow of the Gulf Stream.

Computer models vary greatly on what will happen. The latest results vary from almost 50% slowing of the MOC to no slowing from anthropogenic causes. The

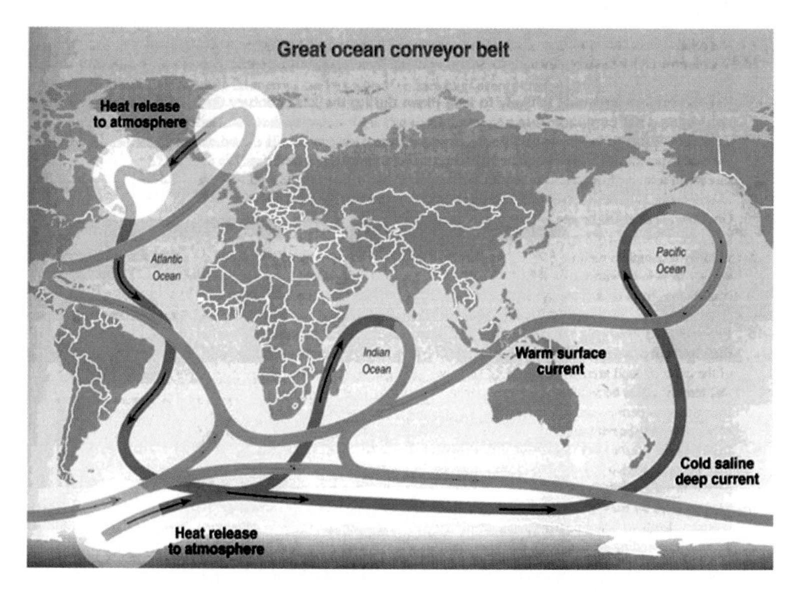

Fig. 1.15 The Great ocean conveyor belt (reprinted with permission from Intergovernmental Panel on Climate Change [3])

problem is complicated by two other known effects, the Atlantic Multidecadal Oscillation and the North Atlantic Oscillation, which can, respectively, accelerate or delay the MOC slowing by a few decades. Furthermore, it depends on where the temperature rise is greater. Both the injection of fresh water from the north and greenhouse warming of the North slow down the MOC, while warming in the South will enhance it. The 2007 IPCC report concludes that there is a greater than 90% probability that there will be *some* decrease of the MOC in the next 100 years, but no simulations predict that the MOC will completely stop. There has been no conclusive evidence of changes so far.

The One-Degree Effect

The earth's average temperature has risen 0.74°C (1.3°F) in the last century, with most of this rise, 0.55°C (1.0°F), occurring after the 1970s. Since our local temperature varies by many tens of degrees between day and night and between summer and winter, how can a one-degree change have the dire consequences attributed to it? The one-degree change is only an average over the whole globe and over a whole year. Any particular place can have swings of temperature much larger than this which are compensated by opposite swings at other places. As will be discussed below, there is evidence that extreme weather events like droughts and floods are occurring more frequently, and these can cause disasters like wildfires, though the causal relation cannot be proven.

In some instances, the effect of even one degree is clear. Much of the permafrost in Greenland is near the melting point. A one-degree warming can cause it to unfreeze, allowing plants to grow. These, in turn, absorb much more sunlight than ice does, triggering accelerated warming by positive feedback. The loss of ice and snow where the temperature is near 0°C has affected the lives of polar bears and their prey, the monk seals. The permafrost under the Arctic Ocean has trapped methane from decaying vegetation ages ago. Bubbles of this gas, with 26 times the warming potential of CO_2, have recently been observed to come out in increasing amounts, though the connection with global warming has not yet been established. On land, the pernicious effects are more subtle but have already been observed. The tree line on mountains has moved upwards. Birds have found their usual food sources diminished during nesting season. Spring seems to arrive earlier. Annual migrations of birds and butterflies are sensitive to small changes in the timing and location of their food sources. Examples of quantitative data are given in Box 1.1.

But are mankind's activities the cause of these changes? Figure 1.11 showed the temperature rise over both land and ocean. The change in air temperature over land is shown more clearly in Fig. 1.16. We see that there was a warming trend from 1890 to 1940, a change of about 0.5°C which was probably due to natural causes. This was followed by a period of global cooling. Wildlife has in the past adapted to

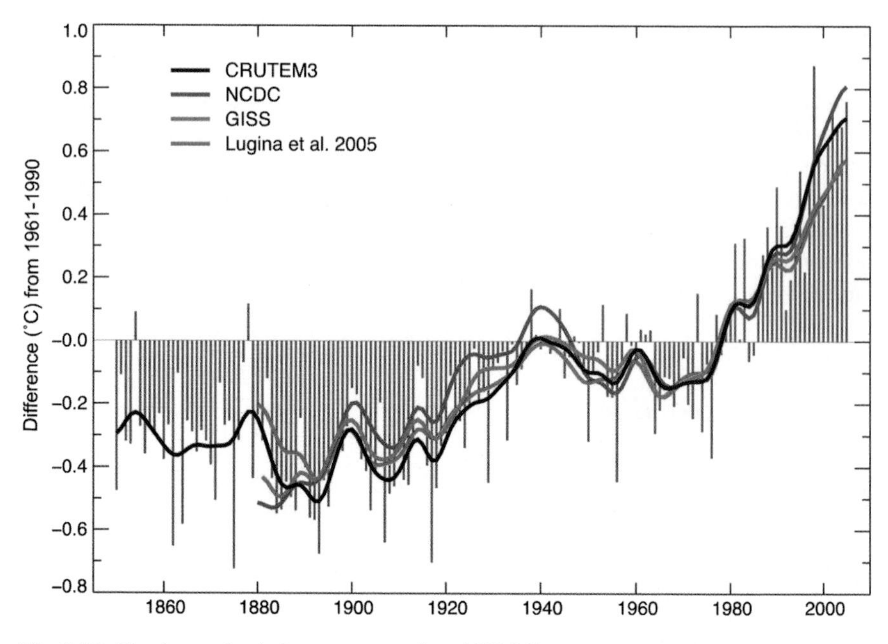

Fig. 1.16 The data on land air temperature since 1850 [6]

such changes. It is natural for some species to become extinct occasionally, just as the dinosaurs became extinct. What is new is that the current temperature rise is noticeably faster and can be related to the emission of GHGs. It is not so much the one-degree (°F) change of the past but the six-degree (°F) change predicted for the next century (Fig. 1.12) that is worrisome. Natural evolution is being driven at an increasing rate by mankind with unknown consequences unless CO_2 emissions can be controlled.

Floods and Droughts

When we talk about rainfall, global averages are of no use; rain and snow occur locally. This can be seen in Fig. 1.17, which shows how precipitation varies from region to region. What global warming does is to increase the occurrence of extremes: severe floods or severe droughts. It is hard to see the long-term trend because of large periodic weather events such as the El Niño Southern Oscillation (ENSO) or the less well-known North Atlantic Oscillation (NAO), which is a modulation of the westerly winds into Europe. Nonetheless, the IPCC 2007 report states that the wet-dry differences (the color depth in Fig. 1.17) have been increasing from 1900 to 2002.[9]

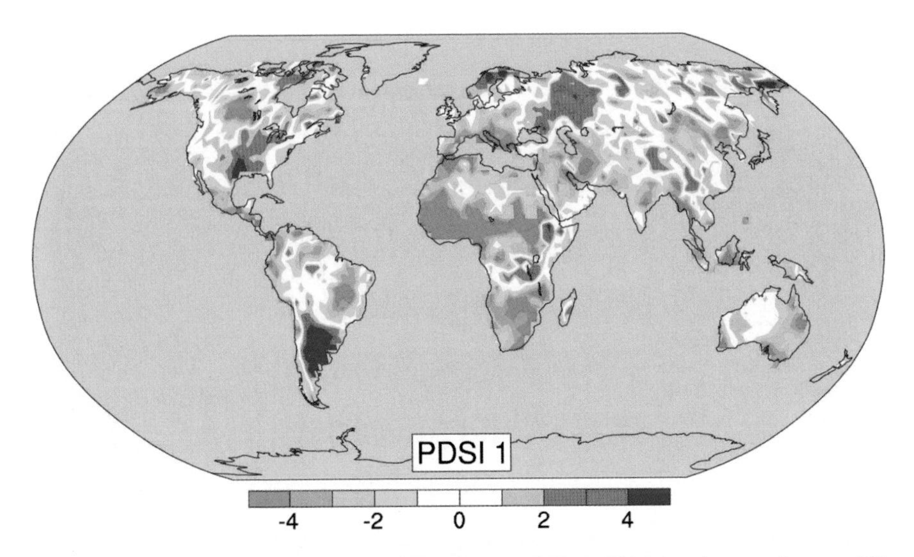

Fig. 1.17 Reddish regions have gotten much drier since 1900, and *bluish* regions much wetter [6]. PDSI stands for the Palmer Drought Severity Index

Box 1.1 Effect of Temperature Rise on Birds and Flowers

The Audubon bird count has been going on for 109 years, and there are 35 million bird records in the database. A 2008 study by California Audubon [12] analyzed the shifts in ranges of 312 species in the last 40 years as the January temperatures rose by 2.5°C (4.5°F). For most species, the shift is northward toward cooler climates and can be over 400 miles. A few examples are shown in Fig. 1.18.

The range over which birds can find sufficient food and nesting sites can be shortened by their geographic displacement (Fig. 1.19). This can be computed using scenarios which assume different rates of anthropogenic carbon emissions and different degrees of mitigation. Consequently, for some birds such as the California gnatcatcher, the predictions can vary widely from model to model.

The migration of birds is also affected by higher temperatures as their nesting period and food sources occur earlier in the spring. Jenni and Kéry [13] have studied the time of migration through Western Europe of 65 species in a 43-year period. Long-distance migrants migrate sooner, but short-distance migrants and multiple-brood birds may delay or not change their migration times.

Flowers also have been blooming earlier as temperatures rise. Using Henry David Thoreau's notes on flowering dates in the 1850s and comparing with his own measurements in Massachusetts, Primack [14][10] has been able to show that the mean flowering date for 43 species has moved up seven days, while the May temperatures increased by 2.9°F (1.6°C) between 1855 and 2006. Some plants were found to bloom 20–30 days earlier.

(continued)

Box 1.1 (continued)

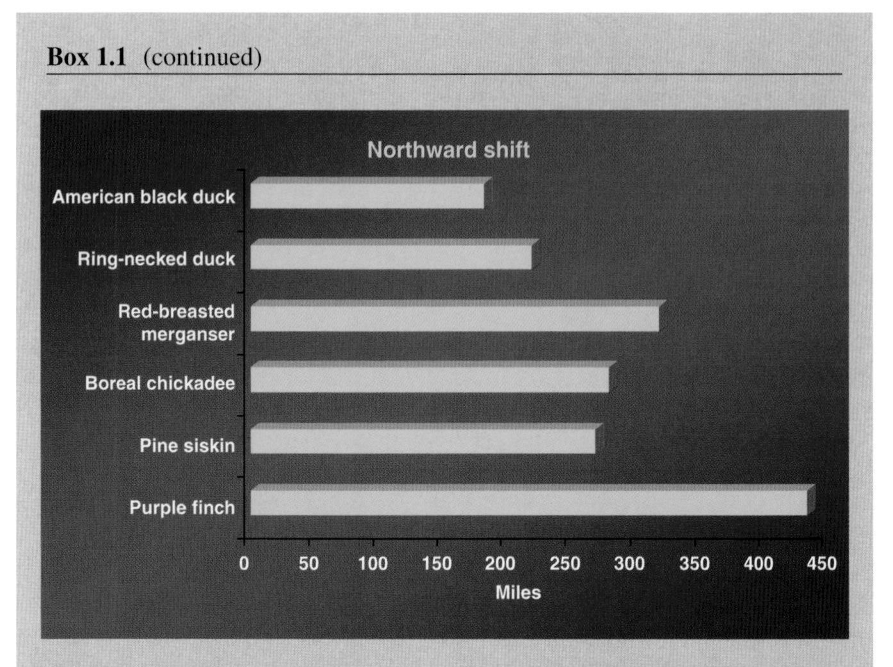

Fig. 1.18 Northward movement of the ranges of representative bird species [12]

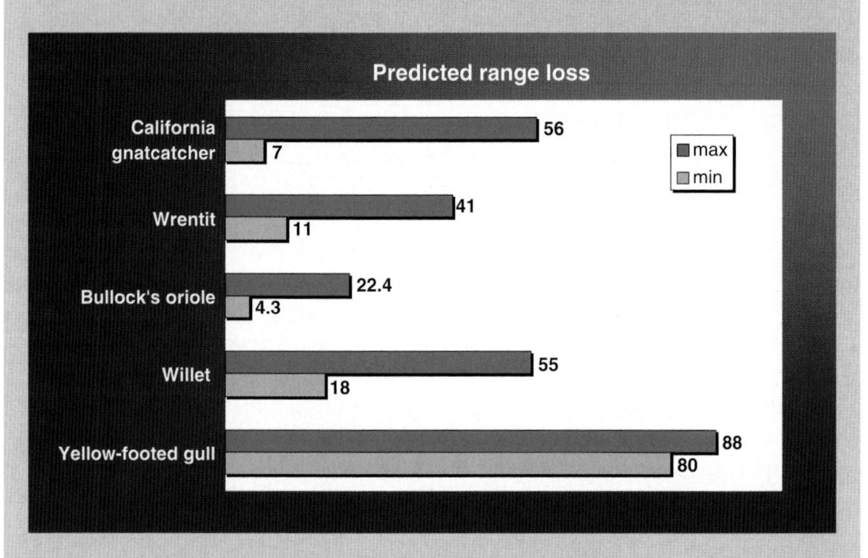

Fig. 1.19 Shortening of bird ranges (in percent) as predicted by computer simulations using scenarios such as those in Fig. 1.12. The *blue* and *red bars* show the minimum and maximum range loss percentages forecast by different climate model [12]

The underlying physics involves the fact that warmer air holds more moisture: 7% more for every degree Celsius rise in temperature. In places where it rains, the larger moisture content in the atmosphere makes it rain harder. At the same time, evaporation is a cooling process, so the transfer of surface moisture into the air tends to cool the surface. Where it does not rain, this cooling effect does not occur, and the land gets hotter and drier. This raises the possibility of heat waves and forest fires, the latter injecting more CO_2 into the air. This is conjectural, but there is an immediate impact of global warming on drought which comes from timing. Earlier summers mean that the snowpack on mountains melts sooner, releasing water before it is needed and causing reservoirs to overflow. The loss of water means drought in the summer.

Effect on Oceans

The oceans are a vast reservoir of CO_2, taking up about two billion tons a year. The rate of this uptake is slowing down, although the yearly amount is increasing simply because there is more and more CO_2 in the atmosphere. From 1750 to 1994, 42% of CO_2 released went into the sea; but from 1980 to 2005 this figure decreased to 37% as a result of the extra CO_2 that we are producing. Carbon dioxide is the gas that bubbles out of soda pop and forms a weak acid when dissolved. The oceans can absorb much more CO_2 than by simply dissolving it, however. That gas reacts with H_2O to form positive hydrogen ions (H^+) and negative carbonate (CO_3^{2-}) and, mainly, bicarbonate (HCO_3^-) ions in a "buffering" process. This increases the uptake of CO_2 by almost an order of magnitude. The possible increase is quantified by the so-called Revelle buffer factor, which depends on the partial pressure of CO_2 at the ocean surface. That is, the more CO_2 that is pushing back into the atmosphere, the less CO_2 the ocean can absorb out of the air. It takes about a year for these pressures to equalize, and it takes thousands of years for carbon in different forms to circulate in the ocean. The CO_3^{2-} ion can also combine with calcium to form calcium carbonate ($CaCO_3$), the material of coral and some shells. These solids sink into deeper water and stay there for millions of years. If we were to stop producing CO_2, it would take 4–10 thousand years for the ocean's partial pressure of CO_2 to get back to normal.

The buffering effect injects much more H^+ ions into the ocean than would be created by dissolving CO_2 into carbonic acid, and this makes the ocean much more acidic. The ocean is naturally mildly alkaline, with a pH value of 7.9–8.3, and anthropogenic CO_2 has decreased it by 0.1 since 1750. This does not sound like a lot, but the number of H^+ ions has increased by 30%. Furthermore, computer models predict a decrease between 0.14 and 0.35 in the 21st century. Acid dissolves carbonate matter such as coral and shells of sea animals, and it can slow or prevent their creation. We have all read about dead or dying coral reefs, though the relation to global warming is conjectural. Phytoplankton, at the bottom of the food chain, absorb almost as much CO_2 as plants on land,[11] and they are consumed by larger organisms which are the food source of all fishes and whales. Most crustaceans such as krill have chitin rather than carbonate shells, but those that are carbonate-based would suffer from increased acidity. The entire food chain can be upset by acidification of the oceans. However, there is so far no scientific evidence that this is happening. The 2007 IPCC report states that the effect of increased acidity on marine organisms is poorly known.

Weather Extremes

Hurricane Katrina leveled New Orleans in 2005 and is most often cited as an example of the effects of global warming. The hurricane season in 2005 had the largest number of hurricanes, and the strongest ones, on record. It is of course not possible to ascribe any single local event, or even a season of events, to a slowly changing general condition. It takes a concatenation of unusual local conditions to produce extreme weather. More far-fetched is the linking of the 2009 wildfires in Australia to global warming.[12] Yes, the tinder may have been dry, but there have been droughts before, such as that in Southeast Asia in 1998–2003, that in Australia in 2002–2003, and that in Western North America in 1999–2004. Other events named in connection with global warming are the floods in Europe in 2002 and the heat wave there in 2003. Eleven of the 12 warmest years have occurred in the past 12 years. The opposite extremes are never mentioned. The winter of 1962–1963 in Europe was so cold that the Seine froze, and oil deliveries could not reach Paris. The European winter of 2008–2009 was the coldest in 20 years. Does global warming really cause heat waves, cold spells, floods, droughts, fires, and storms?

Fortunately, extreme weather events can, and have been, documented statistically. In many regions of the earth, good temperature and rainfall records have been kept and published. Alexander et al. [15] have compiled these data and produced graphs from which trends can be seen. For example, Fig. 1.20 shows maps and graphs of the occurrence of temperature extremes. The figure requires some explanation. At the upper left, the graph below the map in panel (a) shows, for the period 1951–2003, the number of days per year when the night temperature was very cold, when compared with the average number of such days in the period near the center of the graph. We see that the number of *cold* nights has been decreasing recently. The map above the graph shows where these cold nights occur, averaged over the entire period, with blue showing a lesser number of cold night and red a greater number. By contrast, we can look at the number of *warm* nights in panel (c) at the bottom left. We see that the number of very warm nights has increased a lot recently. The map shows, for instance, that western Africa and Latin America have suffered from this the most. In panels (b) and (d), the number of unusually cold and unusually hot *days* is shown. These show the same trend as the nights, but not as strongly. Remember that these data are not about the general warming trend but are about the occurrence of extreme hot and cold spells. These show a trend toward fewer cold spells and more hot spells as we move into the 21st century.

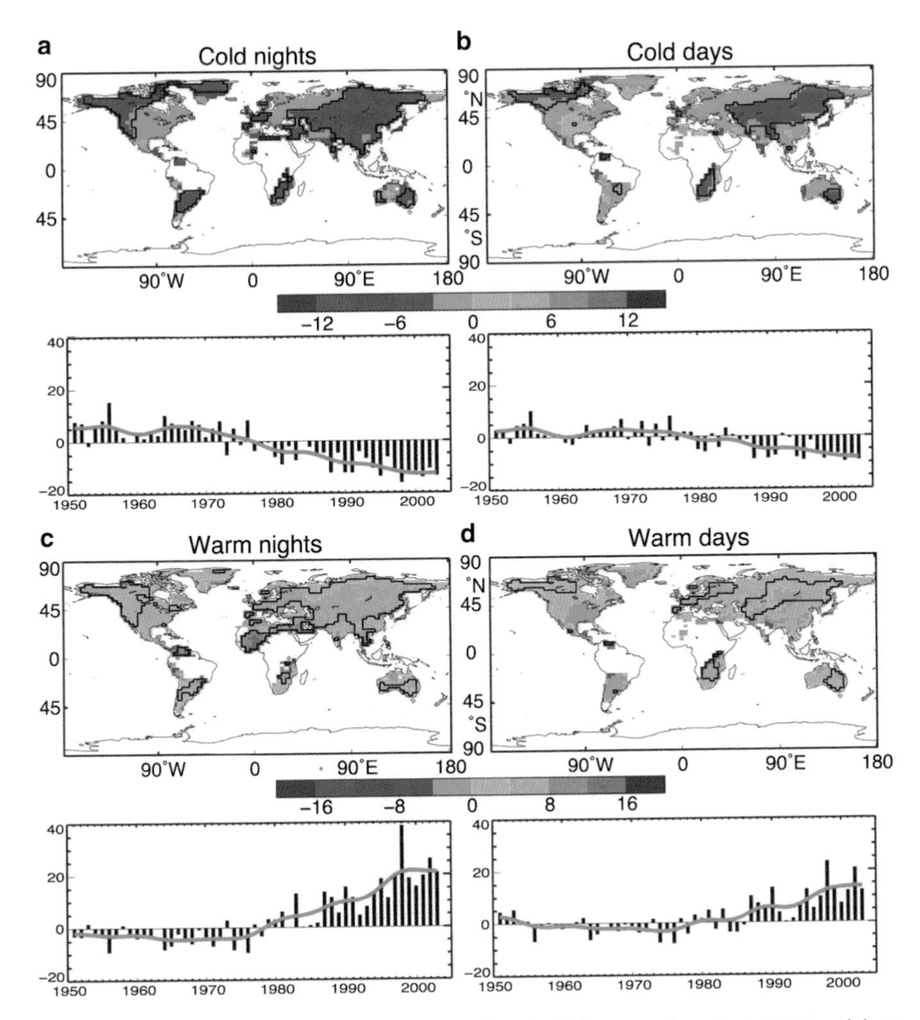

Fig. 1.20 Nights (**a**) and days (**b**) per year colder than the 10th percentile, and nights (**c**) and days (**d**) per year warmer than the 90th percentile, from 1951 to 2003. The *maps* above the *graphs* show the distribution of these extremes over the globe for the entire period. The *heavy lines* show the regions where the data are particularly accurate [6]

The shift of cold and hot spells can be seen more clearly in the bell-shaped probability curves in Fig. 1.21. The blue curves are for the 1950s–1970s, and the red curves for the recent period. The horizontal axis is the number of days per year that have the probability corresponding to the height of the curve. Thus, the peak of the blue curve in panel (a) says that there was a probability of 0.12 (12%) that there were 11 unusually cold nights in any year in that period. The plots (a) and (c) of Fig. 1.21 show the red curves to the left of the blue ones, meaning that there are

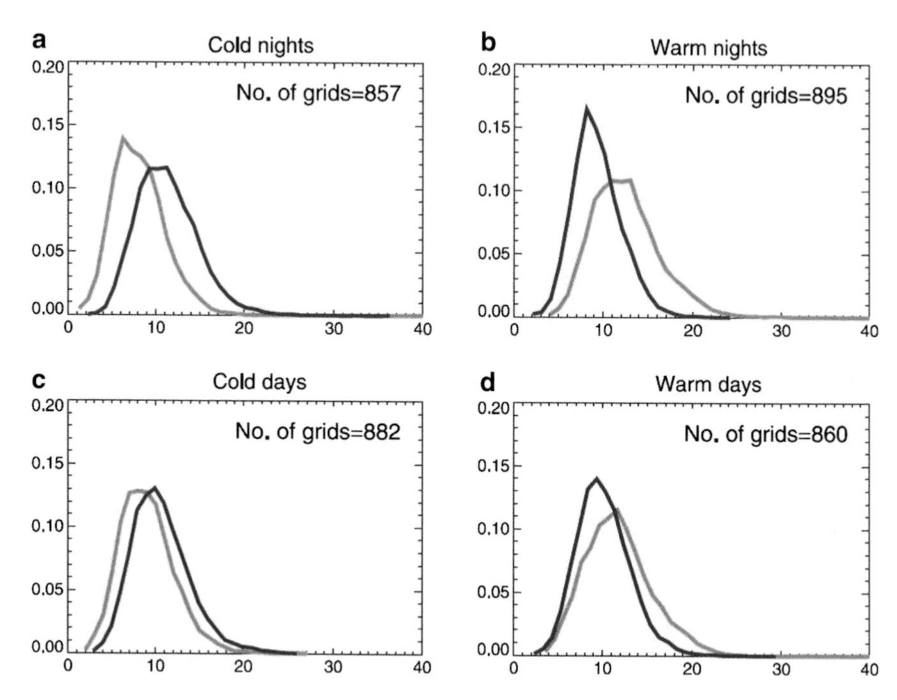

Fig. 1.21 *Bell-shaped curves* showing the probability of having the number of days per year (plotted on the *horizontal axis*) with unusually cold nights (**a**), warm nights (**b**), cold days (**c**), and warm days (**d**) [6]. The *blue curves* are for 1951–1978, and the *red curves* are for 1979–2003 [15]

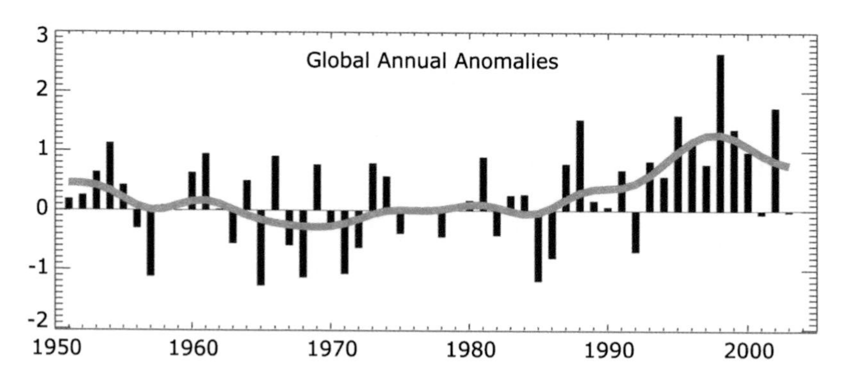

Fig. 1.22 Change in percentage contribution to annual rainfall from very wet days (95th percentile), with 0 representing 22.5%, which is the average percentage in the period at the *center* of the graph [15]

now fewer cold spells; and the plots (b) and (d), with the red curves shifted to the right, show that there are more hot spells in recent years.

The occurrences of unusually heavy rainfall have also been recorded. These extremes are shown in Fig. 1.22. Though there is considerable variation from year to year, a trend toward more rain falling in big storms since 1990 can be seen.

Hurricanes and Typhoons

Extreme events like hurricanes cannot be predicted, and even the statistics are less certain because it is hard to define what constitutes a hurricane, a cyclone, or a typhoon. A useful definition is ACE (accumulated cyclone energy), an index which takes into account both the wind velocities and how long they persevere. The ACE value can be used to tell what is a hurricane and what is just a bad storm. Statistics are gathered for each region and year. Perhaps the most interesting are the data for the Atlantic region. In the 1970–1994 period, there were on average 8.6 tropical storms, 5 hurricanes, and 1.5 major hurricanes; and their average ACE value was only 70% of normal. By contrast, the period 1995–2004 had 13.6 tropical storms, 7.8 hurricanes, and 3.8 major hurricanes, with an average ACE value 159% of normal [6]. In fact, only two years in that period, 1997 and 2002, had fewer hurricanes than normal, and those were El Niño years. It is well known that El Niño produces more severe storms in the Pacific but the opposite in the Atlantic.

Although these statistics show an increase in destructive storms, no direct cause-and-effect relation with global warming can be proved. Nonetheless, there are physical reasons why hurricanes arise, and these are being used in attempts to model hurricanes. When the sea surface temperature rises, more moisture is evaporated into the atmosphere. The water vapor has a greenhouse effect that increases the temperature further. The heated air rises, creating an upward flow of air. When the temperature reaches 26°C (79°F) locally, the air current is strong enough to create a hurricane. Whether this happens or not depends on the wind shear in the atmosphere. If the cross-winds are weak, the upward air currents become very strong in one place, seeded by some random fluctuation there. By Bernoulli's Law, a flowing fluid has less pressure than one that is not moving. This is the same effect that causes a baseball to curve if given a spin such that the air flows on opposite sides of the ball are not equal. The incipient hurricane then has less pressure, and air flows into the column from all sides. The Coriolis force then causes the column to spin and develop into a cyclonic vortex. We described the Coriolis force briefly in Footnote 8. How this force causes winds and spins is interesting and often misunderstood, so we have added a detailed explanation in Box 1.2.

Tropical storms have a cooling effect on surface temperature. Evaporation of seawater cools the surface just as the evaporation of sweat cools our skin. Eventually, the moisture in the atmosphere condenses into rain, reversing the process and carrying the heat back into the ocean; and there is no net cooling. Storms, however, stir up the atmosphere so that this heat is carried up to higher altitudes, where it can be radiated into space before it comes back to earth. This may be a way for nature to stabilize the ocean's temperature. Lightning-lit forest fires renew our forests by burning the undergrowth and allowing new trees to grow. Hurricanes and forest fires may be natural mechanisms that stabilize the present conditions on the planet. Both are catastrophic for mankind, but humans are only a minuscule part of life on earth.

Box 1.2 Why Do Northern Hurricanes Rotate Counter-Clockwise?

Hurricanes have been observed to rotate clockwise in the Southern Hemisphere and counter-clockwise in the Northern Hemisphere, and this has been attributed to the Coriolis force, illustrated in Fig. 1.23. The earth is shown rotating from west to east, causing the sun to rise in the east and set in the west. Several latitude lines are shown. Since these circles are smaller at higher latitudes, the ground speed of the rotation is highest at the equator and diminishes as one moves toward the poles. The atmosphere is dragged by the ground, and therefore the air has a different speed at each latitude, as shown by the lengths of the orange arrows at the left. Nothing happens until the air masses move north–south. Looking at the northern hemisphere in the left diagram, we see that if the air mass at the equator, say, moves northward from A to B, the large velocity of the air at A is brought into a region where the normal velocity is smaller. This motion is indicated by the wiggly blue arrows. The difference between the velocities is shown by the thick blue arrow. The people at latitude B, therefore, feel a wind blowing from west to east. The same happens in the Southern Hemisphere if the air moves south out of the tropics. Now suppose the air flow is *toward* the tropics, southward in the north and northward in the south. This is shown in the right diagram. Then the air masses move into regions where the normal velocity is *larger*. This slowing down of the normal speed appears as a wind going in the opposite direction, namely westward. This is shown by the thick blue arrows in the right diagram. The Coriolis force is the imaginary force that causes that wind.

Whether air moves north or south depends on other conditions, such as temperature or barometric pressure differences at different latitudes. It turns out that for latitudes between 30° and 60° N the motion is northward, as at B,

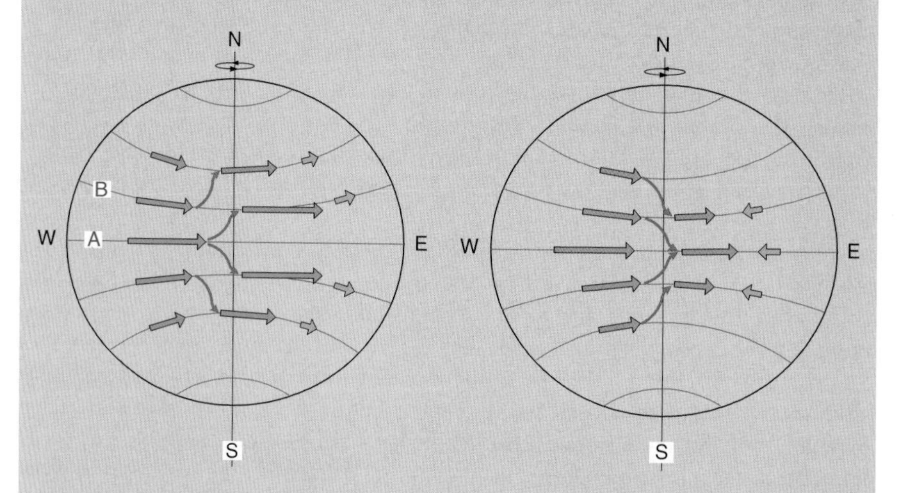

Fig. 1.23 Illustration of Coriolis force causing westerly (*left*) and easterly (*right*) winds

(continued)

Box 1.2 (continued)

giving rise to the Westerlies. These are the winds that cause the flight from New York (41° N) to Los Angeles to be an hour longer than the return trip. At lower latitudes, the N–S motion is *toward* the equator, driving an Easterly. These are the "trade winds" giving the Hawaiian Islands (21° N) their cool.

Now we finally come to hurricanes. The center of a hurricane is a low-pressure area, so air rushes inward. The air mass therefore moves in opposite directions on opposite sides of the eye. This is shown in Fig. 1.24. If this is in the Northern Hemisphere, the Coriolis force pushes the N–S flow toward the west, as shown by the thick blue arrows on the right side of Fig. 1.23. The S–N flow is pushed to the east, as in the left diagram of Fig. 1.23. The E and W flows, of course, do not have a Coriolis effect. The result is that the hurricane rotates counter-clockwise. A hurricane in the Southern Hemisphere would have the arrows reversed, thus causing hurricanes to rotate clockwise.

Is the Coriolis force large enough to do this? A typical hurricane has a diameter of about 500 km (300 miles). If it is located at a latitude of 20°, the difference in the earth's rotation speed between the north and south edges of the hurricane turns out to be about 25 km (28 miles) per hour. This is probably enough to start the rotation, which picks up speed as the hurricane grows. No, the direction of the swirl in a bathtub drain does not depend on hemisphere! A bathtub drain is 25 million times smaller than a hurricane!

All explanations of the Coriolis force assume a spinning object. How do we know the earth is rotating? If we look "down" to the earth from a synchronous satellite, it just sits there; nothing is moving. There is no friction against the vacuum of space to tell that the earth is rotating. Relative to what is it rotating? Actually, it is rotating relative to an inertial frame set by the sun and stars. We can tell that it is rotating because the centrifugal force is palpable. It gives a boost to satellites that are launched in the direction of rotation, which is why so many of them are launched near the equator, and so few have a polar orbit. If the frame of the earth and synchronous satellite were the only frame of reference, the satellite would fall directly down to earth.

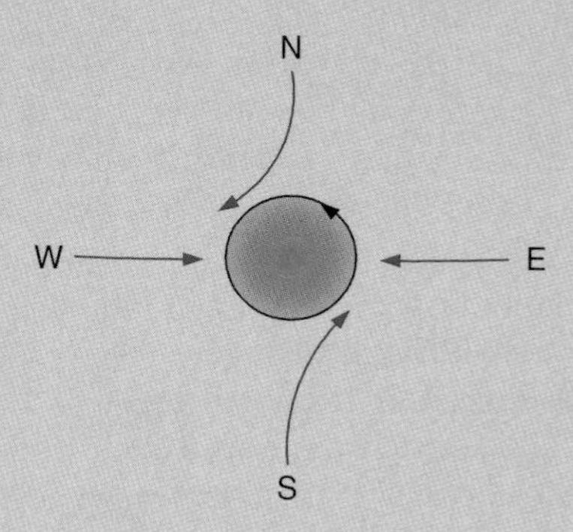

Fig. 1.24 The counter-clockwise torque on a northern hurricane

Slowing the Inevitable

Regardless of the scientific basis of climate change, what can be done about it is a political and economic problem. What makes money is what will happen, but this can be influenced to some extent by laws and subsidies enacted by a savvy government. This well-publicized subject falls outside of the scientific tenor of this book, and only a brief summary is given here. Since the ways to combat global warming depend so much on the way of life and the political setup of each country or community, even the IPCC Working Group 3's voluminous report [16] on mitigation gives few substantive conclusions or recommendations. There is disagreement about the predictions of the IPCC report. Some say that it is too pessimistic, and we need not over-react to the forecasts; others say that the report is not strong enough, and we should act faster than we are doing now. In any case, it is known that the anthropogenic climate change (the only part we can control) is mostly due to GHG emission, particularly of CO_2, and that much of this will persist in the atmosphere for hundreds of years. We can hope to slow down the increase in warming potential, but we cannot expect to recover from our profligate habits for at least half a century.

Mitigation consists of three steps: adaptation, conservation, and invention. Adaptation means taking immediate steps to protect ourselves from impending disasters, such as sea-level rise and violent storms. This means building seawalls, raising bridge heights, strengthening and raising structures near the shore, and so forth. Conservation requires no new technologies or expenses, and many organizations are already promoting this. Lights can be turned off by infrared or motion detectors when no one is in the room. Electronic equipment can be made to draw no current when off. Gasoline can be saved by driving slowly, by carpooling, and by bicycling, for instance. Thermostats can be turned higher in summer and lower in winter. Recycling programs are already in place to save fossil energy used in mining and refining. Everyone is familiar with this list, and many books have been written on "green" living. Along with conservation is efficiency: switching to more energy-efficient appliances which have already been invented. The change from incandescent lamps to fluorescent and LED is being widely implemented. Every time an appliance like a refrigerator has to be replaced, it should be a new, efficient model. Gas–electric hybrid cars and upcoming plug-in hybrids will cut fossil fuel usage, but unfortunately their popularity rises and falls with gasoline prices. The worldwide use of computers has become a large consumer of electricity from fossil fuels. Energy efficiency of computers are increasing all the time, but computers cannot be recycled. New computers all have a large fossil footprint. Houses can be built with better insulation and use of solar energy. Power plants can greatly increase efficiency by co-generation, in which waste heat from electricity generation is recaptured for heating and cooling. Conservation and efficiency are relatively easy to implement, and there is a public will to do this.

The third step in mitigation is the invention of new devices, a longer-term objective. Foremost among these are new ways to generate energy that do not emit CO_2, and these are the subject of Chap. 3. Controlled fusion, the topic of this book, fits into this category of long-term solutions. Shorter-term needs are, for instance, the invention of better batteries or new chemistries for making synthetic fuels. Energy storage is a problem both for transportation and for intermittent energy sources such as solar or wind power, and there has so far been no great breakthrough on batteries. Paradigm-changing inventions may require going back to basics. Forward thinking in the US Department of Energy's Office of Basic Energy Sciences led to a series of ten workshops on Basic Energy Needs such as electrical energy storage, solar energy utilization, and catalysis for energy. The resulting Energy Challenges Report *New science for a secure and sustainable energy future* summarizes the basic scientific advances needed in the long term.[13]

The magnitude of the long-term problem – controlling or reversing global warming in the next 50–100 years – can be seen from the following graphs. We have seen at the beginning of this chapter that anthropogenic forcing of global warming comes mainly from the emission of GHGs, of which CO_2 is the main culprit. Figure 1.25a shows that the major part of this comes from the burning of fossil fuels, so that we must either develop new energy sources or find ways to eliminate the CO_2 pollution. Figure 1.25b shows the distribution of GHG emissions from various human activities worldwide. These activities are so varied among different countries that general methods of mitigation cannot be applied.

From 1970 to 2004, the CO_2 concentration grew by 80%, and the total GHG warming potential increased by 70%. About half of this comes from highly developed nations representing only 20% of the world population. Aggravating the problem is the growth of both population and production. Figure 1.26 shows predictions of population and gross domestic product (GDP) growth and calculations in different scenarios, some 400 in all, without intervention by mitigation techniques. A large divergence of results can be seen, since human behavior has to be assumed in addition to the physics effects considered in climate simulations. Pre-2000 computations are shown by the blue shading, while more recent ones, using different methods, are shown by the lines. Population growth rate has slowed recently, so that the lines give a more optimistic view. Third-world countries will increase their GDPs rapidly as they become industrialized. China has already overtaken the USA as the world leader in CO_2 emissions.

When mitigation is added to the scenarios, different assumptions have to be made for each economic sector in each country or region, and even larger divergence of results is produced. To make sense of the mass of data from some 800 different scenarios, the IPCC has grouped them according to the GHG concentration level or, equivalently, the radiative forcing that each scenario ends up with and has plotted the range of mean global temperature increase above the preindustrial level as predicted by all these models. This is shown in Fig. 1.27. Each category

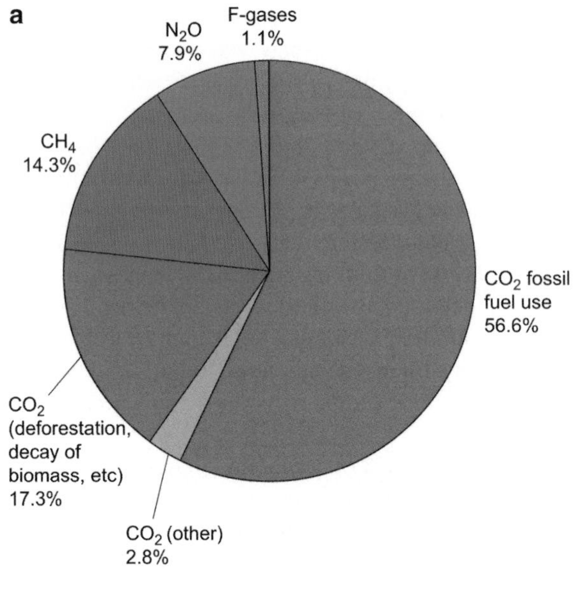

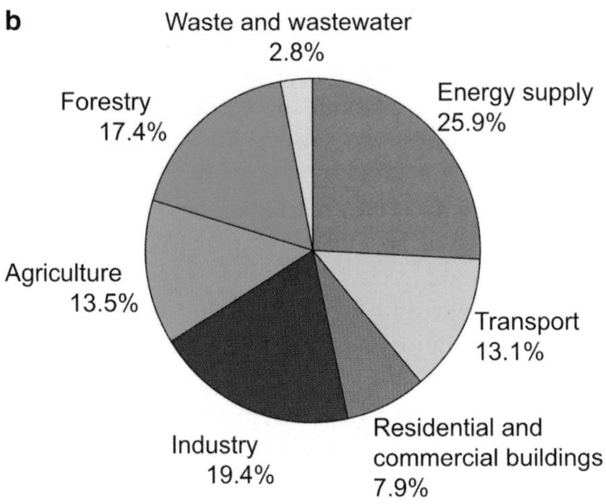

Fig. 1.25 (**a**) Major constituents of anthropogenic GHGs; (**b**) GHG emission by various sectors. Here, F-gases are the ozone-depleting fluorinated gases [16]

from I to VI lumps together scenarios resulting in an increasing range of GHG levels, and the curves show the range of temperature rises that the scenarios in that group predict. The results are also shown in Table 1.1. Here, it is seen that the CO_2 level can be made to peak at some time in the next century and then go down. The larger the CO_2 level, the later this peak will occur. Category IV has the most scenarios; apparently, this is the most anticipated range.

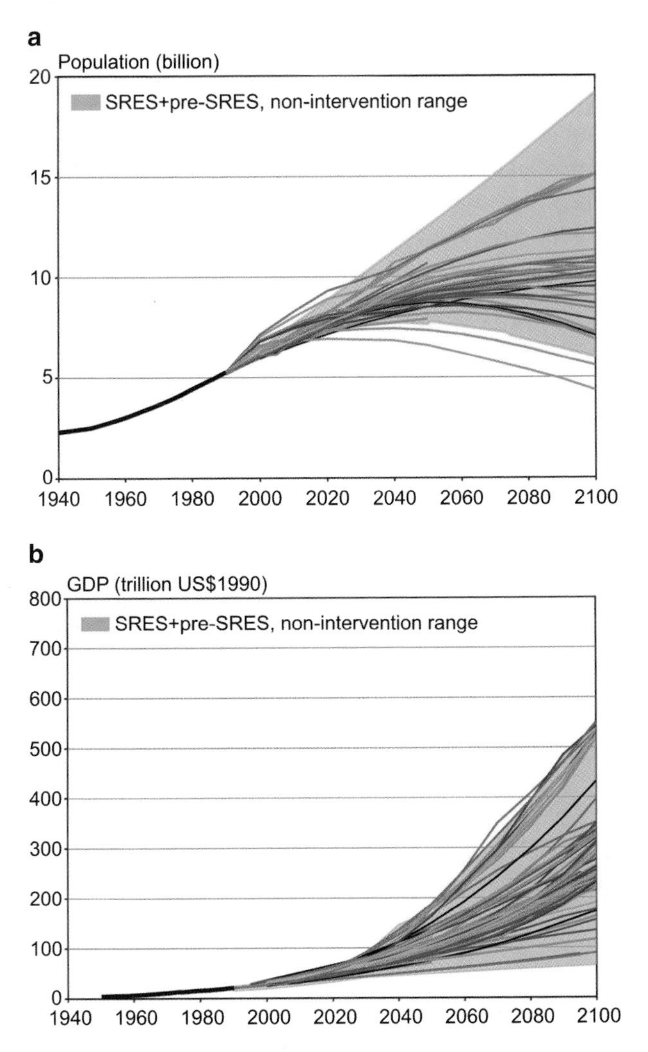

Fig. 1.26 (**a**) World population growth up to 2100 as predicted by various scenarios; (**b**) Predictions of GDP growth in trillions of 1990 US dollars [16]. Here, SRES stands for the IPCC Special Repeat on Emission Scenarios (2000). Both are "baseline" scenarios without mitigation

As complicated as these computations are, they do not tell us how to achieve the stabilization levels specified. No one method of mitigation will do the trick. A simple and attractive way to analyze the problem has been given by Socolow and Pacala [17–19]. They address the intermediate term of the next 50 years, relying on existing methods of conservation and efficiency enhancement but not counting on any new inventions which may come later. Since CO_2 is the dominant GHG, only that gas is considered here to simplify the problem. In Fig. 1.28, the wiggly line shows the data for yearly carbon emissions measured in billions of tons (gigatons) per year (GtC/year). The dashed line is the current path that we are on, and it will lead to a tripling of our current level of about eight GtC/year by the end of the century. The horizontal line is the desired goal of maintaining emissions at the present level. The yellow triangle between these lines represents, then, the reductions in emissions that we have to make to achieve this goal. This triangle is enlarged in Fig. 1.29.

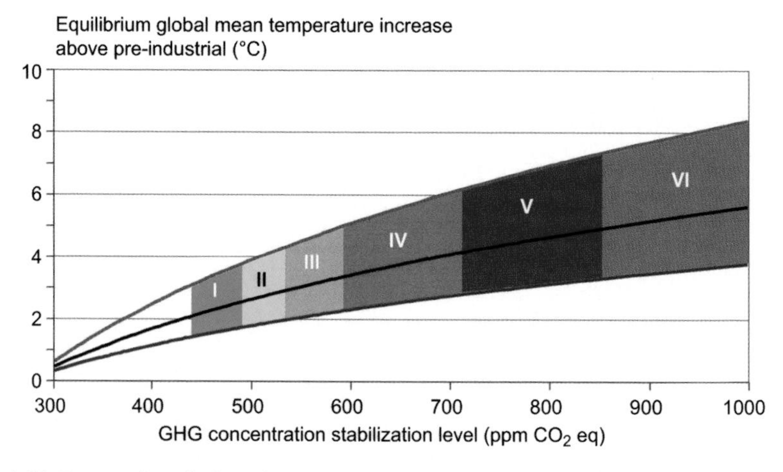

Fig. 1.27 Range of predictions for global temperature rise according to scenarios sorted into Groups I–VI according to the GHG concentration level achieved with mitigation methods [16]

Table 1.1 If the target CO_2 level in column 3 (or the equivalent CO_2 level of all GHGs in column 4) is achieved, the year in which the GHG peaks is given in column 5, and the percentage change in emissions is in column 6 [16]

Category	Additional radioactive forcing (W/m^2)	CO_2 concentration (ppm)	CO_2-eq concentration (ppm)	Peaking year for CO_2 emissions (year)	Change in global emissions in 2050 (% of 2000 estimations) (%)	No. of scenarios
I	2.5–3.0	350–400	445–490	2000–2015	−85 to −50	6
II	3.0–3.5	400–440	490–535	2000–2020	−60 to −30	18
III	3.5–4.0	440–485	535–590	2010–2030	−30 to +5	21
IV	4.0–5.0	485–570	590–710	2020–2060	+10 to +60	118
V	5.0–6.0	570–660	710–855	2050–2080	+25 to +85	9
VI	6.0–7.5	660–790	855–1130	2060–2090	+90 to +140	5
Total						177

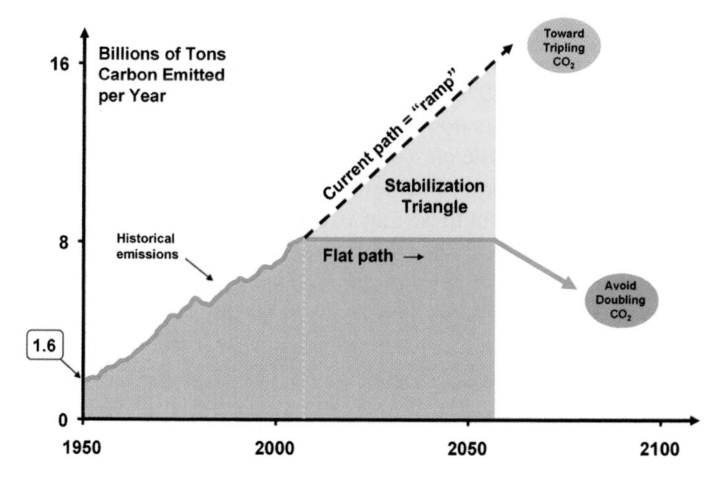

Fig. 1.28 Socolow–Pacala diagram showing the amount of mitigation (*yellow triangle*) needed to keep CO_2 emissions constant at the present level [17–19]

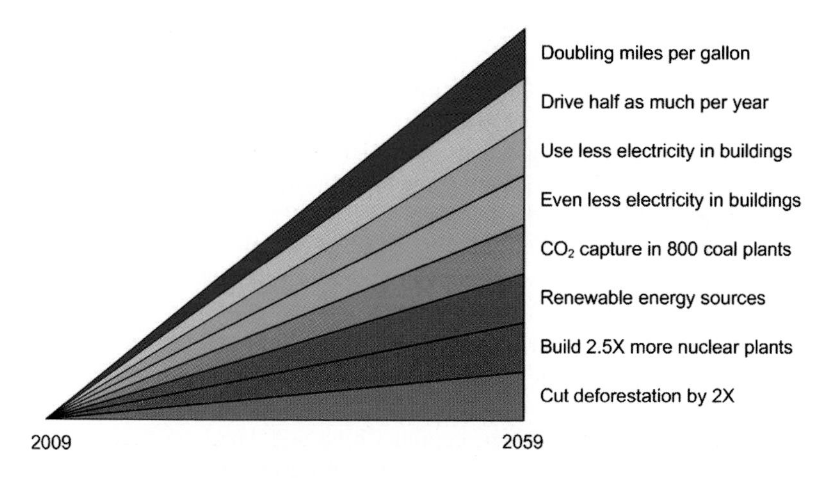

Fig. 1.29 Division of the stabilization triangle into wedges, each representing a cut of one billion tons of carbon emission per year. (Design originated by the Carbon Mitigation Initiative, Princeton University, and replotted from the data in refs. [17–19])

The triangle can be divided into eight "wedges," each representing the contribution of one stratagem to these carbon reductions. Each wedge represents a reduction of one GtC/year in carbon emissions. Together, these wedges would hold carbon emissions to eight GtC/year instead of the 16 GtC/year expected by 2058. Looking at this way, the problem is not so overwhelming. Each sector simply needs to focus on that amount of reduction in its activities. The lines, of course, are not exactly straight; they have been straightened to simplify the idea and make it understandable to all. In fact, the idea is now so simple that the authors have made it into a game that can be played in the classroom, with each student or group of students responsible for finding out how to achieve the goal in one sector. There are numerous ways to make a wedge, but these may overlap. For instance, building 700 fewer coal plants in the next 50 years is one wedge, and so is building 2.5 times more nuclear plants than exist now; but these are the same wedge if the coal plants are replaced by nuclear plants. The wedges in Fig. 1.29 are a few examples chosen so as not to overlap.

From top to bottom: one wedge can be gained if cars averaged 60 miles per gallon (3.9 liters/100 km) instead of 30 mpg (7.8 liters/100 km). Hybrid technology already exists for this. Driving 5,000 miles per year instead of 10,000 would give another wedge. Bicycling, ride sharing, and public transportation could achieve this, but at the expense of personal time. Buildings use 60–70% of all electricity produced, and much of this is unnecessary. Cutting this in half can yield *two* wedges. Requiring 800 coal plants to sequester their CO_2 output would yield one wedge. Building more renewable energy sources such as wind and solar could give one wedge without inventing new technologies. Replacing coal plants by nuclear plants up to 2.5 times their current number would yield one wedge. Cutting in half the area of forests destroyed per year yields one wedge. With so many ways to tackle the problem, this way of dissecting it makes the problem not as mind-boggling as it first seems. It is easier to evolve a strategy. Holding the line is achievable with effort and government incentives. As under-developed countries increase their use of electricity and fuels for cooking, the number of wedges needed will increase, but only by one-fifth of a wedge [17–19].

You may wonder how billions of tons of carbon can get into the air when it goes up as CO_2, which is just a gas weighing no more than the bubbles coming out of a carbonized beverage. Box 1.3 explains how.

Most nations have taken action to do their share in reducing its carbon emissions. With Chancellor Angela Merkel (a physicist) at the helm, Germany leads the way, and other nations have followed. It is the largest market for solar cells and is the third largest producer, behind China and Japan. A feed-in tariff of about 0.5 euro per kilowatt-hour is paid for electricity fed back into the grid. Germany is also a major user of wind power. Its renewable energy sources produce 14.2% of its power, compared with the European Unions' target of 12.5% by 2010.[14] The program is funded by adding 1 euro to monthly electric bills, and the worry is that this will increase with the rapid growth of solar energy deployment. Tony Blair has set emissions goals of a 50% cut by 2050 for the UK. In the USA, California leads the way under Governor Arnold Schwarzenegger, who has introduced ambitious legislation to reduce CO_2 emissions to 1990 levels by 2020 and to 80% below 1990 levels by 2050. The USA, however, has a history of dragging its feet on energy and environment issues since it has more fossil

Box 1.3 How Can CO_2 Weigh So Much?

Here, we are talking about billions of tons of CO_2, a gas as light as the air we breathe. Can our cars and factories actually emit that much weight in a gas? Indeed they can, and here is how. First, a billion is such a large number that it is hard to visualize even though we know that it is a thousand million in the USA and a million million in the UK [A *gigaton* (Gt) is a US billion.] So let us bring it down to something more palpable. There are about a billion cars in the world, so each car emits about a *ton* of pollutants a year, or almost the car's own weight, on average. That is still an unbelievable amount.

The weight of gasoline is mostly in carbon, since gasoline molecules are hydrocarbons with a ratio of about two hydrogen atoms (atomic weight 1) to one carbon atom (atomic weight 12). So 12/14th of the weight of gasoline is the weight of the carbon in it. Gasoline is lighter than water; one liter of it weighs 0.74 kg, compared with the standard weight of 1 kg per liter of water. Of the 0.74 kg, 0.63 kg (six-seventh of it) is carbon. How much does a tankful of gasoline weigh? Say it takes 45 liters (12 gallons) to refill a tank. The weight of a tankful is then about $45 \times 0.74 = 33$ kg (73 lbs), containing $45 \times 0.63 = 28$ kg of carbon. When the gasoline is burned, the carbon and hydrogen combine with oxygen from the air to form CO_2 and H_2O, respectively. Since oxygen's atomic weight is 16, a molecule of CO_2 has atomic weight $12 + (2)(16) = 44$, and the weight of the carbon is multiplied by $44/12 = 3.7$ by picking up O_2 from the air! So when a whole tankful of gasoline is burned, it emits $28 \times 3.7 = 104$ kg (228 lbs) of CO_2 into the air. Suppose a car refuels once every two weeks or 26 times a year, its CO_2 emission is then $26 \times 104 = 2{,}700$ kg of CO_2. This is 2.7 metric tons per year or about 3 US tons! The carbon footprint of driving is even larger, since it takes a lot of fossil energy to make the gasoline in the first place.

The discussion about wedges used units of gigatons of *carbon*, not CO_2, per year. To get back to carbon, we have to divide by 3.7, so our example car can emit $2.7/3.7 = 0.73$ tons or almost a ton of carbon a year. If we increase miles per gallon by a factor of 2, we would save 0.5 ton per year per car or 0.5 GtC/year for one billion cars. By 2059, we expect to have two billion cars and that doubles the savings back to one GtC/year. Hence the top wedge in Fig. 1.29. While we are merrily driving along the highway, the car is spewing out this odorless, colorless gas in great quantities the whole time!

reserves than most countries outside OPEC. The USA did not sign the Kyoto Protocol because it would have cost too much. The 2008 climate-change strategic plan by the Department of Energy called for $3 billion in energy research, which is the same amount as in 1968 in adjusted dollars. Under the Bush administration, the USA failed to live up to its commitment to ITER for two years. ITER is the international project to develop fusion power and is described in Chap. 8. President Obama has appointed Steve Chu as Secretary of Energy and John Holdren (formerly a plasma physicist) as Science Adviser. This administration has already taken steps to move forward aggressively in protecting the environment. For instance, $777 million has been allocated to establish 46 Energy Frontier Research Centers in US universities and laboratories, and a new ARPA-Energy program has been started in the Advanced Research Projects Agency to stimulate new ideas for energy efficiency and curbing of carbon emissions.

The first step that is usually taken for economic reasons is to install a Cap and Trade system, in which companies with large carbon emissions can buy credits from other companies that have emissions below the legislated level. This does not directly reduce overall emissions unless low-carbon companies are new ones using clean energy. Coal plants will find it cheaper to buy carbon credits than to install equipment to capture and sequester their CO_2. A carbon tax would be about $100–$200 per ton of carbon emitted, equivalent to $60 per ton of coal burned or $0.25 per gallon of gasoline [17–19]. Perhaps in anticipation of this tax, which will raise electricity bills, it is encouraging that large companies like Walmart and Google have installed solar panels on their roofs.

Enlightened legislation has succeeded in protecting the environment in the past: CFCs have been eliminated to cure the ozone-hole problem, and lead has been taken out of gasoline, paints, and plumbing. We can succeed again with global warming.

Legislation is also necessary because mitigation involves entire communities, not just individuals. "Greener than thou" is not the right attitude. Here is an example. There was a television program showing the construction of a "green" skyscraper in New York. It was noted that the high building intercepts 40 times as much sunlight as would normally fall on that area. By using partially reflecting windows, the heat load on the building could be reduced, with substantial savings in the energy required for air conditioning. Erecting a building, however, does not change the amount of heat that the sun deposits on each square meter of the earth. What happens is that the building throws a shadow, thus cooling the buildings behind it. This benefit accrues regardless of window design. Reflecting windows would heat the buildings in front, thus increasing *their* air-conditioning energy. Thus, whether total energy is saved or not depends on the energy efficiency of the neighbors' equipment. Market-driven savings are necessarily selfish, and one has to be wary of such profits.

This discussion of mitigation is about the near term of the next 50 years. In the latter half of the twenty-first century, the world will be quite different. New technologies will exist that we cannot imagine now. We went from the Wright brothers to the Boeing 747 in only 67 years.

In 2050, the remaining supplies of oil and gas will be prohibitively expensive. Local power by solar and wind will be commonplace. Coal and nuclear will supply base power in spite of the problem of storing their wastes and the cost of mining. Controlled fusion, which has neither problem, will be coming online as the primary power source. Much of the expense of developing and commercializing new energy technologies can be spared if we finish the development of fusion sooner.

Notes

1. Subsequent letters and rebuttals published in this journal and in APS News showed that a number of physicists believed that variations in solar radiation could have caused the earth's temperature rise. Their proposal to mitigate the American Physical Society's strong statement that climate change is caused by humans was overwhelmingly rejected by the Society.
2. http://www.ipcc.ch or just google IPCC AR4.
3. Note that this is not the half-life of CO_2 concentration in the atmosphere, which is 30 years. CO_2 molecules go in and out of the ocean, and four years is the recycling time. Courtesy of R.F. Chen, University of Massachusetts, Boston, who read this chapter critically.
4. For instance, Hegerl et al. [10], countered by Schneider [11]. Also, Scafetta and West [4] who elicited seven letters to the editor in Physics Today, October 2008, p. 10ff.
5. Not exactly, since fresh water is about 2.5% less dense than seawater.
6. National Geographic News, December 5, 2002.
7. A. Gore, *An Inconvenient Truth*, DVD (Paramount Home Entertainment, 2007).
8. The eastward motion is the result of what physicists call the Coriolis force. The earth rotates west to east (making the sun move east to west daily), and the air picks up the large "ground speed" near the equator. As the air moves northward, it goes into a region of lower ground speed and moves faster eastward than the ground does.
9. What this IPCC graph (FAQ 3.2, Fig. 1.1) means in detail is too complicated to explain and is shown here only to illustrate the large local variations in rainfall data.
10. An impressive graph of the changes in several species appeared in Audubon Magazine, March/April 2009, p. 18.
11. The Ocean Conservancy newsletters, Spring 2008 and Winter 2009.
12. National Geographic Video Program, *Six Degrees Could Change the World* (2009).
13. http://www.sc.doe.gov/bes/reports/list.html.
14. New York Times, May 16, 2008.

References

1. A.B. Robinson, N.E. Robinson, W. Soon, J. Am. Phys. Surg. **12**, 79 (2007)
2. J.M. Inhofe, US senate environment and public works committee minority staff report, http://www.epw.senate.gov/minority
3. Intergovernmental Panel on Climate Change, Assessment report 3, *Working Group 1* (2001)
4. N. Scafetta, B.J. West, Phys. Today **3**, 50 (2008)
5. K.A. Emanuel, *What We Know About Climate Change* (MIT, Cambridge, MA, 2007)
6. Intergovernmental Panel on Climate Change, Assessment report 4, *Working Group 1: The Physical Science Basis* (2007)
7. W.F. Ruddiman, Sci. Am. **292**(3), 46 (2005)
8. A. Scaife, C. Folland, J. Mitchell, Phys. World **20**, 20 (2007)
9. National Research Council, *Surface Temperature Reconstructions for the Last 2000 Years* (National Academies Press, Washington, DC, 2006)
10. G. Hegerl et al., Detection of human influence on a new, validated 1500-year temperature reconstruction. J. Climate **20**, 650 (2007)
11. T. Schneider, Nature **446**, E1 (2007)

12. National Audubon Society, *Mapping Avian Responses to Climate Change in California* (2008)
13. L. Jenni, Kéry Marc, Proc. Roy. Soc. Lond. B **270**, 1467 (2003)
14. A.J. Miller-Rushing, R. Primack, Ecology **89**, 332 (2008)
15. L.V. Alexander et al., J. Geophys. Res. **111**, D05109 (2006)
16. Intergovernmental Panel on Climate Change, Assessment report 4, *Working Group 3: Mitigation of Climate Change* (2007)
17. S. Pacala, R. Socolow, Science **305**, 968 (2004)
18. R. Socolow, R. Hotinski, J.B. Greenblatt, S. Pacala, Environment **46**, 8 (2004)
19. R. Sokolow, S. Pacala, Sci. Am. **295**(3), 50 (2006)

Chapter 2
The Future of Energy I: Fossil Fuels*

There are three different types of power: backbone power, green power, and mobile power. *Backbone power* is the primary energy source that is always there when we need it. *Green power* comes from renewable energy sources which do not pollute. *Mobile power* drives our cars, planes, and other vehicles and has the special requirement of transportability. We will discuss each of these in turn.

Backbone Power

Only 40% of the world's energy use is in the form of electricity; the rest is used for heating and manufacturing. But it is the electric power that governs our way of life in developed countries. During a hot summer day, you have probably experienced a rolling blackout. Night falls and you light a candle. So far so good, and it might even be romantic; but it is too dim to read by. You turn on the radio to find out what the problem is. It does not work. You want to watch TV or play a disk, but those do not work either. You try to call your neighbor to talk about it, but the phone does not work either. Now, where is that phone that connects directly without a power brick? Well, I have all this time to surf the web, you think. The computer is dead as a door nail, and so is the modem. A cup of hot tea would calm your nerves, but… oops! The stove is electric, and so is the hot water heater. Maybe we can take a drive in the moonlight until the power comes back on. But the garage door would not open. There is nothing to do. During the 10-h New York blackout in 1965, people did what came naturally; and the maternity hospitals were jammed nine months later…or so it was reported. This story has been debunked since then.

Heating of homes uses mostly oil and gas, but reliable electric power is still needed in a pinch. Mrs. Johnson, a widow, lives alone in her house in suburbia. The snow is so deep that oil trucks have difficulty in making deliveries. The electricity goes out when a large generator goes down in the public utility. A fierce storm rages outside, and there is no sun. The gusting wind does not provide enough wind power to make up for the

*Numbers in superscripts indicate Notes and square brackets [] indicate References at the end of this chapter.

F.F. Chen, *An Indispensable Truth: How Fusion Power Can Save the Planet*,
DOI 10.1007/978-1-4419-7820-2_2,

shortfall. The inside temperature falls to below zero. Mrs. Johnson has an electric heater, but there is no power. She cannot cook without electricity. After two days, she unfreezes a can of soup by lying next to it in bed. On the third day, she looks at a picture of her grandchildren on her nightstand and wonders if she will ever see them again. Then, on the fourth day, the power goes back on. Yes, she will see them again. Thank goodness for backbone power! This is a dramatization, but loss of backbone power can have deadly consequences. Fortunately, most hospitals have emergency power systems that run on fossil fuels. This is one use of fossil fuels that is defensible.

Renewable energy sources are absolutely necessary for limiting greenhouse gases, but the ones that most people know about – wind, solar, and hydro – are not sufficient or dependable enough to be the primary energy source. Great strides are being made to increase the fractional contribution of these sources, but they can only supplement the primary source. That is because we cannot store energy from intermittent sources or transport that energy from where it is produced to where it is needed. Backbone power has got to be available at all times. This means that reserve generating capacity has to be built to supply power when all else fails. Backbone power keeps people alive and functional in their normal activities. Green energy can save on fuel cost, but not on capital costs, because backbone power plants still have to be built to supply the necessary standby capacity. This will be quantified in the section on wind power. *Only three energy sources fulfill the requirements of backbone power: fossil fuels, fission, and fusion. Of these, only fusion energy has the prospect of being backbone, green, and safe.*

The Energy Deficit

Energy Units

Before we talk about energy, let us be sure we know what it is. If you turn on a 100-W light bulb, it will use up 100 W of energy, right? Not exactly! Watts measure the *rate* at which energy is used, which is called *power*. Energy is something we can store, and power is how fast we use it up. A toaster takes about 1,000 W, or 1 kW, of electricity to run. If we turn it on for an hour, it will consume 1 kWh of *energy*. A 200-W light bulb left on for 10 h would use up 2,000 Wh, or 2 kWh of energy. On a more personal note, suppose you ate a 200-calorie hamburger (a small one). That's energy which you store. Suppose it takes you 2 h of exercise to burn off that energy, then you are using up 100 C/h, which is the average *power* you put out during the workout. What confuses most people is that the well-known electrical unit, the watt, is a unit of power, not energy. You have to multiply by time to get energy.

To compound the confusion, articles about the energy crisis do not use the same units for energy. There are British thermal units (BTUs), terawatt-years, millions of barrels of oil equivalent (MBOE), megatonnes of coal equivalent, and so forth. In this book, we convert all the data to metric units; namely, watts and joules and their multiples. The conversion factors among the most common units are given in Box 2.1.

Box 2.1 Conversion of Energy Units

	Equals this many of these units			
One of these units	kJ	kWh	BTU	BOE
Kilojoule	1	2.8×10^{-4}	0.95	1.6×10^{-7}
Kilowatt-hour	3,600	1	3,412	5.6×10^{-4}
British thermal unit	1.055	2.9×10^{-4}	1	1.7×10^{-7}
Barrel of oil equivalent	6.1×10^{6}	1,700	5.8×10^{6}	1
Tonne of oil equivalent	4.5×10^{7}	1.2×10^{4}	4.3×10^{7}	7.33

	Equals this many of these units				
One of these units	TJ	TW-year	MBtu	Quad	MBOE
Terajoule	1	3.2×10^{-8}	948	9.5×10^{-7}	1.6×10^{-4}
Terawatt-year	3.2×10^{7}	1	3.0×10^{10}	30	5,200
Million British thermal units	1.1×10^{-3}	3.3×10^{-11}	1	1.0×10^{-9}	1.7×10^{-7}
Quad	1.1×10^{6}	0.033	1.0×10^{9}	1	172
Million barrels of oil equivalent	6.1×10^{3}	1.9×10^{-4}	5.8×10^{6}	5.8×10^{-3}	1
Million tonnes of oil equivalent	4.5×10^{4}	1.4×10^{-3}	4.3×10^{7}	0.043	7.33

The first table shows the basic units, the most familiar of which is the kilowatt-hour (kWh) used for electrical energy. A joule is the metric unit for energy; but the joule-per-second, a unit of *power* called the watt, is better known. A kilowatt is then 1,000 W or a kilojoule (kJ) per second. Since there are 3,600 s in an hour, a kilowatt-hour is 3,600 kJ. The BTU, used well before the metric system was established, is still widely used and is conveniently close to 1 kJ. A tonne is a metric ton, equal to 1.1 tons. For energies outside the laboratory, industrial people often use barrels of oil equivalent (BOE), which is obviously imprecise, since it depends on the kind of oil and how efficiently it is burned; but it has been defined by the US Internal Revenue Service as 5.8 million kilojoules.

The second table shows the scaled-up units that one has to use to measure energies on a national or global scale. A terajoule (TJ) is a trillion (10^{12}) joules or a billion kilojoules. In scientific notation, the exponent (the superscript above the "10") is simply the number of zeroes after the "1." Here are the prefixes corresponding to the various multiplication factors:

Thousand: 1,000 (10^{3}), kilo-
Million: 1,000,000 (10^{6}), mega-
Billion: 1,000,000,000 (10^{9}), giga-
Trillion: 1,000,000,000,000 (10^{12}), tera-
Quadrillion: 1,000,000,000,000,000 (10^{15}), peta-
Quintillion: 1,000,000,000,000,000,000 (10^{18}), exa-

(continued)

Box 2.1 (continued)

A terawatt-year is 32 million terajoules, since there are that many seconds in a year. A large power plant generates about 1 GW of power, and thus a GW-year of energy per year. A terawatt-year is the annual output of 1,000 power plants. Since 1 BTU is about 1 kJ, a million BTU (MBtu) is about a billion joules or about 1 GJ. This size unit is used for partial energies. A Quad is a quadrillion (10^{15}) BTU or a billion MBtus, a unit appropriate for worldwide production. It is equal to 172 MBOEs, a unit often used in magazine articles as well as technical journals. We shall convert all graphs to Quads and MBtu's here, the saving grace being that they are close to the modern metric units.

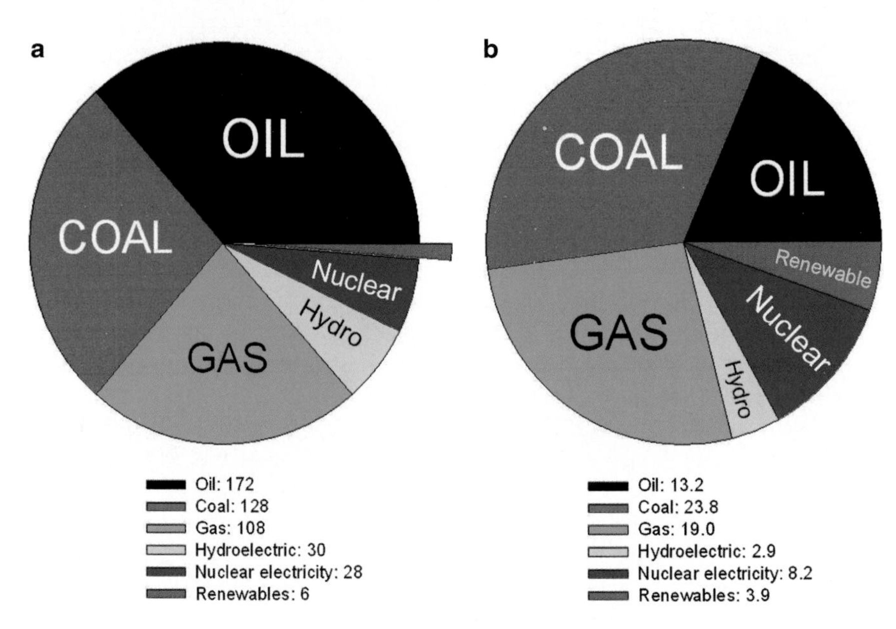

Fig. 2.1 Sources of energy consumed in (**a**) the world[1] and (**b**) the USA[5]. Data are for 2006 and are in units of Quads per year

Energy Consumption

The consumption of energy in the world and in the USA is shown in Fig. 2.1. For the world, the total of 472 Quads is dominated by oil, with all fossil fuels accounting for 79% of the total. For the USA, the total of 71 Quads is dominated by coal, with fossil fuels accounting for 86% of the total. Renewable energy, mainly from wind, solar, and biomass (wood and waste), amounted in 2006 to only 1.3% of the total in the world and 5.5% in the USA.

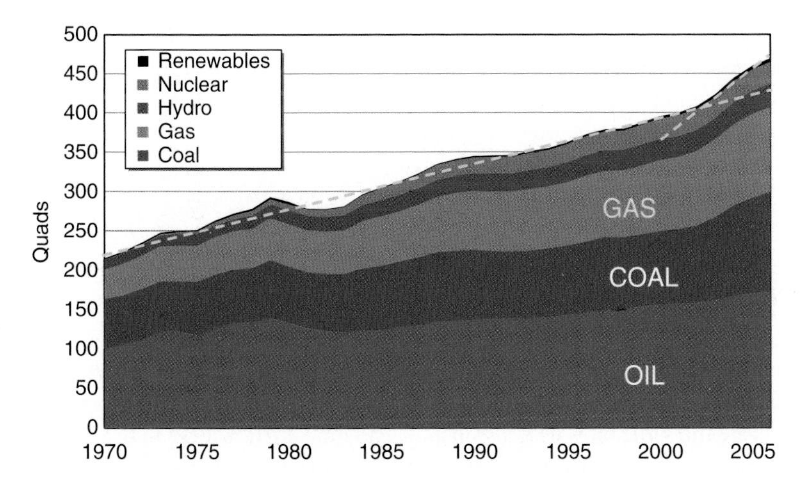

Fig. 2.2 36-Year history of the world's annual consumption of energy from various sources. The *dashed lines* show the average rate of increase from 1970 to 2002 and from 2002 to 2006 (Data from footnotes 1 and 5)

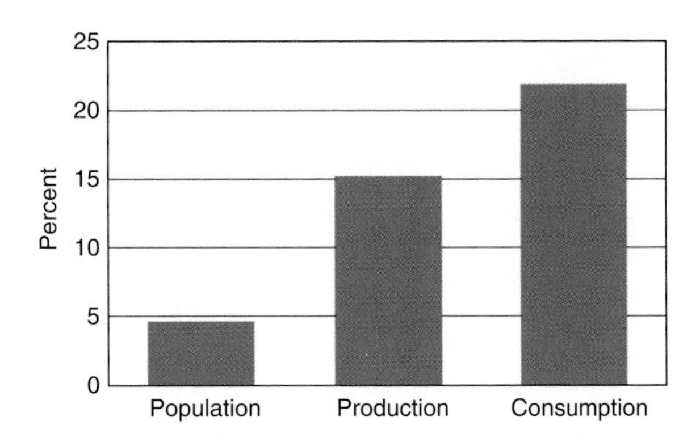

Fig. 2.3 The US share (in percent) of the world's population and energy production and consumption in 2005[1]

The growth of the world's energy consumption over the last 36 years is shown in Fig. 2.2, organized by source. The total dominance of fossil fuels is evident. The contribution of renewable sources is only the thickness of the black line at the top. The dashed lines show that the rate of increase of total annual energy was rather steady from 1970 to 2002 at about six Quads per year. However, the rate seems to have increased since 2002 to about 16 Quads per year.

Figure 2.3 shows the fraction of the world's resources that the USA consumes. We can see at a glance that the USA, with less than 5% of the world's population, consumes 22% of its energy. It is noteworthy that most of this energy, 15% of the total, is produced within the USA, as shown by the middle bar in the graph. The rest is imported. The USA is relatively rich in fossil deposits, and this explains why it has been lagging in the race to develop alternative sources. Countries like France, Germany, and Japan are more dependent on imports and have taken the lead in developing fossil alternatives.

Energy Forecasts

Estimating the energy the world will need in the future is risky business. We have to depend on computer simulations, as we did for climate change. Some of these models are the same ones used in Chap. 1, and they differ widely in the assumptions made in each scenario. Results up to 2030 are shown in Fig. 2.4. The middle bar in each group is the reference scenario, in which policies and laws remain unchanged. The low and high bars in each group are the minimum and maximum predictions across all scenarios. As expected, uncertainty increases with time, and so does the range of predictions. For the case of high economic growth, we see that the present consumption of some 470 Quads will grow to 760 Quads by 2030. By the end of the century, the level will be above 1,200 Quads. The problem is obvious: this doubling and then tripling of energy demand will occur while oil and gas reserves are being completely depleted.

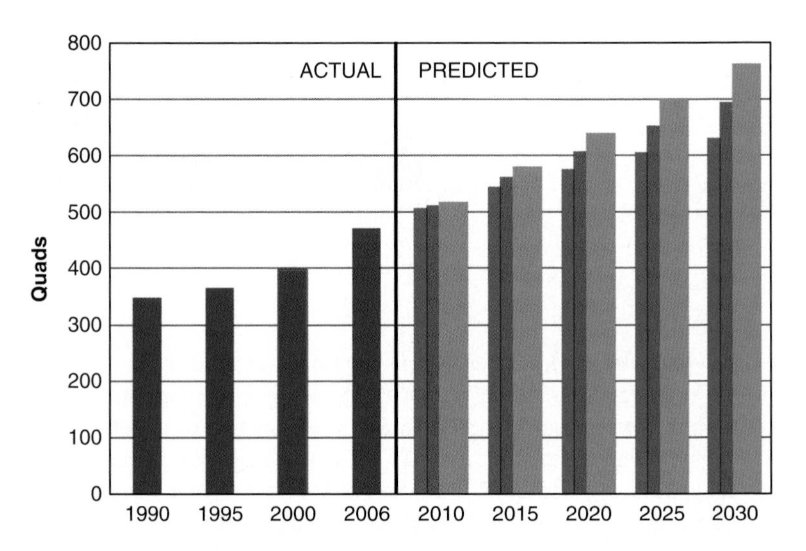

Fig. 2.4 Predictions of the world's annual energy needs up to 2030. The *triple bars* show the minimum, average, and maximum values computed using different scenarios (Data from footnote 1)

What Drives the Increasing Demand?

Population increase is one cause, but not the main one. The projections are shown in Fig. 2.5. In developed countries, the scenarios generally predict a slowing population growth peaking around 2040, followed by a slow decline to the end of the century. The underdeveloped countries in Africa and Latin America are responsible for the continuing increase to 2100. Experts believe that population will stop growing at 10 billion people, the most that the earth can support. After that, we will have to start colonizing the moon and Mars.

It is the productivity of man that drives the need for more and more energy, as shown in Fig. 2.6. One measure of this is the gross domestic product or GDP. This can be evaluated for a single, developed country; but to do this for the whole world requires dealing with different currencies and ways of accounting. For this reason, the GDP for the world is estimated differently by different sources; and the data for the past are not necessarily accurate. Nonetheless, projections for *growth* can be calculated using a consistent system. In Fig. 2.6, we have reduced the GDP data to US dollars of the year 2000. There we see that the GDP is expected to grow exponentially. This is in spite of the fact that the GDP per person in developed countries is expected to decline slightly. It is the industrialization of the rest of the world that drives energy demand.

To illustrate this, Fig. 2.7 shows energy demand in the high economic growth case of Fig. 2.4, broken down between OECD and non-OECD countries. The

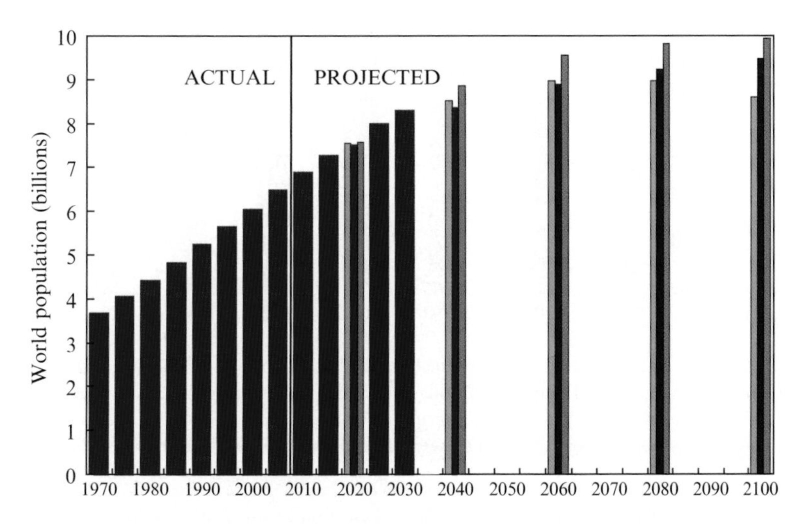

Fig. 2.5 Projections of population increase. The *triple bars* show the predictions of three different scenarios where this information is available (Data from Scenarios of Greenhouse Gas Emissions and Atmospheric Concentrations, US Climate Change Science Program, Synthesis and Assessment Product 2.1a, 2007 and footnote 1)

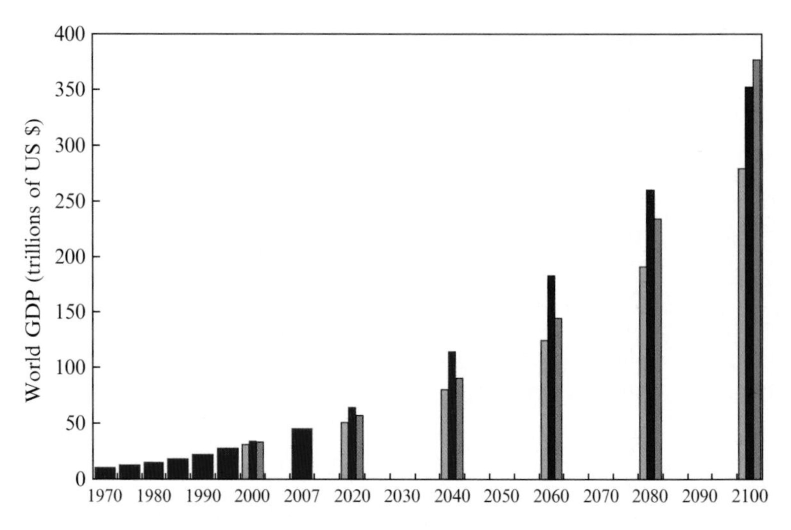

Fig. 2.6 Projections of gross domestic product increase. Units are in trillions of US dollars of year 2000. The *triple bars* show the predictions of three different scenarios from Scenarios of Greenhouse Gas Emissions and Atmospheric Concentrations, US Climate Change Science Program, Synthesis and Assessment Product 2.1a, 2007. The 2007 point is from the CIA World Fact Book

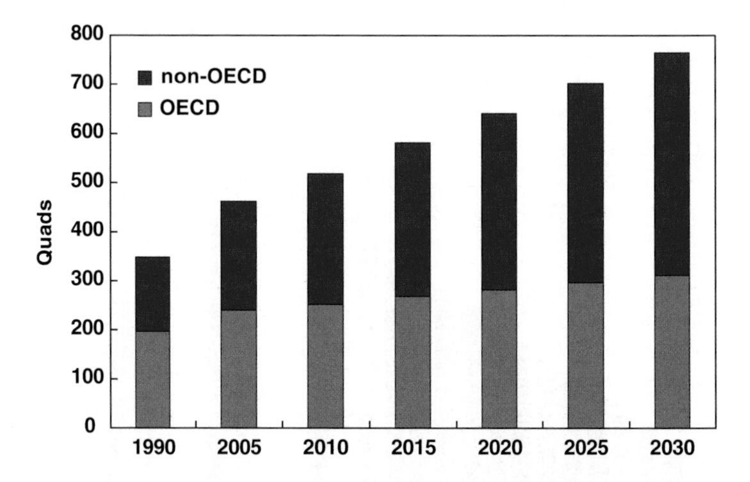

Fig. 2.7 Current and projected energy demand by OECD and non-OECD countries, in Quads (Data from International Energy Outlook 2008, Energy Information Administration, US Department of Energy. See also *World energy, technology, and climate change policy outlook 2030*, Directorate-General for Research (Energy), European Commission, Brussels (2003).)

Organization for Economic Cooperation and Development consists of some 30 industrialized countries mostly those in Europe and North America, plus Japan, South Korea, and Australia. The non-OECD countries include Russia, China, India, Africa, the Middle East, and Central and South America. It is clear that most of the growth is in the non-OECD countries up to the year 2030, and the projections of

GDP in Fig. 2.6 show even greater dominance of the non-OECD countries in the second half of the twenty-first century.

Can we believe these predictions? We may not trust what unseen scientists do with their computers, but this is the best information we have if we are to plan for the future. Doubters and naysayers are usually single persons who act on their intuitions without doing the homework. By contrast, the scenarios shown here are worked out by large groups of experts using massive amounts of data. The ground rules of a scenario are decided at the beginning, and widely differing approaches are taken to cover the spectrum of possible results. For instance, in predicting the path of underdeveloped countries, one scenario assumes that different localities modernize in isolation, following their own customs and ways of life, while another scenario assumes that communication is so good that all countries are connected by the internet and can share methods and economics with the rest of the world. The different regions have GDPs that increase at vastly different rates, but they tend to average out over the world. This leads to the scenario results of Fig. 2.6, which differ greatly by the year 2100 but nonetheless show a definite trend. In the energy projections of Fig. 2.4, these vastly different scenarios still agree within ±10% in the year 2030.

Where Does the Energy go?

Not all energy is the same. Electricity is the form of energy that governs the way we live in modern society; we depend on it in ways that we do not always appreciate. Much of the energy needed by underdeveloped countries will be for building an electricity infrastructure. The next four graphs will show where electricity comes from and where it goes. The readily available data here are for the USA.

Figure 2.8a shows that total energy use in the USA is shared almost equally by the transportation, industrial, residential, and commercial sectors, but they use

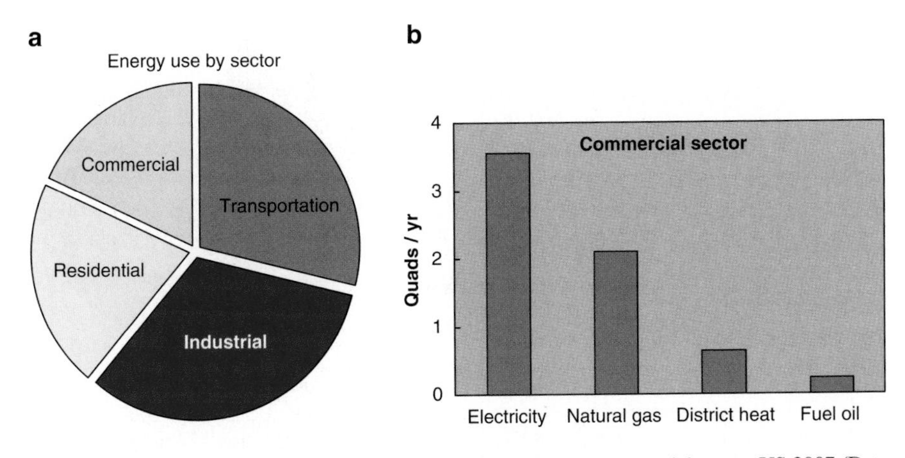

Fig. 2.8 (**a**) Energy use by sector and (**b**) energy sources for the commercial sector. US 2007 (Data from Annual Energy Review 2007, Energy Information Administration, US Department of Energy.)

different sources. Transportation energy comes almost entirely from petroleum (loosely called "oil" here). Industry burns most oil and natural gas ("gas"). In the commercial and residential sectors, electricity and gas are equally important, but electricity is fast overtaking gas, as seen in Fig. 2.8b for the commercial sector. In this sector, lighting and air conditioning in buildings use large amounts of electricity, much of which can be saved by strict conservation practices. In the residential sector, 31% of the electricity is used for space heating, cooling, and ventilation; and 35% for kitchen appliances and hot water. Lighting, electronics, laundry, and other uses take up less than 10% each.[2] Each household in the USA uses 1.2 kW of electricity steadily when averaged over day and night, winter and summer. There being 2.6 persons in each household on average, each person is responsible for about 470 W of electricity consumption.[2] The peak load is, of course, many times that; and power plants have to be built for peak demand.

To make things worse, making electricity is very inefficient. The losses are shown in Fig. 2.9a, and the sources of energy for electricity are shown in Fig. 2.9b. Two-thirds (69%) of the fossil energy used for electricity is lost in production! The main loss is in converting heat into electricity. The raw material, such as coal, has to be prepared to be burned. It then burned to produce steam, and the steam is used to drive a turbine (an electric motor in reverse) to generate electricity. Each of these steps takes energy. The main loss comes from an old thermodynamics principle called Carnot's theorem, which states that the best that any engine can do in converting heat to mechanical energy is to suffer a fractional loss equal to the initial temperature divided by the final temperature. For instance, if the steam is heated to 500°C (932°F) and cooled to 100°C (212°F) to drive the turbine, the absolute temperatures are about 770 and 370 K, with a ratio of about 0.48 or 48%. This is the part that is lost, leaving 0.52 for the part that can be used. So even if all is ideal, the efficiency cannot be more than 52%. Modern heat engines can exceed this figure,

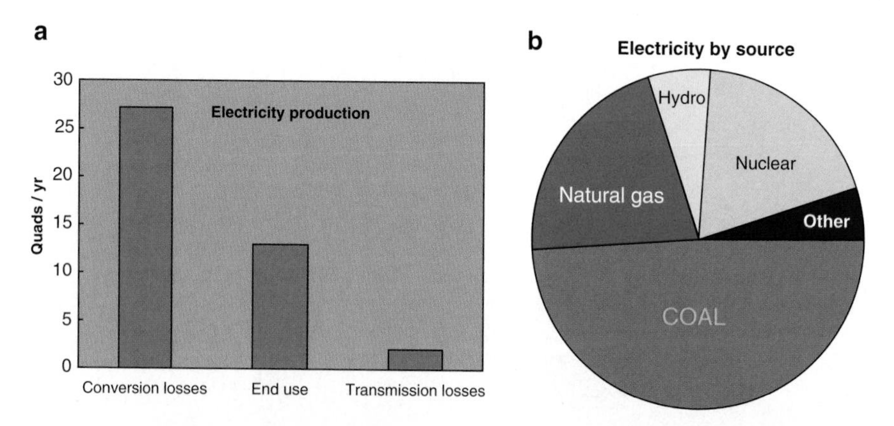

Fig. 2.9 (**a**) Losses in electricity production, expressed in Quads/year rather than a percentage; and (**b**) relative contribution of different sources of electric energy. US data for 2007 and 2003, respectively, from Annual Energy Review 2007, Energy Information Administration, US Department of Energy

but then the turbine is not perfectly efficient either. This conversion loss is shown in Fig. 2.9a. To this we have to add the losses in transmission and distribution, including the heating of the high-voltage cables and the transformers to step the voltage down to wall-plug values. These losses are given by the last column in Fig. 2.9a. What is left for use is the middle column there.

Our thirst for electricity comes at great cost. We are using precious fossil fuels very inefficiently. Systems that produce electricity directly without going through a heat cycle make much more sense. These are hydroelectricity, wind, and photovoltaic solar cells. Solar, unfortunately, has its own physical limits on efficiency, as will be seen later in this chapter. We see in Fig. 2.9b that by far the largest fraction of electricity comes from coal, the dirtiest of all fossil fuels! And we have not yet counted the fossil energy expended in mining, transporting, and refining coal. It is encouraging that the slice labeled "other," which includes wind and solar, appears larger than the splinter seen in our other pie charts. This is because they can produce electricity directly, without going through a heat cycle.

Energy Reserves

Here is the bottom line: how much fossil fuel the world has left, and how long it will last. The data are for 2007, and the heat equivalents have been reduced to Quads.[3] First, let us look at coal, the largest resource, shown in Fig. 2.10. The regions are as follows: Asia Pacific includes China, India, Japan, Korea, Australia, and other nations on the Pacific Rim. Europe and Eurasia include West and East Europe, the Former Soviet Union, Greece, and Turkey. North America is the USA, Canada, and Mexico. South and Central America is self-explanatory, and so is Middle East. Proven reserves are known deposits that can be mined using existing techniques.

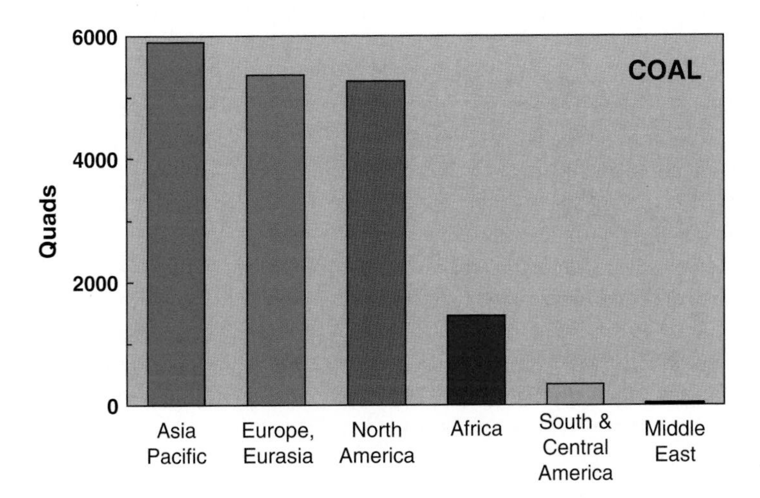

Fig. 2.10 Proven coal reserves by region (Data from BP Statistical Review of World Energy 2008.)

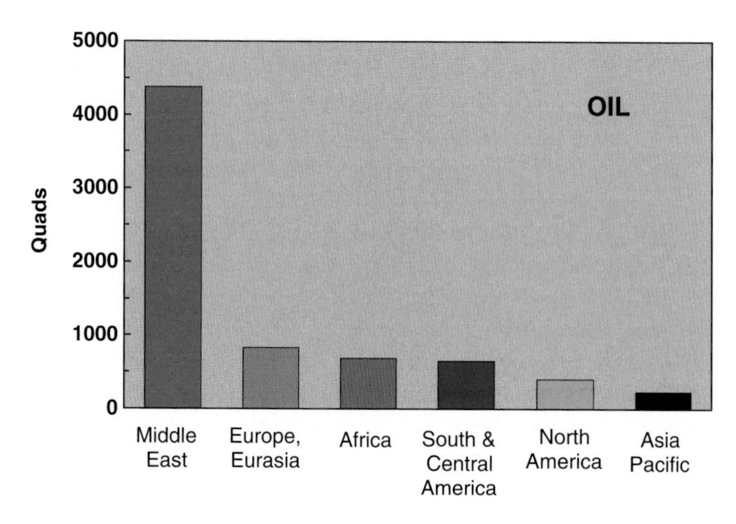

Fig. 2.11 Proven oil reserves by region (Data from BP Statistical Review of World Energy 2008.)

We see that coal deposits are concentrated in the first three regions and are practically nonexistent in the Middle East.

Petroleum, of course, is a different story. Figure 2.11 shows what we already know: most of the world's oil is in the Middle East. In addition to normal oil, there are reportedly large amounts of oil trapped in oil sands and shale oil in Canada. However, this oil is extremely hard to extract, and known methods are energy intensive. This oil is not included here because it would take a new technology to get a large net energy gain.

Natural gas reserves are shown in Fig. 2.12. The Middle East leads here also, but note the difference in scales. The amount of energy in gas is small compared with coal and oil.

The dominance of coal is more clearly seen when we put these reserves on the same scale, as done in Fig. 2.13. For oil, we see from the red columns that we will still be dependent on the Middle East for our main transportation fuel.

Now we come to the crux of the problem: how long will fossil fuels last? This is estimated by the R/P ratio, the ratio of Reserves to Production. Hubbert's Peak, mentioned in the Prologue, has been estimated numerous times, but more exact information is now available in the R/P ratio, shown in Fig. 2.14. If we take the fossil energy available in known deposits in each region and divide by the annual production of energy in that region, we can get the number of years the supply will last *if there is no trade*. Clearly, some regions will be more self-supporting energywise than others. In the real world, we import and export fuels; and the number of years the world's fossil reserves will last is shown at the right of the figure. Oil will be depleted in 42 years, natural gas in 60 years, and coal in 133 years. Note that the consumption rate has been *assumed to be steady at the 2007 level!* With the predicted increase in consumption shown in Fig. 2.4, these reserves will be gone in a much shorter time.

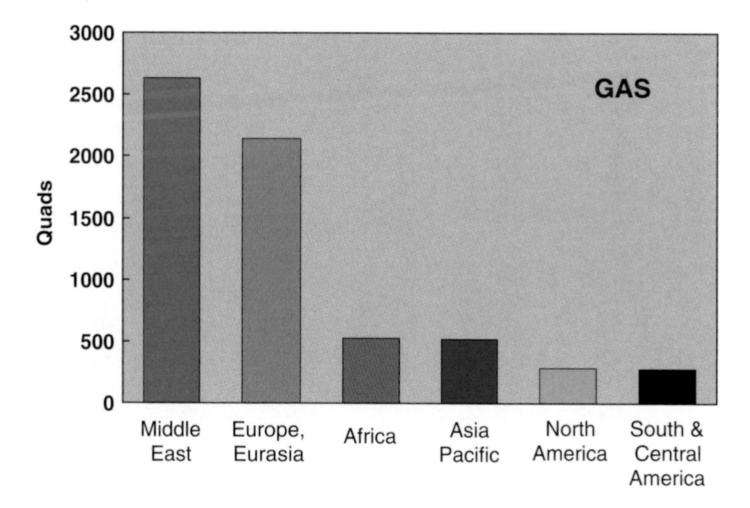

Fig. 2.12 Proven natural gas reserves by region (Data from BP Statistical Review of World Energy 2008.)

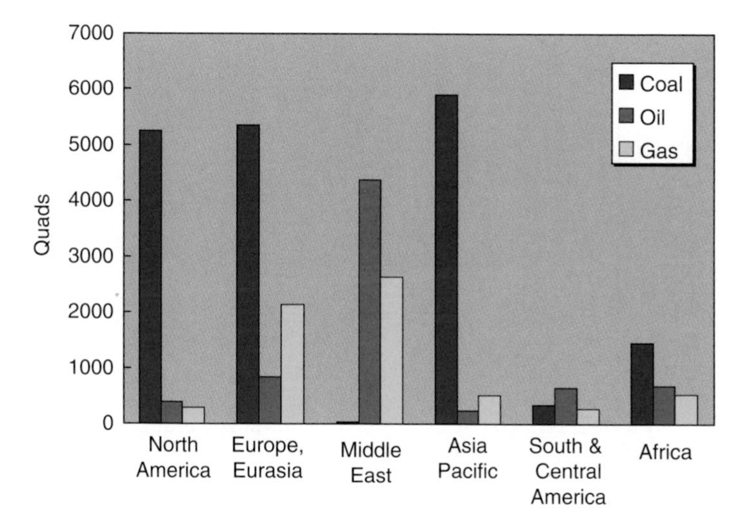

Fig. 2.13 Summary of the world's proven fossil reserves (Data from BP Statistical Review of World Energy 2008.)

Let us examine the case of oil, which is critical for gasoline and all our travels. In the Prologue, we mentioned Hubbert's peak, a prediction by M. King Hubbert in 1956 about the eventual decline of production as we run out of fossil fuels. The shape of the peak is usually shown as a smooth, symmetric curve like that in Fig. 2.15. The dots there are yearly data on oil production in the USA since 1900, expressed in Quads per year of equivalent thermal energy. We see that indeed the data lie on a Hubbert-type curve, and the peak was reached in 1973, the year of the oil crisis.

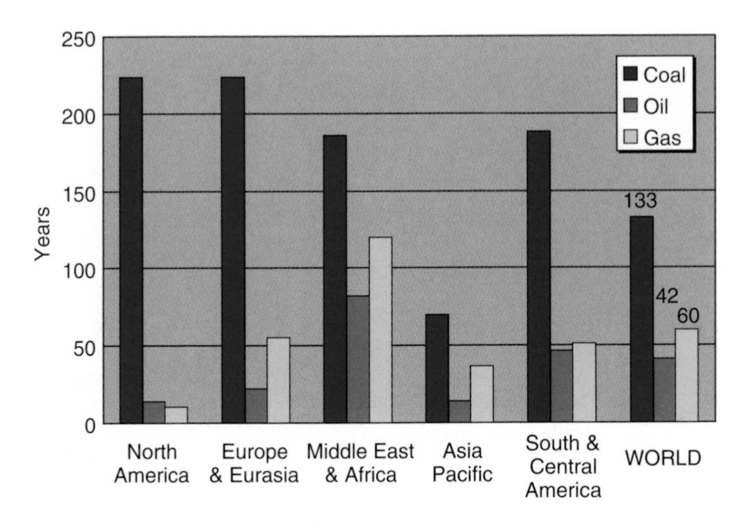

Fig. 2.14 Reserves-to-Production ratio for different regions and for the world (Data from BP Statistical Review of World Energy 2008.)

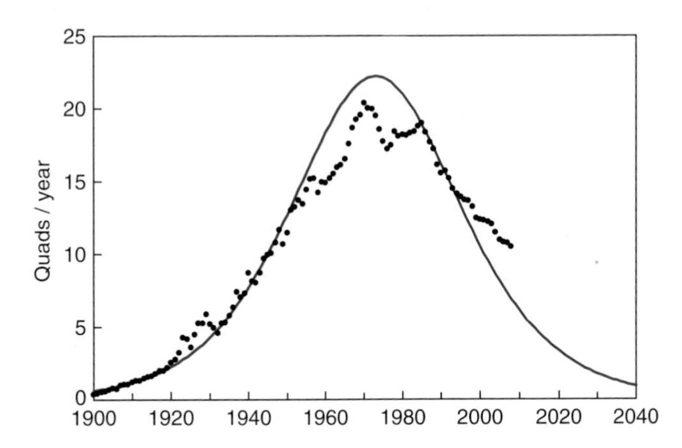

Fig. 2.15 Production of oil in the USA from 1900 to 2008 (*dots*), fitted to a Hubbert-type curve (*line*). The area under the curve is the total amount of oil in conventional deposits (Data from Energy Information Administration website)

US oil production has been declining since then, but clearly this is not true for the whole world. Figure 2.2 showed that use of all fossil fuels, including oil, is still increasing. What the USA lacks, it is importing from the Middle East. We are not changing our habits in airplane and car travel, or in the transport of food and merchandise in trucks. This means that the consumption curve will not be symmetric. It will keep going up and then crash rapidly when oil becomes more and more difficult to find.

When will this come? Figure 2.4 showed predictions of the world's fossil fuel consumption up to 2030. We can get specific predictions for oil for that period from the Energy Information Administration's Reference Case.[1] Using the average rate of increase of 1.2% per year, we can predict the annual consumption beyond 2030. Then, knowing the total amount of oil reserves in conventional deposits in 2007 (7,180 Quads) from footnote 3, we can calculate how those reserves decrease year by year. This is shown in Fig. 2.16. *The oil reserves in the world will be depleted by 2040.* Though this seems to agree with Fig. 2.15, it is different. First, this is for the whole world, including the Middle East, not just the USA. Second, the drop will be much sharper, as shown by the dotted line, since the consumption rate keeps going up until the price of oil becomes prohibitive. It will become imperative to use alternative fuels, so complete depletion of reserves can be avoided. Oil consumption (the same as production when the whole world is involved) will decrease much faster than it rose, giving an asymmetric Hubbert curve. There are unconventional sources which can be tapped at great cost, but this would extend the curve only slightly. The point here is that the *world's* oil will soon be depleted. The world cannot import oil from elsewhere the way the USA can.

The need for petroleum can be mitigated several ways. Cars can be made much more efficient if they are, for instance, made of carbon fiber instead of heavy steel. Current gasoline engines are terribly inefficient. Only 1% of the energy is used to move the driver, and only 10% to move the car; the other 90% is lost in heat.[4] Gas–electric hybrids are already marketed and can double gas mileage. Electric plug-in vehicles use no gas, shifting the burden to the more abundant fuel, coal, which

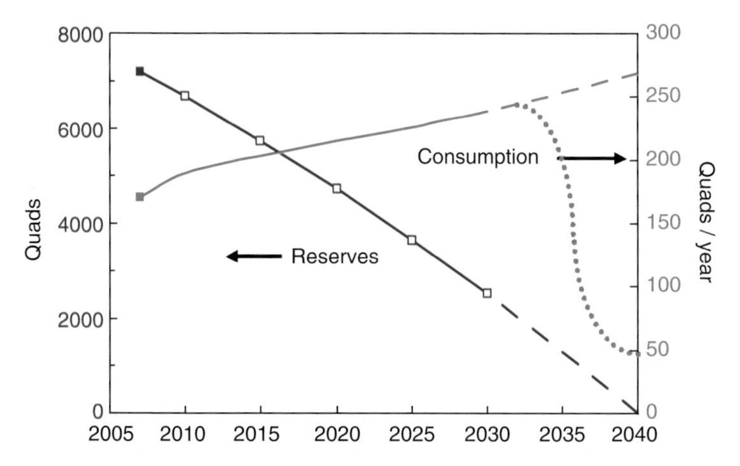

Fig. 2.16 Predictions of world oil reserves (*blue line, left scale*) and annual consumption (*red line, right scale*). The 2007 points (*solid squares*) are actual data. The *hollow squares* are from computer simulations; the *dashed lines* are extrapolations; and the *dotted line* is a conjecture (Data from BP Statistical Review of World Energy 2008 and International Energy Outlook 2008, Energy Information Administration, US Department of Energy. See also *World energy, technology, and climate change policy outlook 2030*, Directorate-General for Research (Energy), European Commission, Brussels (2003).)

can be burned more efficiently at high temperature at central power plants. Alternate fuels such as hydrogen and ethanol have their own problems, which will be discussed in Chap. 3. The buying public's preference for horsepower and speed has to change to one for fuel efficiency. It will take many decades to change the manufacturing infrastructure from one making steel parts and gas engines to another making carbon parts and efficient new types of engines. Changing the infrastructure of fuel distribution (gas stations and pipelines) will also require decades. Thus, the oil problem is already upon us.

We have stressed oil as the most imminent problem, but all fossil fuels will soon have to be replaced. It has been said that there is no shortage of coal, which may be true for North America and Europe, but not for the entire world. China is building a new coal plant every week, thus depleting its reserves rapidly. In Fig. 2.14, the Asia Pacific region already has the lowest amount of reserves compared with its consumption rate. Eliminating greenhouse gases from all coal plants will be very costly, if at all possible. As for oil, it does not make sense to burn up this precious resource when it should be saved for special applications, such as making plastics. By the time, oil and gas run out by mid-century, their entire energy slices in Fig. 2.2 will have to be filled by nuclear, fusion, and renewable energy. Renewables like wind, solar, and biofuels would have to expand a 100-fold to make up the difference. Nuclear energy can do it by expanding 17 times, but it has environmental problems. These sources are discussed in the next chapter. They would be needed, together with continued use of coal, to fill the energy gap in the first half of this century. *If fusion can be online by mid-century, it will help. It will definitely be needed for the second half of the century. By 2100, with even coal and uranium running out, fusion should become the main source of backbone power*. How fusion works and its difficult development will concern us in Part II.

Coal and Carbon Management

Coal is the major problem. It supplies 27% of the world's energy and 40% of its electricity. In the USA, coal supplies 23% of all energy and a whopping 49% of electrical energy.[5] Coal is also the worst CO_2 emitter. In 2007, CO_2 emissions from coal burning amounted to 2.65 billion tons in China and 2.20 billion tons in the USA.[6] No other nation was responsible for more than 0.54 billion tons. No wonder, since China and the USA produced 41.1 and 18.8% of the world's energy from coal because of their large deposits.[3] It is easy to see why coal is so dominant: it is cheaper than oil or gas; there is a large supply of it; and it is easy to transport by rail. The mines are not remote; no pipelines need to be built; and there is no need for tankers which occasionally crash and foul our beaches.

Coal is bad news also because it causes deaths in mining accidents, it destroys the environment when whole mountains are dug up, and it emits many pollutants such as sulfur. We all remember stories of families waiting in vain for news about their loved ones trapped miles deep in the earth with no hope of rescue. In the USA alone, 100 million tons of coal ash and sludge are stored in 200 landfills annually, and these contain dangerous contaminants such as arsenic, lead, selenium, boron, cadmium, and cobalt.[7] The problems are exacerbated by the rapid development of China, where coal plants are being built at the rate of one large one in a week, while the USA has stopped building them as of 2007. Let us concentrate on this biggest problem: the unstoppable industrialization of China and India. In China, 74% of energy comes from coal, and this will increase to 90% with continued growth, though efforts to develop renewables may hold the line at 70%.[8] China has about 30,000 coal mines, 24,000 of which are small ones which use antiquated equipment and are not regulated for safety. In 2006, 4,746 miners died in China, versus only 47 in the USA; both numbers are down from those in earlier years. Chinese coal generates every year 395 billion cubic meters of methane, SO_2, and black soot, all of which have larger warming potential than CO_2. Furthermore, the methane is what causes explosions in mines, and the SO_2 causes acid rain. Of the million people in China suffering from black lung disease, 60% are miners. This disease increases the coal mining death total by 50%.[8] It is not likely that other energy sources can replace coal any time soon, but we can try to mitigate its effect on global warming.

Cap and Trade

The coal industry will not do anything that lowers its profits without government intervention. What is being done in most developed countries is to legislate a decrease in carbon emissions by a certain deadline. The Cap and Trade system allows large utilities to meet these standards without a sudden capital expenditure. However, Cap and Trade does not directly lower total CO_2 emissions. It works as follows. An emissions cap is legislated for each industry, and this cap is divided into credits, in terms of tons of CO_2, that that sector is allowed to emit. Credits are then auctioned off. Heavy emitters, such as a large utility, may find it less expensive to buy credits than to build equipment to reduce emissions, while light emitters, such as a modern, efficient plant, can sell the credits that they do not need. Both utilities would gain financially. To make this work, the government has to establish a fraud-proof monitoring system and assess severe penalties for noncompliance.

Unfortunately, Cap and Trade does not actually decrease carbon emissions because, in the example above, both utilities would emit the same amount of CO_2 that they would without trading credits. It actually allows the large utility to delay investing in the equipment for capturing CO_2, when it should be forced to do it as soon as possible. New power plants using solar or wind energy can sell their credits to coal plants, but these producers of green power are being built *anyway* because they are profitable, not because of Cap and Trade. Cap and Trade does not force industries to lower their emissions if they are already taking steps to do this because of societal concerns or because it is profitable publicity-wise. Only *additional* low-carbon plants should be counted, not those that already exist or are planned.

Loopholes in the scheme allow accounting tricks to get around doing anything constructive. The only advantage of Cap and Trade is to make large polluters aware of what is coming and begin to worry about it.

Carbon Sequestration

To continue using coal, we have to capture the emitted CO_2 and bury it. This is called carbon capture and storage (CCS), but we will continue to avoid acronyms when possible. There are three steps: first, CO_2 has to be separated from the flue gas out of a coal burner; second, the CO_2 has to be transported to a burial site; and third, it has to be injected into a geological formation that can hold it forever. The last part is of course highly debatable; but it is the first part, capture, that is the most expensive. There are three basic ways to do this.[9] In the first method, the flue gas is mixed with a liquid solvent called MEA into which the CO_2 dissolves. The MEA's chemical name is not always spelled the same way, but it is a corrosive liquid found in household products such as paint strippers and all-purpose cleaners. When the MEA is heated to 150°C, pure CO_2 is released, and the MEA is cleaned up with steam to be reused. This method can be retrofitted to existing plants, but there is a huge penalty. The heating and steam production takes up to 30% of all the energy produced by the power plant! The cost of this step can be as much as four times higher than that of the other two steps. At the moment, other absorbers are being tried to lower this cost.[10]

In the second method, the flue gas mixture is controlled by burning the coal in a specific way. When it is burned in air, which is 80% nitrogen and 20% oxygen, there is a lot of nitrogen in the mix, and N_2O is a greenhouse gas. A better way is to remove the nitrogen from air at the outset and burn the coal in pure oxygen. What comes out is water and pure CO_2, ready to be sequestered. However, separating the nitrogen from the air to get pure oxygen requires 28% of the power plant's energy, still a steep penalty. This method is being tested by Vattenfall, Sweden's energy company, in the town Schwarze Pumpe in Germany. The experiment is fairly large – 30 MW – but not of electric utility size. A novel feature was added to this "oxyfuel" process: the flue gas is recirculated into the burner with the oxygen. This keeps the temperature low enough to prevent melting the boiler walls, as would happen with pure oxygen. In effect, the CO_2 in the flue gas replaces the nitrogen in air, diluting the oxygen without using nitrogen.

The third method is coal gasification: the coal is heated to a high temperature with steam and oxygen, turning the coal into a gas, called syngas, which is a mixture of carbon monoxide (CO) and hydrogen (H_2), plus some nasty contaminants. After the syngas is purified, it is the fuel for generating electricity in an "integrated gasification combined cycle," or IGCC, an acronym that seems unavoidable in this case. Coal gasification has been tested in fairly large power plants, but the IGCC sounds like a Rube Goldberg type contraption that has yet to be verified on a large scale. An air separation unit to get pure oxygen is still required both for syngas generation and for burning the syngas later. After the pollutants are taken out, the gas goes into a chamber where the CO combines with steam (H_2O) to form CO_2 and H_2. Pure hydrogen is separated out through a membrane, giving carbon-free fuel. The rest of the gas, containing CO_2, CO, and H_2, is burned with oxygen in successive turbines, a gas turbine and a steam turbine, to generate electricity. The pure hydrogen separated by the membrane can be sold or burned to generate more electricity cleanly. The IGCC can be 45% efficient, compared with 35% in ordinary coal plants limited by the Carnot theorem that we described earlier. Meanwhile, the CO_2 generation is lower, and it comes out in pure form to be stored. This separation system adds only 25% to the cost of electricity. An even more efficient method called chemical looping is under development.[9] New chemical structures for capturing CO_2 are described in Chap. 3 under Hydrogen Cars.

In 2003, the FutureGen Alliance had proposed a plan to test IGCC on a large scale by building a $1 billion plant in Illinois, finishing in 2013. That project was canceled by President G.W. Bush in 2008 because the projected cost had almost doubled. Unbelievably, this figure was an accounting error; the actual increase was to only $1.5 billion. Under President Obama, Energy Secretary Steve Chu has pledged $1.1 billion of economic stimulus money to restart the project, with the other funds to be raised by FutureGen. There is $2.4 billion of stimulus money slated for CCS research. This is to be compared with $3 billion spent by the Department of Energy for this purpose since 2001.

Now that we have separated out the CO_2, the problem is where to put it. There are three main places: old wells, underground, and undersea. The oil and gas that we mine have been trapped in the earth for millennia, so it is possible that porous rock or underground caverns can hold liquids and gases stably. To carry CO_2 to these sites, the gas has to be highly compressed to a small volume and transported by truck or rail. This step entails a certain amount of danger, should there be an accident causing the container to explode and release tons of CO_2 into the atmosphere. The gas is then injected under pressure into depleted oil or gas wells, where it could stay for millennia if it were not for the leaks made in drilling the wells in the first place. These old wells have to be sealed tightly. The trouble is that carbon dioxide and water combine to form carbonic acid, and the seal has to withstand this acid attack. This storage solution is well tested because it is used to store excess gas and oil mined in the summer for use in the winter. The difference here is that the storage has to be stable essentially forever. The possibility of leaks has to be carefully monitored. Injection of CO_2 into oil wells is actually beneficial, for it helps to push the oil up. Toward the end of life for an oil well, the oil gets quite thick; and gas, which might as well be CO_2, is injected to lower the viscosity. This is what happening in those nodding pumps seen along the California coast.

There are many large subterranean formations that can hold carbon dioxide. These are porous sandstone deposits covered with a cap of hard, impervious rock. For instance, such a depository has been found below a little town called Thornton somewhere south of California's capital of Sacramento. It is estimated that it can hold billions of tons of CO_2 in its pores, enough to store away hundreds of years of California's emissions.[11] Of course, no one knows whether it will leak.

Fig. 2.17 The Sleipner Platform in the North Sea (http://images.google.com)

There are plans to drill into this formation and test it, to the dismay of local residents. The reaction, NUMBY (Not Under My Back Yard!), is a switch from NIMBY (Not In My Back Yard!), an epithet used against wind and solar power.

Large geologic formations under the sea have also been found for CO_2 storage. These are layers of porous sandstone called saline aquifers lying deep below the seabed and capped by impermeable slate. Storage in these aquifers is the only sequestration method that has been tested on a large scale, and this is a story in itself.[9, 11–13] The Sleipner Platform, shown in Fig. 2.17, is a huge oil drilling and carbon sequestration plant located in the middle of the North Sea, halfway between Norway and England. It was built in 1996 by Statoil, Norway's largest petroleum company to produce oil while testing sequestration. Built to withstand the frigid conditions and storms with 130-mile winds and 70-foot waves, it houses a crew of 240 whose jobs are considered the most dangerous in the world. Below Sleipner lies not only a rich field of natural gas but also a saline aquifer called the Utsira Formation lying a kilometer below the seabed (Fig. 2.18). The aquifer is very large: 500 × 50 km in area and 200 m thick. It can hold 100 times Europe's annual CO_2 emissions.

There was a special reason to build sequestration into the plant *ab initio*. The gas from the Sleipner field contains about 9% CO_2, too high to burn properly unless reduced to 2.5%. The gas has to be scrubbed using the MEA solvent described above, thus releasing a million tons of CO_2 a year that has to be stored. The way the CO_2 is injected involves a little physics. It is compressed to 80 atmospheres because at this pressure it turns into a liquid about 70% as dense as water. So it is stored as a liquid. When it is mixed with the salt water in the aquifer, it tends to rise, since it is less dense. One worries how fast it moves and whether the 200-m thick layer of shale above the storage volume can spring a leak. Such leaks can arise from drilling through the cap to inject the gas, and these holes have to be carefully sealed with acid-proof material. Statoil has spent millions of dollars to develop a way to measure the spreading and leaking of the CO_2 using sound waves. Since the system has 25-m resolution and the area is measured in kilometers, the amount of data is many megabytes. These data clearly show that the CO_2 is spreading sideways as well as upwards, and that there are no leaks so far. In the best scenario, the CO_2 will eventually dissolve into the brine (in 1,000 years or so) and thus become a liquid heavier than water. This then moves safely downwards, and on a geologic timescale will turn into a mineral, thus locking the carbon away permanently. All fossil fuels will be but a distant memory by that time.

The Utsira formation is unusual in that it is located at the same place as the gas deposit, so that no transportation of the CO_2 is necessary; but it is not unique as a large burial site. It is estimated that the USA has subterranean reservoirs capable of storing 4 trillion tonnes of CO_2, enough to take care of its emissions until coal runs out. Statoil would not have built the Sleipner plant if it did not have to pay an annual $53M carbon tax imposed by the Norwegian government. Global warming cannot be halted without strong legislation by enlightened political leaders. The cost of separating the CO_2 and burying it is estimated to be about $25–$50 per tonne. Though this may come down as new techniques are developed, it is still a huge expense. Three tonnes of CO_2 is produced for each tonne of coal burned, and a fairly large (1 GW) coal plant gives off 6 million tonnes of carbon dioxide per year. The cost of up to $300M would be passed on to the consumer. That is not even the main problem. It is simply not possible to make a fundamental change in all coal plants or to build enough new-technology plants in a short time. Up to now, except for Sleipner, only small, scattered projects for cleaning up coal have been funded,

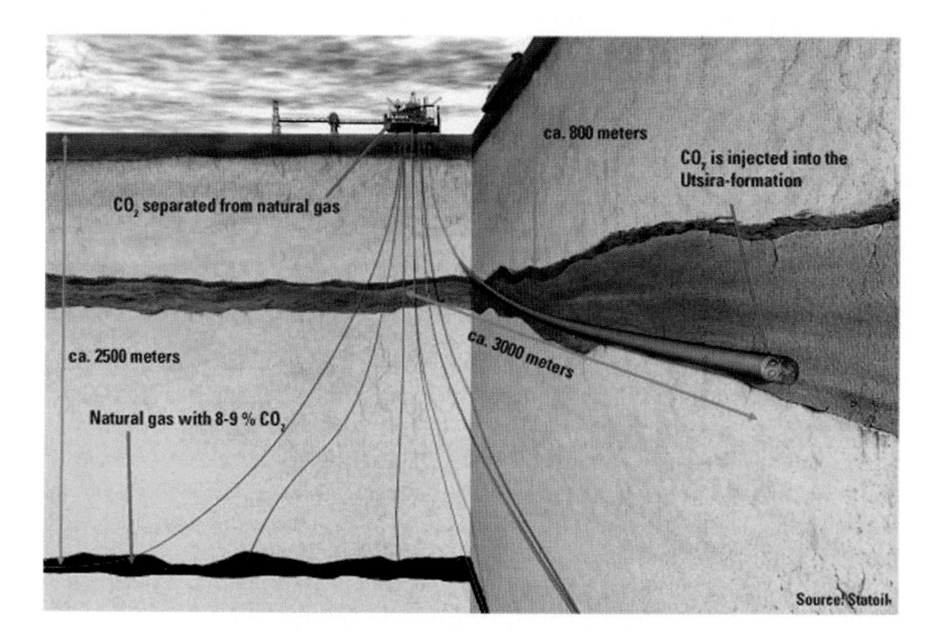

Fig. 2.18 Diagram of the gas field and saline aquifer below Sleipner (http://images.google.com)

with no integrated plan for replacing all dirty coal power with clean coal power. This is in stark contrast to the ITER project for developing fusion power; there, even the political problems of a large international collaboration have been tackled and solved. It may take two or three decades to clean up all coal power, and this is no shorter than the time needed to commercialize carbon-free renewable sources.

Oil and Gas Pipedreams

We discussed the shortage of oil earlier in this chapter but gave short shrift to natural gas, which supplies as much energy as oil, as seen in Fig. 2.1. That is because gas and oil mostly occur in the same places, are mined the same way, and are similarly depleted. We also ignored the minor overlap between oil and gas: oil can be converted to propane and butane gases, which we use for camping and power in remote houses; and gas can be liquefied at low temperatures for more convenient transport as LNG (liquefied natural gas). In this section, we will again consider these fuels together as we consider the various proposals for extending their supplies.

The price of oil can jump wildly, as it did in 2008–2009 from higher than $140 to less than $40 per barrel, and it can jump back. The price of gasoline follows, and this has a great effect on the economy as people travel less and buy fewer large cars. The gas crisis of 1973 even triggered legislation setting the speed limit in the USA at 55 miles per hour. These rapid changes are not our concern here; we are worried about the end of oil and gas altogether. In 2007, BP (British Petroleum) reported that proven reserves are 15% higher than previously thought, so that oil will last another 40 years,[14] 30 more than predicted in Fig. 2.16. There was widespread doubt, however, about this result from a normally reliable source. For instance, the IEA (International Energy Agency) assessed the top 400 oil fields and found them old and in bad condition.[15] They did not see how the present consumption of 87 million barrels per day can exceed 100 million, much less than the 116 million predicted by 2030. Similarly, the six oil fields that produce 90% of Saudi oil were found to be greatly depleted.[16] In the USA, the crunch is already felt as the Alaskan pipeline, built in the 1970s to carry most of our domestic oil and gas, is carrying only one-third as much these days because the wells at Prudhoe Bay are being depleted at the rate of 16% per year. Figure 2.19 shows that discoveries of new oil fields have been declining since 1964.[17]

Russia exports more oil and gas than any other country. It produces 11.8% of the world's oil, compared with 9.9% for Saudi Arabia and 12.4% for Iran, United Arab Emirates, Kuwait, and Iraq combined.[15] Its state utility, Gazprom, is so powerful that it held the Ukraine and other parts of Europe at its mercy by shutting off gas deliveries through its pipeline. The politics of gas and oil are changing. The former holders of power, ExxonMobil, Chevron, BP, and Royal Dutch Shell are being replaced by the new "Seven Sisters": Aramco (Saudi Arabia), Gazprom (Russia), CNPC (China), NIOC (Iran), PDVSA (Venezuela), Petrobas (Brazil), and Petronas

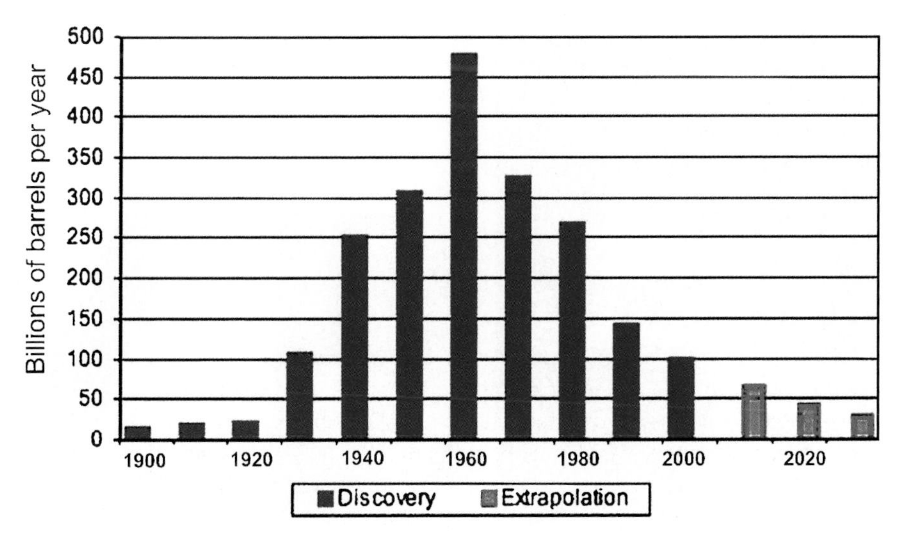

Fig. 2.19 Rate of oil discoveries since 1900 (http://www.theoildrum.com)

(Malaysia).[18] The IEA predicts that 90% of new oil and gas discoveries will come from developing nations. We will next show the different ways in which the industry is trying to explore beyond "proven" reserves.

Deep Drilling

There are new oil fields to be found if one is willing to drill deep enough. In addition to the Caribbean, deeply lying deposits are believed to exist in the North Sea, the Nile River Delta, the coast of Brazil, and West Africa.[19] To see how hard this is, consider Chevron's Jack 2 well, 175 miles offshore in the Gulf of Mexico. The drill goes 1 mile in water down to the bottom, then four more miles down into the ground. To find such large deposits, modern supercomputers are used to analyze seismic signals in three dimensions, requiring the processing of huge amounts of data. A new generation of drilling rigs had to be built to go twice as deep as ever before. These platforms, almost as large and as dangerous as that at Sleipner, cost half a million dollars a day to rent, but they could still be profitable if oil prices stay above $45 a barrel. This large deposit could yield 15 billion barrels, just a drop in the bucket compared with the world's proven reserves of 1,200 billion barrels. New deposits have been found that can be accessed only by *horizontal* drilling.[20] From a central platform, pipelines are drilled down and then horizontally out to deep-lying deposits kilometers away. The oil collected from these wells is then pumped to the mainland in a large pipeline. Figure 2.20a shows what a normal-size drilling rig looks like when the weather is nice. These are ships that go wherever they are needed. Storms and uncontrolled fires make oil drilling a dangerous occupation. An oil platform under less ideal circumstances is shown in Fig. 2.20b.

Fig. 2.20 (**a**) A drilling vessel in the Gulf of Mexico (http://images.google.com); (**b**) The Deepwater Horizon, 2010 (National Geographic Channel, July 2010)

These words, written previously, were brought to a focus when the Deepwater Horizon platform in the Gulf of Mexico exploded on April 20, 2010, killing 11 workers. The huge rig burned for days, and the oil leaking into the Gulf was uncontrolled, contaminating thousands of square miles and disrupting the fishing and shellfish industries in Louisiana. The damage to aquatic and avian wildlife is yet to be determined. Before the leak was capped in August, 4.9 million barrels of oil had been released, exceeding the 3.3 million barrels in the Ixtoc 1 blowout off the Yucatan peninsula in 1979. These numbers overwhelm the 257,000 bbls from the 1989 Exxon Valdez tanker spill in Alaska, whose effects are still felt 30 years later. Energy giant BP, owner of the Deepwater Horizon, suffered severe economic losses. The accident triggered legislation to regulate and restrict deepwater drilling. Aside from ecological concerns, it is becoming apparent that it would be cheaper to develop a substitute for oil than to ferret out the last of the earth's deposits.

Arctic Drilling

There is more oil and gas to be found if you are willing to endure conditions in freezing, inhospitable places. Russia owns a lot of property where no one wants to go. North of Japan lies Sakhalin Island, where they used to send prisoners. The deposits there contain 14 billion barrels of oil and 2.7 trillion cubic meters of gas.[21] Shell and Royal Dutch want to build an 850 km (500 mile) long pipeline to carry LNG to the USA. This is still being contested by Russia. The third largest gas field in the world is at Shtokman, in the Barents Sea near Murmansk, the largest city north of the Arctic Circle. That reserve contains 3.2 trillion cubic meters of gas, compared with 177 trillion in proven reserves. Western companies are bidding to get a part of this. But the cold conditions require the latest equipment and hard-learned techniques. There are icebergs, and shore is 550 km (340 miles) away. A pipeline on the bottom could be scraped by icebergs. Worse yet, antifreeze (glycol) has to be added to prevent the gas from reacting with the water that comes with it to form gas hydrates (more on this later), which can clog up the pipe. The water and glycol have to be separated out later. These arctic mining techniques are being tested in the Snohvit gas field in Norway. It may take $3 trillion to exploit these reserves.[22] It is clear that Russia can afford to do this only with foreign investment.

The Arctic north of Canada contains oil and gas fields made more accessible by the shrinking ice cap and the opening of the Northwest Passage. The US Geological Survey estimated that between 25% (some say 10%) of the world's "undiscovered oil reserves" could lie in the Arctic.[23] This is, of course, an oxymoron. How would you know how much is undiscovered? One deposit was estimated to contain 31 billion BOE in gas, enough to supply the US for four years. The problem here is a political one: no one knows who owns these deposits. North of Canada is also north of Russia, and the Russians planted a Russian flag at the North Pole. Stay tuned.

Shale Oil

Far below sagebrush country where mule deer and sage-grouse roam in Colorado, Utah, and Wyoming, there lie layers of organic marlstone bearing oil. The USA is reported to have two of the world's 2.6 trillion barrels of shale oil locked in the rock there. Of this, 800 billion barrels are deemed recoverable, 2/3 as oil and 1/3 as gas. By comparison, the proven oil and gas reserves of the Middle East total 1.2 trillion BOE. But to get it out, one has to essentially boil rock. Rather than digging up 200 million tons of rock per year to get a million barrels of oil a day, it is less destructive to get the oil out *in situ*, by drilling rather than digging. In western Colorado, Shell Oil has drilled 1,000-foot deep holes to test the feasibility of this process, which works as follows. Three holes a few feet apart are drilled into the shale. In two of them, electric heaters, like toaster wire, are inserted in pipes to heat the rock to some 700°F (370°C). It takes months or years for the rock to reach this temperature, and it has to be kept there for the life of the well, say 10 years. Fortunately, earth is a good insulator. The gas and oil are boiled out of the rock and can then be pumped out conventionally in the third pipe and sent in a pipeline to a processing plant. Mentioning electricity used for heat should raise your hackles because electricity is much more efficient for mechanical work than for heating. That is why your microwave or toaster runs on 1,000 W while a large window fan uses only 100 W. *In situ* mining uses electricity generated by a conventional power plant that loses 69% of the fossil fuel energy that it consumes, as can be seen above in Fig. 2.9a.

Will mining shale oil produce net energy? It is marginal. Let us do a back-of-the-envelope calculation to see if shale mining can be in the right ballpark. One ton (2,000 lbs) of shale will yield 25 gallons of oil.[24] It is easier to use metric units: 1 ton is about 0.91 metric tons (tonnes). At 42 gallons per barrel, we get 0.65 bbls of oil per tonne of shale. If we were to heat water, that would take 1 C/g/°C, so 1 tonne (a million grams) would take a million calories per degree centigrade. One calorie is about 4 J, so a million calories is 4,000 kJ. It is easier to heat rock, however. The specific heat of rock is only about 0.2, so now we only need 800 kJ per tonne per degree centigrade. We have to heat it by 700°F, which is about 380°C. To heat one tonne of rock by 700°C then requires 800×380 or about 300,000 kJ. From Box 2.1, we see that 1 kJ equals 1.6×10^{-7} BOE, so 3×10^{5} kJ equals about 0.05 BOE. But we get only 0.65 bbls from 1 tonne of rock, so 1 bbl of shale oil requires 0.08 BOE of electrical energy for heating alone. If the electrical plant is 30% efficient, 1 bbl of shale oil needs 0.25 barrels of *real* oil for heating. To this we have to add the energy to run the refining plant. Using microwaves to heat, as proposed by Raytheon,[25] would be even less efficient.

In addition to all this, it is planned to build a "freeze wall" to keep oily liquids from seeping into the groundwater. This would be a wall of existing rock and water 1,800 feet deep and 20 feet thick surrounding the drill sites. By drilling more pipes, a cold ammonia solution is circulated to keep the wall at freezing temperature. Since refrigeration is even more inefficient than heating, this could double the electrical cost, and we would get only two barrels of shale oil for each barrel of oil equivalent in, say, coal used to generate the electricity. There would be an advantage in that oil is a liquid and much more valuable than coal for transportation. Destruction of the environment is a high price to pay for this marginal fossil resource.

Estonia provides an example of what happens if shale is strip-mined.[26] There, shale oil provides 70–90% of the electricity. Shale is crushed to 6–10 mm size (about 0.5 in.) and burned in boilers topped by 250-m high chimneys. The ash and pollutants are blown up the chimneys, and the large particles of shale are re-burned when they fall back down. CO_2 emission is 10 million tonnes a year. Only after new boiler technology from Foster-Wheeler of the USA was adopted did the SO_2 and N_2O emission fall below acceptable levels. Solid slag is piled up 100 m high. Five million tonnes of ash are produced annually. This is pumped with waste water into a huge lake formed by a surrounding levee 30 m high. This blue-green lake looks nice but is a toxic stew containing potassium, zinc, sulfates, and hydroxides.[26]

Tar Sands

If you thought shale oil was bad news, you should see what tar sands are like. At least shale is in one solid piece. Tar sands are a mixture of oil, sand, water, and worse yet, clay, made of very fine particles. To get the oil out is harder than cleaning up an oil spill on a beach. The huge deposits of tar sands, or sand oil, in western Canada are often in the news as examples of untapped energy reserves, estimated to be around 1.7–2.5 trillion barrels of crude oil.[27,28] In the northwest corner of the province of Alberta, the Athabasca River starts from Athabasca Lake and meanders all the way to Jasper National Park. At the northern end, near Fort McKay, is the largest of three oil sand deposits in Canada. Alberta's sands yield a million barrels a day and have a proven, economically recoverable reserve of 173 billion barrels, perhaps extendable to 315 billion, compared to 264 billion in Saudi Arabia.[27] The USA gets 10% of its imported oil from these sands. That's the good news. The bad news is that it takes a lot of energy to get the oil out, and there is a huge environmental impact in doing so. For deep deposits, the *in situ* method described above for shale can be used, with wells drilled down to the tar sand layer and then horizontally along the deposit. Steam is injected in one well to melt the tar, which drips down into a lower well and is then pumped up to the surface. Open-pit mining uses less energy but still requires heat. For each barrel of oil produced, *in situ* mining of tar sands emits 388 lbs of CO_2 and open-pit mining 364 lbs, compared with 128 lbs in conventional oil mining.[27] Eighty percent of the deposits lie deep enough to require energy-intensive *in situ* mining.

Here is how it works. Tar sands contain oil in the form of bitumen, which is as thick as molasses in the summer and as hard as a hockey puck in the winter. Roughly speaking, the sands consist of 10% bitumen, 5% water, 20% clay, and 65% sand.[28] To get at them, the forest has to be cut down first; then, to dig down to the sands, 100 feet of earth has to be removed: 4 tons of earth for each barrel of oil. Huge shovels then scoop the sands into monstrous trucks three stories high, carrying 400 tons at a time. The ore is dumped into crushers and then into tanks where warm water at up to 80°C (175°F) is added to form a slurry. The slurry is pumped in a pipeline up to 5 km (3 mi.) long to a separation tank. The pipeline serves an important function. The lumps rub against its walls during the transport in such a way that the bitumen is separated from the sand and becomes attached to air bubbles. The air and bitumen form a froth which rises to the top and can be separated out, while the sand and clay fall to the bottom. Some of the bitumen is still in the mix, which can be recirculated to get more bitumen out in a secondary froth. It takes time for the air bubbles carrying the bitumen to rise to the top, because they collide with the heavy stuff going in the opposite direction. A faster way is to put the mixture between two parallel plates which are inclined at an angle to vertical. The bubbles then rise along the top of the gap while the water and sand fall at the bottom plate, and they do not have to collide. Really high tech. The air in the froth is then boiled off, leaving an emulsion of water (30%), bitumen, and clay. An emulsion is a mixture of immiscible fluids, such as vinegar and oil in salad dressing. If the water droplets in this emulsion would coalesce, the water would sink to the bottom and the oil to the top, as in salad dressing left standing around. In a draconian twist, the water droplets are coated with a fine layer of particles from the clay, which keeps the droplets from coalescing. Solvents have then to be added to get the water out. The bitumen ends up with 2% water and 0.8% clay. The chemicals in these contaminants will corrode the pipelines later on, so the oil next goes to an upgrading plant, where it is heated to 480°C (900°F) and compressed to 100 atmospheres. The energy cost of this would be excessive if the heat were not recovered for the initial heating of the oil sands.

There are other energy costs not mentioned above. The shovels that do the digging have steel teeth each made of a ton of steel and wear out in a day or two. The energy used to mine and refine that steel is usually ignored. The trucks use 50 gallons of diesel per hour, and their huge tires last only six months. The sands have to be near a river, because lots of water is needed for washing, 200,000 tons of which has to be heated every day at Athabasca. What happens to the sand and clay that were removed in the first step? They go into tailing ponds, which are the worst news yet. The tailings are a thick sludge consisting of the waste water and 30% sand and other solids from which the bitumen had been stripped. It also contains toxic chemicals. One pond can cover four square miles (10 km^2), and there are 50 square miles (130 km^2) of these ponds in Canada, about a third of the area despoiled by tar mining. A sand dike 300 feet (100 m) high around each pond contains the tailings, but some suspect that the toxic chemicals have leached into rivers and lakes. Fish have been found with unusual red spots on their skins. Once 500 ducks landed on the oily brew and died. Self-operated, flapping mechanical falcons have been installed as scarecrows, insufficient for the purpose. It takes 1–2 years for the clear water to rise to the top, from where it can be reclaimed to supply half the water for mining. What is left at the bottom, however, is still liquid and is difficult to solidify to restore the forest land. So far no tailings pond has been reclaimed.

The mines operate day and night, winter and summer, to supply the demand for oil. The large reserves are there, but the price is steep. The cost of mining is many times the cost for conventional oil, and this does not include the cost of carbon capture and sequestration, which has not started. Tar mining emits CO_2, and more CO_2 is emitted when the oil is converted to gasoline and burned in cars. The environmental impact alone makes this a poor choice for stretching our oil supply. Perhaps the most poignant argument is that the energy used in tar mining is mostly natural gas, the cleanest burning fossil fuel. This is wasted to produce a low-grade oil because liquid fuels are so valuable for transportation.

Oil from Algae

We know that photosynthesis in trees uses sunlight to convert CO_2 into oxygen. Could it also produce oil? It turns out that fast-growing algae, considered scum that chokes up ponds, can contain both biodiesel oil and carbohydrates that can be fermented into ethanol. Funded by venture capitalists, hundreds of startup companies are scrambling to develop a process that can be commercialized to compete with fossil oil. The Center for Algae Biotechnology was founded in 2008 in San Diego, CA, and 200 companies have been set up in that area. In the Imperial Valley to the east, there are 400 acres (81 hectares) of algae farms.[29] Algae, half of whose weight is in oil, are expected to produce 10,000 gallons of oil per acre per year, compared with 650 gallons from oil-palm trees.[29] Growing algae needs carbon dioxide, which can be from the exhaust of coal-burning plants, and light, but not necessarily sunlight. This is because most of the sunlight is of the wrong frequency (color). Algae can be grown in acres of tanks lined with plastic sheets and given the right amount of CO_2 and water at the right temperature. Under the best conditions, algae of the right species can double their weight in 1 day.

Small-scale experiments at OriginOil, Inc.[30] have shown an efficient way to grow algae, harvest it, and extract the oil. Efficient LED lamps of the right frequency are used instead of sunlight to grow the algae, and CO_2 is fizzed in. After harvesting, the cell walls of the algae have to be broken up to release the oil. CO_2 is first added to change the pH. Then pulsed microwaves are applied whose frequency, intensity, and pulse rate are feedback-controlled. The mix is then moved to a settling tank, where gravity causes the oil to rise to the top, the biomass to the bottom, and water in between. The biomass can be used for feedstock. The separation occurs in a single step with no further input of energy. Whether oil from algae will be worth it is not yet known, since the process is still in the research stage.

Gas Hydrates

Gas hydrates are solids like ordinary ice, but they exist only under high pressure, typically below hundreds of meters of ocean. They contain methane bubbles trapped by H_2O molecules and will burn if ignited in air. The methane is believed to have been created by bacterial action ages ago. Gas hydrates are found on continental shelves and under the tundra in the Arctic. Figure 2.21 shows why. The dotted vertical line shows the freezing point of water; it is around 0°C and does not change much with pressure. Water is liquid on the right of the line and is ice on the left side. Gas hydrates, however, can exist only at great depths, where the pressure is high; and the depth is greater if the temperature is higher. The possible temperature-depth combinations for gas hydrates are shown by the yellow part of the diagram. In the ocean at temperatures above freezing, gas hydrates lie below 500 m of seawater. In the Arctic, they are closer to the surface because the temperature is lower.

The US Geological Survey (USGS) has led the exploration of gas hydrates under coastal waters such as the Carolina Trough, in the North Slope of Alaska, in the Gulf of Mexico, and even in the Bay of Bengal in India and the Andaman Sea of Thailand. Drilling projects in the Gulf of Mexico were carried out in 2005 and 2009 to obtain cores of the layers where hydrates are found. Detailed data have been obtained not only on the concentration of gas hydrates but also on the nature and stability of the sand layers through which the drill goes. It has been estimated that the amount of fossil energy in these hydrates can exceed the energy in all other fossil fuels on earth, but this is highly speculative. One estimate is between 100,000 and 300 trillion cubic feet of gas in hydrates, which translates to the same number of Quads, since 1 trillion cu. feet contains approximately 1 Quad of energy. This compares to the world's proven conventional gas reserves of 6,385 Quads given in Fig. 2.12. The highest estimate is 47,000 times this number and should be taken with a grain of salt. More accurate information became available more recently.[31]

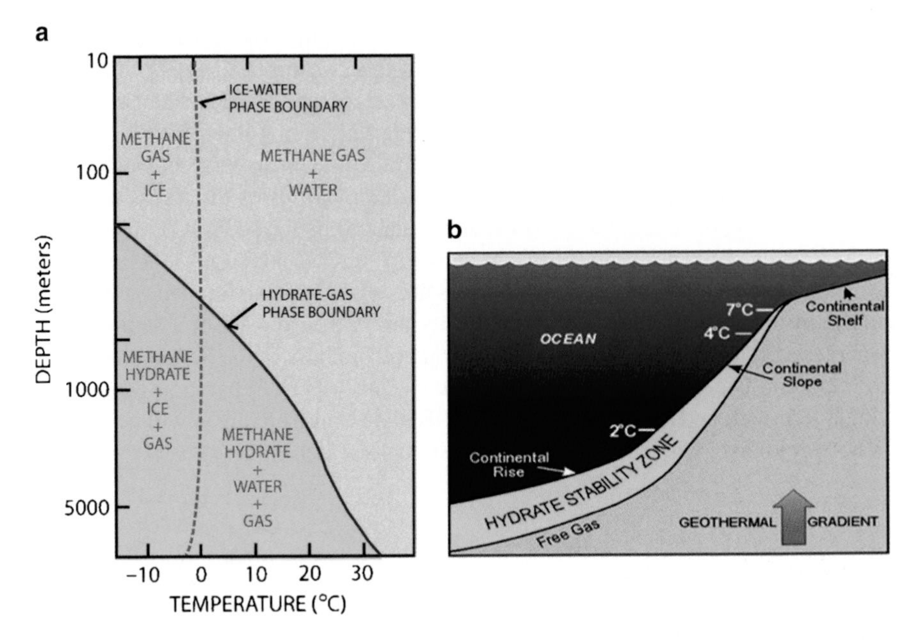

Fig. 2.21 (**a**) The depths in the ocean where gas hydrates can be found (*yellow region*), depending on the temperature (USGS website http://energy.usgs.gov/other/gashydrates); (**b**) relation of this region to the continental shelf (W.F. Brinkman, US Department of Energy, *FY 2011 Budget Request to Congress for DOE's Office of Science*)

We are clearly still in the exploration state on this resource. There is no proposed method to mine gas hydrates safely and distribute the methane. The problem is that methane is a greenhouse gas ten times as powerful as CO_2, and it is released as soon as the hydrates are relieved from the pressure they are under. Leakage of a small fraction of this gas into the atmosphere would be catastrophic. Methane can also be released from sand layers that the drills have to go through. Although methane is a clean-burning gas and emits less CO_2 than other fuels, it is still a fossil fuel, so CO_2 is emitted when it is burned. Even if it's true that the gas reserves in hydrates are huge, it is dangerous to exploit this source when completely carbon-free energy sources can be developed. These are the subject of the next chapter.

To conclude this chapter philosophically, we refer back to Fig. 2.2 in the Prologue. There we saw that the use of fossil fuel occupies only a thin slice of human history. For millions of years, solar energy was stored in trees which decayed and were stored deep underground as carbon compounds. This fortuitous treasure was discovered by man and is being recklessly consumed to advance our civilization without regard for the future. Mother Nature's endowment, however, was not meant to be wasted. The endowment was sufficient for humans to develop enough intelligence to find an unlimited resource: fusion energy. First, she showed us the enormous power available by leading us to develop the hydrogen bomb. She is now gently leading us to the next step. In Chap. 7, we will see evidence of her helping hand.

There have been totally unexpected bonuses in our attempts to control the reaction. It is a way to continue the benefits of the one-shot legacy of fossil fuels without destroying all the species of birds, fish, and animals that she has created.

Notes

1. International Energy Outlook 2008, Energy Information Administration, US Department of Energy. See also *World energy, technology, and climate change policy outlook 2030*, Directorate-General for Research (Energy), European Commission, Brussels (2003).
2. Energy Information Administration, US Department of Energy website.
3. BP Statistical Review of World Energy 2008.
4. Car of the future, NOVA on public television, May 2009.
5. Annual Energy Review 2007, Energy Information Administration, US Department of Energy.
6. Los Angeles Times, November 18, 2007.
7. IEEE Spectrum online, May 2009 (Institute of Electrical and Electronic Engineers).
8. Yang Yang, Woodrow Wilson International Center for Scholars, Princeton University, report (April 2007).
9. Physics World, September 2006 and July 2007.
10. A new method has recently been found in the laboratory to use porous solids instead of liquids to capture CO_2. These metal-organic frameworks (MOFs) are like molecular sponges that selectively trap CO_2 in microscopic pores and then release it much more easily than MEA can. A liter of MOFs can hold 83 L of CO_2. Their use in real smokestacks has yet to be tested. [L.A. Times and Scientific American website, June 30, 2009.]
11. Janet Wilson in the Los Angeles Times (date unknown).
12. Los Angeles Times, September 9, 2006.
13. IEA Greenhouse Gas R&D Programme, Gloucestershire, UK.
14. Time Magazine, July 9, 2007.
15. Wall Street Journal, May 22, 2008.
16. Money Magazine, February 2005.
17. http://www.theoildrum.com
18. Los Angeles Times, March 19, 2007.
19. Business Week, September 18, 2006.
20. Video, *Deep Sea Drillers*, National Geographic, July 14, 2009.
21. IEEE Spectrum, September 2006.
22. Business Week, June 30, 2008.
23. Time Magazine, October 1, 2007.
24. Audubon Magazine, March–April 2009.
25. IEEE Spectrum, December 2008.
26. IEEE Spectrum, February 2007.
27. National Geographic, March 2009.
28. Physics Today, March 2009.
29. L.A. Times, September 17, 2009 and Popular Mechanics, March 29, 2007.
30. http://www.originoil.com
31. *Realizing the Energy Potential of Methane Hydrate for the United States*, ISBN: 0-309-14890-1 (National Academies Press, 2010).

Chapter 3
The Future of Energy II: Renewable Energy*

Introduction

Many governments are providing support and subsidies for the development of renewable sources of energy. As a result, thousands of companies, some funded by venture capital, have been founded to tackle this problem. The incentive, however, is always commercial. The world runs on money, and nothing gets done without the possibility of profit. This incentive, however, is artificial. What matters more is the *fossil footprint* of each technology. That is, how much fossil energy is used in manufacturing and maintaining the equipment, including the mining of the raw materials, their transportation, and the assembly and installation of the power units.[1] After all, the goal is to *replace* fossil fuels, not to buy more of it to buildup a new business. "Green" energy has to be self-sustaining energy-wise. This seems obvious, but only the wind people have been brave enough to calculate their fossil footprint and publicize the results. This chapter also describes new inventions and ideas which give hope for the future but are as yet untested on a large scale.

Electricity is the kind of energy that our modern lifestyle depends on. Making electricity from fossil fuels requires going through a heat cycle. As explained in Chap. 2, the thermodynamics of heat cycles puts a limit on efficiency. Power plants have to be carefully designed to approach even 40% efficiency. Sixty percent of the energy in the fossil fuels that we burn up is lost in the production of electricity. Most of the renewable energy sources, however, can generate electricity *directly*, without going through a heat cycle, thus avoiding that 60% loss. This is the case for hydroelectricity, wind power, and solar power. The bad news is that these sources are local, or intermittent, or have their own inefficiencies. Hydro is well established, but not everyone has it. The realities of wind and solar will be covered next. The possible backbone energy sources, fission and fusion, still have to go through a heat cycle. The second or third generation of fusion reactors, however, could possibly produce electricity directly in so-called "mirror machines." These advanced systems will be covered in Chap. 10.

*Numbers in superscripts indicate Notes and square brackets [] indicate References at the end of this chapter.

F.F. Chen, *An Indispensable Truth: How Fusion Power Can Save the Planet*,
DOI 10.1007/978-1-4419-7820-2_3, © Springer Science+Business Media, LLC 2011

Wind Energy

Windmills have been used for energy long before there was electricity. We are now returning to this source by building wind farms. Wind is actually a kind of solar energy, since it is produced by sunlight heating different parts of the earth differently. Figure 3.1 shows a typical modern wind farm. The original concept was that these farms can be built on open land where it is usually windy and, consequently, where not many people live. Farmers can lease the land to power companies for $3,000–6,000 per turbine per year and still let their cattle graze among the towers. This seems ideal, but people began to object. The wind farm at Altamont Pass near San Francisco is notorious for the number of birds that its 5,000 turbines were killing every year. The Elk River Wind Farm in Kansas was built on a pristine prairie, the home of the sage grouse and the lesser prairie chicken.[2] This habitat is now cut up by roads, transmission lines, and power stations. To get enough wind power to make a difference, the environment does have to suffer, but the benefits of this free energy far outweigh the disadvantages. China hopes to get half its electricity from wind by 2020, thus cutting its carbon emissions by 30%.[3] The scenery will surely suffer, but there the objectors have less of a voice. Wind power is not free of technical problems, but these seem to be less severe than with other

Fig. 3.1 A modern wind farm (This is a publicly accessible photograph shown on many websites. The location is not identified)

green technologies. In some places, like Texas, the cost of wind energy is already competitive with that from oil.

The Birds and the Bats

In spite of its economic efficiency, wind power has encountered considerable opposition. Initially, many bats and raptors were found to be killed in wind farms. At Altamont Pass, the count was 1,300 raptors a year, including more than 100 golden eagles.[4] This wind farm was located on a bird flyway, and the obvious solution was to avoid these flyways. Apparently, the raptors would land on top of the turbine and look for rodents on the ground. Once they saw one, they would dive right through the whirling blades. There was such concern that the state of California issued guidelines for the treatment of birds in the development of wind power.[5] This report did not say how to avoid bird kills, but did outline the procedures for licensing and monitoring. Bats are not the most lovable creatures, but they do eat a lot of insects. Golden eagles have a regal name, but they have practically hunted the island fox of California's Channel Islands to extinction. Wind power's impact on wildlife is monitored by various organizations.[6]

This problem has not surfaced with modern turbines such as those shown in Fig. 3.1. These are much taller than first-generation turbines and turn at much slower speeds. But the clinching argument lies in the numbers. Ten to 40 thousand birds and bats are killed per year in wind farms. Compared to this, 100 *million* are killed per year by cats, and 60 million by cars and windows (which they fly into).[4] It is just that no one goes around counting these carcasses the way they do on wind farms. If global warming is not controlled by eliminating fossil fuels, many more birds and animals will die and even become extinct, as we saw in Chap. 1.

There are other environmental objections. Wind farms cannot always be built where there are no people. The noise can be bothersome, and the effect on scenery, even of offshore turbines, often cannot be tolerated. There is a NIMBY (Not In My Back Yard) sentiment. Objectors have their own website.[7] The technical problems have to do with time and place. Since wind speed fluctuates, the excess energy generated in periods of strong wind has to be stored, and there is no easy way to store that much energy. Wind farms are usually built far from population centers where the energy is needed. This involves modifying the power grid with new transmission lines. This presents a chicken-and-egg problem: neither the wind farm companies nor the transmission line companies want to proceed without the other.

The Growth of Wind

Being the most economical of the renewable technologies, installation of wind turbines has grown rapidly in the last few years. In Fig. 3.2, we see that Europe

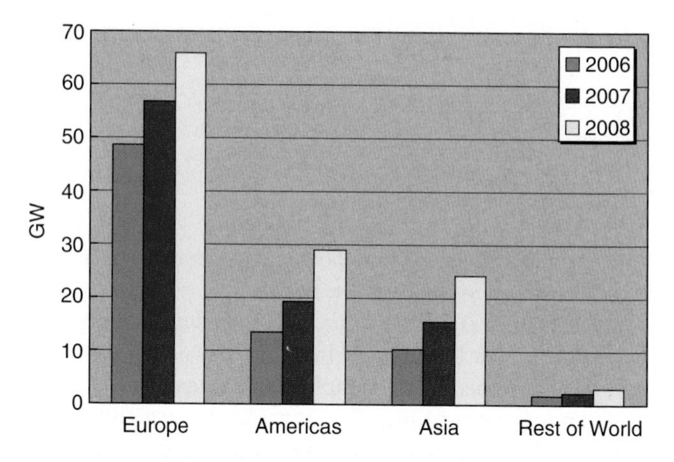

Fig. 3.2 Accumulated installed wind power from 2006 to 2008 in three continents. The scale is in gigawatts (GW), which are millions of megawatts. Redrawn from Vestas Wind, No. 16, April 2009. Original data from BTM World Market Update 2008 (March 2009)

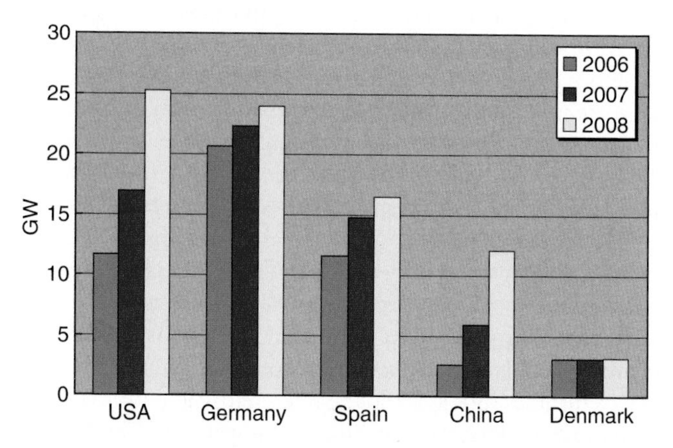

Fig. 3.3 Installed wind power in the top four countries plus Denmark [Vestas Wind, No. 16, April 2009. Original data from BTM World Market Update 2008 (March 2009)]

leads in this field, being more dependent on foreign oil than other continents. It has also had a head start, but other nations have been advancing more rapidly. Between 2006 and 2008, wind capacity has more than doubled in America and Asia. The units in Fig. 3.2 are in gigawatts (GW) or millions of megawatts. A large coal plant generates roughly 2 GW of heat, giving 1 GW of electricity. So the 65 or so GW of *peak* wind power in Europe in 2008 would replace, roughly, 65 coal plants. We will see later that the *average* power of wind turbines is much less.

The installed wind capacity of the top countries is shown in Fig. 3.3, again in gigawatts. We see here that the head start of the European countries is being rapidly overtaken by the USA and China. Wind power has more than doubled in the USA and more than quadrupled in China in the two years. Denmark's wind capacity is typical of many other small European countries, and it is shown here because Denmark has been a leader in developing the technology of wind turbines and their deployment onshore and offshore. Currently, 20% of the electricity in Denmark is supplied by wind.[8] It is estimated that by 2013 electricity from wind will cost $0.055/kWh, compared with $0.05 from coal or gas, $0.06 from nuclear, and $0.22 from solar.[8]

At one time, after the Chernobyl accident, Germany wanted to eliminate all its nuclear reactors, replacing them with wind and solar plants. A feed-in law has been in place since 1990, requiring utilities to buy energy from green sources that feed into their grid.[9] The plan was to install 500 MW of offshore wind in the North Sea by 2006 and 2,500 MW by 2010. The major players are the large utility companies E.ON Netz, REpower Systems, and the giant Swedish firm of Vättenfall. However, this was harder than they thought, the subsidy was too small, and the enviromentalists were too vocal. Only a few offshore turbines have been installed. Chancellor Angela Merkel lowered the costs by shifting the burden of new transmission lines to the power grid operators from the wind developers. Now 900 MW of turbines have been ordered, and E.ON Netz will spend $254 million (€180) to build a cluster of turbines in the North Sea, using some of the huge 5-MW turbines from REpower (later in this chapter). Nonetheless, wind is so capricious that it can supply only a small fraction of the energy now generated by nuclear reactors.[9]

In the USA, installed capacity was close to 30 GW by the middle of 2009, providing 1.4% of the country's electricity. The states with the most wind power are Texas (7.1 GW), Iowa (2.8 GW), and California (2.5 GW). The largest wind farms are the Horse Hollow, Capricorn Ridge, and Sweetwater farms in Texas; Altamont, Tehachapi, and San Gorgonio in California, and Fowler Ridge in Indiana.[10] Wind supplies 5% of the renewable energy in the USA, compared with 1% for solar; and renewables account for 7% of total energy. The Great Plains states, like Kansas, have great potential for further development. The current rate of buildup (Fig. 3.3) is on track to attain the Obama administration's goal of doubling clean energy by 2012. Little has been done so far about offshore wind capture. There are plans to try this on the East Coast. The technology will be far behind that of the Danes, who have been researching this for many years. The economic crisis may slow down the investment in this field. For instance, T. Boone Pickens had planned to spend $10 billion to build the largest wind farm in Texas, but the plans were scrapped when the price of oil dropped to the point where wind became too expensive. Ideology is again the slave of economics.

For the far future, the proponents of wind power have no such reservations. Figure 3.4 shows the predictions of the experts at Vestas Wind Systems of Denmark. The blue part of the curve shows the 16-fold increase of the world's wind turbine capacity from 1997 to 2008. The red part of the curve shows the expected future growth up to 1.3 *trillion* watts by 2020. Whether this will actually happen is problematical. As we shall see, this would require a large amount of backbone power to back up the wind power, and too much fluctuating power may make the power grid unstable. The good news is that wind installations have a very small fossil footprint.

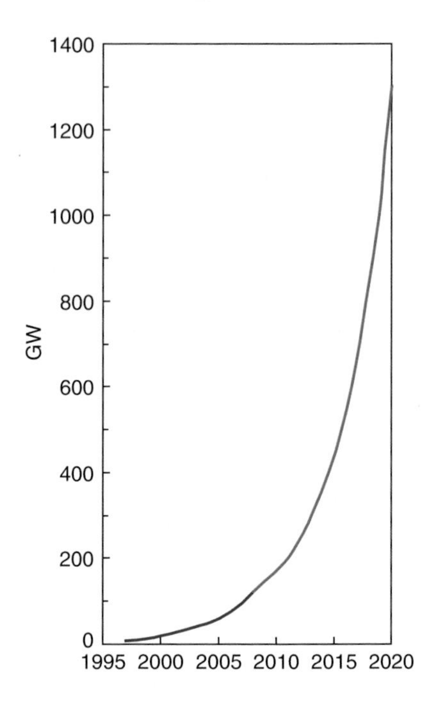

Fig. 3.4 Actual (*blue*) and predicted (*red*) wind capacity, in gigawatts, from 1997 to 2020 [Vestas Wind, No. 16, April 2009. Original data from BTM World Market Update 2008 (March 2009)]

When is a Megawatt Not a Megawatt?

When they talk about wind turbines, the quoted power of a turbine is the *peak* power, the most it can generate when the wind is strongest. The power output of a turbine varies as the *cube* of the wind speed. That means that if the wind drops from 20 miles per hour (9 m/s) to 10 mph (4.5 m/s), the electric power produced goes down by a factor of *eight*. The average number of megawatts generated is then much lower than the maximum that the turbine is built for. Figure 3.5 gives an idea of how variable wind power is. The data are from the area controlled by E.ON Netz, a large company that controls 40% of Germany's wind capabilities. During the year, this example shows that the power varied from 0.2 to 38% of the peak grid power! For this reason, a turbine built to generate 5 MW actually produces much less than that on the average. Just how much is shown in Fig. 3.6. This graph shows how much time during the year the wind power generated in a certain area was the number of gigawatts shown on the vertical scale. The time is measured in quarter-hours. We see that the maximum installed capacity of 7 GW was never reached, and even 6 GW was produced for a very short time. The average power over the year was less than one-fifth of the installed power capability. For half the year, less than 14% of the installed capacity was usable.[11] So 7 MW can mean only 1.3 MW!

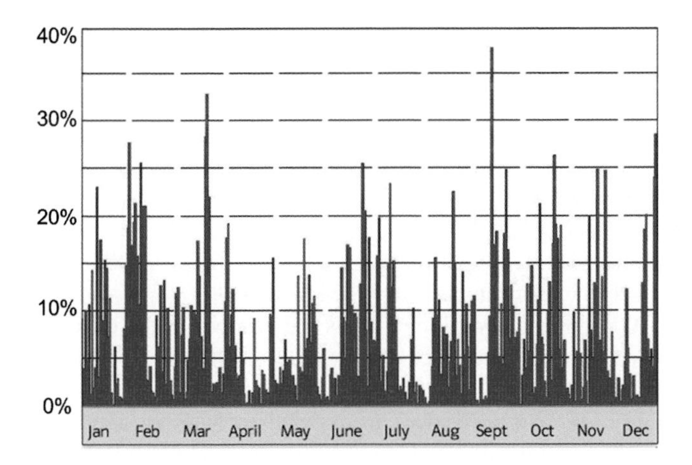

Fig. 3.5 Daily fluctuations of wind power in 2004 in the E.ON Netz control area. The scale gives the contribution of wind power to the peak grid load. Adapted from E.ON Netz Wind Report, 2005

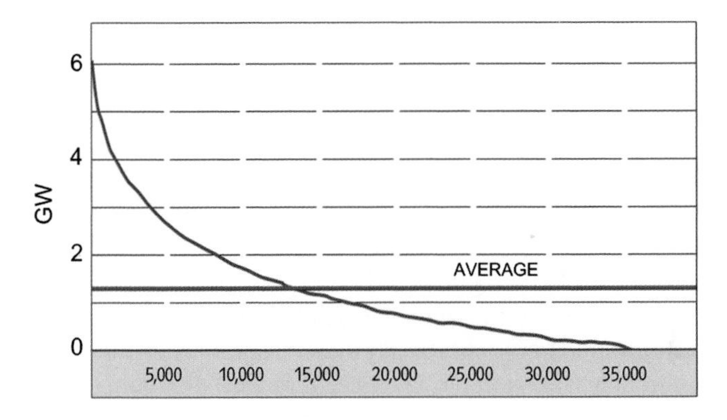

Fig. 3.6 The number of quarter-hours in 2004 in which the wind power generated by E.ON Netz was the number of gigawatts plotted on the vertical scale. (There are 35,000 quarter-hours in a year.) For instance, there were about 5,000 quarter-hours in which the power was 3 GW, and about 17,000 quarter-hours when the power was 1 GW. The average was 1.3 GW. Adapted from E.ON Netz Wind Report, 2005

We often see statements like "The 5-MW titan [in Denmark]…will average enough power for 5,000 homes,"[8] or "The 108 [1.5-MW] turbines…in the Colorado Green project…produces roughly enough electricity each year to supply more than 52,000 homes[4]." The first averages out to 1 kW (peak) per home, while the second works out to be 3.1 kW (peak) per home. Clearly, this number will depend on the amount of wind at each locale as well as the lifestyle there in terms of electricity use.

In 2001, the yearly average electricity consumption in the USA was 1.2 kW per home[12] or 0.47 kW per average person. This is on a steady basis, averaged over a whole year. Now, 1.2 kW goes into 1 MW (=1,000 kW), 833 times. So if average wind power is only 20% of the peak power, as we found above for Germany, 1 MW would supply only a fifth of 833 or 166 homes. This is a little less than the number of 250–300 homes quoted in footnote 4, but the discrepancy can be accounted for if steadier winds were assumed. The Colorado example above works out to give an average-to-peak wind factor of 38%, just twice the number for Germany. This means that 1 MW of peak power in Colorado would supply 320 homes, in good agreement with the quoted number of 250–300 homes for the US average. In the Denmark example above, 1 MW of *peak* power would provide *average* power for 1,000 homes, about three times the number in the Colorado case. It is quite possible, however, that electricity is used much more sparingly in Denmark than in the USA.

In summary, the average power from wind turbines is only 19–38% of the installed power capability, depending on the location. The number of homes a wind farm can power also depends on the energy usage pattern in that location. Consequently, claims about the efficiency of wind farms can vary widely and cannot always be believed without checking the facts.

Size Matters

The Nørrekær Enge wind farm in Denmark is replacing 77 old-style turbines with 13 large new ones. At 2.3 MW peak, these few modern turbines can produce twice the power of the old ones. Since winds are steadier higher off the ground, the *average* power over the year will be *four times* larger. Germany is planning to add 25,000 MW by 2020 by *repowering* their wind farms with the new turbines.[13] Why is it worth the trouble? Not only is the first generation of turbines getting old, but wind is stronger and steadier at higher altitudes. Doubling the height of the turbine will increase the wind velocity by 10%. Since the power varies as the *cube* of the velocity, this 10% translates of a 34% increase in wind power. The trend is to build fewer very tall towers with very long blades.

These new turbines are huge. The largest so far is Enercon's E-126, shown in Fig. 3.7. Its rotor diameter is 126 m (413 feet) and its total height is 198 m (650 feet)! This is like the length of two football fields stretching up into the sky. Compared to this, the height of the Statue of Liberty is only 93 m (305 feet); of the Washington Monument, 169 m (554 feet); and of the Eiffel Tower, 324 m (1,063 feet). Those who have been up the Eiffel Tower can testify to the winds up there! Unlike these other structures, the turbine cannot be built step-by-step from the ground up. The blades and the *nacelle* (the housing holding the blades and the generator) have to be preassembled and lifted up by cranes. The cranes themselves are so tall that they have to be assembled with smaller cranes. It is a very dangerous operation: the slightest wind can cause everything to come crashing down.

Each blade is 200 feet (60 m) long, and the blades catch so much wind that they have to turn at only five revolutions per minute, or once every 12 seconds. No birds

Fig. 3.7 The Enercon E-126 turbine and cranes (http://www.metaefficient.com/news/new-record-worlds-largest-wind-turbine-7-megawatts.html, February 2008)

would be so slow as to be struck. This turbine is rated at 6 MW but is expected to produce more than 7 MW peak power. Calculations like the ones we did above show that a single E-126 turbine can power 5,000 European households or 1,776 American households. Of course, the power cannot go directly to houses because wind power varies. The power is fed into the electrical grid as a small fraction of the power there, and the wind replaces only some of the fossil fuel or nuclear energy that the grid has to supply.

Figure 3.8 shows the interior of the nacelle of a smaller turbine made by Germany's Siemans. Inside the nacelle, there are motors and controls that change the pitch of the blades as the wind varies, generators that convert the rotating motion into electricity, and a gearbox that connects the rotor to the generator. These nacelles can be the size of a conference room. Figure 3.9 shows an offshore array of 5-MW turbines made by REpower of Germany. A closeup of the nacelle can be seen in Fig. 3.10. Since the turbine is not easily accessible, the nacelle is made to accommodate workers lowered to the platform from a helicopter.

Fig. 3.8 Detail of the nacelle on a Siemans offshore turbine (http://www.powergeneration.siemens.com/press/press-pictures/)

Fig. 3.9 Erecting an offshore array of REpower's 5-MW turbines (http://www.repower.de/)

Fig. 3.10 The large nacelle of REpower's 5-MW turbine (http://www.repower.de/)

Offshore Wind Farms

In Europe, the emphasis is on offshore turbines because of the lack of space and objections to the noise and aesthetics of onshore wind farms. It is more expensive to build towers in the sea, and there are problems with storms, icebergs, and salt water, raising the cost of operation and maintenance. However, the wind can be steadier and stronger at low altitudes so that the towers do not have to be quite so high. Denmark had the most installed offshore wind power as of 2005 (Fig. 3.11) and has led in the development of the technology.

As Fig. 3.12 shows, there are different ways to mount the towers in the sea depending on the depth of the water. If the installation is kilometers offshore, the turbines have to be floated and tethered to the bottom. This is much harder than for floating oil rigs because the towers have to be kept from turning, leaning, or tipping over. Except for experimental trials, no floating turbines have yet been installed, though Germany envisions placing them as far as 40 km offshore.[9] In September 2009, Vestas Wind Systems of Denmark announced its V112-3.0MW turbine specifically designed for offshore use.[14] This turbine incorporates new technology for increased efficiency, reduced noise, and resistance to the severe conditions, including a heating system to keep the parts from freezing. The power curve for the V112 is shown in Fig. 3.13. The turbine cuts in at a wind speed of

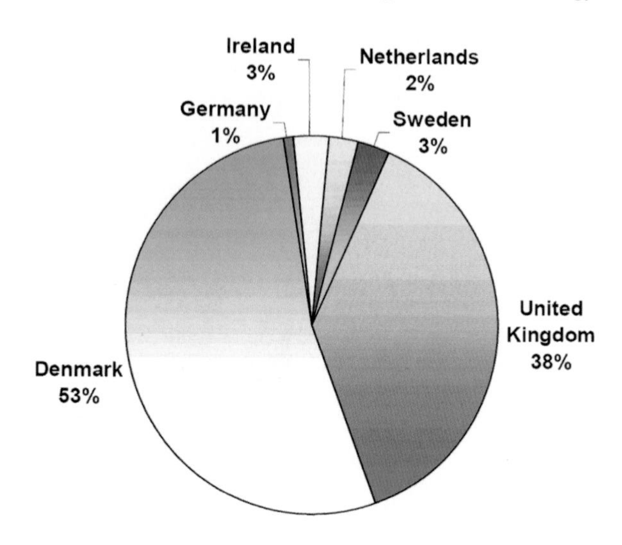

Fig. 3.11 Distribution of offshore windpower in Europe, as of 2005 (*Energy from Offshore Wind*, US National Renewable Energy Laboratory, NREL/CP 500-39450, February 2006. *Engineering Challenges for Floating Offshore Wind Turbines*, NREL/CP 500-38776, September 2007)

Fig. 3.12 Methods for installing offshore turbines (*Energy from Offshore Wind*, US National Renewable Energy Laboratory, NREL/CP 500-39450, February 2006. *Engineering Challenges for Floating Offshore Wind Turbines*, NREL/CP 500-38776, September 2007)

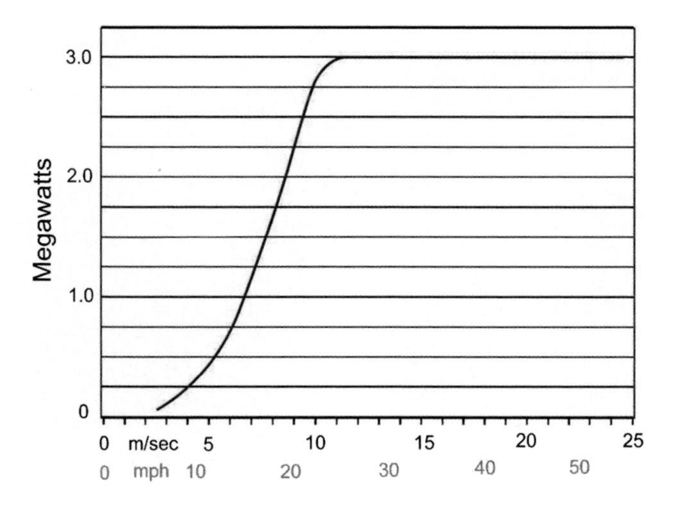

Fig. 3.13 Power curve for the Vestas V112 turbine. The wind speed at the bottom is listed both in meters per second and in miles per hour (*red*). Adapted from Vestas brochure *V112-3.0 MW, One Turbine for One World*

3 m/s and achieves its maximum output at 12 m/s. It can maintain this output up to 25 m/s. The steep dependence of power on wind velocity can be seen from this curve.

Blade Design

The picture of the V-112 in Fig. 3.14 shows that the blades of modern turbines have been designed with special shapes to maximize efficiency at all wind speeds and to minimize turbulence.[15] Such shapes are also seen in newer airplanes (Fig. 3.15). Blades may evolve further to incorporate scalloped edges, as these have been found to reduce drag on a humpback whale's flippers.[16] As the wind speed varies, each blade's pitch is changed with a motor to capture the most energy. In very strong winds, the blades are feathered as in airplanes. The fiberglass blades are much thinner than on windmills and there are only three of them per rotor. This design is driven by cost.[17] More blades will not only be too expensive but will also require sturdier towers to support in strong winds. The blades are so long that even at only 5 rpm, the tip of a 200-feet (60-m) blade travels at 170 miles per hour (75 m/s).

The diameter of the rotors is very large because these catch more wind when the speed is low. This is explained in Fig. 3.16,[18] which is drawn for a situation when the average wind speed is 7.5 m/s (17 m/h). The smooth, peaked curve at the left shows how often each wind speed occurs. The speeds are on the bottom scale. Notice that most of the time, the speed is between 2 and 12 m/s. The rightmost of

Fig. 3.14 Blade design of the Vestas V90 turbine

Fig. 3.15 Blade design of the Aerospatiale ATR-42 A2-ABP

the rising curves shows the turbine's output power for a 50-m diameter rotor. The power is limited by the size of the generator. The curve labeled 50 m-3.0 MW, therefore, rises as the wind speed increases but stops rising and stays flat when the curve reaches 3 MW (at around 16 m/s). Increasing the generator's capability to

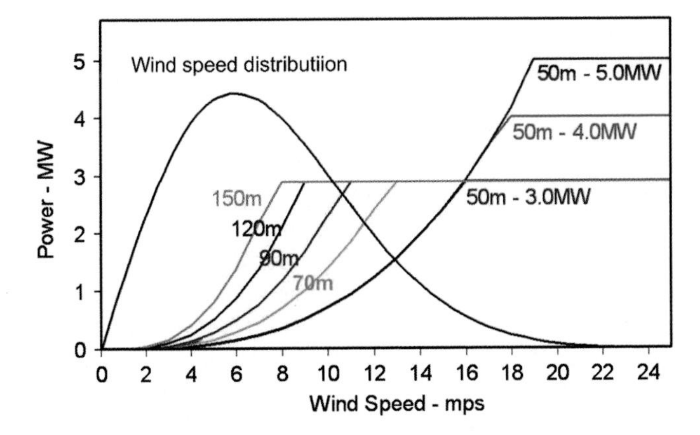

Fig. 3.16 Curve on *left*: distribution of wind speeds (arbitrary units) when the average speed is 7.5 m/s. Curves on *right*: turbine power in megawatts as generator size is increased. Curves in *middle*: turbine power of a 3-MW generator as rotor diameter (m) is increased (adapted from a presentation by Chris Varrone, Chief Strategist, Technology R&D, Vestas Wind Systems). The horizontal scale is wind speed in *meters* per second. To convert to miles per hour, see Fig. 3.13

4 or 5 MW permits capturing the energy of the strongest winds, as shown by the uppermost curves on the right. However, these occur only a small part of the time. If, instead, we keep the generator at 3 MW and increase the rotor diameter, we get the colored curves labeled 70, 90, 120, and 150 m. These rotor diameters utilize the slow wind speeds more efficiently, even though the 3-MW generator cuts the curves off when the available power reaches 3 MW.

Even larger rotors would capture more of the slow winds under the peaked curve, but then the towers would have to be even taller than the monsters that we now have. High hub heights also contribute to a turbine's efficiency. Winds are stronger away from the ground, where the trees, grass, hills, and structures impart a drag. This was a rather technical discussion, but it shows why it pays to tear down old turbines and replace them with fewer large ones.

How Turbines Work

As we stated at the beginning of this chapter, wind turbines produce electricity efficiently without going through a steam cycle. The generators in the nacelles are basically electric motors run in reverse, so that instead of electricity causing something to turn, the turning of the blades causes electricity to be generated. Of course, it is not that simple, and this gets a little technical. The pitch of the blades is varied to keep them turning at the same speed as the wind varies. The rotor is connected to the generator through a gearbox. The gears are needed to increase the rotational speed of the rotor (about 5 rpm, say) to the speed of the generator (about 1,000 rpm, say). The gearbox tends to wear out before anything else, and new turbines are being developed to do the switching electronically, without moving parts. Since it takes a second or two to change the pitch of the blades, gusts of wind will make the rotor turn faster, and the generator has to handle that.

The next problem is to match the electric output from the generator to the AC grid. Though there are different kinds of generators, it is not always possible for them to turn out AC at the same frequency as the grid. The generator's output will vary with the wind and may be nowhere near the 50- or 60-cycle frequency of AC power. It will also be *reactive*. That is, the voltage and current of the output will not be in phase, varying nicely together as they should. To manage this, the generator's output is processed by a *converter*. The AC is first converted to DC, and then the DC is converted to 50- for 60-cycle AC so that it can be sent into the grid. We commonly use converters on a small scale. The power bricks that charge our cell phones and laptop computers convert AC to DC. There are small devices for cars which can convert the DC from the cigarette lighter into AC to run a portable household appliance. But for a 5-MW turbine, the electronics and capacitors to handle this conversion would fill a small factory. Basically, a sizable electric substation has to be built at the base of the tower or inside it. Five megawatts is a lot of power; it is equivalent to 6,700 HP. The switching of this much power requires some heavy-duty transistors, and there is a proposal to develop silicon carbide (SiC) switches, which can handle this better than ordinary silicon.[19] These large components needed to convert wind power to grid power are a part of the cost and environmental impact that people do not usually know about.

The Fossil Footprint

Wind contributes less than 1% to the world's energy. The planned buildup in wind power will have to use mostly fossil fuel energy and thus contribute to CO_2 emissions. Fortunately, wind is one renewable energy source that can payback this energy in months instead of years. Careful analyses of energy use in wind energy generation have been made by Vestas Wind Systems in 1997[20] and 2006.[21] Vestas is a large Danish manufacturer that has installed 38,000 turbines, about half the world's total. The bottom line is that the fossil energy used can be recovered in about four months for a 600-kW turbine in 1997 and in about 6.8 months for an offshore 3-MW turbine in 2006.

These so-called life-cycle analyses are interesting because they give a good idea of all that is involved in building a wind farm. We'll take the 2006 study as an example. The study begins with the description of a fictitious power plant to be built.[21] This plant will consist of 100 Vestas V90-3.0MW turbines built 14 km (9 miles) offshore in water 6.5–13.5 m (about 33 feet) deep. Each turbine will produce 14 GWh/year for a total of 1,400 GWh/year for the whole plant. That's 1.4 billion kWh/year of electricity, compared to the 2,300 kWh that an average Danish household uses per year. That is enough power for 600,000 homes! It turns out that large plants require less energy per kilowatt produced than small ones.

The energy used to build this plant is divided into four parts: (1) manufacture of the components, (2) transport, construction, and installation of the turbines, (3) their operation and maintenance, and (4) their dismantling and disposal at the end of life. The lifespan is assumed to be a conservative 20 years. The components consist of the foundations, the towers, the nacelles, the blades, the transformer station, the transmission lines up to the grid, and even the boat dock for offshore plants. The foundation if offshore would be a steel tube 30 m long 4 m in diameter, and 40 cm thick. The transition piece to the tower is of concrete. The tower is made of steel, and all the energy used in making the steel from ore, fabricating the tower, and sandblasting and painting the surface is counted. The nacelles contain the gearbox, the generator, the transformer, a switchboard, a yaw system, a hydraulic system, and the cover. When these components are made by subcontractors, all the energy used in those factories is accounted for. The blades are made of 60% fiberglass and 40% epoxy, and the spinner on which they are mounted is plastic.

Transporting these components to the site by truck or boat uses gasoline or diesel, and the large cranes used for installation use more fuel. A transformer station for the offshore plant is to be built on three concrete piles 14 m above the water. The steel structure is 20 m × 28 m in size and 7 m high, with a helicopter platform on top. To carry the power to land, two 150-kV underwater cables are used up to a cable transition station 20 km away. From there, 34 more kilometers of dry cables carry the power to land. For maintenance, it is assumed that half the gearboxes and generators in the station will have to be replaced or repaired during the 20-year life cycle. Each turbine will be inspected four times a year, and the energy used to transport the inspectors by car, helicopter, and boat is also counted. A resource one usually does not know about is the use of sacrificial aluminum anodes for cathodic protection against the attack of parts by salt water. Since the aluminum cannot be reclaimed, the energy in mining is lost.

At the end of life, the turbines, towers, and foundations have to be dismantled and disposed of. Metals can be 100% recycled, with 90% recovery, and 10% going to landfill. Materials like fiberglass, plastics, and rubber can be burned; and the heat can be captured for use. Energy is actually recovered in the dismantling stage. When all this is added up, each turbine's energy cost over 20 years is 8.1 million kWh, while it is producing 14.2 million kWh/year. Dividing these two numbers gives the 0.57-year or 6.8-month energy payback time quoted above. This is for an offshore plant. An onshore plant produces only half as much energy, but it also takes half as much energy to build and maintain. Amazingly, the energy recovery time is almost the same, at 6.6 months. As for the carbon footprint, such a plant generates about 5 g of CO_2 for every kilowatt-hour (kWh) of electricity generated. By comparison, normal European power plants emit 548 g/kWh. Wind is indeed a very clean way to generate energy, but it has other problems.

Energy Storage

Since wind is so variable, one problem is how to smooth out the fluctuations. This means storing the energy. The method depends on how long the energy has to be stored. The capacitors in the nacelles and the turbine's transmission station need to store energy only from one AC cycle to the next, and those capacitors are already very large. Storing enough energy to last an hour would do a lot of smoothing.[22] Batteries can do this, but they are too expensive. The best batteries are lithium-ion (as in laptop computers) and sodium-sulfide (NaS). Enough Li-ion batteries to service a large turbine would cost as much as the turbine itself. A 1-MW bank of NaS batteries would be the size of three shipping containers. A 34-MW NaS demonstration plant in Rokkasho, Japan, occupies 16 large buildings.[22] This does not seem practical either. Storing mechanically in large flywheels is not yet taken seriously.

There is also day–night storage for 8 hours or longer. If there is a hill, pumped hydro can be used. The excess energy is used to pump water into a reservoir uphill. The energy is then regained quite efficiently by hydroelectric power. A scheme that is being taken seriously is compressed-air storage. Excess wind energy is used to pump air into underground salt domes or porous sandstone topped by shale. These sites, also usable for CO_2 storage, can be found over 85% of the USA.[23] The energy is recovered by bringing the compressed-air backup to help spin a natural-gas turbine driving an electric generator. The scheme is shown in Fig. 3.17. The turbines there are *gas* turbines, not *wind* turbines. A gas turbine is shown in Fig. 3.18. When natural gas is burned, the expanding air blows through the fan blades and turns the shaft, which then turns an electric generator. The compressed air from underground can add to this push, increasing the efficiency of the turbine by 60% or more. However, there is a heat cycle involved. When the air is first compressed, it heats up, and that heat is lost to the rock. Then when the air is decompressed, it cools down and has to be reheated to help drive the turbine. If you examine Fig. 3.17 closely, you will see that the heat for this is recovered from the hot air leaving the turbines after it has done its work. The loss in efficiency is more than 50%.[22] Nonetheless, large projects for such storage are planned in Iowa, Minnesota, Texas, and the Dakotas.[23]

Meshing with the Grid

Wind power rarely occurs where it is needed most. Conversely, you would not want to live where the wind is always fierce, like the west side of the Falkland Islands. New transmission lines are necessary, and this obstacle is preventing wind power from developing as fast as planned. In Germany alone, it is estimated that 2,700 km (1,700 miles) of extra-high-voltage lines will be needed by 2020 to carry an expected 48 GW of wind power. These lines run at up to 380 kV, compared to high-voltage lines at 110 kV, which are scary enough, and will cost over 3 billion

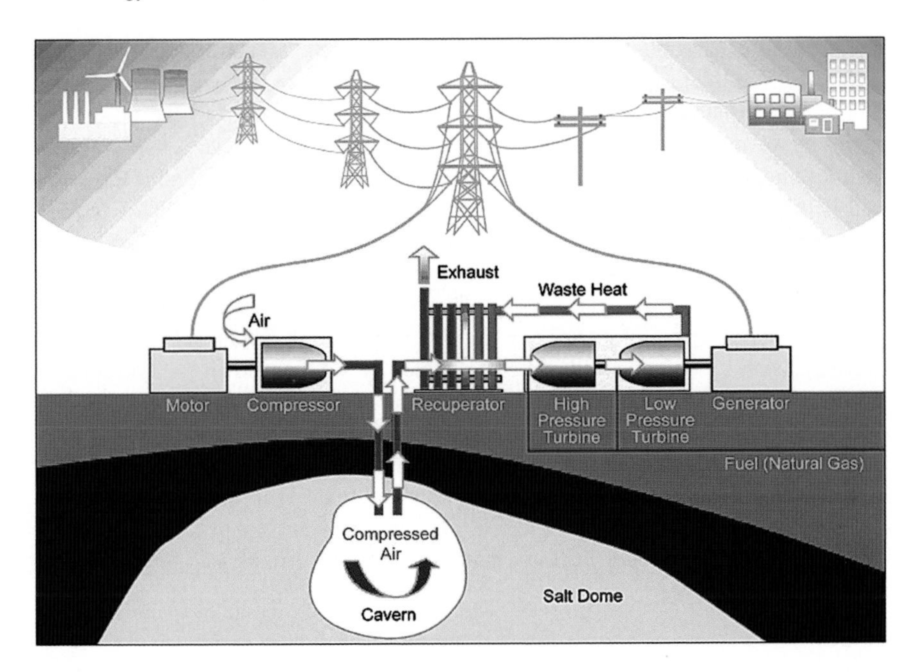

Fig. 3.17 Compressed air energy storage scheme for wind power (Vestas Wind, No. 16, April 2009)

Fig. 3.18 A gas turbine (http://www.powergeneration.siemens.com/press/press-pictures/)

euros.[11] Traditionally, power plants are built near population centers, so the transmission lines are short. Distributing wind power will require new rights-of-way, some of it underground. These lines cost 7–10 times as much as standard lines.[11] There will be political, legal, and social problems in addition to the large cost. Germany, and even all of Europe, is small compared with the US. Transmission lines are an even bigger problem for wind power in the USA.

Load distribution is another big problem. If the wind input to the power grid varies by as much as 10%, the grid can become unstable. However, if several wind sources are connected to the same grid, load variation can be avoided if the power can be switched in and out fast enough from each of the sources. This requires accurate forecasting of the wind speed and close collaboration among grid operators. The Nordic countries of Sweden, Norway, and Denmark are close enough to pool their resources for load leveling.[18] They can exchange wind and hydro energy. For instance, when wind power is excessive in Denmark, it can sell the power to Norway. Norway can accommodate the power by slowing down its hydroelectric power, storing the energy in the reservoir above a dam. The hydro energy can be sold back to Denmark when the wind dies down.

Wind is so variable that it can never be a large fraction of the total grid power. Not only that, but it must be backed up by conventional fossil fuel or nuclear plants. Estimates vary from 90%[11] to 100%.[24] *That is, for every megawatt of new wind power installed, one megawatt's worth of new coal, oil, gas, or nuclear plants have to be built.*

The Bottom Line on Wind

Wind is an attractive source of free energy. It generates electricity directly. It does not pollute, and it can generate enough energy to cover itself in half a year. The technology is well developed and is actually rather interesting. *But wind can never be a primary source of power.* It is too variable, and the problems of energy storage, transmission, and load leveling are overwhelming. Wind power is suited for islands such as the Galapagos[25] and Hawaii,[26] where all other energy must be imported. *Wind is the best of the renewable sources of auxiliary power, but it cannot supply backbone power.*

Solar Energy

The Nature of Sunlight

If we take a solar cell a square meter in size, put it on top of the atmosphere, and face it directly toward the sun, it would receive solar radiation energy of 1.366 kW/m^2. Take it down to the surface of the earth, and the light will be attenuated by the air's absorption and scattering. The net result is the convenient figure of 1 kW/m^2. Over the whole earth, there is enough sunlight in an hour to supply all the energy use in the world for a year! If you find this hard to believe, as I did, we can do a back-of-the-envelope calculation in Box 3.1.

Box 3.1 How Much Sunlight Does the Earth Get in 1 hour?

The radius of the earth is about 6,400 km (4,000 miles). Replace the earth with a disk 6,400 km in radius, and the disk would get the same amount of sunlight. We do not count the back side of the disk, and that takes care of the fact that there is no sunlight at night (Fig. 3.19).

The area of the disk is πr^2, as you well remember. That works out to be some 130,000,000 km^2. In square meters, the area of the disk is a million times that, which is 130,000,000,000,000 m^2. (Those who are meter-challenged can think of a square meter as a square yard.) Each square meter gets 1 kW, so the total power over the earth is that large number of kilowatts. The number is too long to write, but we can use shorthand and write it as 1.3×10^{14} kilowatts, where the 14 stands for the number of decimal places after the "1." (This is scientific notation, which was explained in Chap. 2.)

To compare this with our energy consumption, we have to convert kilowatts into Quads per year. We can use Table 2.1 in Chap. 2 to make this conversion. It takes several steps, but 1.3×10^{14} kW is the same as 440 Quads per hour. Also in Chap. 2, we found that our civilization consumes about 500 Quads per *year*, almost the same number. So indeed, sunlight hitting the earth every hour carries about the same energy as we consume in a year!

Fig. 3.19 The same amount of sunlight falls on the earth as on a flat disk of the same diameter

Figure 3.20 shows the annual variation of sunlight. This shows that the earth is tipped relative to the plane of its orbit. Consider a location in the northern hemisphere, on the upper red line, say. In the summer, the sun would be on the left, so as the earth rotates, more of that red line is in the sunlit region, and less in the blue night region. Days are longer than nights. In the southern hemisphere, the opposite is true. When the earth moves to the opposite side of its orbit, the sun appears to come from the right. The blue region is then sunlit, and less time is spent in there than in the yellow night region. Days are shorter then nights in the northern hemisphere. Furthermore, the sun never gets high above the horizon at high latitudes. Since we cannot easily store solar energy from summer to winter, solar power is inequitably distributed.

(continued)

Box 3.1 (continued)

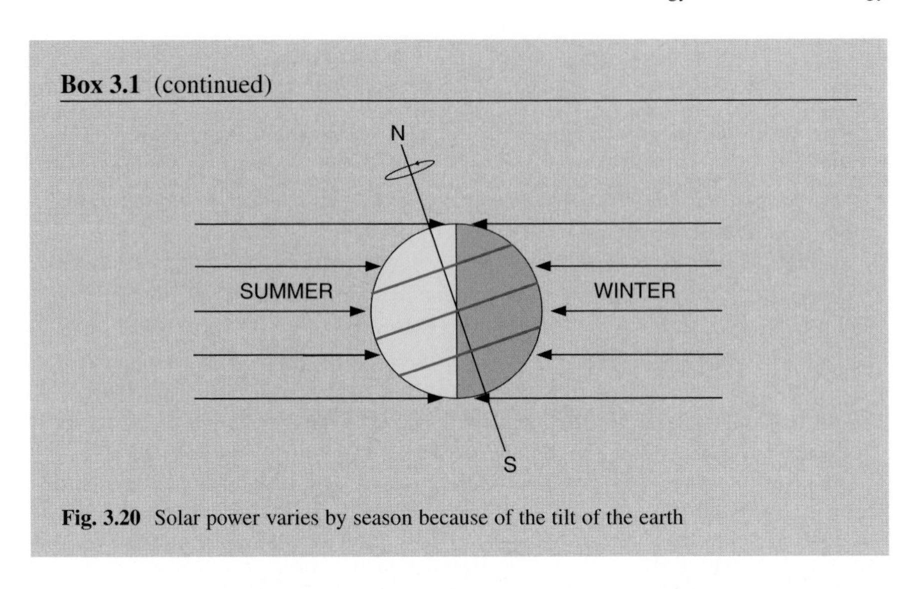

Fig. 3.20 Solar power varies by season because of the tilt of the earth

So why aren't we all fried by the sun? First, there's the factor of 2. Except at the equator, sunlight comes in at an angle, not from overhead, so that the power is spread out over a larger area. To figure out how many kW/m^2 there are at any given latitude and longitude is a long exercise in spherical trigonometry, but we can average. The area of a hemisphere is $2\pi r^2$, happily just twice that of the disk. So the average insolation over the earth is only 0.5 kW/m^2. Since the earth's axis is tilted with respect to its orbit, there are seasons; and people living at high latitudes have a bigger difference between winter and summer. They also get less sun altogether. Figures 3.21 and 3.22 show this. The number 0.5 kW/m^2 is averaged over latitude and seasons. Then there are clouds and storms and smog which prevent the sun from shining. That cuts the average to below 250 W/m^2, and it is not available everywhere. Even so, it is a lot of energy, if we could only learn how to capture it efficiently. The average person in the USA uses about 500 W of electricity, averaged over 24 hours. Two square meters of solar cells in a good location could generate this if they were 100% efficient. Right now, it is hard to get 10% except in the laboratory.

Ways to Use Solar Power

Although articles on solar power appear often in public media, it is not always made clear that there are many ways to capture that energy, and that these methods are quite different from one another. First, there is local solar vs. central solar. Locally, sun falls on every rooftop, and there is no excuse not to use this free energy. Centralized solar power plants are another matter. These take up large areas and have to transmit the energy from sparsely populated to densely populated regions. The plants also have to compete with coal and nuclear plants on cost.

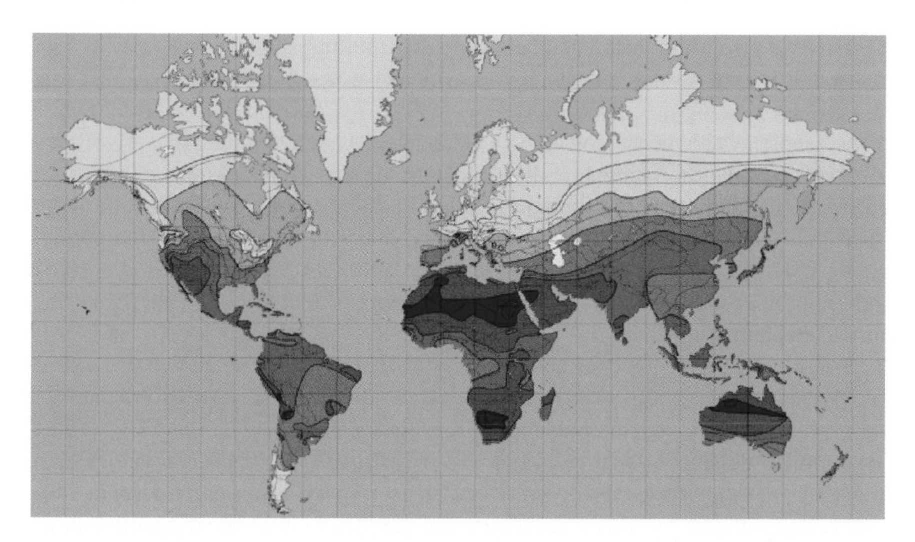

Fig. 3.21 Distribution of average solar energy incident onto the earth, with the darker colors indicating more sunlight (http://images.google.com). This shows that solar power is most abundant in the least populated regions of the earth

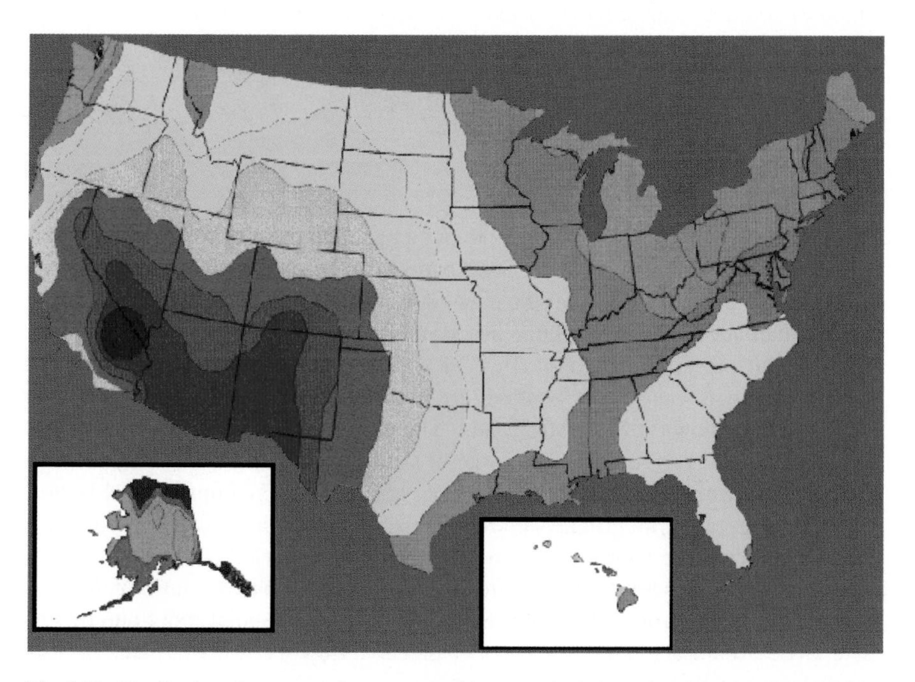

Fig. 3.22 Distribution of average solar energy incident onto the USA, with the *red* colors indicating more sunlight (http://images.google.com). This shows how difficult it would be to transport solar energy from the southwest to where it is needed in the northeast

There is also a big difference between solar thermal and solar electric. In solar thermal, sunlight is used to heat a liquid, typically water, and that heat is either used directly for heating or is used to generate electricity. *Local* use of solar thermal is very simple: water heated on the roof can directly reduce one's gas or oil bill. *Centralized* solar thermal is literally done with mirrors. Acres of mirrors motorized to follow the sun focus the sunlight into a boiler on top of a tower. There a liquid such as water or liquid salt is rapidly heated and stored in tanks on the ground. Since heat is hard to transport long distances, the hot liquid is used in a steam generator to produce electricity. Most of the energy is then lost in the thermal cycle, as was explained in Chap. 2.

Solar electric is commonly called *photovoltaic* or PV. There are two main kinds of PV: silicon and thin film. Solar cells made of silicon are expensive, and there are several kinds of these: polycrystalline, amorphous, and microcrystalline. Polycrystalline silicon solar cells can be very efficient, but these are so expensive that they are used where cost does not matter, as in space satellites. Amorphous silicon is less efficient but much less expensive, and they could be competitive in the market. The new microcrystalline silicon cells under development may turn out to be a good compromise. The fastest growing segment, however, is in *thin-film* solar cells. These are much cheaper than silicon ones, use very little material, and can be used for both local and central power. Although thin-film cells are the most inefficient of all, the possibilities for their deployment are tremendous. For instance, windows could conceivably be coated with thin-film cells. The following sections will tell how these various solar energy methods work.

Panels on Every Rooftop

The easiest way to use solar energy is to put a panel on the roof to heat water. This is already done in many countries. Such panels can be seen as one rides on a train in Japan. In a place like Hawaii, the panel does not have to be very big at all; 1 m^2 is more than adequate. A panel can be just a flat box with a glass top and a black bottom to absorb all the sunlight (Fig. 3.23). The panel is connected to the usual water heater with two pipes. A small pump circulates the water up to the solar panel and back down to the water heater. The gas or electricity driven heater then does not have to turn on as often to keep the hot water at the set temperature. No fancy electronics are needed, so the cost is low. Solar swimming pool heaters are even more economical. The same pump used for the water filter can pump the water up to panels on the roof, from where the water siphons down without further pumping energy. Since the temperature rise in each pass is only a couple of degrees, no high-temperature materials are needed. Black plastic panels, about one by two meters, are used. Each has many small channels to flow the water in parallel. Such panels have lasted over 30 years.

The fossil footprint of rooftop solar thermal collectors has been analyzed by the Italians [1]. As with the life-cycle analyses described in the previous section on

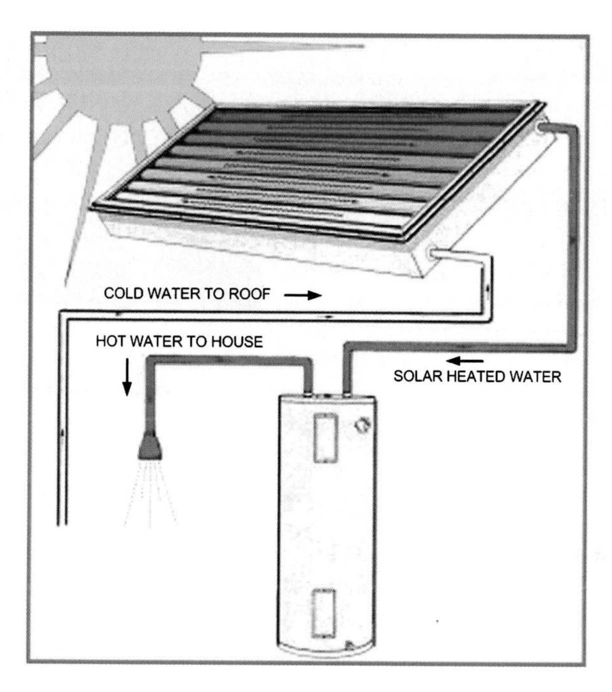

Fig. 3.23 The simplest implementation of a solar water heater (http://images.google.com)

Wind, all the energy used in producing the materials used and in installation, operation, and maintenance is added up; and the energy recovered in the recycled materials at the end of life is subtracted. The energy comes from conventional sources, mainly fossil fuel plants. This is then compared with the solar energy produced during the lifetime of the equipment. The resulting energy payback time lies somewhere between 1.5 and four years. However, the systems considered include an insulated tank on the roof, and this is the main contributor to the weight of the galvanized steel component, which accounts for 37% of the energy used. For systems without a rooftop tank, the energy payback time should be closer to the lower limit of 1.5 years. All the solar heating collected after that is real "green" energy. There is really no reason for every house *not* to collect the solar energy that falls on its roof.

Photovoltaic (PV) solar panels on the roof are another matter. These are expensive, but they provide electricity, not just heat. It costs about $5 a watt to have PV installed on the roof. Since the electricity use per home in the USA is about 1.2 kW averaged over the whole year, one would need about 5 kW to cover the peak hours. The cost is then 5000 × $5 = $25,000. People usually pay between $20,000 and $40,000 for their systems, but there is a 30% federal rebate and sometimes also a state rebate in the US. PV systems are usually guaranteed to lose no more than 20% of their efficiency after 25 years. States with net metering will allow the electric meter to count only the external energy used and to run backwards if the solar cells produce more energy than is used. The savings in electricity

Fig. 3.24 A 4.4-kW photovoltaic roof installation (http://www.californiasolarco.com)

bills can payback the PV cost in about 15 years without rebates or about eight years with rebates.[27] This presumes that there is a large roof area with an unobstructed view to the south (in the northern hemisphere) (Fig. 3.24).

Whether PV solar can pay for itself of course depends on where you live. The number of Peak-Equivalent Hours per Day is a measure of how much usable sunlight is available in a given place. The average in the USA is 3.5–6.5 hours. Winter in the Northwest would give only 1.5–2.5 hours, while summer in the Southwest can give 8 hours.[27] At 2 hours of intense sun equivalent, a 5-kW PV system would yield 10 kWh of electricity. Remembering that the average use per home is 1.2 kW, amounting to $1.2 \times 24 = 28.8$ kWh/day, we see that a large system can supply about a third of the electricity requirements even in the Northwest. The good news is that even on cloudy days, 20–50% of solar energy can still be obtained.

Of course, the sun does not shine when we need electricity the most; namely, at night when the lights are on and we are watching TV. The energy has to be stored. In the Southwest, the peak power is so large that it cannot be used right way; it *has* to be stored. This requires batteries, which increases the cost of solar energy beyond that for the panels themselves. The most economical batteries available today are the lead-acid batteries used in cars. A whole bank of them will have to be installed in the house. There are larger, more compact lead-acid batteries available. These are used, for instance, in African safari camps in case diesel fuel for their generators cannot be delivered. A 20-feet (6 m) row of these can supply the minimal needs of a camp for three days. PV power, stored or otherwise, cannot run appliances because they produce direct current (DC) power. An inverter has to be used to convert the DC to AC at 60 cycles/s in the USA and 50 elsewhere. This is an additional expense that must be counted.

There are other impediments to local solar power that are not widely known. Shadows, for instance, can completely shut off a solar panel. This is because each solar cell produces only 0.6 V of electricity. The cells in a panel are connected in series to buildup the voltage to at least 12 V, which the batteries and inverters need.

If one cell is in shade, it cuts off the current from all the cells. This is like the old strings of Christmas tree lights which were connected in series instead of parallel. If one bulb burns out, the entire string goes out.

Dangers

In spite of its low voltage, rooftop solar may actually be the *deadliest* source of energy! This is because the panels get dirty and need to be cleaned of dust, dirt, wet leaves, and bird droppings in order to maintain their efficiency. People will naturally climb up to the roof to clean their panels. Statistics on people falling off the roof and ladders are not readily available, but here are some figures for accidental deaths from falls. The Center for Disease Control and Prevention[28] reports that 15,800 adults above age 64 died from unintentional falls. Another branch of CDC shows 19,195 total accidental deaths from falls in 2006.[29] Data from the US Census in 2000 show that deaths from falls from one elevation to another were 3,269 in 1996.[30] If we conservatively take the smallest number, about 3,000, and say that maybe 10% of those were falls from a roof or a ladder going to the roof, then 300 US deaths occur annually from such falls. Now if rooftop solar becomes widespread, this number may grow by an order of magnitude to 3,000 deaths per year. Compare this to the annual average of 32 coal-mining deaths in the USA from 1996 to 2009![31] Even the 4,000–6,000 coal-mining deaths in China is comparable to the number of USA fatalities if local solar power expands as intended.

Factories are usually large, single-story building with flat roofs. These would be ideal for solar installations. Forward-looking companies like Walmart and Google have already installed solar power on their roofs. Covered parking lots are also good candidates, and some are already being converted. These installations would be serviced by professionals, not homeowners. No doubt measures will be taken to make solar systems for homes safer. Panels can be designed with this in mind.[32] Perhaps a cottage industry of panel cleaners will arise, the way chimney sweeps have been reinvented. Rooftop solar is needed, but its dangers must be minimized.

Central-Station Solar Power

Solar Thermal Plants

We next consider large power plants that collect solar energy. There are two main kinds: solar thermal and solar electric. Solar thermal is more straightforward and easier to understand. It's done with mirrors. In one type, a large area of ground is covered with mirrors which reflect the light onto a "boiler" on top of a tower. One such installation is shown in Fig. 3.25. It is also called a *solar concentrator*. To keep the cost down, the mirrors are usually flat; but that means that they have

Fig. 3.25 A solar power tower in the Mojave Desert, USA (http://ec.europa.eu/energy/res/sectors/solar_thermal_power_en.htm)

to be controlled to follow the sun. The immense heat impinging on the boiler from all these mirrors brings a liquid to a very high temperature. This liquid is then piped down to storage tanks, where it can be kept until used, thus solving the day–night storage problem. The liquid can be water, oil, or molten salt. Water can only be heated to 100°C before it turns into steam, but molten salt can be heated to 1,000°C. It can be held at 600°C without damaging its container. However, it has to be used or drained before it cools into a solid, never to be melted again. To produce electricity, the hot salt is piped to a heat exchanger, where it turns water into steam, and the steam is used to run a standard steam turbine to generate electrical power. The heat cycle is at most 30–40% efficient, so there is a 70% loss in addition to the losses in focusing onto the boiler all the sunlight that falls onto the ground.

Parabolic mirrors, harder to make, can bring the sunlight onto a focus as the sun moves vertically in the sky. This method is used in linear systems like the one shown in Fig. 3.26. A long pipe fixed at the focus of the mirrors carries the fluid to be heated down to the end of each row, where it is transferred to storage tanks. Cheaper flat mirrors could be used this way if they are controlled to tilt so as to keep the reflected energy onto the pipe as the sun moves in the sky. This kind of system is shown in Fig. 3.27. The flat mirrors are remotely controlled to pivot around the circles, keeping the sunlight on the overhead pipe. The mirrors can also be set at different angles, like a Fresnel lens, to simulate a parabolic mirror which focuses on the tower regardless of the sun's position.

The *fossil footprint* of these systems can be found in several life-cycle analyses of central-tower and parabolic-trough solar thermal power plants. We give here representative figures from studies of two installations in Spain [2]. The first is a

Fig. 3.26 A parabolic trough system (http://thoughtsonglobalwarming.blogspot.com/2008/03/solar-thermal-company-says-it-could.html)

Fig. 3.27 A linear array of mirrors that rotate to follow the sun (http://www.instablogsimages.com/images/2007/09/21/ausra-solar-farm_5810.jpg)

central-tower type producing 17 MW from 2,750 mirrors totaling 265,000 m^2 in area and occupying a land area of 1.5 km^2 or 0.58 square miles. The parabolic trough system produces 50 MW from 624 mirrors of 510,000 m^2 area and occupying a land area of 2.0 km^2 or 0.77 square miles. The tower generates 104,000 MWh of electricity per year, while the larger trough system yields 188,000 MWh. Though these two systems seem very different, their other numbers, including their fossil footprint, are quite similar. Both are assumed to have a 25-year lifetime. Both use molten salts for energy storage, the tower having a 16-h storage capacity, compared to the trough's 7.5 h. Both systems are about 46% efficient in gathering the sunlight on their grounds, and a thermal efficiency of about 37% in converting that to electricity. The overall efficiency is about 16%. This is about twice as good as that of current commercial photovoltaic systems.

As in the case for Wind, the life-cycle studies here consider the amount of material used in building the installation and the energy used in mining, refining, and transporting each type of material. More energy is used in constructing and installing the mirrors, the buildings, the heat storage equipment, and the electrical generation plant. Gas and electricity from conventional sources are used in operating the plant. Decommissioning includes tearing down the plant and returning recyclable materials. This usually nets a negative energy cost. The bottom line is that both plants will have an *Energy Payback Time of 12.5 months*. This is shorter by at least a factor of 2 than that of photovoltaic systems. Instead of parabolic lenses, one can use Fresnel lenses. These are lenses that are collapsed into a flat sheet such that grooves in the sheet have the same angle as the lens would have at the same position. Fresnel lenses are the flexible plastic sheets that one can buy to magnify reading material. If such lenses can be manufactured on a large scale, the energy payback time would go down to *6.7 months*, comparable to that of wind turbines. The downside of solar thermal plants is the large amount of real estate they use A normal coal or nuclear plant produces 1,000 MW, 20 times that of the parabolic trough plant described here. Since 50 MW required 2 km^2 of land, 1,000 MW would require 20 times that or 40 km^2, an area two-thirds the size of Manhattan Island in New York!

Nonetheless, solar concentrators, especially the linear kind, are gaining steam, so to speak. There have been dubious pronouncements that 9% of the area of Nevada could provide enough solar electricity to supply the entire USA.[33] New mirror materials are being invented, with thick glass replaced by thin glass. The mirrors have to withstand the harsh desert environment and not fade with time. They are front-surface mirrors, not like the back-surface ones at home. A thin mirror would have at least six layers: a substrate of stainless steel or aluminum, a layer of adhesive and a layer of paint, then a copper back layer, and finally a silver reflection layer covered with a thin protective glass layer.[33] Solar thermal plants are capital intensive. Their electricity costs about $0.16/kWh, hopefully halved by 2012, compared to $0.06/kWh for conventional power. Grand plans are being made for 200–1000 MW size plants in sunny places like western USA, Spain, Israel, Egypt, and Mexico.

In all life-cycle studies, the *carbon footprint* is also calculated. This is the amount of CO_2 emitted in the life cycle of the installation. We have omitted this information because it is harder to understand, and the resulting payback time is about the same as for energy if fossil energy was used. Use of renewable energy for manufacture would, of course, decrease the carbon footprint.

Solar Photovoltaic Plants

If one were to build a solar power plant to compete with coal or nuclear plants, a number of problems have to be overcome: cost, transmission, storage, and energy payback time. Solar shares with wind the problems of transmission and storage, but wind is cheaper. Solar *thermal* has an easy way to store energy for short periods, but this is not available for solar photovoltaic. Let us first consider the problem of cost. Solar cells made of silicon are expensive, but nonetheless 90% of installed cells are made of silicon because those were invented first. The fastest growing market nowadays is in *thin-film* solar cells, which are much cheaper. Led by First Solar of the USA, rapid buildup of solar power in Germany, China, and the USA is being done with thin-film cells.

To compete with standard energy sources in cost, the magic number of $1/W of peak installed power is sometimes quoted. Silicon cells have been working their way down in cost but are still far from this goal. Thin film, however, may have already reached "grid parity." Where does this magic number come from? A rough calculation is given in Box 3.2 to show that it is quite reasonable.[34,35] The diluteness of sunshine means that central solar power would require lots of acreage. Box 3.3 shows that a solar plant generating the same power as a coal plant would occupy at least 100 km^2 (10,000 hectares or 24,700 acres). Figure 3.28a shows what a solar farm looks like. It is a 100-hectare, 14-MW plant opened in 2008 in southern Spain. The 120,000 panels can handle 23 MW of peak power. Figure 3.28b shows an aerial view of the area, which was cut out of sunny wine-growing country. This amount of land is necessary to supply electricity for a small town of 20,000 homes.

Box 3.2 Price of Solar Cells for "Grid Parity"

This is a little complicated because "$1 per watt" refers to watts, which are not units of energy. We have to take into account that *kilowatts* give instantaneous *power*, while electricity costs are given in units of *kilowatt-hours*, which are *energy* units. A kilowatt-hour is the amount of energy generated by a 1-kW source of energy in an hour.

As deduced earlier, one peak kW of solar power yields an annual average of about 200 W as sunlight varies from day to night and summer to winter. This is the same as saying that the Peak-Equivalent Hours per Day is about five. So at $1 per peak watt, 1 kW of peak power would cost $1,000; and 1 kW of average power would cost about five times as much or $5,000. For this much investment, how many kilowatt-hour do we get? Well, that depends on the life of a solar cell. There are 8,766 hours in a year; and if we assume a lifetime of 20 years for the cells, they will last about 175,000 h. Dividing $5,000 by this, the cost of solar electricity would be $0.03/kWh, compared with $0.10/kWh for average electricity cost in the USA in 2009,[34]

(continued)

Box 3.2 (continued)

which is three times higher. However, $1/W is the cost of the solar cells only. The cells have to be mounted, transported, and installed, and substations have to be built to convert the low-voltage DC from the cells into high-voltage AC for the grid. Some mechanism must be built to store the energy for nighttime use, and long transmission lines have to be built to carry the electricity from the desert to population centers. The price for thin-film cells is reported to be as low as $1.18/W in 2009.[35] However, First Solar executives estimate that the price of $1/W may have to be halved before grid parity is achieved.

Box 3.3 Covering the Desert with Glass

A typical large coal or nuclear plant produces 1 GW of electricity. How much area would a comparable central solar photovoltaic plant take up? Using the figure of 200 W/m^2 given above for average solar radiation, we multiply by a solar cell efficiency of 8% to get a net power of 16 W/m^2 from thin-film solar panels. More power is lost in the electronics and the inability to tilt the panels economically to follow the sun. A more realistic estimate for net power may be 10 W/m^2 for a power plant. At this rate, a 1-GW power plant would require 100,000,000 m^2 of space, the area of a square 10 km (6.25 miles) on a side. How much does it cost to cover such an area with solar cells? At $1 per peak watt or $5 per average watt, 1 GW would cost $5 billion for the cells alone. Compare this with covering the desert with other materials. Cheap plywood costs about $20 for a 4×8 foot sheet, 3/4-in. thick. This works out to $6.73/m^2, or $670 million for 100 million m^2, only about seven times less. Cheap window glass costs about $58/m^2 or $5.8 billion for 100 million m^2. This is more than the $5 billion for solar cells! To produce photovoltaic cells at $1/W would be a remarkable achievement. Solar cells, which are glass coated with multiple delicate layers of semiconductor material, with electrodes, have to be manufactured at less cost than the retail price of ordinary glass!

With prices near grid parity, industrial investment in solar panels is expanding so fast that the numbers of dollars and megawatts given now will change rapidly. China is the largest manufacturer of solar panels, 99% of which are exported. China had only 140 MW of photovoltaic cells installed in 2009 but has plans to expand to 20 GW (gigawatts or thousands of MW) by 2020.[36] The USA plans to have 5–10 GW installed by 2015. Spain added 2.3 GW in 2008, catching up with Germany's 5.8 GW already in place.[37] First Solar has ramped production to 192 MW/year, but at this rate many manufacturers will have to participate in the growth of central-station solar photovoltaic.

Fig. 3.28 A large solar farm in Jumila, Spain (http://ourworldonfire.blogspot.com/2008/08/worlds-largest-pv-solar-farm-opens.html; http://technology4life.wordpress.com/2008/01/31/the-world%C2%B4s-largest-pv-solar-plant-open-in-southern-spain/)

Storage and Transmission

With the price of solar cells under control, the next problem is to find a way to store the energy collected during the day for use at night. *Storage of energy is not the same as storage of fuel.* For instance, gasoline does not take much space, but after it is burned, the energy can only be stored in batteries and such, which are large and expensive. Storage is not a big problem for rooftop solar because that energy is only a small supplement to the electrical grid, and large power plants are still needed to supply nighttime energy. If solar farms are to provide backbone power, however, storage is needed to cover nights and cloudy days. The same methods described above for wind power are also available for solar. Capacitors or batteries to serve a GW-size solar farm would be prohibitively large and expensive, and making them would greatly increase the fossil footprint. Pumping uphill to get hydro at night is not practical, since deserts have few hills. For lack of a better idea, what is usually proposed is the unproven concept of compressed air energy storage (CAES), as shown in Fig. 3.17. Excess solar energy is used to force compressed air into underground caverns or salt domes. Unlike CO_2 storage in such natural structures, the gas does not stay there. It is taken out at night, and its pressure is used to drive turbines to generate electricity. As explained under Wind, there is a large energy loss due to the heating of air when it is compressed.

If energy is so hard to store, what about transmitting it from the southwestern USA to the east coast? A Smart Grid for the USA is under discussion for distribution of renewable energies. This is a huge project that cannot be carried out in less time or for less money than developing fusion reactors. The electrical grid is a complex network of high-voltage lines, ranging from 115 to 765 kV, connecting power generators to user sites. It has to respond to sudden changes in power needs, and its reliability is tightly regulated. Even without the special needs of renewable energies, it has to be modernized in any case because of aging equipment and the especially stringent requirements of digital circuits [3]. Another publication from the Electric Power Research Institute proposes superconducting transmission lines cooled by liquid hydrogen, which would not only lower transmission losses but also supply hydrogen for cars [4]. Even if it makes sense, it will take many years for such a new idea to reach the design, costing, and building stages. Rights-of-way will be legal roadblocks for new transmission lines. Carrying central-station solar power straight from Arizona to New York or from North Africa to Paris requires changing the whole infrastructure.

Is Large-Scale Solar Power Really Feasible?

Proponents of solar power have calculated what it would take for a sizable fraction of the world's energy to be provided by sunlight. Jacobson and Delucchi [5] estimated that the world will need 16.9 TW (terawatts or billions of kilowatts) of energy by 2030. If we were to use only Water, Wind, and Sun power, only 11.5 TW would be needed, since these sources can generate electricity directly, without going through a thermal cycle. This amount can be generated by WWS in the proportion shown in Fig. 3.29. Water energy (1.1 TW) is to come from hydroelectric and geothermal plants, and from tidal turbines yet to be developed. Wind power (5.8 TW) will come from 3.8 million wind turbines and from machines driven by ocean waves, which arise from wind. Solar power (4.6 TW) will require 89,000 300-MW power plants and 1.7 billion rooftop collectors. These three sources would have to work together to cover the daily and annual fluctuations. More than 99% of these numerous installations have still to be built.

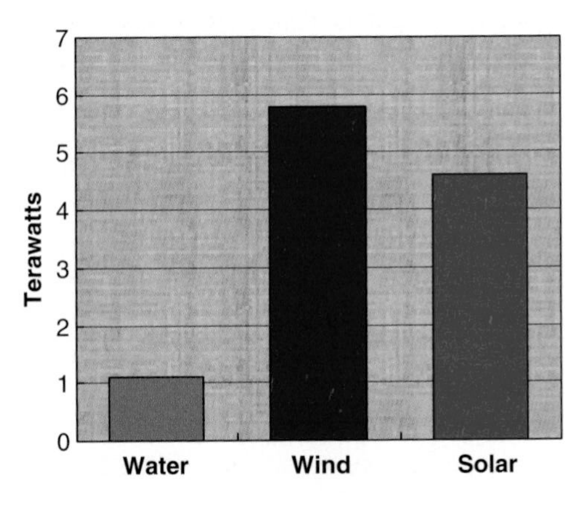

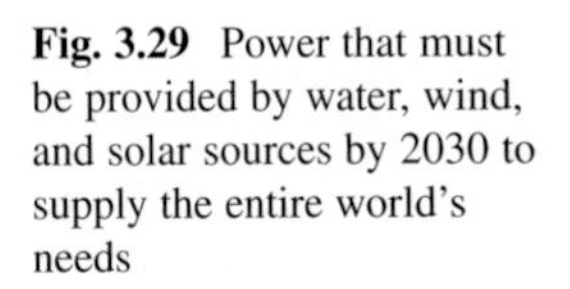
Fig. 3.29 Power that must be provided by water, wind, and solar sources by 2030 to supply the entire world's needs

The solar part of this has been evaluated in great detail by Fthenakis et al. [6]. They estimate that plants located in the Sahara, Gobi, or southwestern US deserts can produce photovoltaic electricity at \$4/W and \$0.16/kWh. This includes the entire plant, not just the panels themselves. Since residential electricity costs closer to \$0.12/kWh than the average of \$0.10/kWh, and since there are rebates, the cost of solar is already competitive with standard sources. The authors point out that electricity storage and transmission have still to be developed, and this has to be done using conventional fuels, since solar energy is still small. However, the energy payback time is of the order of two years (as will be shown here later); and once solar grows to 10% or more of total energy, further development could be done without the use of fossil fuels. These studies seem to be realistic, since the authors point out that there are many problems that still need to be treated in detail: the availability of rare materials, the sites for compressed-air storage, the transmission problem, the commercial problems in scaling up, and ecological damage to land and wildlife. If 10% solar cell efficiency is achieved and 2.5 times more land area than cell area is required, then 42,000 km^2 of desert area could supply 100% of the electricity for the USA (if it can be stored and transported). This seems like a large area, but it is less than half the area of the lakes produced by dams for hydro in the USA, and solar produces 12 times the energy. Lakes like Lake Mead have drastically changed the landscape. The change may have been welcomed by boaters, but not by the fish.

At this point, it is becoming clear that WWS (water, wind, and solar) sources have some large problems to overcome: storage of intermittent energy; transmission over large distances; use of large land areas; ecological damage to land and wildlife; unsightly encroachment on the landscape and seascape; and legal, political, and environmental objections to these intrusions. Overcoming these obstacles may take longer than developing compact power centers, like nuclear fusion, which avoids these problems. Replacing the power core of a coal or nuclear plant with a fusion reactor would retain the electrical generators, transmission lines, and real estate already in place. There would be no noticeable difference to the public except that all CO_2 emissions and fuel costs would be eliminated. The great advantage of WWS, however, is that feasibility is already proven; and further improvements in technology can be tested on a small scale, privately financed by industry. By contrast, each step in the development of fusion is so costly that the expense is best shared among nations.

How Photovoltaics Work

A solar cell is an electronic device made of semiconductors in layers, just as computer chips are, but much larger and simpler. Since each cell produces less than 1 V, cells have to be connected in a series to give a useful voltage, like 12 V. Flashlight batteries generate 1.5 V, and we use two of them in series to get the 3 V required by the bulb. Solar *panels*, about half a square meter in size, contain many cells connected together by transparent wires. The difference among conductors (like metals), insulators (like glass), and semiconductors arises from quantum mechanics, which mandates that energy levels in a solid are *quantized*. That means that electrons cannot have any old energy but must have an energy on one of the allowed levels. Furthermore, no two electrons can be on the same level. This situation is shown in Fig. 3.30. Energy levels occur in bands, two of which are shown, each containing seven energy levels. There are, of course, zillions of levels in actuality. In an insulator, the levels in the lower band are all filled, one electron in each level. This material cannot conduct electricity, because the electrons cannot move. To move, they would have to gain a little energy, but there is no level close enough for them to move up to. In a conductor, the lower band is filled, but the material has some electrons in the upper band, which is not full. Those electrons *can* conduct electricity because there are levels above that they can move up to. In a semiconductor, the lower band is full, but the bandgap is small, so if the topmost electron gets a big enough kick (from sunlight, for instance), it can jump up to the upper band, where it can move. So a semiconductor conducts sometimes.

The most common semiconductor is silicon. The bandgap in silicon is 1.1 eV. It is not important at this point to know how much energy an eV is; it will be explained amply in Chap. 4. The "kick" that the electrons get from sunlight to cross the bandgap depends on the color of the light that hits it. Sunlight contains a range of colors, as we know by separating them with a prism (Fig. 3.31), giving rise to the proverbial sequence violet, indigo, blue, green, yellow, orange, and red. Light can be considered as a stream of photons, which are particles with energy but no mass. No, they do not follow $E=mc^2$! Each color corresponds to photons of a certain energy. Those at the blue end of the spectrum have more energy, and those at the red end have less. For a photon to make a semiconductor conduct, it must have an energy of at least 1.1 eV. That means that the part of sunlight redder than that will be lost. For silicon PV, the idea is to add semiconductors with other bandgaps that can capture the other parts of the solar spectrum.

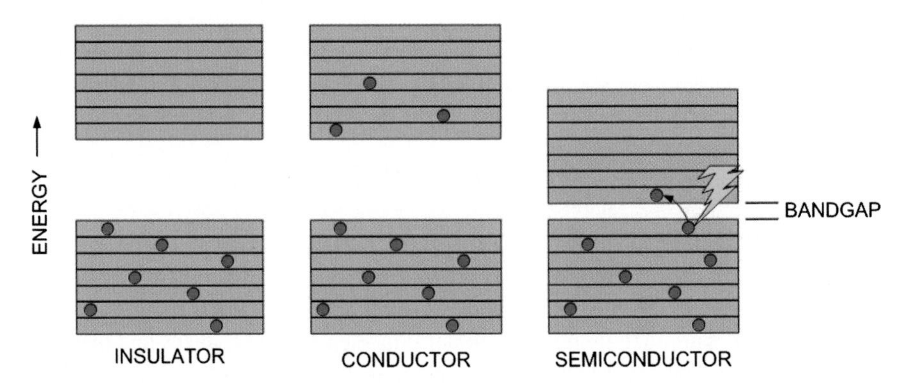

Fig. 3.30 How semiconductors differ from other materials

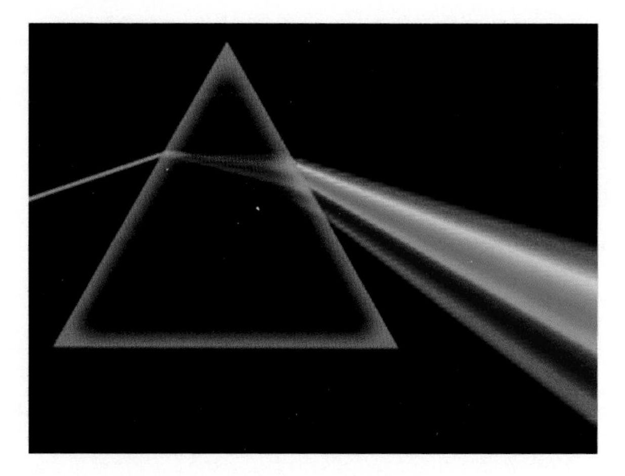

Fig. 3.31 The colors of sunlight (http://images.google.com)

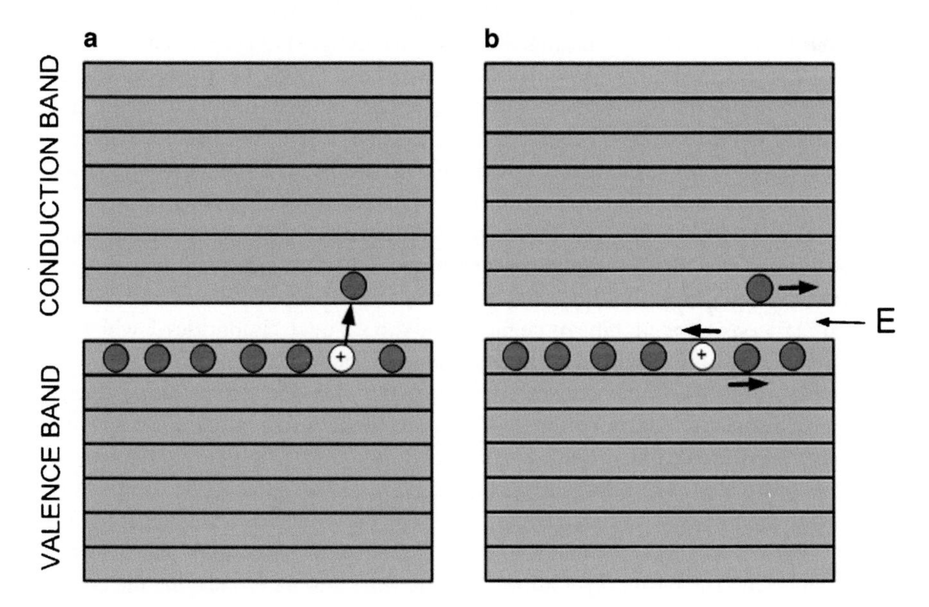

Fig. 3.32 Creation of an electron–hole pair and how a hole moves

After a photon kicks an electron into the conduction band, what happens next? This is shown in Fig. 3.32. This is the semiconductor part of Fig. 3.30, but showing only the electrons on the top level. After an electron is kicked into the conduction band, it leaves a hole in the valence band. What we have not shown is that the electrons actually belong to atoms consisting of a positive nucleus surrounded by enough electrons to make the whole atom uncharged. These atoms are locked into a crystal lattice. In Fig. 3.32a, an electron has been knocked out of one atom into the conduction band. It leaves behind an atom with a missing electron and therefore

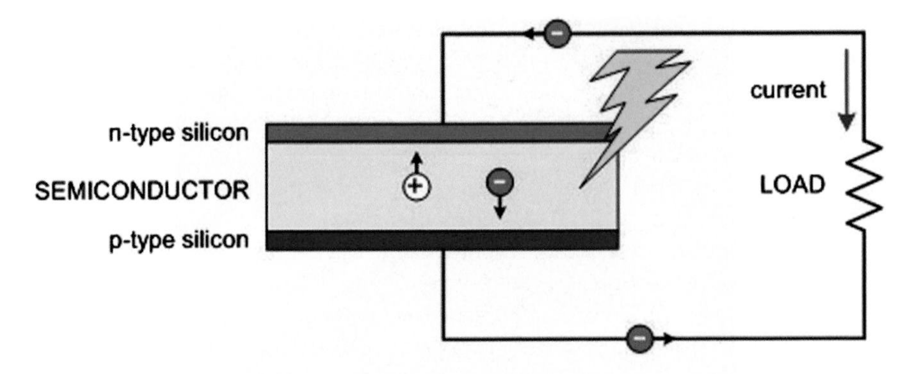

Fig. 3.33 Basic element of a solar cell. The electric current carried by the electrons is opposite to their motion

has a charge +1. That atom, shown in white, has a "hole" in it; that is, a place where an electron should fit but is missing. An electron can then jump from a neighboring atom, thus filling the hole but leaving a hole in the neighboring atom. As shown in B, the hole can move like a *positive* electron! If an electric field is applied, the electron in the conduction band will move one way, and the hole in the valence band will move the opposite way. These electron–hole pairs will conduct electricity, and now we have to see how the current is collected.

The electrons and holes cannot be collected directly with a copper plate connected to a wire because these charges cannot cross the interface between these very different materials. A buffer layer has to be added. These buffer layers are made of "doped" silicon. Here, doping is legal. By adding a few "impurities," which are specially chosen atoms with one more or one less electron than silicon has, we can make *n-type* or *p-type* highly conductive semiconductors. The former has a net negative charge, and the latter a net positive charge. We can then make a sandwich of three layers to form the basic unit of a solar cell (Fig. 3.33). Opposite charges attract, so when solar photons create electron–hole pairs in the silicon, the electrons are attracted to the p-type layer at the bottom, and the holes to the n-type layer at the top. Since they are negative, the electrons carry a current in the opposite direction to their motion. The buffer layer allows them to flow into wires carrying the current to the load (the appliance or battery that uses the juice). When the electrons reach the n-type layer, they fill the holes that had migrated there. The voltage generated is the bandgap voltage. The larger the bandgap, the higher the voltage. This makes sense, since only the energetic photons can push an electron across a large bandgap.

Silicon Solar Cells

By far the most common type of solar cell because of their long history, silicon solar cells are fast being overtaken by thin-film cells, which are much less complex and costly.

Crystalline silicon is expensive and takes a lot of energy to make. It also absorbs only part of the solar spectrum and does it weakly at that. Only those photons that have more energy than silicon's bandgap can be absorbed, so the red and infrared parts of sunlight are wasted. That energy just heats up the solar cell, which is not good. The blue part of the solar spectrum is also partly wasted for the following reason. Each photon can release only one electron regardless of its energy as long as it exceeds the bandgap. So a very energetic photon at the blue end of the spectrum uses only part of its energy to create electric current, and the rest of the energy again is lost as heat. To capture more colors of sunlight, cells made with other materials with different bandgaps are used in the basic cell instead of silicon. These other semiconducting materials are called III–V compounds, and they are explained in Box 3.4.

Box 3.4 Doped and III–V Semiconductors

The way semiconductors can be manipulated is best understood by looking at the part of the periodic table near silicon, as shown in Fig. 3.34. The Roman numerals at the top of each column stand for the number of electrons in the outer shell of the atom. The different rows have more inner shells, which are not active. The small number in each cell is the atomic number of the element. Silicon (Si) and germanium (Ge) are the most common semiconductors and are in column IV, each with four active electrons. They share these with their four closest neighbors in what is called covalent bonds. These are indicated by the double lines in Fig. 3.35. These bonds are so strong that the atoms are held in a rigid lattice, called a crystal. The actual lattices are three-dimensional and not as simple as in the drawing. The crystal is an insulator until a photon makes an electron–hole pair by knocking an electron into the conduction band, as we saw in Fig. 3.32.

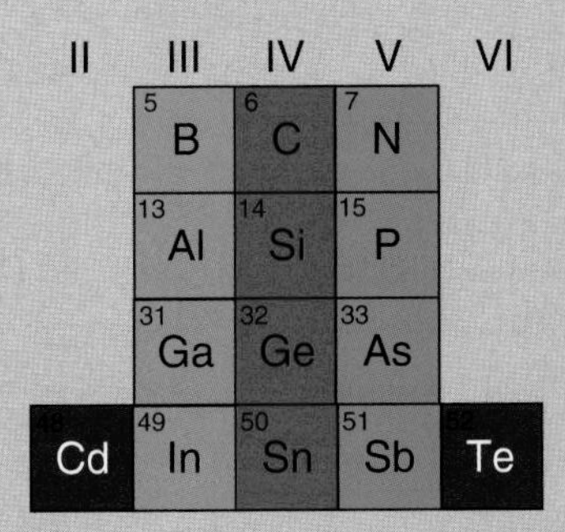

Fig. 3.34 The periodic table near silicon

(continued)

Box 3.4 (continued)

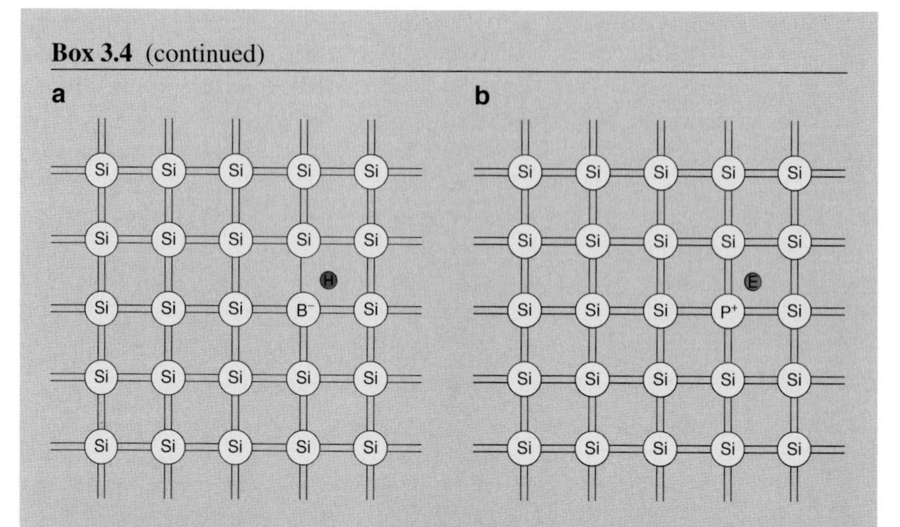

Fig. 3.35 A silicon lattice doped with (**a**) boron and (**b**) phosphorus

However, there is another way to make Si or Ge conduct. We can replace one of the silicon atoms in Fig. 3.35a with an atom from column III, for instance, boron. We would then have a "hole." That's because boron (B) has only three active electrons and leaves a place in a covalent bond where an electron can go. Since holes can move around and carry charge as if there were positive electrons, this "doped" semiconductor can conduct. We can also dope Si with an atom from column V, such as phosphorus (P), as shown in Fig. 3.35b. Since phosphorus has *five* active electrons, it has an electron left over after forming covalent bonds with its neighbors. This is a free electron which can carry current. Note that the P nucleus has an extra charge of +1 when one electron is removed, so the overall balance of + and – charges is still maintained. The conductivity can be controlled by the number of dopant atoms we add. In any case, only a few parts in a million are sufficient to make a doped semiconductor be a good enough conductor to interface with metal wires. Any element in column III, boron (B), aluminum (Al), gallium (Ga), or indium (In), can be used to make a p-type semiconductor (those with holes). Any element in column V, nitrogen (N), phosphorus (P), arsenic (As), or antimony (Sb), can be used to make an n-type semiconductor. When the doping level is high, these are called p^+ and n^+ semiconductors.

Now we can do away with silicon! We can make compounds using only elements from columns III and V, the III–V compounds. Say we mix gallium and arsenic in equal parts in gallium arsenide (GaAs). The extra electrons in As can fill the extra holes in Ga, and we can still have a lattice held by covalent bonds. We can even mix three or more III–V elements. For instance, $GaInP_2$, which has one part Ga and one part In from III and two parts P from V. There are just enough electrons to balance the holes. This freedom to mix

(continued)

Box 3.4 (continued)

any of the III elements with any of the V elements is crucial in multijunction solar cells. First, each compound has a different bandgap, so layers can be used to capture a wide range of wavelengths in the solar spectrum. Second, there is *lattice-matching*. The lattice spacing is different in different compounds. Current cannot flow smoothly from one crystal to another unless the spacings match up. Fortunately, there is so much freedom in forming III–V compounds that multijunction cells with up to five compounds with different bandgaps have been matched. Figure 3.36 shows how the three layers of a triple-junction cell cover different parts of the solar spectrum.

At the bottom of Fig. 3.34, we have shown a *II–VI* compound, cadmium telluride (CdTe). Each pair of Cd and Te atoms contributes *two* electrons and holes. This particular II–VI material has been found to be very efficient in single-layer solar cells. It is one of the main types of semiconductors used in the rapid expansion of the thin-film photovoltaic industry.

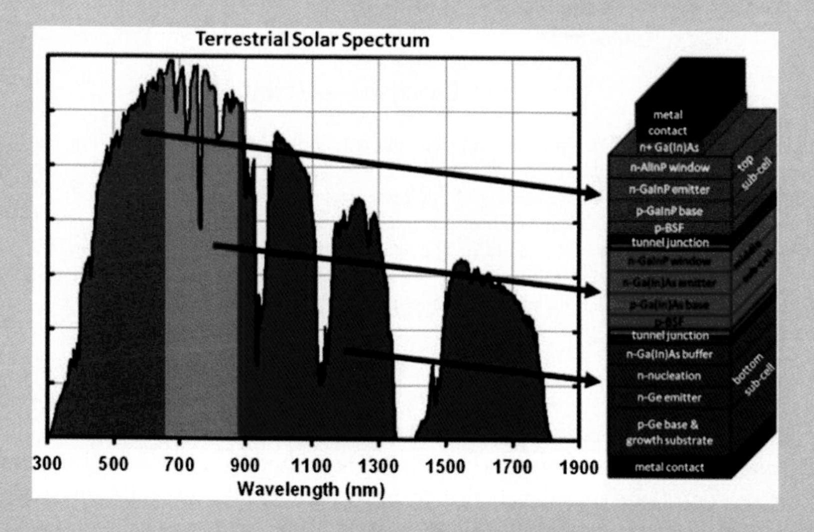

Fig. 3.36 The parts of the solar spectrum covered by each subcell of a triple-junction solar cell (http://www.amonix.com/technology/index.html)

By adjusting the compositions of these III–V compounds, their bandgaps can be varied in such a way as to cover different parts of the solar spectrum. This is illustrated in Fig. 3.37. The spectrum there will be explained in Fig. 3.40. The different cells are then stacked on top of one another, each contributing to the generated electric current, which passes through all of them. There are many layers in such a "multijunction" cell. The layers of a simple two-junction cell are shown in Fig. 3.38. The top cell has an active layer labeled n–$GaInP_2$ and is sandwiched

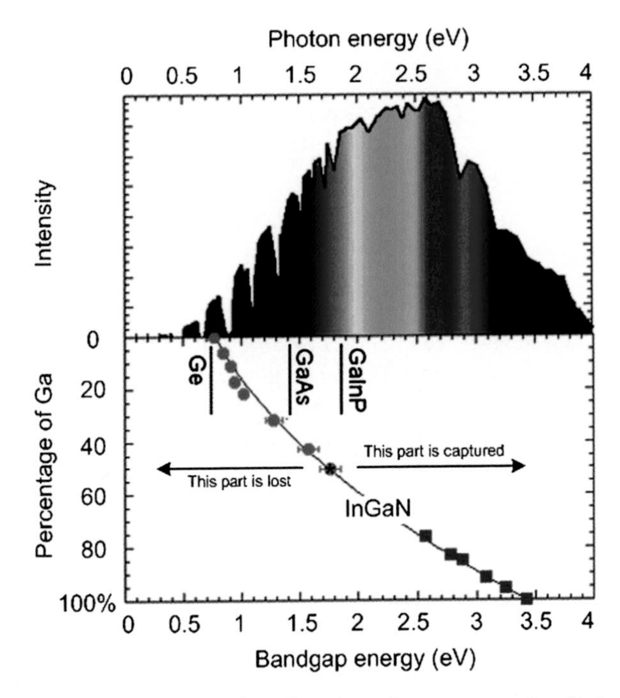

Fig. 3.37 *Top*: the solar spectrum plotted against photon energy in eV. Long (infrared) wavelengths are on the *left*, and short (ultraviolet) wavelengths are on the *right*. The visible part is shown in the *middle*. *Bottom*: bandgaps of various semiconductors plotted on the same eV scale. The bandgaps of Ge, GaAs, and $GaInP_2$ are fixed at the positions marked. In InGaN, half the atoms are N, and the other half In and Ga. The bandgap of InGaN, given by the data points, varies with the percentage of Ga in the InGa part. As illustrated for the marked point, the part of the spectrum on the *blue side* of its bandgap is captured, and the part on the *red side* is lost (adapted from http://emat-solar.lbl.gov)

between the current-collecting buffer layers labeled n–$AlInP_2$ and p+GaAs. This is the basic cell structure shown in Fig. 3.33. The bottom cell has an active element labeled n–GaAs surrounded by buffer layers. Connecting the two cells is a two-layer tunnel diode, which ensures that all the currents flow in the same direction. Up to five-cell stacks have been successfully made,[38] yielding efficiencies above 40%, compared with 12–19% for single-silicon cells. Each cell in a stack has three layers plus the connecting tunnel diode. However, not all the layers are equally thick as in the diagram, and the entire stack can be less than 0.1 mm thick! Pure crystalline silicon needs at least 0.075 mm thickness to absorb the light and at least 0.14 mm thickness to prevent cracking [7], but this does not apply to the other materials.

The semiconductor layers are the main part of a solar cell, but they are thin compared with the rest of the structure. A triple-junction cell is shown in Fig. 3.39. The support layer could be a stainless steel plate on the bottom or a glass sheet on the top. The top glass can also be grooved to catch light coming at different angles. At the bottom is a mirror to make the light pass through the cell a second time.

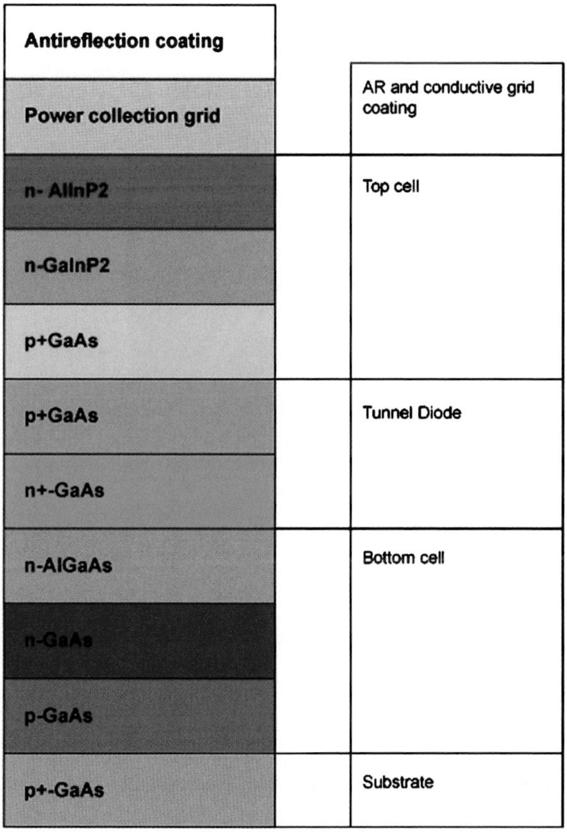

Fig. 3.38 The parts of a two-cell stack using gallium–indium–phosphide ($GaInP_2$) and gallium arsenide (GaAs) (http://www.vacengmat.com/solar_cell_diagrams.html)

At the top is an antireflection coating such as we have on camera or eyeglass lenses. The current is collected by a grid of "wires," formed by a thin film of conducting material. The top grid has to pass the sunlight, so it is made of a transparent conductor like indium-tin oxide, which is used in computer and TV screens for the same purpose. The photovoltaic layers have to be in a specific order. At the top is material with the largest bandgap, which can capture only the blue light, whose photons have the highest energy. The lower energy photons are not absorbed, so they pass through to the next layer, labeled "green" here. This has a lower bandgap and captures less energetic photons. Last comes the "red" layer, which has the smallest bandgap and can capture the low-energy photons (the longest wavelengths) which have passed through the other layers unmolested. If the red layer were on top, it would use up all the photons that the other layers could have captured, but it would use them inefficiently, since the voltage generated is the same as the bandgap voltage.

The voltage generated by each cell is only about 1.5 V, so cells are connected into chains that add up the voltage in series to form a module. Modules giving a

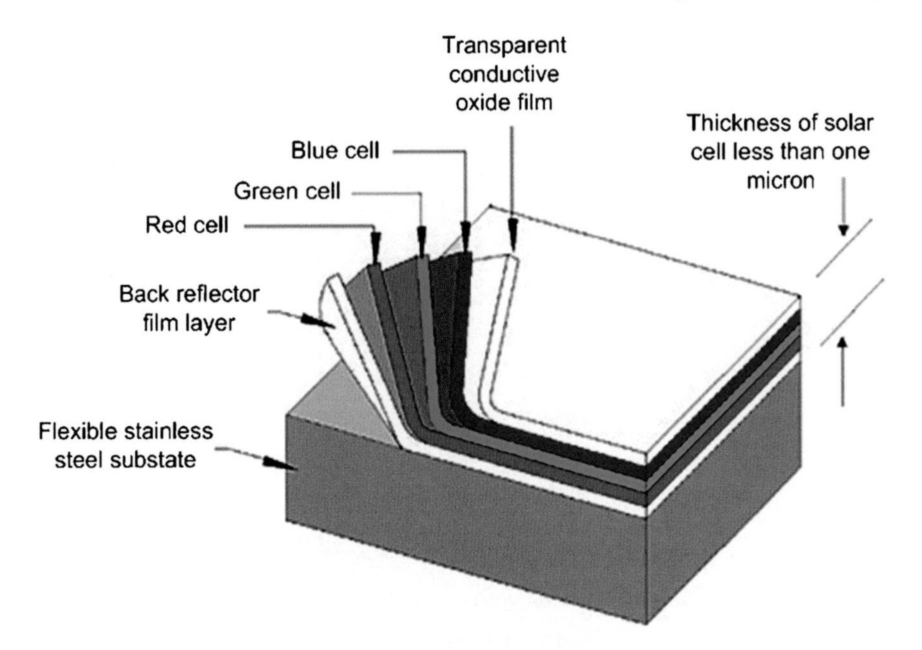

Fig. 3.39 A typical multijunction solar cell assembly. All the layers in the active part of this cell are less than 1 μm (1/1,000th of a millimeter) thick (http://www.solarnavigator.net/thin_film_solar_cells.htm)

voltage of, say, 12 V are then grouped into arrays, and thousands of arrays make a solar farm. Modules and arrays generally need to be held in a frame, adding to the cost, and the frames have to be supported off the ground. There is a problem with the series arrangement of the cells. If one cell fails, the output of the entire chain is lost, since the current has to go through all the cells in a chain. Similarly, if one of the layers in a cell fails, there can be no current going out of that cell. Fortunately, the failure rate of commercial units is known and is not bad. Solar cells can still produce 80% of their power after 25 years or more, at least for single-junction cells.

Solar cell efficiency is degraded by another effect: the colors to which a cell responds is fixed in the design of the photovoltaic layers, but the color of sunlight changes with time and place. At sunset, the light is redder and yellower. This means that the blue cell cannot put out as much current. Since the same current flows in series through the whole stack, the red cell's larger current cannot all be used; its excess current turns into heat. The atmosphere alters the solar spectrum more than you might think. This is shown in Fig. 3.40. In space, the spectrum is almost exactly like that of a classical blackbody. In the visible part of the spectrum, about 30% the intensity is absorbed by the atmosphere. In the infrared region, large absorption bands are caused by gases in the atmosphere. This spectrum is further degraded by the atmosphere during the day as the sun goes lower in the sky.

Multijunction and crystalline silicon solar cells are so expensive that they are not suitable for solar farms, but they have two good applications. First and foremost, these are used where cost is not a prime concern: in space satellites. The ruggedness of

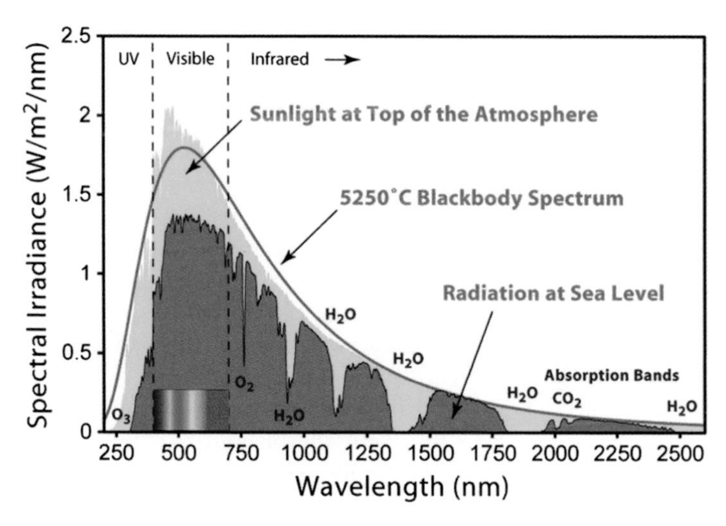

Fig. 3.40 The solar spectrum in space (*yellow*) and on the earth's surface (*red*). The visible region is shown by the small spectrum at the bottom. Parts of the spectrum are heavily absorbed by water vapor, oxygen, and CO_2 (http://en.wikipedia.org/wiki/Image:Solar_Spectrum.png)

silicon and the efficiency of multijunction are needed out there. The sunlight is stronger, and cooling has to be considered because there is no air. Missions to the moon and Mars will no doubt have the most expensive solar cells made. On the earth, expensive solar cells can be used in *concentrator PV systems*. Since multijunction cells are so expensive, it is cheaper to make large-area Fresnel lenses to catch the light and focus it onto a small chip. The solar intensity can be increased as much as 500 times ("500 suns"). The solar cell will be very hot, but cooling on earth is not a problem. This idea has attracted commercial interest. The Palo Alto Research Center of Xerox Corp. has developed a molded glass sheet with bumps like bubble-wrap. Each bump contains two mirrors configured like a Cassegrain telescope to focus sunlight onto a small cell. The amount of PV material needed is reduced by at least 100 times. Making high-quality silicon is very energy-intensive, but some forms of it can be used for terrestrial solar cells. More on silicon is given in Box 3.5.

Box 3.5 The Story of Silicon

Oxygen and silicon are the most abundant elements on the earth's crust, oxygen mostly in the form of water (H_2O) and silicon in the form of rock (SiO_2). These molecules are prevalent because they are very stable; it takes a lot of energy to break them up. The solar cell business got a head start because the semiconductor industry had already built up the infrastructure for producing pure silicon. Without a source of silicon, the expense of making a silicon solar cell would have been prohibitive.

(continued)

Box 3.5 (continued)

The integrated circuits that make computers, cell phones, iPods, and other electronic devices work are made of 99.9999% pure silicon. These chips are made of single-crystal silicon. First, pure silicon is produced from quartz. It is then melted in a crucible by heating to above 1,400°C (2,600°F). This requires a lot of energy: think of the molten rock flowing from the Kilauea caldera in Hawaii into the sea. A seed crystal is then dipped into the liquid and slowly drawn upwards, carrying some silicon with it. As the silicon solidifies, it takes on the crystalline structure of the seed; and a large cylindrical ingot is formed. The entire ingot, 400 mm (12 in.) in diameter, is a single crystal. This is then sliced into wafers about 0.2 mm thick. The "sawdust," or kerf, takes up 20% of the silicon, and it cannot be re-used because of contamination by the cutting tool. To make computer chips, a wafer is processed to make hundreds of chips at once, each containing millions of transistors. The wafer is then sliced into the individual chips, each no larger than 1 cm^2 in size. The cost of the silicon wafer is minor, since the chips are worth a million dollars. For solar cells, however, the large areas required mean that the silicon is the main expense, even when off-grade material rejected by the semiconductor industry is used. Silicon shortages cause large fluctuations in price. Note that to form solar cells, the silicon has to be re-melted, using more energy.

Single-crystal solar cells are the most efficient because electrons and holes flow easily along the lattice. However, silicon made of small crystals is cheaper and easier to make. The silicon can be poured into a crucible without the slow drawing-out process. Depending on the crystal size, this is called multicrystalline, polycrystalline, or microcrystalline silicon. In these materials, electron flow is interrupted by their bumping into grain boundaries. This causes a higher resistivity and hence loss of energy into heat. Most silicon solar cells are made of polycrystalline silicon.

There is also amorphous silicon, which is really a thin-film material. The silicon atoms are not in a lattice at all but are randomly distributed. The production process is entirely different. A glass substrate is exposed to silane (SiH_4) and oxygen (O_2) in a plasma discharge, where the hydrogen latches on to the oxygen to form water, and the silicon is deposited onto the glass. The electrical conductivity of amorphous silicon is very poor, and it has to be improved by adding hydrogen in a subsequent hydrogenation process. The result is called a-Si:H. Its power output decreases about 28% at first use, so it has to be "light-soaked" for about 1,000 h before it stabilizes. It is also less efficient in the winter, when the temperatures are lower. The efficiency is only about 6%, but amorphous silicon is much cheaper than any crystalline form and can be used in large installations. Crystalline silicon, on the other hand, is suitable for space applications but not for large solar farms.

Thin-Film Solar Cells

We have already seen that multijunction solar cells use thin films made of III–V or II–VI materials. The problem with crystalline silicon is that it is what is called an *indirect bandgap* material. We need not go into the physics of this. What it means is that a palpable thickness of silicon (about 0.1 mm) is needed to absorb photons, and we saw in Box 3.5 how hard it is to make pure silicon. Thin-film materials, on the other hand, have direct bandgaps. The absorption is so good that thicknesses are measured in microns,[39] typically 1 μm, which is a thousandth of a millimeter. By comparison, the thickness of an ordinary piece of paper is about 100 μm (0.1 mm or 0.004 in.), the same as a human hair. Thin films that can absorb 98% of sunlight are only 1% of that thickness. No wonder that even a thin layer of sunscreen spread on the skin can protect against sunburn. Since crystalline silicon in a solar cell has to be over 100 μm thick, thin-film solar uses 100 times less semiconductor material than silicon.

However, the small amount of material required for thin-film solar cells is not the main reason for their success. It is because manufacturing techniques developed by First Solar, Inc. of the USA and other companies have reduced the cost so that solar power is commercially viable. Development advances much more rapidly when support moves from the government to private industry, where the monetary incentive is strong. First Solar became dominant in its field by optimizing the use of CdTe (cadmium telluride). This material, with a bandgap of 1.45 eV, combines the best combination of voltage and current for the higher power output from a single layer. First Solar started with a plant in Ohio with 90 MW/year of production capability, then added a 120-MW/year plant in Germany and a 240-MW/year plant in Malaysia. It has contracted with China to produce 30 MW in 2010, then 100 and 870 MW by 2014, and finally a total of 1,000 MW by 2019. The entire production process, from deposition of all the layers to assembly and to testing, takes only 2.5 h on their automated production line. Benefiting from economy of scale, First Solar has lowered the cell cost to below $1/W and the module cost to $110/m^2. The goal is to bring this down to $0.50/W or $1.50/W including balance-of-system. The cost of electricity would be 6–8 ¢/kWh.[40] Producing 1 GW/year in solar cells would give the company one-sixth of the world's share.

The layers of a CdTe solar cell are shown in Fig. 3.41. The layers are deposited on a 60 cm × 120 cm glass superstrate 5 mm thick. This is about the size of a quarter-sheet of 4 × 8-feet plywood and will yield many cells. Below that is a thin SiO_2 layer for insulation, followed by a transparent conducting layer of SnO_2, which is the top electrical contact. A thin layer of CdS (cadmium sulfide) follows. Only about 0.1 μm thick, it serves as the n-doped layer in Fig. 3.33. It must be thin to allow the light to reach the absorbing layer of CdTe. Sulfur is a Column VI element, which has been left out of Fig. 3.34 to avoid clutter, so CdS is a II–VI compound. It turns out that CdS is naturally slightly n-doped in production, and CdTe is slightly p-doped [8], so the other layers in Fig. 3.38 are not necessary to separate the electrons from the holes, greatly simplifying the device. The main CdTe layer

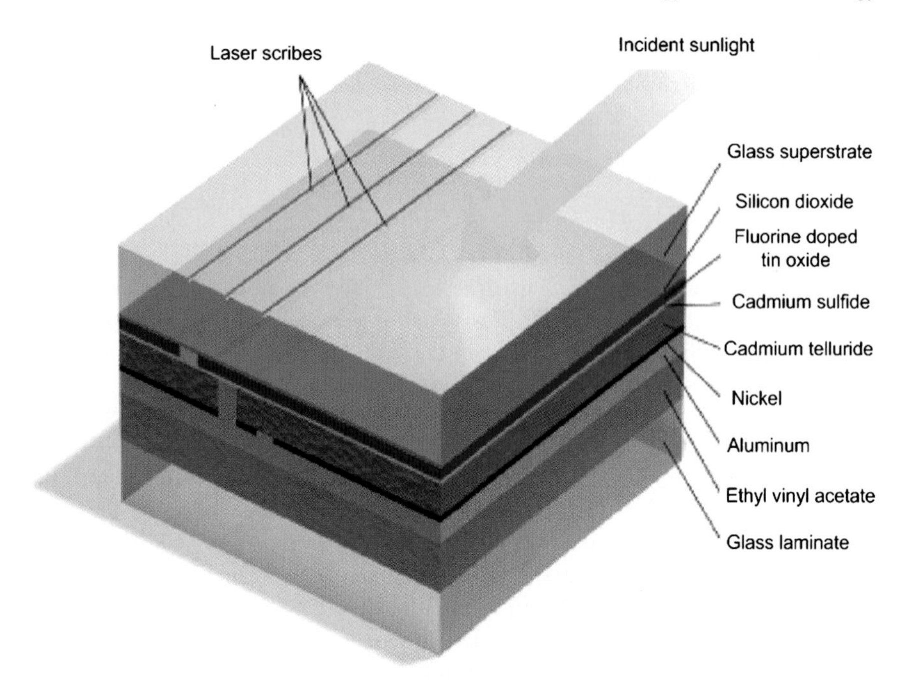

Fig. 3.41 Schematic of a CdS/CdTe solar cell (IEEE Spectrum, August 2008)

is 1–5 μm thick; Gupta et al. [9] have shown that the performance does not improve much beyond 0.75 μm. At the bottom is the other electrode, made of gold, nickel, or aluminum, followed by a plastic binder and a glass protector. Laser scribing is used between the deposition of the various layers to divide the cell into smaller cells and to connect them in a series to raise the voltage to 70 V. After all this, the whole sheet is annealed between 400 and 500°C in $CdCl_2$ gas to improve the efficiency by as much as a factor of 2.[41] The reason for this is not well understood. Such a cell puts out about an ampere of current and up to 75 W of power at 10.6% efficiency.[41] Improvement to 12% may be possible.

The record efficiency achieved in the laboratory is 16.5%. To do this, the transparent conductor at the top, usually tin oxide, was replaced by cadmium stannate, which has higher conductivity and is more transparent. A buffer layer of zinc stannate was then added below it.[42] As current flows through the cell, its internal resistance causes energy to be lost as heat. This loss is measured by the *filling factor*, which is the percentage of the ideal power that is actually usable. The best that can be achieved is 77%.[42] Although the general production process is well known[41] (see ref. [8]), the know-how details are closely guarded secrets. For instance, the bottom contact tends to be unstable, and adhesion is affected by the annealing step.

Thin-film materials competing with CdTe are amorphous silicon (a-Si:H) and copper indium gallium diselenide (CIGS). Amorphous silicon has a low efficiency of 6–7%, but it has had a head start because the manufacturing equipment had been developed in the semiconductor industry. This material loses the red part of the solar spectrum, and there are attempts to add a 2-μm layer of microcrystalline silicon to add the blue part. The efficiency might then go up to 11% to compete with CdTe. CIGS has a laboratory efficiency of 19.5% vs. 16.5% for CdTe. In modules, the efficiencies are 13 and 11%, respectively; and in production they are 11.5 and 9% [8]. CIGS is harder to make, but it is being pursued because of the possibility of 25% efficiency. Currently, it has only a 1% market share, compared with 30% for CdTe and 60% for a Si.[43]

Fossil Footprint and Environmental Issues

Many life-cycle analyses have been made of both silicon and thin-film solar cells. In 2007, Raugei et al. [10] published a careful study of the environmental effects of both silicon and thin-film solar cells using actual production data from Europe. For polycrystalline silicon (the most common kind), one had to decide where it came from. If it came from the electronics industry, even if it is the off-grade rejected material, the energy cost is very high, as shown in Box 3.5. On the other hand, if the solar industry grows to the extent that it can build its own factories to produce solar-grade silicon of lower purity, the energy cost would be much lower. Both the worst-case and best-case scenarios for silicon were compared with CdSe and CIS (copper indium diselenide) thin-film systems. (CIS is similar to the CIGS mentioned above.) The results for energy payback time are shown in Fig. 3.42.

Here is what went into these calculations. First, the materials used were listed. For thin film, these were glass, plastic, water, and the electronic layers. Glass was

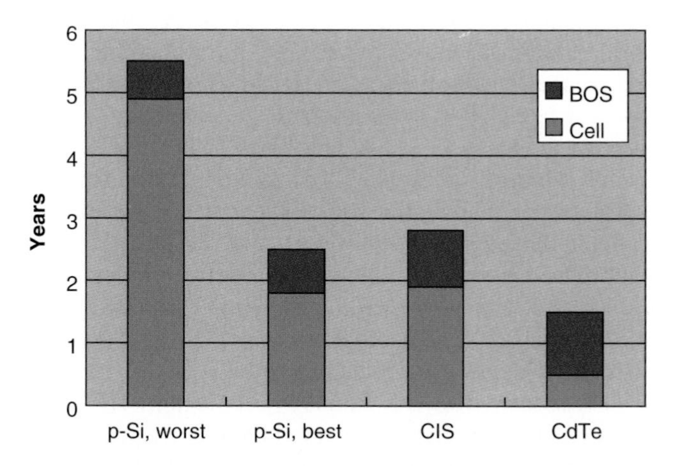

Fig. 3.42 Energy payback time, in years, for polycrystalline silicon (p-Si) and thin-film (CIS and CdTe) solar modules. For p-Si, the worst-case and best-case scenarios defined in the text are given. The *bottom part* of each column is for the bare cell, while the *top part* is for the balance-of-system (BOS), which includes the frame, supports, and electrical equipment for converting DC to AC [10]

by far the largest part, and the thin films the smallest. Then the energy to make these materials and the electricity to fabricate the cells in the factory were evaluated. That is for just the bare cell. To this must be added the balance-of-system; namely, the parts needed to complete functioning modules and arrays. These include aluminum for the frame, steel for the supports, cables and connectors, and the electric equipment for converting DC to AC at the grid voltage. There is also fuel oil used in installation. The energy cost of decommissioning and recovery of materials at the end of life was not included, but these were considered by Fthenakis [11]. The energy used was assumed to be the mix of fossil and hydro energy typical of Europe, with 32% average efficiency in generating electricity. As for the solar energy output, the assumptions were quite conservative. The sunlight available was 1,700 kWh/m^2 per year, typical of southern Europe, not a desert. A 25% efficiency loss was assumed to account for dust accumulation and electrical equipment. The lifetime of the system was taken to be only 20 years.

The calculated energy payback times are shown in Fig. 3.42. As expected, it is very long for silicon in the worst case, when it is obtained from the electronics industry. However, if special factories are built to produce solar-grade silicon in ribbon form, the payback time is competitive with thin film. CdTe is the clear winner in this study, its payback time being only 1.5 years. The graph also shows the breakdown between the energy costs of the bare cells and the balance-of-system or BOS. Note that for CdTe cells, it is the BOS that takes the most energy to make. The global warming potential (CO_2 emissions) of these systems is usually also calculated in these studies, and of course it is much smaller than that of fossil-fuel energy sources. After initial greenhouse gas (GHG) emissions during buildup, a solar plant produces electricity with almost no emissions for up to 30 years.

Cadmium is a very toxic element. In 2009, there was an uproar because some toys imported from China were found to contain cadmium in ingestible form. However, that does not mean that a compound like CdTe is toxic. Salt, NaCl, is certainly not dangerous although sodium and chlorine are themselves very toxic elements. In the case of CdTe, one worries that Cd could be emitted into the environment during manufacture and operation, even though the cells are encapsulated in glass and Cd is very stable, with almost zero evaporation. Unknown to most people, incidental emission of Cd also occurs in coal and oil plants. Raugei et al. [10] estimated the emission of Cd from a solar plant and found that it is 230 times smaller than from a coal plant for the same energy output! Detailed evaluations of dangers from toxic substances have been done by Fthenakis et al. [12, 13].

The amount of land used in solar power and the environmental impact on it has been compared with other energy sources by Fthenakis and Kim [14]. Not surprisingly, these solar proponents find that solar energy requires the *least* amount of land and biomass energy requires the most. The use of land in coal and nuclear power includes the land destroyed in mining and waste storage. Hydroelectricity uses dams which convert land into lakes. However, the usage of the area may actually be improved, and wildlife may only be changed from animals to fish. A large area covered with solar arrays may still allow desert animals, birds, and tortoises to live if some plants are allowed to grow under the panels. However, the reflectivity (albedo) of the desert will be decreased by the absorbing black panels. A very large area of these may affect cloud formation and the entire climate in the region.

The life-cycle studies of solar power are less complete than those of wind power and seem to be optimistic. The wind studies included replacement of parts as they wore out and the energy costs of inspection and maintenance, including the gasoline usage by the inspectors. Dust will cover the solar panels and should be washed off. In the desert, there is no water. The glass covers of the panels will be blasted in sandstorms. In temperate climates, plants will grow and have to be pulled out before they get too high. With no weeding, a solar farm will be immersed in a dense forest in 10–20 years. Space must be left open between rows of panels for machines to do this. The Mars rovers have experienced what happens to solar panels without maintenance. Dust accumulated on them, decreasing their power. The Rovers depended on wind storms to blow the dust off. After years of dust accumulation, the power became so weak that communication became difficult. The rovers had to be manipulated onto a crater's edge to tip the panels to face the sun more directly. In solar farms on earth, the panels are fixed.[44] It has been estimated that mechanisms to track the sun would add 25% to the cost of the panels but could increase their capacity by 40%. The cost of energy storage for night time was not included in these studies.

However, the storage problem was addressed in admirable detail by Mason et al. [15]. The only method being considered is CAES, which is described in the Wind Energy section (Fig. 3.17). The electrical energy being stored is used to compress air into these caverns. When the energy is needed, this compressed air is released and used to help drive a gas turbine to produce electricity. CAES has been tested only at two places: in Germany, where a 290-MW plant has been operating since 1978, and in Alabama, where a 110-MW plant has been operating since 1991. These CAES systems were used to store excess electricity produced conventionally in off-peak hours. There are numerous sites in the USA where caverns suitable for CAES exist, but they cannot be close to the solar farms for several reasons. The deserts where there is the most sunlight have few suitable sites and insufficient water needed for cooling. They are also far from population centers. A system of high-voltage DC (HVDC) transmission lines is proposed to connect the solar plant to the storage plant. The energy capacities of the two plants also have to be matched.

The Mason study [15] considered a storage plant that provides peak power 10 hours a day, Monday through Friday when it is needed, and another for base-load power 24 hours a day for a future central-station solar farm. The daily solar output during the year was calculated, as well as the storage requirements during each day. The costs of the solar and storage plants were carefully itemized, including such items as maintenance, land preparation, interest during construction, and replacement of parts. The HVDC cost was included, as well as the substations for converting DC to AC. The results for a peak-load PV-CAES system are summarized in Fig. 3.43. The cost of electricity from PV systems with storage is compared with that from an advanced-cycle natural gas plant with carbon sequestration. In the next 10 years, the cost of the PV part is expected to go down, but the CAES part does not go down as much. It accounts for a third of the total cost. Solar electricity for peak loads, it appears, will be competitive with that from natural gas by 2020.

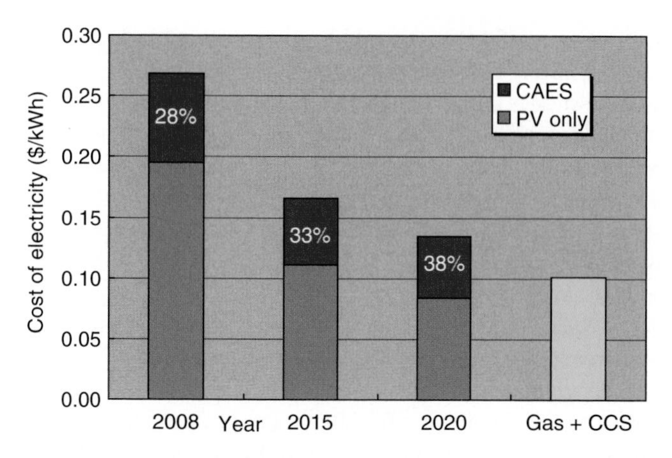

Fig. 3.43 Cost of electricity in 2007-dollars per kilowatt-hour for thin-film photovoltaic (PV) plants with and without compressed air energy storage (CAES). The *yellow bar* is for advanced gas turbines with carbon capture and sequestration (CCS) [15]

For base loads, however, PV-CAES electricity will cost \$0.118/kWh, considerably more than the \$0.076 and \$0.087, respectively, from gas and coal plants, both with CCS. This is all conjectural, however, since the cost, safety, stability, and legal problems of underground storage have never been tested on a large scale.

Ideas on the Horizon

There is no dearth of ideas on new ways to make solar cells, but these are not yet practical. Solar power has a great advantage in the development stage over other technologies such as wind, nuclear, or fusion. New ideas can be explored on a small scale. No large machines or wind turbines have to be constructed. Experimental solar cells can be as small as 1 cm^2. This means that new ideas can be developed profitably by small companies, thus shifting the research burden to the commercial sector. Large, government-funded installations are still needed for commercial viability, but not for testing new ideas. These ideas fall into the category of Generation III solar cells, as shown in Fig. 3.44.

In this graph, the efficiency of solar cells is plotted against their cost per square meter and per peak watt. The three elliptical areas are where Generations I, II, and III lie. Generation I comprises the single-junction silicon cells, costing more than \$3.50 per peak watt and achieving efficiencies no higher than 18%. Generation II contains the thin-film and organic cells, which are much cheaper but have low efficiencies. Generation III includes multijunction cells with efficiencies above 40% and new ideas which are still in the thinking stage. The efficiencies of these solar cells can go above the 31% of the theoretical maximum known as the

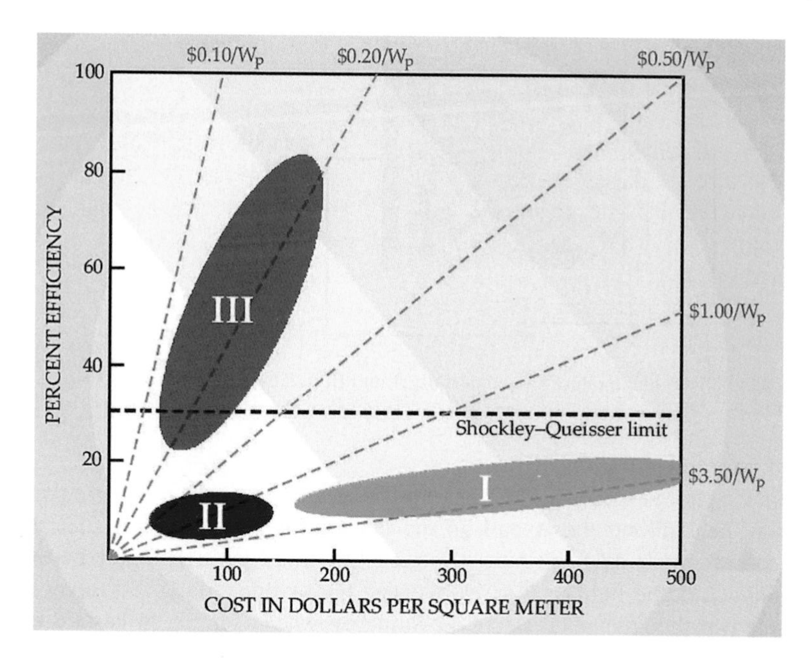

Fig. 3.44 The three generations of solar cells, plotted according to cost and efficiency [16]. The horizontal axis is in dollars per square meter, while the diagonal lines give the cost in dollars per peak watt. The horizontal dashed line is a theoretical limit explained in the text

Shockley–Queisser limit. The limit applies to single-junction cells in unconcentrated sunlight whose photons produce only one electron each and whose excess energy is lost as heat. Generation III cells go higher by violating these conditions. For instance, concentrating the sunlight can give more than one electron per photon, and new nanomaterials can capture the excess energy as current [16].

Organic Solar Cells

Organic solar cells have been invented which are cheaper and easier to make than thin film and which have great promise in small, personal applications. The best of these are made of polymers (a general name for plastics) with long chemical names abbreviated as P3HT and PCBM. They have different bandgaps and different affinities for electrons and holes. Rather than separating them into layers as in CdTe cells, these two polymers are mixed completely together to form what is called a *bulk heterojunction* material. The mixture melts at a temperature below 100°C and, in liquid form, is easily coated onto a substrate, where it solidifies. The substrate can be a piece of cloth! By cooling the mixture at a particular rate, it *self-organizes* into connected clumps where the P3HT and PCBM are separated. A cartoon of this is shown in Fig. 3.45.

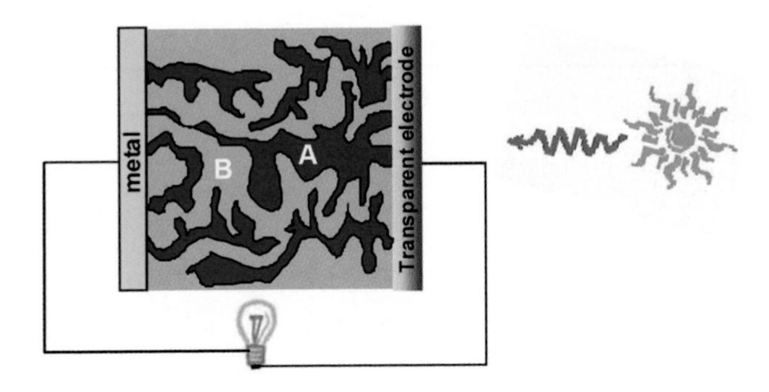

Fig. 3.45 Self-organization of two materials, A and B, in a bulk heterojunction organic solar cell [17]

When a photon strikes a P3HT region (A), it creates an electron–hole pair. The electron then follows the A path to the top transparent electrode. (*Electrode* is defined in footnote 45.) The hole is attracted to the PCBM (B) region because of the natural electric field that arises between the two materials, and the hole follows the B path to the metal electrode. Similarly, when a photon strikes a B region, the electron jumps into the A region, the hole stays in B, and both charges move to their respective electrodes following the strands of A and B. When the two electrodes are connected through a load, the electron current provides the solar power. The fortuitous way these polymers organize themselves avoids all the complicated layers of silicon or CdTe in conventional cells, but the trick is to get the right self-organization by slowly cooling the mixture with careful temperature control.[46]

The first experiments used a polymer layer less than a quarter of a micron (1/4000th of a millimeter) thick and less than a tenth the size of a postage stamp. A sunlight-to-electricity conversion efficiency of 4.4% was achieved [18], together with a high filling factor (defined above) of 67%. Many efficiency claims are deceptively high because small samples collect sunlight from the edges as well as the top, but in this case a proper test was done at the National Renewable Energy Laboratory to avoid this. Further improvement was made in 2009 using a polymer called PBDTTT, whose chemical name would take up two lines. The partner material was not a polymer but carbon in the form of fullerene, commonly known as buckyballs, the familiar spherical carbon lattices made of triangles and named after Buckminster Fuller. This organic solar cell was 6.77% efficient, had high output voltage, and captured more of the infrared energy than the previous model [19]. The current was also reasonable in spite of the crooked paths that the electrons have to follow.

With efficiencies comparable to those of amorphous silicon cells, organic solar cells have great possibilities because they are inexpensive and can be put into almost anything, such as hand-held electronic devices and fabrics. They have already been built into backpacks to charge iPods and cell phones. They are not suitable for large installations, however, because the polymers are attacked by oxygen and last only one or two years. However, they will last almost indefinitely in an oxygen-free environment such as the inside of double-glazed windows.[46]

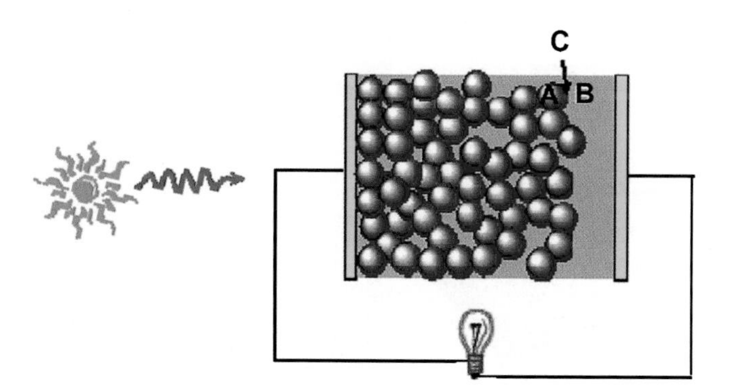

Fig. 3.46 Cartoon of a dye-sensitized solar cell [17]. A is a nanoparticle, B is a conducting liquid, and C is a layer of dye on each particle

Further in the future are such inventions as dye-sensitized and quantum-dot solar cells. *Dye-sensitized cells*, also called Grätzel cells, consist of nanoparticles of titanium dioxide (TiO_2), each only about 20 nm in diameter, coated with a layer of dye, as depicted in Fig. 3.46. (The prefix nano indicates sizes measured in billionths of a meter or millionths of a millimeter.) TiO_2 is a large bandgap semiconductor, so by itself it would absorb only ultraviolet light. The dye, however, is excited by sunlight of any desired color and can inject an electron into the nanoparticles. The electron then hops from one particle to another to get to one electrode. This leaves the dye with an electron missing, so it has to grab one from the electrolyte (a conducting liquid containing iodine) in which the particles are immersed. Efficiencies of 11–12% have been observed in the laboratory, but what it would be in production is unknown. Since a part of the cell is liquid, it has to be sealed, which is rather inconvenient. Solid or gel electrolytes have been tried, but their efficiencies are very low, 4% or so [17].

Since the electrons have to jump numerous times to get to the positive electrode, the motion can be speeded up by using nanowires or nanotubes instead of nanoparticles. Figure 3.47 shows how this would work. The nanowires are heavily coated with dye, and electrons can readily flow along them right to the electrode at the bottom. In this case, the wires are made of zinc oxide (ZnO) instead of TiO_2. Carbon nanotubes have also been used. The tubes, 360 nm long, have a surface area 3,000 times that of a flat surface [21], but of course no amount of surface area can collect more sunlight than falls on the surface facing the sun. Efficiencies of 12% have been observed in the laboratory.

A further improvement can be obtained by replacing the dye with *quantum dots* (QDs), which are nanocrystals of InP (indium phosphide) or CdSe (cadmium selenide). These are *really* small, only about 3 nm in diameter. They can be coated onto TiO_2 or ZnO nanowires to replace the dye coating in Fig. 3.46 or 3.47a. By varying the size of the dots, different colors of the solar spectrum can be absorbed. When a photon hits a QD, an electron–hole pair is created, and the electron falls into the nanowire and is carried straight to an electrode, as in a dye cell. QD cells

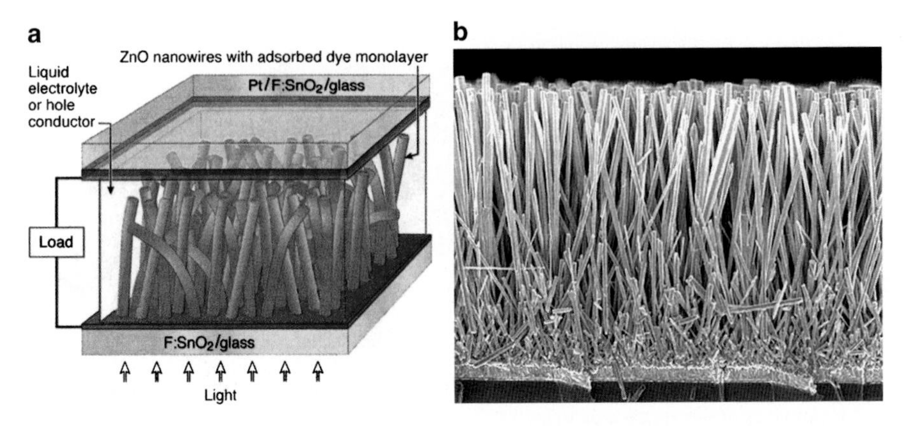

Fig. 3.47 (**a**) Diagram of a dye-sensitized cell using ZnO nanowires [20]; (**b**) microphotograph of actual nanowires [17]. This figure is turned 90° relative to Fig. 3.46

can have higher efficiency than dye cells because they can violate the theoretical limit shown in Fig. 3.44. They can give both higher voltage and higher current [22]. Normally, when a photon has more than enough energy to push an electron across the bandgap into the conduction band (Fig. 3.32), the extra energy goes into the electron. These "hot electrons" then cool and drop down to the bottom of the conduction band, so the output voltage is only the bandgap voltage. In QDs, the hot electrons cool much more slowly and can get into the circuit before losing all their energy, so the cell's output *voltage* can be higher than assumed by the simple theory. Furthermore, the hot electrons can have enough energy to create more electron–hole pairs by themselves, without photons. This increases the cell's *current* over the theoretical limit.

Though quantum-dot solar cells are still in the experimental stage, the way to make nanowires [23] and QDs [24] is well documented. They share all the advantages of organic solar cells in small applications and have the prospect of much better efficiencies. They have not been proved to be suitable for solar farms.

Heat can drive electric currents directly by the Seebeck effect, giving rise to *thermoelectric power*, which is illustrated in Fig. 3.48. If we apply heat to one side of a thermoelectric material, the hot particles at the top move faster than the cold particles at the bottom, so particles tend to drift from top to bottom. Now if on the right side, we have an electron-rich (n-type) material, the electrons will be driven from the top electrode to the bottom electrode. To close the circuit, we put an electron-deficient material (p-type) on the left, where the holes will drift downwards, and we connect the two bottom electrodes to a load. The electrons will then flow through the wire from right to left to fill the holes. Since the electrons are negative, the electrical current goes from left to right. A working arrangement might look like that in Fig. 3.49. Solar concentrators are used to increase the heat applied to the thermo-photovoltaic (TPV) cell, and the bottom of the cell has to be kept cool by water or air flow.

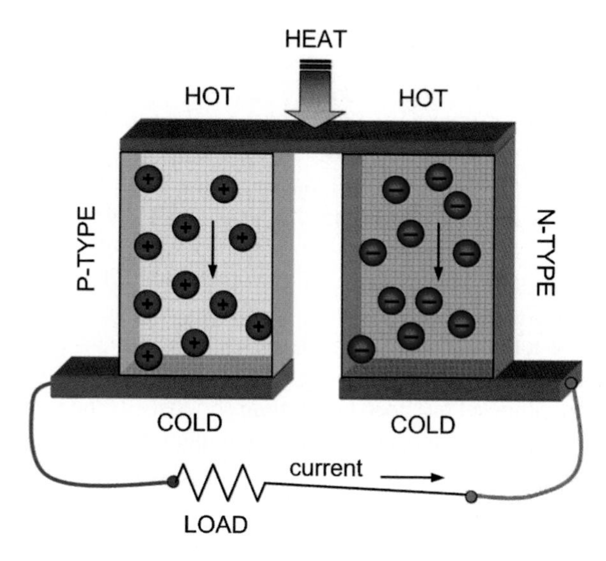

Fig. 3.48 Illustration of direct production of electricity by heat (J.P. Heremans, adapted from an invited paper to the American Vacuum Society, November 11, 2009)

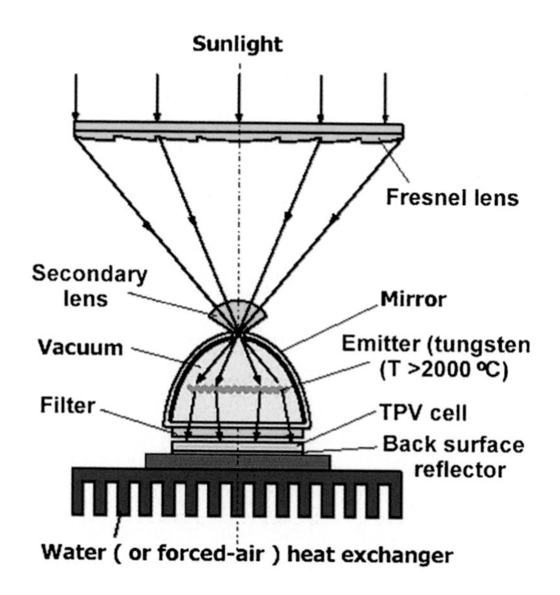

Fig. 3.49 Illustration of thermo-photovoltaic solar cell (*Basic Research Needs for Solar Energy Utilization*, US Department of Energy Office of Science workshop, April 2005)

This idea is still in the initial stages of testing the thermoelectric efficiencies of compounds like PbTe, Bi_2Te_3, $AgSbTe_2$, and $AgBiSe_2$ and formulating new ones. Note that the latter two are type I–V–VI semiconductors [25]. Research is also proceeding on using nanowires and quantum well structures for this purpose [26, 27].

Geoengineering

Articles in the popular press have intrigued the public with wild ideas, some of which have even been legitimized under the rubric *geoengineering*. For instance, instead of reducing GHGs, why don't we shield the earth from getting so much sunlight? This could be done by sending zillions of small plastic sheets up into orbit to reflect sunlight over large areas of the earth. It has also been suggested to use natural plant spores which have large area for their weight. This would not ride well with the resort business! More seriously, such a large-scale, uncontrolled experiment would have unpredictable consequences for our climate and for life itself. It may even trigger an ice age. Such proposals are, of course, science fiction.

The following proposal has been taken more seriously. If the sun does not shine all the time on terrestrial solar panels, why not put them in space? In a geostationary orbit, 22,000 miles (36 km) above the earth, the panels will receive the whole 1.366 kW/m^2 of sunlight instead of the 1 kW/m^2 that reaches the earth, and the weather is always clear. That's only 37% more, but nights will be shorter since the satellite is so high that it will not always be in the earth's shadow when it is nighttime on earth. Gyroscopes can keep the panels always pointed at the sun. If expensive multijunction silicon solar cells are used, the efficiency could be 40%. How much area would be required to produce the power of a coal or nuclear plant, say 1 GW (1,000 MW)? (There are thousands of such plants in a large country.) For the sake of argument, let us assume that the satellite panels get an average of 1 kW/m^2. To generate 1 GW at 100% efficiency would require 1 million square meters or 1 km^2 (0.39 square miles). At 40% efficiency, it would require 2.5 km^2 or just about 1 square mile of panels. That is a lot to send into space! The panels would not last may years because they would be damaged by micrometeorites and solar flares. The moon's gravity would make the satellites drift from their geosynchronous orbits, so a supply of propellant is necessary to make corrections. This supply cannot last many years either.

Then there is the problem of getting the power back to earth. It is proposed to transform the solar energy into microwaves and beam the energy back to the earth at a wavelength that is not absorbed by the atmosphere. Of course, this would be in a desert area with few storms and clouds, and that means building transmission lines to population centers. Microwaves are strongly absorbed by water vapor in the atmosphere. Low frequencies, like the 2.45 GHz (gigahertz) used in microwave ovens are well absorbed by water, which is why microwave ovens work in the first place. To get good transmission, the frequency has to be high, like 100 GHz. Such frequencies can be generated by *gyrotrons*, and the most advanced of these are being developed for the large fusion energy experiment ITER, which is described in Chap. 8. In the laboratory, a gyrotron has produced 1.67 MW at 110 GHz for 3 μs, and 800 kW at 140 GHz for 30 min [28]. Though continuous operation at such powers is expected to be attainable on earth, it may not be possible in space because of the lack of air and water for cooling. Gyrotrons are large devices containing heavy magnets into which energetic electron beams are injected. The magnets help convert the electron energy into microwaves, but not all the energy can be extracted because as the electrons slow down, they get out of sync. The best efficiency that can be hoped for is about 50%. The rest of the energy goes into a beam dump, which has to be cooled. One can build a heat engine that generates electricity from that heat to accelerate more electrons, but that would make the device even more complicated than it is already. There is a further loss at the receiving end in converting the microwave energy into AC power. Even worse, high-power microwaves are known to break down air and make plasma that can scatter or reflect the microwaves. Solar panels in space may gain a factor of 2 in available sunlight over those on land, but more than this is lost in transmission even if the technology can be developed. Regardless of the cost, this is a really bad idea!

The Bottom Line on Solar Power

We started with the fact that the sun gives the earth 1 kW/m^2, enough energy in 1 hour to supply the earth for a whole year. Now we understand why it is so hard to capture that energy. The atmosphere absorbs part of the sunlight. The sun does not shine at night and does not rise high in the winter. There are cloudy and stormy days. There is little sunlight at high latitudes, where the power is most needed. Solar cells can capture only part of the solar spectrum and are not efficient at that. The peak efficiencies quoted apply only when the sun is directly overhead. The color of sunlight changes near sunset and no longer matches the color the solar cells are optimized for. Solar panels cannot economically be turned to follow the sun as it moves across the sky. We are lucky to capture a few percent of solar energy, but even that is a lot of energy that should not be wasted.

Local solar panels on rooftops and exterior walls should be popularized and widely accepted as standard for new structures. These can contribute a few percent to the grid's power, but no more because solar power is intermittent and cannot economically be stored. Selling excess power back to the power station is just a gimmick; the utilities could care less about this small disturbance.[47]

The advances in thin-film technology have made photovoltaic solar power competitive with conventional power sources. The energy payback time will fall below one year, which is short enough, though not a short as for wind power. To use this technology for large solar farms to provide central-station power, however, is fraught with problems. The main problem is that the sun does not shine at night, the time when people turn on their lights. There is no cheap, proven method for storing that much energy. Alternatively, one can build high-tech transmission lines to carry the electricity across time zones from daylight to moonlight, but this will take many decades to implement.

Solar power is an important supplement to grid power, but it is not suitable as a primary central-station power source. Fifty years from now, only coal, fission, and fusion are capable of supplying the dependable, steady backbone power that the civilized world can count on.

Energy for Transportation

After electricity, the form of energy we would miss most is that in gasoline. Our dependence on oil leads us into wars in the Middle East. The price of oil disrupts our economy. The oil crisis of 1973 was so severe that a speed limit of 55 miles per hour was legislated in the USA. (But it had the beneficial effect of increasing government funding for controlled fusion research!) Train buffs will remember the times when trains carried coal, and this was shoveled into steam engines to drive the huffing and puffing trains across the country. Nowadays liquid fuel is at a premium. Gasoline, diesel fuel, and liquefied natural gas are used for transport by cars, buses, trucks, trains, airplanes, and ships. Half of all the world's oil is used for transportation. How can this be replaced by clean energy? Wind and solar produce electricity, which is not easy to carry around. We cannot all drive nuclear submarines.

Hydrogen has been hyped as a promising candidate for a nonpolluting fuel. It is surprising how many people still think that hydrogen is a source of energy! In fact, it takes a lot of energy to produce hydrogen. Water is one of the most stable elements on earth, which is why we have a blue planet. To take H_2O apart into hydrogen and oxygen requires a large energy source to supply the world's transportation needs. Cars run on hydrogen emit only water, but hydrogen is currently produced from natural gas. This not only depletes our precious reserves, but also carbon dioxide is emitted in the process. Even though we still use fossil energy to make hydrogen, transportable hydrogen still has a role to play in reducing pollution. To clear up popular conceptions on hydrogen, we will consider this topic first.

Hydrogen Cars

A Hydrogen Economy?

If we were to replace gasoline with hydrogen to fuel most of our cars, here is how it might work.[48] Until nonfossil energy sources are available on a large scale, hydrogen will be made from natural gas. Gas stations would be replaced by hydrogen stations to which natural gas will be delivered. Hydrogen would be generated locally and stored in underground tanks under pressure. Cars will have plastic tanks in their trunks to hold enough hydrogen for 200–300 miles. These tanks have to be under at least 300 atm pressure, but hose connections can safely handle the filling of the tanks. Hydrogen does not explode unless it is first mixed with oxygen. Inside the car, a *fuel cell* combines the hydrogen with oxygen from air to produce electricity. There is an electric motor, and the car then runs as an electric car, with H_2O as the only emission. The fuel cell–electric motor combination is much more efficient than a gas engine, and less energy is used than if the natural gas or hydrogen is burned directly. Wind or solar power can produce electricity to use directly in the electric motor, but batteries need further development and in any case need a long time to charge. Hydrogen serves as a way to carry the energy. *It is not burned directly in hydrogen cars*. The main problems are the fuel cell, which is very expensive, and the sequestration of the CO_2 if natural gas is used. Discussion of these subjects will follow. Right now it is not clear whether hydrogen cars or plug-in electrics will ultimately win out as the better solution for clean mobile power.

How to Carry Hydrogen [29]

Pound for pound, hydrogen carries three times the energy of gasoline, and fuel cells can use this energy much more efficiently than can an automobile engine. However, if we carry hydrogen as a gas in a 20-gallon gasoline tank, there is only enough energy to drive the car 500 feet. There are two ways to carry more hydrogen: liquefy it or compress it. Hydrogen turns into a liquid at −253°C or just 20° above absolute zero. Needless to say, it takes a lot of energy to run the cryogenic equipment to cool to this temperature. Even if the tank in the car is very well insulated, the hydrogen will boil off slowly overnight. In use, it has to be heated rapidly to feed the fuel cell at a rate depending on the speed of the car. On top of this, each liter or gallon of liquid hydrogen has only 30% the energy of an equal volume of gasoline. It makes more sense to compress the hydrogen.

Scuba tanks and laboratory gas cylinders are heavy. For cars, light-weight tanks made of carbon fiber composites have been developed to hold pressures as high as 10,000 psi (pounds per square inch) or 69,000 kilopascals, which is 700 times atmospheric pressure. Normal would be about 5,000 psi, which is higher than in scuba tanks. The cost of such a tank would be at least ten times higher than for a gasoline tank of equal volume. Regardless of this, can the tank be large enough to power a car for 300 miles (480 km)? A back-of-the-envelope calculation of this is given in Box. 3.6. Squeezing the hydrogen takes energy, most of which shows up as heat of compression. The compressing has to be done beforehand, since the hydrogen has to cool. Otherwise, not enough can fit into the tank. When the hydrogen is released for use, it will be too cold for the fuel cell and has to be heated up.

Under development are ways to store hydrogen in solids. Metal hydrides can absorb hydrogen like a sponge under pressure and then release it under heat when the pressure is relieved. As shown in Fig. 3.50a, the hydrogen molecules go between the atoms of the solid, so it can hold 150% more hydrogen than an equal volume of liquid hydrogen [29]. Unfortunately, the chemicals found so far are either too heavy, react too slowly, or require too high a temperature. Some can absorb only 2% of their own weight in hydrogen, even without the pressurized container. The fuel stored this way for a 300-mile trip would weigh half a ton [29]. Magnesium hydride can store 7.6% of its weight, but needs an inconvenient temperature of 300°C. The most promising ones are complex hydrides combined with a "destabilizer." For instance, lithium and magnesium borohydrides ($LiBH_4 + MgBH_4$) will combine into two other hydrides and release hydrogen at a low temperature [30]. This can hold 8.4% of hydrogen by weight (Box 3.6). The reaction is reversed when hydrogen is added under pressure at a filling station. Unfortunately,

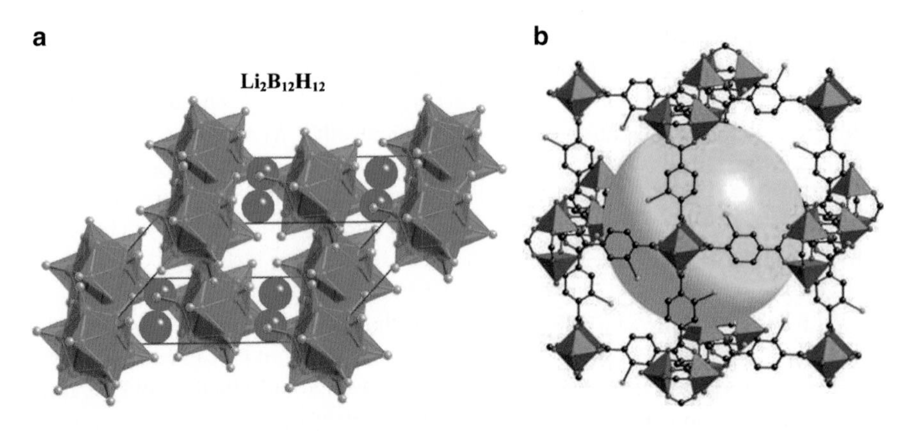

Fig. 3.50 Schematics of (**a**) a destabilized hydride [30] and (**b**) a metal-organic framework [31] for trapping and storing hydrogen

Box 3.6 Carrying Compressed Hydrogen in a "Gas" Tank

The energy content of a gallon of gasoline is about the same as that of 1 kg of hydrogen, so 1 gallon $\approx$ 1 kg H_2. (In metric units, it is not as convenient: 1 L $\approx$ 0.12 kg H_2.) Say it takes 20 gallons of gasoline to go 300 miles. Since fuel cells are more efficient, it would take not 20 kg but only, say, 8 kg of H_2 to drive a car that far. From high-school chemistry, we remember that a mole of gas occupies 22.4 L, so there are 2 g of H_2 in 22.4 L. The density at standard conditions is then 2/22.4 = 0.089 g/L. At 10,000 psi (700 atm) compressed hydrogen at room temperature would have a density 700 times higher or 63 g/L. Eight kilograms would then occupy 8,000/63 = 128 L or about 34 gallons. So to go as far as a normal car, a hydrogen car would need a 70% bigger tank, not including the mechanisms for handling the compressed gas.

There is also the question of weight. The US Department of Energy has set a goal that the weight of a tank should weigh no more than 17 times the weight of the fuel. (The fuel weight is more than 6% of the tank weight.) Hydrogen tanks so far are 25–50 times as heavy as the fuel.

the reaction rate is too slow to be useful even though the material is in the form of a fine powder to expose large surface area and reduce the path for heat conduction.

A promising new material called metal-organic frameworks (MOFs) has been invented by Yaghi [31]. These are extremely light-weight chemical structures that act as nets to trap larger molecules, as illustrated in Fig. 3.50b. Just like a net, a MOF has struts linked together with strong bonds, forming a scaffold to enclose a large space.

This atomic net has the largest area per unit weight ever obtained: 4,500 m^2/g. That means that a paper clip's weight of material can cover a football field. With considerable chemical derring-do, hundreds of different kinds of MOFs have been created for different purposes. For storing hydrogen in cars, one liter of the compound MOF-177 can store 62 g of H_2, exceeding the 6%-by-weight rule in Box 3.6. However, this has been done so far only at 77 K (kelvin: degrees centigrade above absolute zero). This is liquid-nitrogen temperature, easy to get in the laboratory but hard to maintain in a car, though much easier than the 20 K of liquid H_2. A MOF that works for hydrogen at room temperature could get to 5% H_2 by weight, but it is not easy to manufacture on a large scale. Another type of compound called COFs *is* suitable for that, and research is proceeding to make those work at room temperature. COFs can also help with carbon capture in coal plants. A tank filled with MOFs can hold nine times as much CO_2 as one without MOFs [31]. Other compounds called ZIFs can actually selectively capture the CO_2 going up a smoke stack.

Chemical storage of hydrogen is an active research area in laboratories, but nothing works well enough so far to proceed to the next step of engineering large-scale production.

Anatomy of a Fuel Cell

The heart of a hydrogen car is the fuel cell, whose parts are illustrated in Fig. 3.51. Hydrogen is forced into the channels in the anode plate and is then spread out uniformly in the diffusion layer. This layer has been described as a wet rag whose moisture content must be carefully controlled to keep the proton exchange

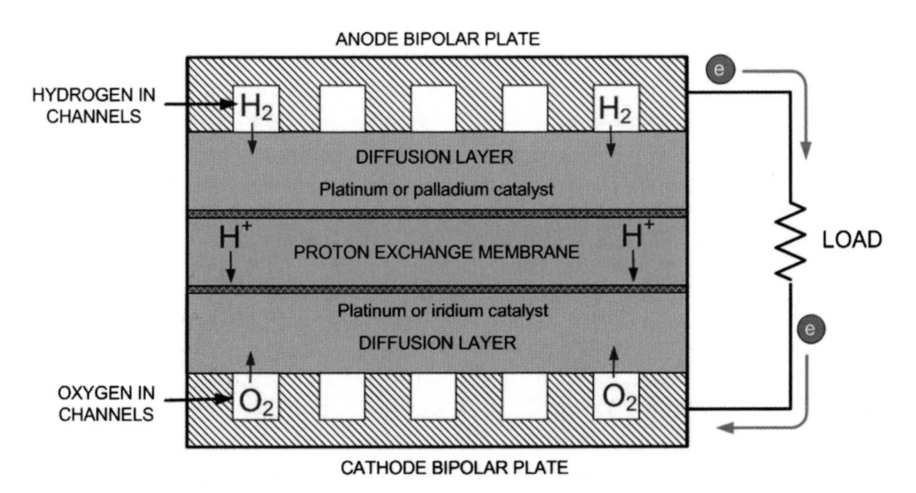

Fig. 3.51 Schematic of a fuel cell. It is not to scale. The catalyst layers and the PEM are only 10's of microns thick, while the diffusion layers are 100's of microns thick. The bipolar plates are of macroscopic dimensions

membrane (PEM) from drying out without dripping. The PEM is a plastic layer like plastic wrap made of a special material called Nafion® made by Dupont Chemical. It has the magical property of allowing hydrogen ions (H^+) to pass through but not electrons. It is the platinum catalyst layer that dissociates hydrogen gas (H_2) and ionizes it into two hydrogen ions (H^+). This is an even more magical property. The catalyst layer consists of platinum nanoparticles thinly deposited on carbon paper which has to be rough to present a large surface area and porous to let the water through. The electrons, being blocked by the PEM, are drained off into a wire to form the electric current that is the output of the cell. When the H^+ ions reach the other side, they encounter another catalyst layer, which could be platinum or iridium. Meanwhile, oxygen (O_2) from air is pushed into the cathode plate and diffusion layer to meet the hydrogen ions in the catalyst layer. Therefore, the O_2 is dissociated into atoms (O) and picks up electrons from the wire that has gone through the load to become negative ions (O^-). Each O^- then combines with two H^+s to form H_2O. Hydrogen and oxygen have been combined to form water and electricity. All in all, the fuel cell is a serendipitous invention, but it has problems.

Each fuel cell generates only 0.6–0.7 V, so as many as 100 of them have to be connected in a series to form a *stack* with a useful voltage output. Platinum is a precious metal used in jewelry and in catalytic converters. Its price drives the price of fuel cells to about $73/kW, twice the commercially viable value.[49] Cyclic operation of PEMs degrades their performance. PEMs have to be heated to at least 60°C from a battery before they can even start, and they need about 100°C to operate reliably. The water in the cell must not boil or freeze under all driving conditions. Corrosion of the bipolar plates is a problem; they cannot be made of a metal that can corrode and contaminate the system with iron or chromium. A carbon compound has to be used. Besides the electric motor, the car has to have a system to pressurize the gases. And the fuel cell has to last for 300,000 miles.

Currently, the whole shebang is too large to fit inside a car but can be used in trucks. No large-scale production and testing has been done. What can be gained is a fuel-cell efficiency of 80% times another 80% efficiency of the electric motor, giving a maximum efficiency of 64% in the conversion of hydrogen energy to mechanical energy. This compares favorably with the efficiency of gasoline-driven cars, about 15%, but the energy in producing the hydrogen has not yet been counted. If that part is 40% efficient, the net efficiency is $64 \times 40 = 26\%$, still higher than burning natural gas in a gas engine. However, the real gain will be when hydrogen is produced in fission or fusion plants with no use of fossil fuels or emission of GHGs.

Sources of Hydrogen

Only minute quantities of hydrogen occur naturally in the atmosphere, but it can be produced efficiently in a process called *steam reforming*. When methane (CH_4), the main constituent of natural gas, is heated up to 700–1100°C in the presence of water (H_2O), two reactions occur. First, CH_4 and H_2O combine to form hydrogen (H_2) and carbon monoxide (CO). Then, the CO reacts with more H_2O to form CO_2

and more H_2. The net result is that methane and water are made into hydrogen and carbon dioxide. The second reaction is exothermic (it gives off heat), so that heat can be used for part of the heat needed to drive the first reaction. The rest of the heat comes from burning some of the methane. The CO_2 has to be sequestered using one of the methods discussed in Chap. 2.

Large factories for steam reforming already exist in the petroleum industry because hydrogen is needed for taking the sulfur out of gasoline and for producing ammonia and fertilizers. These sources supply the hydrogen for initial tests of hydrogen cars. There are other possible ways to produce hydrogen. The classical way is direct hydrolysis of water. An electrolyte is added to the water to make it conduct electric current. Two electrodes[45] in the form of plates are then put into the solution, and a DC voltage is applied between then. Water molecules are broken up into hydrogen and oxygen, and they bubble out separately at each electrode. The efficiency of the process depends on the electrolyte and electrode design, but in any case is quite low. If the energy used to produce the electricity is counted, the energy content of the hydrogen is perhaps a third of the energy used to produce it by electrolysis. Even that may be worth it if the original energy source is nonpolluting, such as a fission or fusion power plant. Pricewise, it is estimated that 1 kg of hydrogen costs \$7–\$9 to make by hydrolysis, compared with \$4–\$5 by steam reforming. The nuclear industry has plans to demonstrate hydrogen production at \$1.50/kg by 2015.[50] One kilogram of hydrogen has about the same energy as 1 gallon of gasoline, but these prices cannot be compared directly with the price of gasoline because cars use and carry hydrogen and gasoline in completely different ways.

There are several new ideas on hydrogen generation without producing CO_2 also. One is to use dye-sensitized solar cells plus a catalyst to get hydrogen directly from sunlight. Another is to perform artificial photosynthesis by growing algae. The most advanced is a system to run a hydrogen fuel cell *backwards*, using solar electricity to make hydrogen rather than using hydrogen to make electricity. In the Compagnie Européenne des Technologies de l'Hydrogène (CETH) in France, a machine called the GenHy5000 Water Electrolyzer has successfully done this [32]. About the size of a refrigerator, the hydrolyzer produces H_2 at the rate of 5,000 L/h at atmospheric pressure using electricity with 62% efficiency. It has run continuously for 5,000 h, but efficiency will drop with intermittent use. When powered by rooftop solar cells, the hydrogen can be generated and stored at 10-atm pressure for later use. For automobile refueling stations, higher pressures will be required. The hydrogen can be allowed to build up pressure as it is generated. A smaller model has run at 30 atm for a total of 10,000 h. Its other data are: voltage 1.7 V, current 1 A/cm^2, temperature 90°C, and power consumption 4 kWh/m^3 of H_2. The noble-metal content in the catalysts is 1.5–3 mg/cm^2, and the hydrogen is 99.99% pure. Though this is a fuel cell run backwards, years of research have yielded valuable data on fuel cells in general: what materials to use, how to make them, how long they will last, and how they can be contaminated. In particular, it was found that the catalyst layers are best deposited directly on the membrane, and a method was devised to do this using frequency-modulated electric pulses.[51]

In spite of the problems with the fuel cell, prototype hydrogen cars costing millions of dollars have been made. The Honda FCX, for instance, is sleek, normal-looking passenger car with a 100-kW fuel cell stack weighing 148 lbs (67 kg) and occupying 57 L (2 cu feet). Four kilos of hydrogen are stored at 5,000 psi in a 170-L (6 cu feet) tank. A matching 100-kW (134 HP) electric motor runs on a lithium-ion battery charged by the fuel cell. The relation between kilowatts and horsepower (HP) will be found in Box 3.7. The mileage is stated to be 60 miles/kg of H_2, and the range is 240 miles (386 km). The car could be leased at $600/month, but full production is not expected before 2020.

Box 3.7 Kilowatts and Horsepower

Kilowatts and horsepower are both units of energy relevant to electric cars. A kilowatt (kW) is approximately four-thirds of a horsepower (HP), and 1 HP is about three-fourths of a kW. The exact values are as follows:

1 kW = 1.341 HP
1 HP = 0.746 kW
1 W-hr = 4.8 HP-sec
50 W-hrs = 241 HP-sec

Bottom Line on Hydrogen Cars

Hydrogen cars are electric cars whose energy is carried by pressurized hydrogen. The technology is in its infancy, especially on the manufacture of fuel cells at reasonable cost. Right now, hydrogen is made from natural gas, and the only gain, at great expense, is barely a doubling of the efficiency of burning the gas directly in reciprocating engine. Carbon dioxide is still emitted in the generation of hydrogen. Hydrogen is clean energy only when fission or fusion plants supply the energy to hydrolyze water to make it. Other nonpolluting sources such as hydroelectricity and solar and wind farms are not sufficient to replace the 383 million gallons of gasoline we consume per day in the USA [29]. The infrastructure for distributing hydrogen [4] will cost perhaps half a trillion dollars.

Electric Cars and Hybrids

The gasoline engine is a marvelous piece of engineering. Honed over hundreds of generations of models, it fires an explosion thousands of times a second, and yet we can hardly hear it as it smoothly pushes the car through the air. What is wrong with it? It uses gasoline very inefficiently, and it emits carbon at a rate equivalent to throwing a charcoal briquette out the window every quarter mile.

Electric cars are even quieter ... so silent that it has been proposed to put a noise generator in them to warn pedestrians. Electric cars have no emissions, but they get their electricity from power plants that emit GHGs. However, power plants burn fossil fuels much more efficiently than cars do, so the total emissions are lower. It is because power plants run at much higher temperatures than cars can, and the Carnot efficiency (see Chap. 2) is much higher. There is a big difference between 40 and 15% efficiency, and most people do not realize this. The main problem with electric cars is the battery. There is no type of battery of reasonable size and weight that can take a car 300 miles on one charge, and it takes many hours to recharge the battery. If you run out of "gas" in an electric car, you would have to stay in a motel with a plug. But electric cars have great advantages. We will consider these next and hybrids later.

Efficiencies of Gas and Electric Cars

A normal car can use only about 15% of the energy in gasoline, though some say it could be 30%. The breakdown is shown in Fig. 3.52. Most of the energy is lost in heat, 30% in the radiator and 30% in the exhaust from the muffler. A few percent more is lost in the engine and in the transmission line between the motor and the wheels. Fully 17% is used in idling while the car is not moving, such as at a red light. The motor has to be running so that it can start again rapidly. Accessories such as lights and radio take only 2%. That leaves only 12.6% for propulsion of the car.

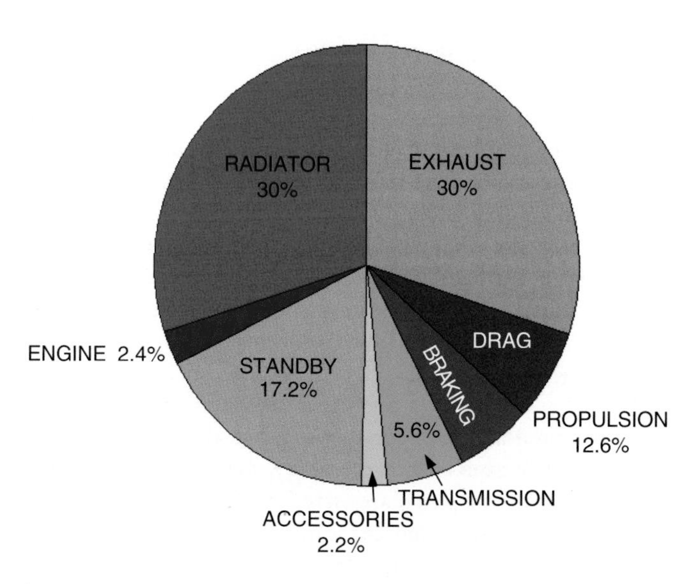

Fig. 3.52 Where the energy goes in driving a car on gasoline. Data from http://www.fueleconomy.gov

About half of this is lost as heat in the brakes to stop the car. The rest, 6.8%, is all there is left to move the car!

Electric cars store energy in a battery bank and use that to drive a motor that drives the wheels. The battery may get a little warm, but the heat energy lost is trivial compared with the 60% in normal cars. The stand-by energy is saved since the motor simply turns off when the car is coasting or stopped. The braking energy is recovered into the battery, though the brakes will get a little hot, and that energy will be lost. The accessories, including the lights, the radio, and the computer, will take a few percent, and so will the transmission, but all the rest is available to move the car. Electric cars can convert about 75% of the energy stored its battery into useful power. The battery is charged with electricity from the grid, and the environmental impact of that process depends on the location. In most places, coal or natural gas is used to generate electricity, and GHGs are emitted. However, this is better than burning oil products in cars for several reasons. Power plants can be 40% efficient, three times better than cars. So less fuel is consumed and less CO_2 is emitted. Furthermore, power plants can be located some distance from cities, thus sparing them from pollution. Electric vehicles emit only water. In locations where hydroelectric or nuclear power is available, the air is even cleaner. Even noise pollution is abated.

Vehicles running totally on electricity are being used successfully in service vehicles and golf carts, which do not have far to travel. The Tesla Roadster has shown that electric cars can have sports-car performance at a price. The big bugaboo is transportable energy. There is no known type of battery that will carry a car 300 miles and recharge in 5 min, as we can get from gasoline. Meanwhile, we can save on gasoline by using hybrids. These will be discussed next, followed by battery prospects.

Gas–Electric Hybrids

The range and charging problems of electric cars are solved by combining an electric motor with a gasoline motor. The most successful of these hybrids has been the Toyota Prius, which approximately doubles the mileage of a normal car. The way it does this, however, is not what most people would imagine. Instead of carrying a large battery, the Prius carries a battery so small that it can be hidden. When we drive, we subconsciously vary the pressure on the gas pedal every second or so as the road curves or rises and falls a little, or because of traffic. Each time the car coasts, its kinetic energy charges the battery, and this energy is re-used in the next few seconds when the gas pedal is pressed to maintain speed. At a stop light, the braking energy is stored and used for startup when the light turns green. Just by saving these small, instantaneous bits of energy, the car can greatly reduce its gas consumption. A dashboard display shows a red symbol every time 50 Wh of energy has been saved and re-used by the car. Fifty watt-hours sounds like a piddling amount of energy. A TV or computer draws 5W when it is *off*, so 50 W-hrs can power only 10 such devices in a home for one hour out of 24. However, as shown in Box 3.7, 50 W-hrs is equivalent to 241 horsepower-seconds, or almost 50 horsepower for five seconds. This allows the car to have fast pickup after a stop. Fifty *kilowatt*-hours (67 HP-hrs) would be more normal for a car that didn't have instantaneous response to small accelerations and decelerations. Indeed, hackers who have modified the Prius by adding a large battery have increased its mileage from 45 mpg (5.2 liters/100 km) to 100 mpg (2.4 liters/100 km), but at great expense. More on the hardware in the Prius is given in footnote 52.

Hybrid cars incorporate many other improvements to decrease fuel consumption. A continuously variable transmission is more efficient than a 4-speed automatic or a 5-speed manual. A switch available in some models turns off the gas motor altogether so that the car runs on electric alone until the battery gets low. When the energy used to climb hills is recovered and the braking energy is stored for use in starting again, an electric car is very efficient in city traffic. In traffic jams when normal cars are burning gas without moving, electric hybrids can get surprisingly high mileage. Driving at high speeds is another matter; the car has to push its way through the soup we call air. In perfect streamlining, the front of the vehicle slices the air apart. The air streams above and below then rejoin each other at the back of the car, pushing the car forward. But there is friction, and heat is lost in the windshield; and there is turbulence, so the stream at the rear is not smooth. There are also protuberances: windows, door handles, tires, and, above all, the rear view mirrors. Sticking your hand out the window at autobahn speeds will show how much energy is needed to push through the atmosphere. Wind drag accounts for 60% of energy use; tire friction, 10%; and engine and transmission line losses account for the rest. In the Prius, sticking to the speed limit can save 10% in gasoline, but over-inflating the tires can save only 1%. Retuning the electronic fuel injection can save 10%. Effective streamlining is measured by the drag coefficient C_d, on which more information is given in footnote 53.

In both hybrids and normal cars, gasoline is used inefficiently when the car is cold. A car rated at 30 miles/gallon (mpg) may get only 12 mpg when it first starts. A Prius which gets 45 mpg when warm drops to 30–35 mpg until the engine and catalytic converter warm up. This loss is avoided when running on electric alone. In hybrids, battery power can be used to heat up the catalytic converter more rapidly. Both motors in a hybrid depend on rare, precious metals. A catalytic converter contains about 5 g of platinum worth about $500. On the other hand, electric motors use permanent magnets made with neodymium. Their batteries may contain more than 10 kg of lanthanum. These materials, however, can be recycled. Many rare-earth elements are used in hybrids, and the concern is that China has a near monopoly on the supply of these elements.

Plug-in Hybrids

Until the battery problem is solved, electric hybrids will continue to evolve. The next step is the plug-in hybrid, in which the battery is charged overnight from the grid. Since most people in cities usually drive no more than 30 miles (50 km) a day, a slightly larger battery will store enough energy for that, so that the gasoline engine need not be started except on weekends. Air quality in cities would be greatly improved. There are actually two types of plug-in hybrids. The usual one works like the Prius: the battery is charged from the grid as well as by the gasoline motor. Two motors drive the car. In a *series* hybrid, a small motor runs only to charge the battery. The propulsion is entirely electric. The savings in fossil-fuel consumption and GHG emission have been estimated in a report by Electric Power Research Institute and the Natural Resources Defense Council (EPRI-NRDC).[54] It turns out that it matters whether the battery is sized to give 10, 20, or 40 miles of electric driving.

The EPRI-NRDC report considers scenarios, nine in all, depending on whether PHEVs (plug-in hybrid electric vehicles) have a low, medium, or high penetration into the market, and whether the power industry makes a low, medium, or high effort to reduce their emissions. Although the nine results vary by a factor of 4, they are all good. The GHG reduction in 2050 is predicted to be between 163 and 612 million metric tons (in the USA). An idea of how they expect PHEVs to take over the market is shown in Fig. 3.53. If no progress is made in battery technology (which is unlikely), PHEVs will take over more than half the car market!

Table 3.1 compares various kinds of hybrids with normal cars.[55] The data are for 12,000 miles of driving in year 2010. The normal hybrid generates its own electricity and therefore uses more gasoline than PHEVs, though less than gasoline cars. PHEV10 is a PHEV that can go 10 miles on one charge. PHEV20 and 40 have bigger batteries to go 20 and 40 miles. All the hybrids are assumed to have

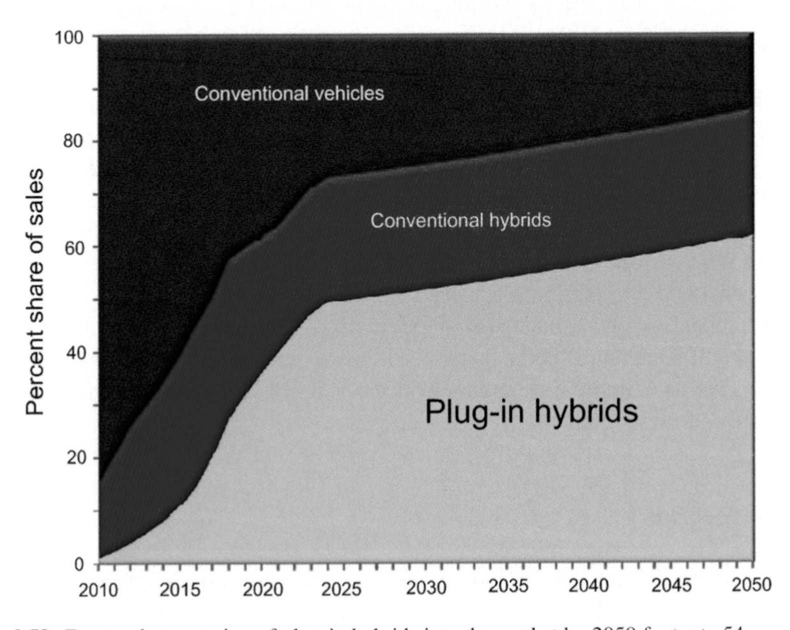

Fig. 3.53 Expected penetration of plug-in hybrids into the market by 2050 footnote 54

Table 3.1 Comparison between normal cars and hybrids of various types

Type of car	Normal gas	Normal hybrid	PHEV10	PHEV20	PHEV40
Gasoline (gallons)	488	317	277	161	107
Electricity (kWh)	0	0	467	1,840	2,477
Fuel economy (mpg)	25	38	38	38	38
Cost of electricity	0	0	$55	$215	$290
Cost of gasoline	$1,464	$951	$831	$483	$321
Total for 12,000 miles	$1,464	$951	$886	$698	$611

a gas motor averaging 38 miles/gallon. The PHEVs use more electricity from the grid and less gasoline, so their carbon footprints are smaller. Remember that electricity generated at a power plant uses less oil than electricity generated in the car. If the power plant uses hydroelectricity or nuclear power, the carbon footprint is more than halved.

How much money will a plug-in hybrid save? This depends, of course, on the battery size in the PHEV and on local prices; but here is an example. The breakdown between electricity usage and gas usage in Table 3.1 is based on some data on driving habits. In electric drive, a Prius-type hybrid uses 150 W-hrs of electricity per kilometer.[56] This works out to be 0.24 kWh/mile. In 2009, the average cost of residential electricity in the USA was 11.7¢/kWh. The cost of 2,477 kWh in the PHEV40 case is then $2477 \times \$0.117 = \$290$. In the PHEV40 column, we see that 107 gallons of gasoline are consumed. If we assume a price of $3.00/gallon, the gas cost is $321 and the total fuel cost is $611. These are the figures in the last column of Table 3.1. The other columns are calculated the same way. As for the "normal" cars, all the energy comes from gasoline, so there is no electricity cost. We see that hybrids save on the cost of fuel, but these savings may not offset the premium one pays for hybrids at present. For the plug-in hybrids, there is a "sweet spot" around the PHEV20, whose fuel costs are much lower than for the PHEV10 but not much higher than the PHEV40. Since most people do not drive 40 miles every day, the extra cost of a large battery is not worth it. However, individuals are not "most people"; they can buy a plug-in suited for their own driving habits.

There has been some concern about the effect of numerous plug-in hybrids on the grid. Since charging a PHEV on household current can take upwards of eight hours, most people would want 240-V service installed. Then charging can be done in 2–3 h. At this rate, however, as much as 6.6 kW of electricity is drawn. Each car that is plugged into that service is like adding three houses to the grid, each house with their lights on and air conditioner working.[57] If every household has a plug-in, the local grid would have to be boosted. However, the EPRI-NRDC study shows that the industry experts are not worried. They show a profile in which 74% of the charging is done between 10 p.m. and 6 a.m., with a small daytime peak between 10 a.m. and 4 p.m. There are minima around 8:30 a.m. and 5:30 p.m. when people are commuting. The grid can handle that load, at least for the present.

Batteries

Electric cars can go a long way toward relieving our dependence on oil, but the bottleneck is the battery. We are spoiled by having cars that can go 300–400 miles (500–600 km) without refueling and can be filled up in 10 min. There has been no path-breaking invention in batteries in the last few decades. Figure 3.54 shows where we are. Each rectangle is the range occupied by one type of battery according to how much it weighs and how big it is compared to the energy it can store. Lighter batteries are to the right, and smaller batteries are near the top. At the bottom left is the old stand-by: the lead-acid battery used in conventional cars. It is heavy and big for the amount of energy it carries. The only improvement over the last 50 years is that they are now sealed, so that we don't have to check the fluid level and add water every week or so. The first experimental electric cars carried a load of lead-acid batteries. One battery is only good for starting a car and keeping its headlights on for a few hours; it cannot move a car very far. The small carbon-zinc and alkaline batteries we use for small appliances and toys are off the chart because they are not rechargeable. Nickel-metal-hydride (NiMH) batteries, however, are successfully used in cars, notably the Prius. These were chosen because they are safer than lithium and have proven reliability. The best we have at present is the lithium-ion battery. As Fig. 3.54 clearly shows, "lithiums" are lighter and smaller for the same amount of energy. They are used to power laptop computers, cell phones, cameras, and other small appliances. Their safety and reliability are, however, worrisome for use in cars. There is hope, however, because electric cars like the Tesla Roadster have shown that, if cost is not a consideration, sport-car performance can be

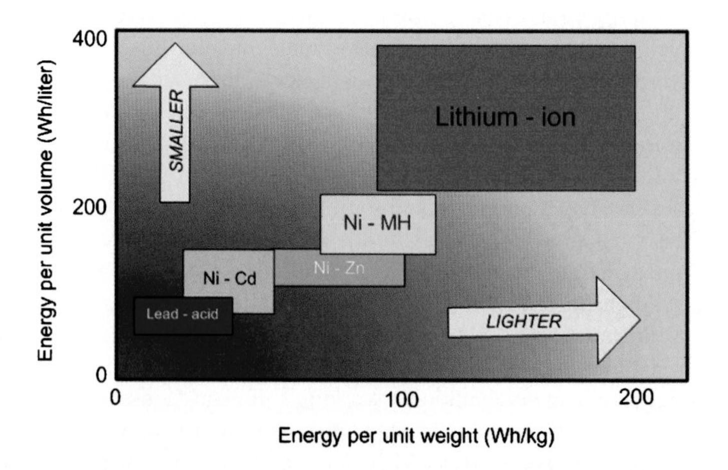

Fig. 3.54 Performance of major types of batteries. For each type, the *horizontal axis* shows the energy stored per unit weight in watt-hours per kilogram, and the *vertical axis* shows the energy stored per unit volume in watt-hours per liter. Adapted from *Basic needs for energy storage*, Report of the Basic Energy Sciences Workshop for Electrical Energy Storage, Office of Basic Energy Sciences, US Department of Energy (July 2007)

achieved with a 6800-cell Li-I battery good for 244 miles. With a 288 HP (215 kW) motor, the car goes 125 mph (200 kph) and accelerates 0–60 mph in 3.7 s. Charging the battery on 240 V takes 17 kW in 3.5 h.

Aside from cost, lithium batteries have two main problems. Safety is the main concern, since lithium batteries have been known to explode, as they did in some laptops a few years ago. When a short circuit occurs in such a battery, the chemicals can burn and cause short circuits in neighboring cells, which release more heat, starting a runaway reaction. Unlike hydrogen, which cannot burn without oxygen from the air, lithium batteries have the oxygen inside. The solution is to divide the lithium battery pack into small isolated units which are then connected together with wires. The second problem is life span, which depends on how often the battery is recharged. Even if it is not used, a lithium battery can lose as much as 20% of its capacity per year [33], as many laptop owners have found to their dismay. The number of charge–recharge cycles is limited to several thousand. For cars, 5,000 cycles would be good for 10 years for most drivers, and this is close to present technology. However, it would be hard to build enough extra capacity for the car to maintain its driving range for 10 years. Charging a lithium battery too fast or overcharging it could cause *plating* of the electrodes, which shortens it life. These problems are slowly being solved as companies move into this rapidly expanding market. The target price set by the US Advanced Battery Consortium for electric car batteries is $300/kWh. Lead-acid batteries cost about $45/kWh, compared with NiMH batteries, which cost $350/kWh for small ones to $700/kWh for ones used in cars. Right now Li-ion batteries cost $450/kWh [33]. Perhaps economy of scale will bring the prices down as electric cars overtake the market.

How Batteries Work

Normal batteries like the AA- and AAA-size ones we use everyday are sandwiches of three materials made into long sheets, as shown in Fig. 3.55a. The anode and cathode materials are separated by a thin insulating sheet, and all three are made as thin as possible and rolled up tightly to fit the largest area into the smallest space. The anode and cathode materials have a chemical potential between them such that the anode is negative and the cathode is positive. They are connected to the contacts at the bottom and top of the battery, respectively. When a light bulb is connected to the contacts, an electric current flows, lighting the bulb, and discharging the built-up charges between the sheets. The chemical potential sets the voltage of the battery, typically 1.5 V, and the area of the sheets determines how much charge they can hold, and therefore the "life" of the battery. Most batteries are not rechargeable.

Lithium-ion batteries are rechargeable. How they work is illustrated in Fig. 3.55b, where the anode and cathode layers are represented by shelves holding Li ions. The anode material is usually graphite (loosely packed carbon) holding some positive lithium ions. The cathode can be made of any of a number of materials, including proprietary ones, which largely determine the performance of the battery. Before the two electrodes are connected together, the chemical potential between them

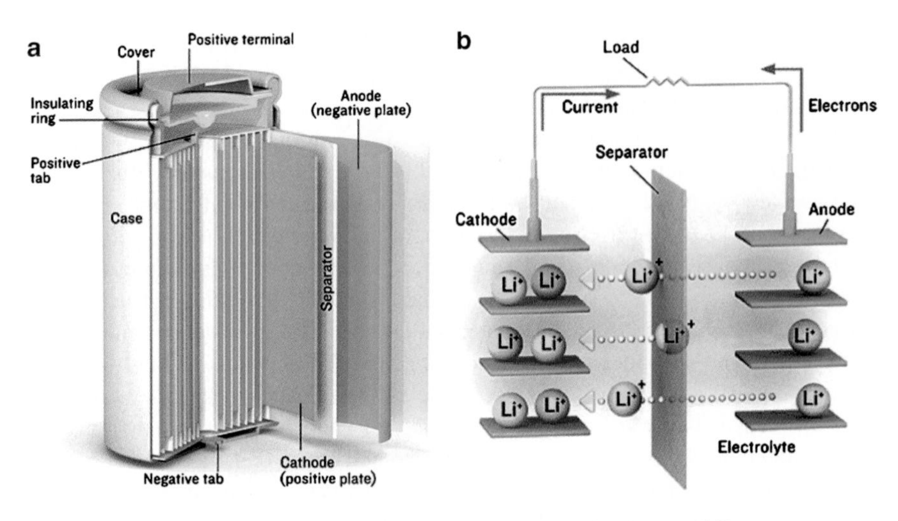

Fig. 3.55 (**a**) Construction of a battery; (**b**) Layers of a lithium-ion battery [33]

draws the lithium ions from the anode to the cathode until the extra positive charge added to the cathode cancels out the chemical potential. The ions travel through an electrolyte, which is a conducting liquid like salt water, only thicker. It is the gooey stuff that leaks out of an old battery. A thin plastic sheet, the separator, prevents the electrodes from touching each other. The separator is thin enough to allow the ions to pass through. A short circuit develops if there is a hole in the separator. Now if the battery is connected to a load, electrons which are attracted by the extra positive charge on the cathode can flow through the load to do useful work. As shown, the electric current is in the opposite direction to the electron motion because the electrons carry negative charge. To recharge, a negative voltage is applied to the anode to draw the lithium ions back. This is what takes hours. A large battery pack could consist of 100 cells, each 5 cm in diameter and 20 cm long (4×8 in.), divided up into modules so that overheating in one module does not spread to others.

As for cathode materials, cobalt-containing compounds such as cobalt dioxide have high-energy density and are commonly used for small Li-I cells, but they are not suitable for cars because of a tendency toward thermal runaway. The best found for cars so far is iron phosphate, which is more stable and less likely to overheat. It gives lower voltage, so that chains of batteries have to be longer to provide a high output voltage. Higher power and longer life are claimed if the cathode is made with nano-sized divots to increase surface area [33]. More on this will come in the next section. The race to make the best iron phosphate battery has already led to patent fights among battery companies.

The long charging times for Li-I batteries have been overcome by Ceder et al. [34] working with $LiFePO_4$ (lithium–iron–phosphate) cathode material.

A123 Systems, a company started in Boston, has expanded into a $91M business in Asia using this material in small batteries for power tools and hobbyists.[59] Employing techniques from ultracapacitors (next section), Ceder et al. form the cathode in such a way that it has large surface area with channels aligned so that Li ions can get in and out of the cathode rapidly. In small samples, discharge times of the order of seconds were observed, more than ten times faster than normal. Critics, including J. Goodenough, an inventor of $LiFePO_4$ cathodes, doubted that charging times could be as short as discharging times.[60] However, Ceder claims that the rates are for both charging and discharging. If we accept that, there is still a problem with charging a car, even a hybrid, in 10 minutes. It requires a lot of power. A plug-in hybrid using 0.24 kWh/mile can go 40 miles (64 km) on about 10 kWh of electricity. To put that much energy into a battery in 10 minutes would require 60 kW of power, enough to run an office building. Charging at home would have to be scheduled so that not everyone on a grid line plugs in at once. However, there is no need to charge that fast at home; overnight will do. Where fast charging is needed is in filling stations en route. To charge nine cars at once would require half a megawatt of power. Probably high-voltage lines and a small substation would be required at each "gas" station. Some people suggest that such stations should have large battery banks to store the energy slowly and continuously so that not so much instantaneous power is needed. In any case, building the infrastructure to support electric cars is worthwhile for saving oil and cleaning up the environment. Ultimately, when oil runs out and fission and fusion plants generate most of the energy for transportation, the electric grid will have to handle the power for all vehicles.

Supercapacitors and Pseudocapacitors

A battery stores a lot of energy in its chemicals, but chemical reactions are slow and cause a battery to charge and discharge slowly. A capacitor, on the other hand, can charge and discharge extremely fast. It stores energy with two electrodes and a separator the way a battery does, but it does not involve chemical reactions. It also can be recycled limitlessly and does not decay with time. Capacitors are used in almost all electronic circuits and come in many sizes. Millions of small ones can be made on a computer chip, and large ones the size of a waste basket (trash bin to Anglophiles) are used by power companies. *Super*capacitors are capacitors that still use no chemicals but can hold much more energy than previously possible. Used in combination with batteries, they help overcome some of the drawbacks of batteries. *Pseudo*capacitors are supercapacitors with reacting chemicals, thus combining the virtues of capacitors and batteries. A few diagrams will show how interesting these new developments in transportable energy storage are.

Figure 3.56a shows a normal capacitor. The positive and negative electrodes are metal sheets separated by an insulator called a dielectric. When the capacitor is charged by applying voltage between the electrodes, the charges move to the inner surfaces of the dielectric, and they attract opposite charges onto the surfaces of the dielectric. There are then sheets of opposite charges on each interface, and they stay

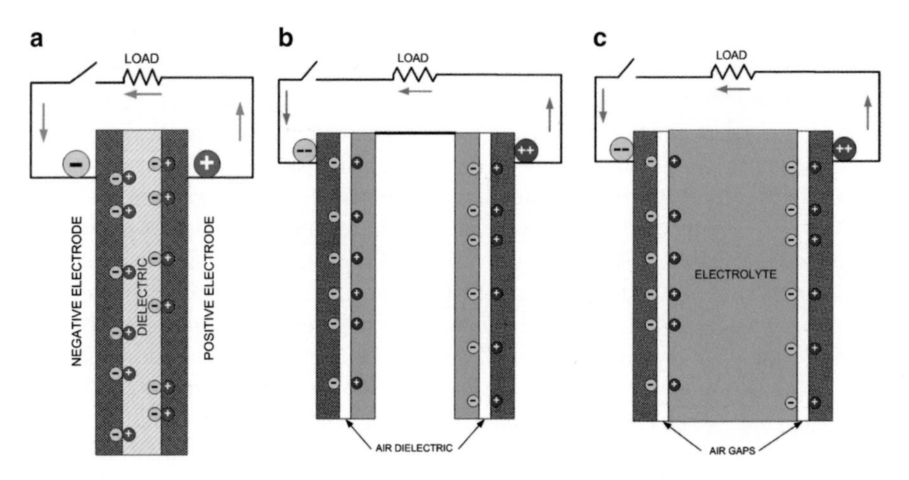

Fig. 3.56 Diagrams of (**a**) a normal capacitor, (**b**) two capacitors with air gaps in series, and (**c**) two capacitors in series joined by an electrolyte

there when the switch is opened. These charges cannot move together to annihilate one another because the dielectric is an insulator. The energy is stored in the dielectric. When the switch is closed to hook up a load, the opposite charges on the electrodes move through the load to combine with one another, thus applying the energy that was stored. The dielectric, which had zero total charge all along, then redistributes its charges to match the charges left on the metal sheets, if any. The energy storage capacity of a capacitor (hence its name) depends on three factors: the area of the sheets, the thinness of the dielectric, and "dielectric constant." The latter is a number varying from 1 (for air or vacuum) to 3 (for plastic), to 5 (for glass), and as high as 80 for water. The higher the number, the more energy the dielectric can hold for a given voltage between the electrodes.

To get more energy into a capacitor, one can work with these three factors. Capacitors are already made as thin as possible and rolled up to get the largest area for their size. Supercapacitors, however, can have much thinner dielectrics and much larger areas by virtue of nanotechnology. This can be explained step-by-step. In Fig. 3.56b, we show two simple capacitors connected in series. The inner electrodes are not metal but conducting liquids (electrolytes). The gaps are filled not with a dielectric but with air. This lowers the dielectric constant to 1, but thickness of the gap is much, much smaller. Now if we connect the two capacitors not by a wire but by simply extending the electrolyte as in Fig. 3.56c, we have a capacitor whose capacitance depends on the thicknesses of the two gaps, and not by the thickness of the electrolyte layer. Next, we can increase the area by roughening up the inner surfaces, as shown in Fig. 3.57a. This is done by coating the electrodes with a layer of "activated" carbon, which consists of fine particles. Special processing techniques make the surfaces of these particles break up into channels nanometers in size, as shown in Fig. 3.57b. The electrolyte goes into these channels but does not actually touch the carbon because of a nanoscopic surface tension effect.

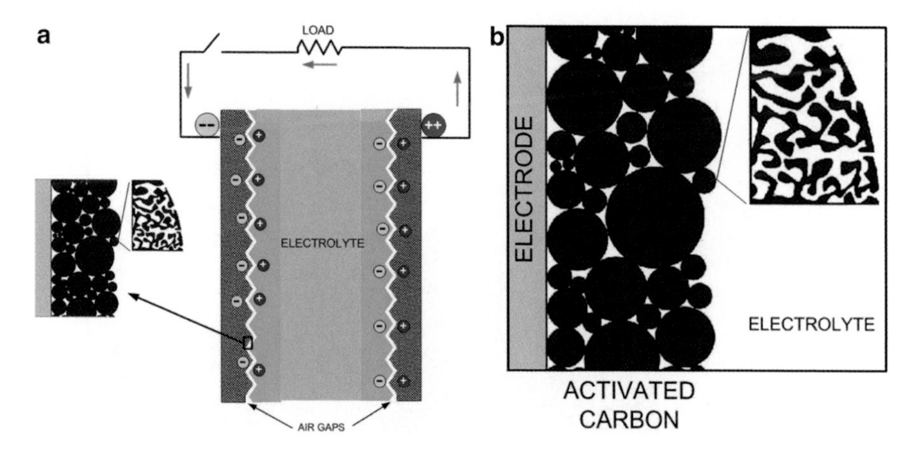

Fig. 3.57 (**a**) Schematic of a supercapacitor; (**b**) enlargement of section shown in (**a**) Adapted from Basic needs for energy storage, Report of the basic energy sciences workshop for electrical energy storage, office of basicenergy sciences, US department of energy (July 2007)

This forms an air gap of nanometers thick. The capacitance is increased to tens of thousands of times.

Capacitance is measured in *farads* (named after Michael Faraday). The energy a capacitor can hold is proportional to its capacitance and the square of the voltage it can take before arcing over. While usual capacitors have capacitances of picofarads to microfarads and a rare one may have a farad, supercapacitors (also called ultra-capacitors) can have 5,000 farads. They can hold 5% as much energy as a automotive Li-I cell in the same size package.[61] They can supplement Li-I batteries in electric cars by storing and releasing braking energy more quickly than the batteries can. They can store enough energy to be used on short trips by buses, garbage trucks, and the like.

Pseudocapacitors add porous electrode structures like those of Fig. 3.57 to a Li-I battery using molybdenum trioxide (MoO_3). The trick is to find a material that can make a chemical battery and yet can be processed in such a way as to have a large area, rough surface. This has been accomplished in the laboratory by Brezesinski et al. [35]. Still in their infancy, pseudocapacitors have the potential to store enough energy fast enough to be useful in smoothing the output of intermittent energy sources such as wind and solar.[62] The development of such electrochemical capacitors will fill the gap in Fig. 3.58 between batteries and capacitors in their abilities to store large amounts of energy and to cycle the storage fast. There are still other types of batteries which lurk in the future, such as metal–air batteries, especially zinc–air and lithium–air batteries. Since the cathodes are air, these could have very large storage per unit weight. They are the only batteries that could approach the energy density of gasoline. However, there are several performance defects, most seriously inability to be recharged completely. The physics of the reversible reaction is still unknown;[62] but, with intensified research, there is hope for a paradigm-changing advance with these new types of batteries.

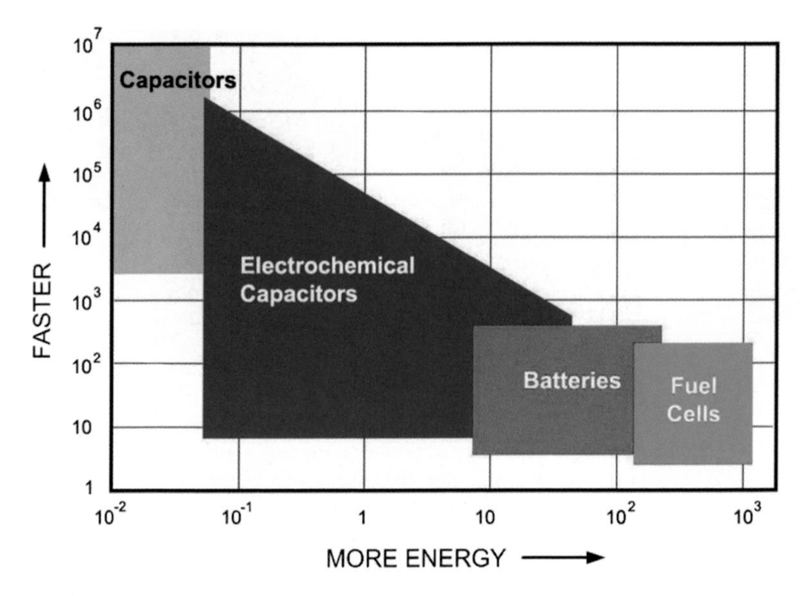

Fig. 3.58 Performance of different types of transportable energy storage, "Fuel cells" here refers to hydrogen storage and use in fuel cells. Adapted from basic needs for energy storage, report of the basic energy sciences workshop for electrical energy storage, office of basic energy sciences, US department of energy (July 2007)

Summary of Electric Cars

Electric cars will be necessary when oil becomes scarce. Electricity to drive them can come from fossil fuel plants or from carbon-free sources like fission or fusion reactors. Even if fossil fuels are used, GHG emissions are greatly reduced if the fuels are burned at a central utility rather than in vehicles. The main problem is the lack of a suitable battery. Recognizing this urgent need, the Obama administration in the USA has allocated \$1.5 billion to the development of advanced batteries. This will greatly expedite this field of research, which was previously hampered by the lack of funding.

Biofuels

Instead of using electric cars, we can lower our dependence on foreign oil by converting plant matter into ethanol. About 10 billion gallons of ethanol were produced in the USA in 2009, a small but growing fraction of the 140 billion gallons of gasoline consumed. Ethanol burns 22% more cleanly than gasoline because it contains more oxygen, but it contains only two-thirds the energy per gallon. Most ethanol is sold as E10, a 10% mixture of ethanol with gasoline. Most cars can run on E10 without modification. E85, which is 85% ethanol, requires modified "flex-fuel" engines, which are installed in many trucks. In Brazil, the leader in biofuels, all cars are so modified because the country is completely independent of foreign oil, having started 25 years ago to produce biofuels from sugarcane.

In the USA at the present time, ethanol is produced from corn, not the stalks but the good part, the ears that we and the cows eat. This has played havoc with the prices of corn and soybeans. The corn is ground up, fermented, and then distilled to evaporate off the alcohol. The beer industry knows this well. What is left is still good for cattle feed. The first distillation yields only 8% ethanol, so it has to be repeated many times to get to 99.5% high octane fuel. This takes energy, at present coming mainly from fossil fuels. More energy is used in planting and harvesting the corn, in making the fertilizers, and in trucking the corn and the fuel. Pipelines cannot be used for ethanol because it is soluble in water, and water in the pipes would cause them to rust. Gasoline does not have this problem. The use of fossil energy also entails GHG emission, negating the cleanliness of ethanol exhaust. There has been controversy as to whether making ethanol from corn actually provides more energy than it consumes, and whether there is any saving of GHG emissions. Early reports in the popular literature were rather negative toward ethanol.[63,64] Much of the pessimism came from papers by Pimentel [36], which indicated that the energy in corn ethanol is 30% less than the energy used to make and transport it. However, other data, mostly recent ones, show a net gain in energy, though much smaller for corn than for cellulosics, which we will describe shortly. Wang's [37] life-cycle analysis shows that to produce one energy unit of corn ethanol, 0.7 energy units of fossil energy has to be used. This means that about 40% (=0.3/0.7) more energy comes out than goes in. When blended with gasoline, E85 of course has better energy savings than E10. As for GHG emissions, E85 saves 29% and E10 26%. Wang also gives a chart showing all the studies made so far on this topic. Twelve of these showed an energy gain, while nine showed an energy deficit. The breakeven is still marginal, but the saving grace is that only 15% of the fossil fuel used is in the form of oil, the scarce commodity that depends on the Middle East. The stance of the US government is that the energy balance is positive, but no firm numbers are given.[65]

How does Brazil do it? Because they have the climate and labor, they can grow sugar directly instead of extracting it from corn. Sugarcane yields twice as much ethanol per acre than corn. Biofuels from sugarcane give 370% more energy than is used in production.[63] The stalk is 20% sugar, and the rest can be burned to generate electricity. One factory is self-supporting; it can generate enough electricity to run the whole operation. This huge plant produces 300 million liters of ethanol and 500,000 tons of sugar per year. Between the biofuel and the electricity, the plant produces eight times the energy that it uses.[64] But there is a big problem: deforestation. An area the size of the state of Rhode Island was razed in half of 2007 to plant sugarcane, and the acreage is to double in the next 10 years.[66] Worldwide, deforestation accounts for 20% of carbon emissions, which is why Brazil ranks fourth in the world on carbon emissions.[66] There is more bad news. Sugarcane has to be cut by hand, and it is hard work in the heat. It is so hard that many workers die at it. To make the cutting easier, the cane is burned every year even though it does not have to be. This releases large amounts of soot and strong GHGs to pollute the air. This sours the sugar business.

The USA cannot grow so much sugarcane, but it cannot grow enough corn either. If all the present corn and soybean crops are used to produce biofuels, there would be only enough to supply 12% of the gasoline and 6% of the diesel oil that we consume.[64] But why use only the sweet part of the corn? We could also use the stalks. The stalks are made of cellulose, as are many other plants. Cellulosics are our best hope for a source of biofuel. Cellulose has a rigid molecular structure that is stiff and can allow plants to grow vertically. This is how corn can grow high as an elephant's eye. The very structure of cellulosics makes them very hard to break down into alcohol. At present, it takes 30% more energy to make the fuel than it gives back [37]. There is an intense effort to find more efficient ways to do this, including using high-speed computers to model the chemical reactions. The Obama administration in 2009 allotted $800 million to the Department of Energy's biomass program, and $6 billion in loan guarantees to start biofuel projects beginning in 2011.[63]

Cellulosics can be found everywhere in corn stalks, wood chips and sawdust, wheat straw, paper, leaves, and specially grown crops of grasses and other fast-growing plants. The Departments of Energy and Agriculture in the USA estimate that 1.3 billion tons of cellulosics can be gathered and grown each year without affecting food crops for either humans or animals. It is possible to produce ethanol, gasoline, diesel oil, and even jet fuel from cellulosics. The amount of cellulosics available equate to 100 billion gallons of gasoline equivalent per year, about half of our needs [38]. To do this, of course, is very hard.

There are three ways to make fuel from cellulosics [38]. At an extreme temperature of 700°C, steam or oxygen can turn the biomass into syngas which is carbon monoxide and oxygen. This is done under pressures of 20–79 atm in the presence of a special catalyst. Coal plants are already set up to produce syngas (see Chap. 2). But a reactor to do this with cellulosics would be so expensive that the capital cost would not be paid back for perhaps 30 years. A second method reproduces the conditions in the earth which made fossil fuels in the first place. At temperatures of 300–600°C in an oxygen-free environment, the biomass turns into a biocrude oil. This crude oil cannot be used directly because it is acidic and would ruin the engine. It would have to be converted to usable fuel. A new idea called *catalytic fast pyrolysis* is being investigated which would convert biomass into gasoline in a few seconds! Fast means that the biomass is heated to 500°C in one second. The molecules then fall into the pores of a catalyst which turns them into gasoline. The whole process takes 2–10 seconds.

The third, more promising way to treat cellulosics is slow and less dramatic; but it could move out of the laboratory into industry. In the ammonia fiber expansion process, the fiber is softened by pressure-cooking at 100°C in a strong ammonia solution. When the pressure is released, the ammonia evaporates and is captured and recycled. The cellulose is than fermented with enzymes into sugar with 90% yield. Distillation then yields ethanol. What is left is lignin, which burns well and can be used to boil water to generate electricity. Of course, burning generates CO_2, but with biomass this CO_2 was taken from the air when it was growing, so there is no CO_2 added to the atmosphere. What spoils this rosy picture? It's the enzyme.

The bacteria that make the enzyme can be found in only a few places, the best of which is in the guts of termites! We know that termites eat wood. They have an enzyme in their stomachs that turns that into something digestible. The enzyme is not easy to reproduce, unlike the yeast that makes yogurt. Presently, they cost $0.25/gallon of ethanol.[67] To mass-produce either the enzyme or the termites is unthinkable. People are finding mushrooms in Guam or other bugs that could make such enzymes.[63]

If we can get over that hurdle, we can think about switchgrass, which you have heard of. A fast-growing source of cellulose, switchgrass needs no fertilizer and little water. It grows in places not suitable for other activity. Its roots grow 8–10 feet down, stabilizing the soil and also drawing CO_2 into the ground.[68] It grows for 5–10 years before reseeding. It has four times the energy potential of corn. The US Department of Energy's goal is to make cellulosic ethanol cost-competitive with gasoline by 2012. The 100 billion gallons of gasoline equivalent per year quoted above will also lower our GHG emissions by 22% relative to our 2002 emissions. Even if switchgrass is grown outside of farm land, it will still take a lot of land. To supply all the transportation fuel for the USA would take 780 billion liters of ethanol per year.[69] At the rate of 4,700 L of ethanol per year per hectare, it would take 170 million hectares or 650,000 square miles. Only Alaska, more than twice the size of Texas, has that much area.

Fortunately, new ideas are coming from people thinking out of the box. James Liao [39] has found a way to make more complex alcohols which contain more energy than does ethanols and, moreover, are miscible with gasoline but not water. Such an alcohol is isobutanol. The enzymes that ferment sugar into isobutanol are more common than those in termites: they are found in *E. coli*. Yes, this is the same bug that causes food poisoning, but its use can be controlled, and it is surely not hard to reproduce. The problem is not entirely solved because biomass has first to be converted to sugar before the process can start. To get around this, Liao has engineered a cyanobacterium [40] that can turn CO_2 and H_2O into a biofuel! Plants do this all the time by photosynthesis, but the result is cellulose. A bacterium has been engineered that can photosynthesize isobutyraldehyde, which boils at a low temperature so that it can be separated from water. That chemical can then be easily converted into isobutanol. To be competitive with current production of bio-diesel from algae, the rate has to exceed 3,420 μg/L/h. The best achieved so far is 2,500, which is promising and can be improved with further research [1]. However, making diesel from algae is very slow and space consuming – only 100,000 L (26,000 gallons) per hectare per year. Two companies, LS9 and Amyris, both in California, are involved in this development.[70] It remains to be seen if this process is economically feasible.

To make transportable fuel, it would seem simpler to make electricity in fission and fusion power plants and develop smaller and lighter batteries for electric cars. Government policy, however, has to take economic stimulus into account. Farmers in Iowa and Nebraska have to be kept happy. The subsidies for ethanol production in Midwest states resulted from strong lobbies. It would seem that our corn is stored not in silos but in pork barrels.

Nuclear Power

Importance of Nuclear Power

Both fission and fusion involve nuclear reactions, but the term "nuclear" usually applies to fission, and we shall use it with that connotation here. Nuclear energy is a mature technology. It is the only time-tested, continuous, dependable source of base-load electricity that does not emit GHGs and can be conveniently located. It has three well-known disadvantages: danger of nuclear accidents, danger of proliferation, and storage of radioactive wastes. We shall treat these one by one. Nuclear power is important for the world's energy needs, but it has been impugned – indeed, attacked – by the press and environmentalists who have not done their homework and studied the risks and costs of the alternatives.[71,72]

France has set an example. It generates 75% of its electricity from nuclear and 15% from hydro, both of which have no CO_2 emissions.[73] There have been no reported deaths. France has led in research on next generation reactors and has begun building them. Other countries which do not have coal have a high percentage of their electricity from nuclear: Belgium (54%), Ukraine (47%), Sweden (42%), S. Korea (36%), Germany (28%), and Japan (25%).[74] Worldwide, the percentage is 15%. Because of its size, the USA's 20% constitutes one-third of the world's total. The supply of uranium will outlast that of oil and gas, and future breeder reactors will generate their own fuel. Fusion reactors will take time to develop, and fission can supply "green" power in the interim. The nuclear waste problem will become more acceptable when the public realizes that fission will eventually be phased out by fusion, so the problem will last only for a few human generations.

How Nuclear Reactors Work

The Cast of Characters

The atomic number of an element is the number of protons in the nucleus. Uranium, element 92, has atomic number 92. Fissionable elements all have atomic number 92 or higher.[75] The mass number is the total number of protons and neutrons in the nucleus. So uranium 235 has 92 protons and 143 (=235−92) neutrons. The atomic *weight* is a loosely used term which is essentially the mass number but differs by a fraction because protons and neutrons do not weigh exactly the same; they are bound with different energies; and energy and mass are interchangeable, according to Einstein. The symbol for uranium 235 is ${}_{92}U^{235}$, but we shall write it is U^{235} because the 92 is already specified by "U." Elements can have the same atomic number but different mass numbers; these are called *isotopes*. Here are a few isotopes of importance in fission:

U^{238}: The normal isotope of uranium in nature.
U^{235}: The fissionable isotope of uranium, with an abundance of only 0.7% in nature.

P^{239}: Plutonium (element 94) is generated in a reactor and fissions easily.
U^{239}: Uranium 239 decays[76] in 23 min.
Np^{239}: Neptunium 239 (element 93) decays in 2.4 days.
Cs^{137}: Cesium 137 (element 55) decays in 30 years.
I^{131}: Iodine 131 (element 53) decays in eight days.

The first group of three contains the isotopes we will be discussing. The next two are intermediate states in the transformation of uranium into plutonium in a reactor. The last two are the most dangerous reaction products when released into the air in an accident. The decay times here are half-lives. Isotopes never completely disappear. Half of what is left goes away in a half-life. Note that only isotopes with odd mass numbers are fissionable.[77] What is not given here is the tremendous amount of energy that nuclei can give. A single-fuel pellet, the size of a AAA battery, can make as much electricity as 6 tons of coal.[78]

The Chain Reaction

When a U^{235} nucleus is joined by a slow neutron, it can split into two nuclei further down in the periodic table, plus two or three neutrons, as shown in Fig. 3.59. In this case, the fragments are Ba^{144} and Kr^{89}. Adding the mass numbers will show that three neutrons are released in this case. A chain reaction occurs when one, and only one, of these neutrons splits another U^{235} nucleus to make more neutrons to keep the chain of reactions going. If two neutrons cause further fissions, the reaction will run away. Figure 3.59 shows another way to continue the chain. As we shall see, there are many more U^{238} nuclei in the fuel than U^{235}s, so a neutron can enter a U^{238} nucleus to form U^{239}, which then beta-decays into Pu^{239}, which is fissionable. A neutron hitting the Pu^{239} will cause fission and keep the chain going.

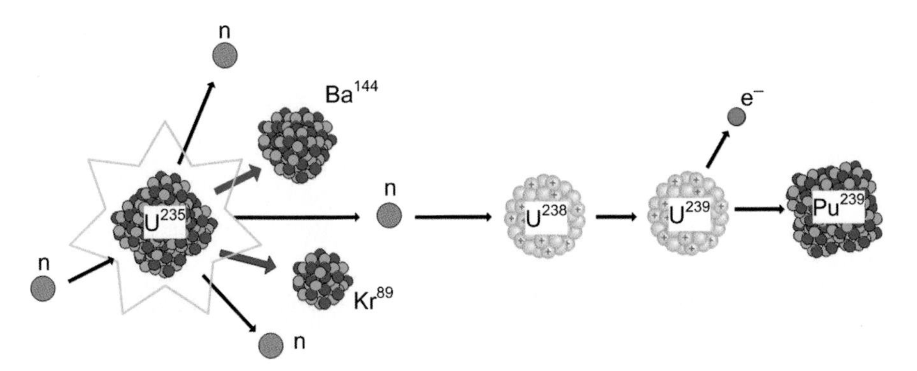

Fig. 3.59 Illustrating the fission of U^{235} into two fragments with the release of three neutrons and a lot of energy. One neutron subsequently enters a U^{238} nucleus, creating U^{239}, which then decays into fissionable Pu^{239}

Moderation is the Key

Of course, things are not this simple. The neutrons from a fission have energies around 2 MeV (about 20 billion°K). (Definition of electron-volt units is given in Chap. 5). They have to be slowed down to normal temperatures before a nucleus will accept them. Room temperature is about 0.025 eV. A *moderator* is used to do this. Here, a moderator is not the chair of a panel discussion; it is an element that slows down neutrons efficiently without absorbing them. The most common moderators are *light* water [ordinary H_2O, *heavy* water (D_2O), and graphite (very pure carbon)]. Only light elements (those with low atomic masses) can be moderators. The reason is that neutrons are light, and they will bounce off a large nucleus without losing much energy. A marble striking a billiard ball will just bounce off. A cue ball striking an 8-ball can come to a complete stop, losing all its energy to the 8-ball, because it has the same mass. Light water is a better moderator than heavy water because the H is closer to the neutron in mass than the D, but it's not twice as good because the H can *capture* the neutron to make a D. A deuteron is less likely to capture yet another neutron to make triply heavy tritium. Carbon has mass 12, so graphite is a weaker moderator than water, but it has other properties, like staying solid at high temperature. Moderators are so important that nuclear reactors are classified according to their moderators.

Isotope Separation

Fresh uranium is mostly U^{238}, with only 0.7% of U^{235}. Unless neutrons are very carefully preserved, there are not enough of them to sustain a chain reaction without increasing the amount of U^{235}. Normally, uranium has to be enriched to 3–5% U^{235} by separating out the U^{235} and adding it to normal uranium. Because the two isotopes differ in mass by only 1.3%, separation is slow; and large installations are needed to fuel power plants. The two main methods are *gas diffusion*, used in the USA and France, and *gas centrifuge*, used in Russia and the rest of Europe [41]. In gas diffusion, uranium hexafluoride (UF_6) is passed multiple times through porous barriers through which U^{235} passes 0.43% faster than U^{238}. Gas centrifuges are tall cylinders spinning at high speeds in vacuum. The centrifugal force pushes the heavier isotope out faster. Though gas centrifuges are more efficient, using only 0.09% of the energy generated by the plant compared with 3.6% for gas diffusion, it is a newer technology and it would be costly for the USA to convert to it. The operative word here is not "convert" but "covert." Centrifuges are discussed further in the Nuclear Proliferation section.

Advanced technology has not overtaken these brute-force methods. Accelerating uranium ions in beams in which the isotopes would have different momenta was tried initially. During WWII, a plasma discharge was tried in the USA, but instabilities arose. This was the origin of Bohm diffusion (see Chap. 6). In the 1970s, a laser method was developed at the Livermore laboratory in California in which a laser beam could preferentially put U^{235} into an excited state, and this could allow it to be extracted separately. At the same time, another laser method was applied to UF_6 at the Los Alamos laboratory in New Mexico. A scheme by John Dawson to use two-ion hybrid cyclotron waves in a uranium plasma was implemented at TRW Inc. in Redondo Beach, California [42]. Though this produced palpable amounts of U^{235}, the project was canceled in favor of the Livermore project for political reasons.

Inside a Nuclear Reactor [41]

In a generic reactor, fuel rods are carefully spaced inside the moderator – water, say – so that each neutron generated inside a fuel rod and slowed down in the moderator produces just one neutron when it causes fission in another fuel rod. The fuel is uranium oxide, UO_2, a black powder created from UF_6, pressed into pellets, sintered, and ground to size. The pellets are slid into thin tubes about the diameter of a pencil and 5 m long. The pellets cannot be large because the heat generated inside has to escape to the coolant. Also, since most of the uranium is U^{238}, the neutrons have to get out of the pellet into the moderator before they are absorbed by the U^{238}. The coolant is usually the same kind of water as the moderator, but it gets hot and carries the output energy. Hundreds of fuel rods make up a fuel assembly, and hundreds of assemblies make up the fuel load, which can weigh 100 tons. The fuel lasts about four years, and one-fourth of it is renewed each year. There have to be enough fuel to make up a *critical mass*, ensuring that at least one neutron from each reaction will find another U^{235} nucleus to split. The fuel assemblies have to be spaced just right inside the moderator for this to happen. When fuel assemblies are renewed, they are shuffled so that the new ones and the half-used ones are distributed evenly. The heat produced is carried away by the coolant and is used to generate electricity at 30% efficiency in steam turbines. One ton of fuel can generate 30 MW of power and 40 GW-days of energy.

Types of Reactors [41]

A boiling water reactor is a light-water reactor (*LWR*) using H_2O as both moderator and coolant. The fuel rods are simply placed in the water, which is allowed to boil under pressure, producing steam directly to the turbines. The water, however, is exposed to radioactive material. A pressurized water reactor (*PWR*) or the European version called *EWR* contains the water under 153 atm of pressure so that it cannot boil at its temperature of 322°C. This water goes to a heat exchanger to transfer the energy to outside water which does not touch any radioactive material. All the reactors in France are PWRs. Standardizing to a single type reduces the risk of accidents.

A *CANDU* (Canadian Deuterium Uranium) reactor was invented because Canada had no enrichment facilities. It burns natural uranium containing only 0.7% U^{235}. With so few fissionable nuclei, the moderator has to be heavy water, D_2O. Hydrogen would absorb too many neutrons. The fuel rods are double tubes, the inner tube contains the fuel pellets and cooling water. Gas in the outer tube insulates the heat from the moderator, which is at room temperature. No thick domed vessel is necessary to contain the reactor. With so little U^{235}, the power output is only 20% of other LWRs, so the fuel has to be replenished often. It is done continuously, going from one end of the rods to the other. There is no proliferation risk due to enriched fuel, but plutonium is produced and comes out with the expended fuel. It can be stolen more easily since it comes out continuously instead of at a fixed time under heavy guard [41].

*AGR*s are early (advanced gas-cooled reactors) developed in England using a graphite moderator and 600°C carbon dioxide as a coolant.[78] Natural uranium could be used at lower temperatures where a low-absorbency "Magnox" fuel cladding could be used, but enrichment is needed for the advanced types. Yet another acronym is the European pressurized reactor (EPR), a safer type being constructed in Finland and France. These two projects have been delayed by cost overruns and safety protests.

Liquid-metal fast breeder (*LMFBR*) reactors are an entirely different breed. *Fast* refers to the fast, or prompt, 2-MeV neutrons emitted in fission. In LWRs, these neutrons have to be slowed down by the moderator before they can cause U^{235} to fission. In breeders, the fuel is U^{238} with 10% Pu^{239}, and U^{235} is not used. Twelve percent of the fast neutrons cause fission in the U^{238}, and the rest are captured. But as Fig. 3.59 shows, the capture of a neutron by U^{238} produces an atom of Pu^{239}, which is a good fuel. Those neutrons that do not get captured immediately eventually slow down and cause U^{238} and Pu^{239} to fission. By covering the chamber with a uranium blanket, which can be made of *depleted* uranium from an LWR, more plutonium can be produced than is used. Breeder reactors can breed fuel from natural uranium.

No moderator is needed; in fact, it is essential not to have any material that will slow down the 2-MeV neutrons. However, there has to be a coolant, and the coolant must not slow down or capture the neutrons either. There are only two elements in the periodic table that can be used: sodium (Na) and lead (Pb). These can be used in liquid form and do not capture many fast neutrons. Sodium, which melts at 98°C, is chosen for convenience in spite of its nasty nature. Although it is harmless when combined with another nasty element (chlorine) in table salt, pure sodium will explode if it touches water. It is the liquid metal in LMFBR. These reactors cannot go critical with normally enriched uranium. A chain reaction requires 10–12% enrichment.

This technology has been well tested in the Superphénix reactor on the Rhône river in France. The 3,000 tons of sodium coolant was in its own closed loop, and heat was exchanged to a secondary sodium loop not exposed to radioactivity. Steam was created in a second heat exchanger. The reactor ran between 1995 and 1997, producing 1.2 GW of electricity between repairs. The sodium ran at 545°C and never boiled, so there was no high pressure. The fuel elements had thicker walls than in LWRs and produced twice the energy per ton. Sodium leaks have been the main problem. A smaller LMFBR, the Monju in Japan, developed a leak in the intermediate coolant loop in 1996. No radioactivity was released, but the sodium fumes made people sick. The reactor was restarted in 2010.[81] LMFBRs are ready for the next generation of reactors. Gas cooling in the intermediate heat loop is the only improvement needed.

Reactor Control

A chain reaction requires active control. The reproduction ratio of neutrons has to be exactly one. Too few neutrons, the reaction dies. Too many, the reaction runs away. The reaction rate depends on the temperature of the moderator (how much it absorbs) and the freshness of the fuel. Fission occurs so fast that it would be impossible to stop a chain reaction except for a lucky circumstance. *A small fraction of the neutrons are delayed.* In uranium, 0.65% and in plutonium, 0.21% of the neutrons from a fission event are emitted only after 10 seconds. Since every neutron is needed, the chain reaction does not proceed instantaneously; there is a time lag. The moderator and coolant in the reactor have high heat capacity, so the temperature inside the reactor changes even more slowly. There can be as much as 20 minutes to react to a temperature change. *Control rods* made of boron carbide (BC), a powerful absorber, are moved in and out of the moderator to control the neutron population. This is normally done automatically, and reactors have run for years without trouble. The few accidents that have occurred are due to human error in response to an abnormal condition. The danger is not only when the chain reaction is going too fast and the temperature rises. If the temperature goes too low, voracious neutron absorbers like Xe^{135} can accumulate and poison the reactor. It cannot be restarted until all the Xe^{135} has built up and then decayed with a half-life of 8 hours [41].

Fuel Reprocessing

France and Japan reprocess spent fuel to recover plutonium and 0.9% enriched uranium out of it; the USA does not. Here is what is involved. The spent fuel rods are cooled for 1 year in water ("swimming pools"). They are then taken apart underwater by remote control. The fuel pellets are dissolved in chemicals to separate out the uranium and plutonium. These are sent to Russia for isotope separation in centrifuges. Their oxides are made into an LWR fuel called "mixed oxide" or MOX. Ceramic MOX is radioactive and expensive. The arguments for reprocessing are that uranium fuel is not wasted, and there is less left-over radioactive waste to store underground. The long-lived part is four times smaller than in stored waste without reprocessing. The arguments against reprocessing are that the plutonium can be stolen for bombs, and that it is simpler and cheaper to just store the spent fuel.

Radioactive Waste Storage

When fuel elements come out of a reactor, they are still generating heat, so much that they would be red-hot if not cooled. They are placed under water in "swimming pools," which are steel-lined concrete pools filled with very pure water. The rods are cooled here for many years under careful surveillance. The heat drops to 1% after one year and is down to 0.2% after five years. The 100+ reactors in the USA are straining the capacities of these on-site pools. A 1-GW plant generates over

20 tons/year of nuclear waste. Before the cooled fuel rods are taken out of the water, they are sliced up and the materials sorted out by remote control. The radioactive material is dried and sealed in steel tubes filled with an inert gas. These are then put into concrete casks for on-site storage. They are cooled by normal air circulation. This is an intermediate, above-ground type of storage. In the USA, there are 66 commercial sites and 55 military sites storing these casks.[82] There are also ten "orphan" sites where the reactor no longer exists but the waste remains. Ultimately, the long-lived "actinides" with half-lives of 300,000 years or more should be stored underground, but there are no definite plans to do this. The temporary solution is the permanent solution so far.

For underground storage, the high-level waste is cast into glass logs and welded into stainless steel canisters. These are to be stored in a large underground tunnel system in a geologically stable environment, like a salt mine or rock formation. The site must be immune to infiltration of water and such disturbances as earthquakes. The waste cannot be moved there until it is cool enough not to heat the rock. You have seen the charts of the decay of radioactivity from each element over 10,000's, 100,000's, even millions of years. In 600,000 years, the radioactivity level is down to that of natural uranium. Reprocessing of fuel in France is estimated to cut this time to 60,000 years.

In the USA, $9B had been spent to characterize a site under Yucca Mountain, Nevada. This project was canceled by the Obama administration in 2009. There are only two projects in the world at this time devoted to underground nuclear storage. One is on Olkiluoto Island in Finland at a place that satisfies the requirements and where no one is likely to build a housing project. Finland has four reactors generating 25% of its electricity. The project is estimated to cost €3B ($4B). In Sweden, two sites have been chosen after a long campaign in which many proposals were considered and open discussions involved both politicians and citizens. This could not be done in the USA because of the military component. No construction has started, but there is likely to be less public opposition than elsewhere.

The nuclear waste problem will become worse since more reactors are being built or planned. The longevity of geological formations cannot be proved. The danger to future generations is a legitimate concern. However, fusion power can help in two ways. First, subcritical fusion reactors can be built to generate neutrons for transmuting actinides into stable elements. Second, if nuclear power can be considered only as a temporary solution, like wind or solar power, until fusion comes online, the buildup of radioactive waste will eventually stop; and underground storage may not be necessary.

Nuclear Proliferation

Plutonium is very good for making a bomb. It does not need enrichment. Uranium has to be highly enriched for explosive purposes, and gas diffusion plants are so large that a terrorist would have to be quite an industrialist to build one. It is easier to steal plutonium. Breeder reactors make plutonium. Recycling nuclear waste also recovers plutonium and makes MOX. Places where plutonium is made or transported have to be heavily guarded.

The development of gas centrifuges has posed a new problem [43]. These devices are relatively small and much more efficient. The separation factor is 1.2–1.5, compared with 1.004 in gas diffusion. Uranium has to pass through a centrifuge only 30–40 times before it reaches weapons grade. It would take many times more in gas diffusion. The uranium is in the form of UF_6 in gaseous form. It has to be under partial vacuum so that it does not solidify and gum up the works. This means it cannot leak out. Centrifuges are small enough that a hundred of them can be installed in a building that looks like any other industrial building. Centrifuges can be connected in series so that the output of one goes into the next for further separation. A cascade of over 100 centrifuges can be designed to optimize the number used at each stage of enhancement. One cannot prevent the construction of such a cascade for peaceful production of 5% U^{235} for power plant. The problem is that the cascades can be reconfigured in a few days to produce weapons-grade uranium. For instance, the output of 5% U^{235} from two-thirds of the cascades in a plant can be sent to the remaining one-third for further enrichment to 90% U^{235}. The power used in either case is only about 160 W/m^2, compared with 10,000 W/m^2 in gas diffusion, so the clandestine activity cannot be detected by the power consumption, which is like that of any well-lighted building.

India and, in response, Pakistan were the first to use gas centrifuges. This is the reason for the recent attention given to Iran for its construction of an isotope separation facility. The danger exists whether or not nuclear power is used for energy.

Nuclear Accidents

In the early days of civilian nuclear power, there have been a number of small accidents in different countries, but usually there was little release of radiation. Workers were exposed to it, and four died, one in Yugoslavia, one in Argentina, and two in Japan.[83] This does not include deaths in Russia. The two well-known, large accidents are Three Mile Island and Chernobyl.

At Three Mile Island in Pennsylvania, USA, the 1-GW Unit 2, a PWR, had a problem in March 1979. A mechanical failure was compounded by operator error. A pump for the cooling water stopped, and the water got hot, increasing the pressure. Automatically, a relief valve opened to let the steam out into the containment building, and control rods dropped in to stop the chain reaction. The relief valve was supposed to close at a certain pressure, but it got stuck open. The hot fuel rods continued to produce steam, and most of the water was lost out the open valve. The operators misinterpreted the signals and thought that there was *too much* water, so they shut down the pumps, making things worse. Only the bottom of the fuel elements was covered with water. The tops got so hot that the cladding electrolyzed steam into hydrogen, and a hydrogen bubble was formed, preventing water from entering for days. The fuel melted, and 700,000 gallons of radioactive water covered the floors of the buildings [41]. Although the people in surrounding areas were scared and were evacuated, only a small amount of radioactive material escaped. *There were no deaths.* Statistically, the amount of radiation could have caused three deaths in 20 years, but none has been reported.

The Three Mile Island accident turned a lot of Americans against nuclear power, but compare its safety with that of other energy sources. In 2010 alone, we have had the methane explosion in a West Virginia mine which killed 25 miners, followed by the Deepwater Horizon disaster in the Gulf of Mexico, which killed 11 workers. In each case, families waited and waited for good news about their loved ones, but in vain. The grief is repeated hundreds of times all over the world. The oil leak following the fiery destruction of the Deepwater drilling platform was far larger than the Exxon Valdez spill in Alaska and covered hundreds of square miles of the Gulf. Both aquatic wildlife and migrating birds suffered from the environmental damage. Compared to the fossil-fuel industry, a well-regulated nuclear industry is a far safer way to get energy.

The Chernobyl accident is another matter. The dire consequences of the accident were caused by the organization of the Soviet Union.[84] Failures were covered by lies. Tight secrecy kept workers from learning from the experiences of others. Those in command made policies without caring about the actual situations they covered. The chief engineer disregarded the protocols anyway. Workers were not well trained to know about the dangers, and they disregarded orders. One of the four reactors at Chernobyl in the Ukraine was being shut down for maintenance. The chief engineer decided to test whether power could still be produced while shutting down. He did not consult the safety personnel or the set rules. The workers turned off safety devices. The control rods were withdrawn to get power while the chain reaction was slowing down due to xenon poisoning. A decrease in cooling water caused the fuel rods to heat up, increasing the power output. The reactor had not been designed to shut down automatically when this happens. There was a runaway reaction and a power surge that ruptured fuel tubes. The hot fuel reacted with water to cause a steam explosion which blew off the 1,000-ton top of the reactor. This broke all the fuel tubes, and a second explosion sent most of the reactor core into the air.

The explosion was like the volcano in Iceland that erupted in 2010, stopping all air traffic in Europe. This time, a radioactive cloud went as high as 10,000 m (30,000 feet), carrying 50 tons of nuclear fuel. The surrounding area was sparsely populated; a nearby village was in great danger. Nonetheless, the man in charge, arriving from Moscow, gave orders not to evacuate because it would create panic. It was a plasma physicist, Evgeny Velikhov, who finally convinced him that people had to get away. Meanwhile a large crew (200,000 in the first year) was trying to clean up the mess. They were walking directly on radioactive material, receiving a lethal dose within minutes. Winds carried the radioactivity all over Europe, but where it landed was random, depending on rain. Most of the volatiles were iodine-131 and cesium-137. The iodine fortunately has a half-life of only eight days, but the cesium lasts for 30 years. The Cs^{137} carried 500 times more radioactivity than created by the bombs on Hiroshima and Nagasaki.

Statistically, health experts calculate that this accident would cause 30,000 deaths in 20 years. However, this is still a small number compared with other types of accidents. It amounts to a probability of 0.6 deaths per 100,000 people per year. For a well-regulated industry with accidents like that at Three Mile Island, the figure is 0.00007 per 100,000 per year. This is to be compared with 16 for motor vehicle accidents, 0.41 for airplanes, and 5.15 for falls [41]. Falls were considered earlier in the Solar Energy section. Chernobyl was a lesson in bad management, but it will never happen again. Nuclear power poses less risk than almost anything we do.

Future Reactors

Generation III reactors will have more efficient use of fuel and better safety features but no radically new designs. Advanced Boiling Water Reactors, Advanced CANDU heavy-water Reactors, and EPRs will be added to the list of acronyms. Generation IV reactors will be of two main types: breeder reactors, either liquid-metal or gas cooled (discussed above); and very high temperature reactors (VHTRs) [44]. Of these, the most interesting is the pebble-bed modular reactor (PBMR), shown in Fig. 3.60.

The "pebbles" are tennis-ball size spheres containing both the fuel and the moderator. The small grain of fuel can be any fissionable material such as enriched uranium, plutonium, or MOX, the mixed oxides of both. The fuel is surrounded by a layer of porous graphite to absorb gaseous products of the reaction. This is covered by a thin layer of silicon carbide, which is an impenetrable barrier that can take high temperatures. The outer layer of the fuel grain is pyrolytic carbon, which

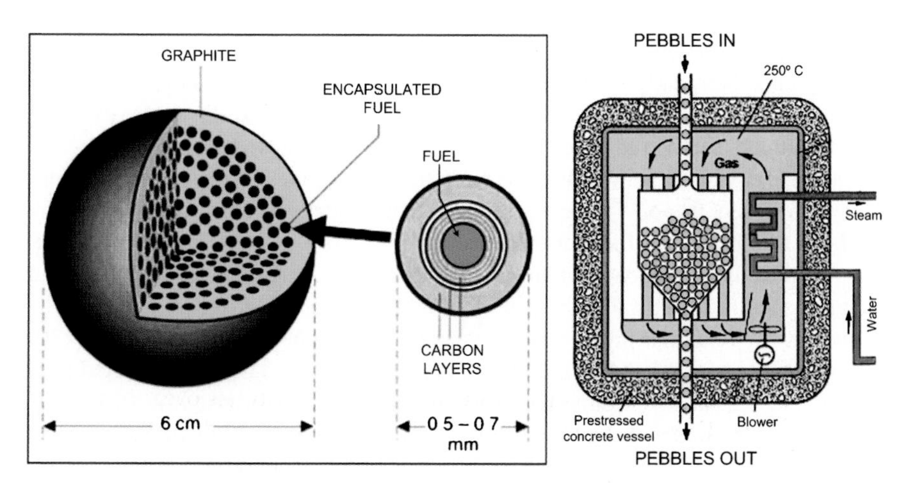

Fig. 3.60 Diagram of a pebble and a pebble-bed reactor vessel (European Nuclear Society: http://www.euronuclear.org/info/encyclopedia/p/pebble.htm)

is dense and can take extremely high temperatures. These tiny fuel grains are dispersed in the graphite moderator, which forms the bulk of the pebble. The reactor core contains some 360,000 pebbles, enough to make a critical mass with the spacing fixed by the spherical pebbles. Helium is circulated through the spaces between the spheres for cooling, and the helium then carries the reaction energy to a heat exchanger.

The design has built-in safety features. The reaction products are contained within the fuel grains and the pebbles. In fact, depleted pebbles can be their own waste containers. The helium is not radioactive even if it leaks out. The reactor can operate at 1,000°C to raise the thermal efficiency to 50%. If the coolant fails, the reactor cannot go critical because the U^{238} part of the fuel absorbs more neutrons at higher temperature, thus slowing down the reaction if it gets hot. The pebbles might reach a temperature of 1,600°C, but the pebbles are still stable at that temperature, and the reactor core will just stay that way until cooling is restored. The pebbles can be dropped in at the top and removed from the bottom of the reactor core. This allows the pebbles to be periodically examined and removed to storage if they have been used up.

Critics of PBMRs cite the possibility that the graphite would catch fire if it contacts air or water at these extreme temperatures. PBMRs are being developed in Germany, the USA, the Netherlands, and China. The automatic safety mechanisms have been tested on a small scale.

Fission–Fusion Hybrids

This subject is logically treated here because of the radioactive waste problem of fission reactors. However, fusion reactors have not yet been described. This section can be best understood if Chap. 9 on fusion engineering is read first. The reason for combining fusion with fission is that it could benefit both systems. Fission reactors can be run subcritically for better safety, and their high-level wastes can be transmuted into fuel and a much smaller amount to be sequestered. Fusion reactors, on the other hand, can be run subcritically also, without producing all the energy of the reactor, greatly accelerating the time for their development. Many plasma theorists have advocated fission–fusion hybrids, notably Jeffrey Freidberg at M.I.T. and Wallace Manheimer at the Naval Research Laboratory in the USA. The idea was first proposed by none other than Hans Bethe. However, their arguments do not include specifics on how a hybrid reactor might be designed. A group at the University of Texas has proposed a reactor based on a spherical torus (see Chap. 10), a new fusion device that has not been extensively tested. The most detailed engineering design has been done by a group at the Georgia Institute of Technology (Georgia Tech) under the leadership of W.M. Stacey. Their subcritical advanced burner reactor [45] will be described here. A diagram of it appears in Fig. 3.61.

Within the D-shaped toroidal-field coils is the plasma of a fusion reactor, shown in yellow. Surrounding that is the fission fuel core, which is divided into four

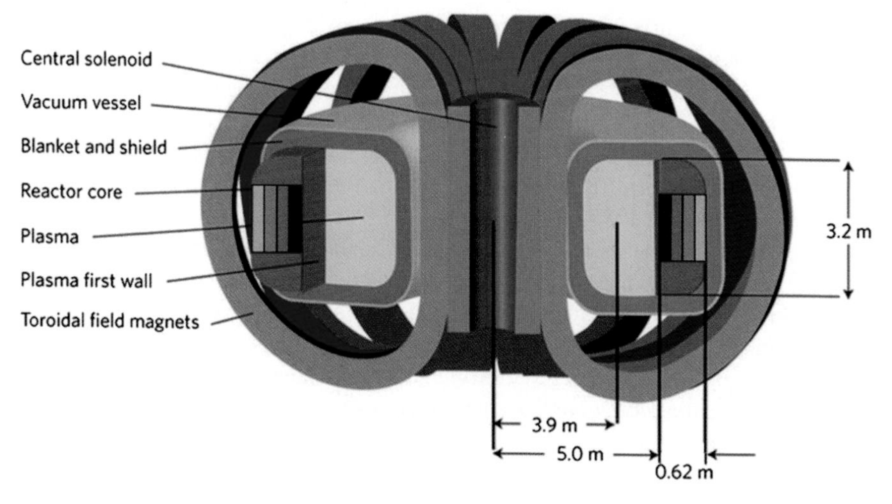

Fig. 3.61 Rough diagram of a conceptual fission–fusion reactor [45]

concentric rings (gray). Surrounding both is a neutron absorbing blanket which breeds tritium from Li_4SiO_4 for DT fuel. The fission part is an LMFBR designed at Argonne National Laboratory. The fuel is 36 tons of transuranic waste from LWRs consisting of 40% Zr, 10% Am, 10% Np, and 40% Pu. It is in the shape of 7.3 mm diameter fuel pins, 271 of which form a fuel assembly. The fuel pins include a channel for the liquid sodium coolant. Their complete design and manufacturing process have been specified [46]. The fuel rings (batches) contain 918 assemblies. The tokamak part is a scaled-down ITER operating with conservative parameters lower than the maximum values needed for energy production. These include factors which will be explained in Chap. 9: the Greenwald limit, normalized beta, big Q, and the bootstrap current fraction.

The operating characteristics of this reactor have been extensively calculated. The fission part will generate 3 GW_{th} (gigawatts thermal). It runs subcritically, generating fewer neutrons than is necessary to maintain a chain reaction. The missing neutrons are generated by the fusion part. Since its mission is not to generate power, it can be designed to contribute only 250–500 MW_{th} of energy. The fission fuel is burned in 750-day cycles. Each batch spends one cycle in each position, for a total exposure of four cycles or 3,000 days. After that, it is removed to storage, and its decay heat over the next million years has been reduced by a factor of 2, and thus the storage facility requirements have been halved. The total time of exposure is limited by the life of the fuel cladding under neutron bombardment, set at 200 dpa (displacements per atom).

This amount of burnup of actinides can be greatly improved by reprocessing. If the fuel from the hybrid after four burn cycles is reprocessed, then mixed with "fresh" waste from LWRs and sent through the hybrid again, the decay heat of the ultimate product can be reduced by 99%. High-level storage facilities can be reduced by a factor of 100. If the 200-dpa limit on neutron damage can be relaxed so that the fuel can be burned for four 3,000-day burn cycles for a total of 12,000 days (25 years), 91.2% of the transuranic waste can be removed after only once through the hybrid reactor. Such a fission–fusion hybrid can treat the waste from four 1,000-MW_e LWRs.

It is possible for the fission reactor to go critical. Zirconium is added to the fuel so that there is negative feedback: when the temperature rises; the reaction slows down. However, if this does not work and there is a runaway reaction, there is less time available for control rods to be inserted than in a normal LWR. Fortunately, there is a simple solution. The reaction cannot run without neutrons from the fusion reactor. The plasma producing these neutrons can be shut off within a second or so by a massive injection of gas.

Proponents of hybrids see that they can make fission safer and at the same time let fusion get online faster. Skeptics see that these would be extremely expensive and difficult reactors to design and construct and would detract from the main objective of developing pure fusion. In any case, this subject is still in its infancy compared with Generation III fission reactors or with tokamak fusion reactors.

Other Renewables

Hydroelectricity

Hydroelectric power is the simplest, most direct way to produce electricity. A dam is built, and water is released to turn large generators. No heat, no complicated equipment, no fuel transport, and no pollution. The power is available in controllable amounts any time. This is an ideal situation that no other source can emulate. Of course, it is not available everywhere. Worldwide, hydro accounted for only 2.2% of total energy consumption in 2006, compared with 6.2% for nuclear.[85] Some countries, such as Bhutan, depend entirely on hydroelectricity, and Bhutan actually exports part of it. Iceland uses hydro for 73% of its energy. The role of hydro in various parts of the world is shown by the blue bars in Fig. 3.62. In the USA, hydro accounts for 7% of electricity generated and 36% of all electricity from renewable sources.[86] Renewables provided 7% of all energy consumed in the USA in 2007. China has the most hydro power. The Three Gorges Dam, completed in 2008, has generating capacity for 26.7 GW of electricity. This is comparable to the output of 25 coal plants.

Construction of dams can change the landscape and displace wildlife, especially fish, but this is a small price to pay for free energy. Dam breaks pose a danger to downstream residents. Climate change can affect the distribution of rain and snow, causing some rivers to increase, and some to decrease their flow rates.

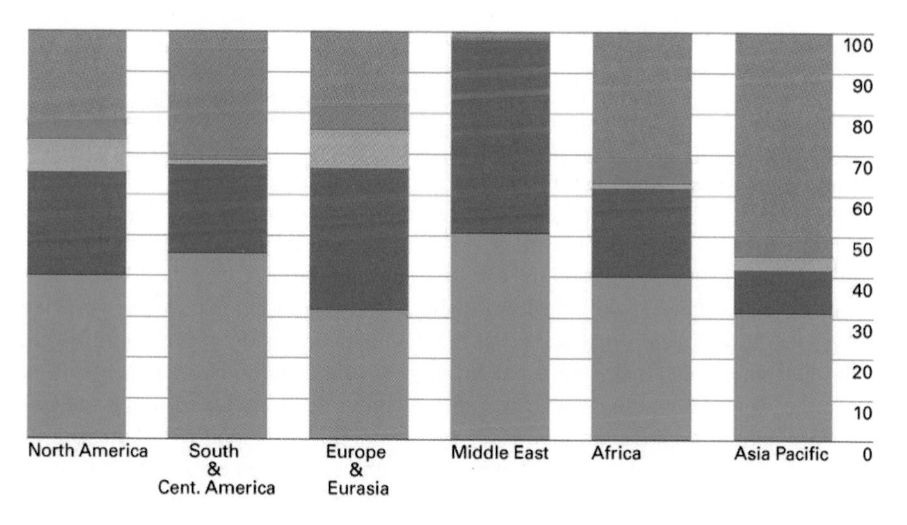

Fig. 3.62 Fuel sources in regions of the world by percent. From the *bottom up*, the sources are oil, gas, nuclear, *hydro*, and coal (BP Statistical Review of World Energy 2008)

However, these drawbacks are minor, and hydroelectricity will continue to be an important part of our energy mix even if most of the best hydro sources are already being used.

Geothermal

Geothermal energy comes from the hot rock deep down that makes geysers and warm pools for spas and mud baths. It mostly occurs at the junctions of tectonic plates. Worldwide, 10.7 GW of electricity is generated geothermally in 24 countries, and another 28 GW is used for heating. The USA produces the most geothermal power, 3 GW, in 77 plants mostly in California. The Philippines is second with 1.9 GW and gets more than a quarter of its energy from geothermal, as does Iceland. These numbers are very small on a world scale, and we need not say much about this energy source.

The capital expense of geothermal plants, used for exploration and drilling, is comparatively large. There is no fuel cost, but electricity is used to run the pumps. Once a bed of hot rock is found, a production well is drilled to extract the steam. If this is hot, above 180°C (360°F), it can be used directly to drive steam turbines to generate electricity. If it is cooler [below 150°C (300°F)], it is used for space or water heating. The used, cooled water is injected back into the rock in an injection well. With the steam, GHGs also came: CO_2, methane, ammonia, and hydrogen sulfide, which smell. Whether these emissions are lower than from a comparable fossil-fuel plant depends on the location. The water also contains undesirable chemicals: mercury, arsenic, antimony, boron, and salt. All in all, geothermal energy is not going to be a solution to the world's problems.

Wave and Tide Energy

The motion of tides, currents, and ocean waves can be used to produce energy. A few places, like the Bay of Fundy, have high tides, and the water rushes through a narrow channel four times a day. If the speed is greater than about 5 knots (2.6 m/s), the current can drive an electric generator, but there are very few such sites. A new method called Vivace[87] is claimed to work at speeds as low as 2 knots (1 m/s). Flexible cylinders are anchored to plates on the sea bottom. Currents flowing back and forth make the cylinders flex and wobble, and this motion is used to generate electricity. How they do that and what the cylinders are made of are not revealed. Tides and waves also make the ocean level go up and down. Several systems using this effect are based on the same principle. A rigid tube is anchored to the sea bottom. A diaphragm inside the tube is driven up and down by a buoy floating on the top. As the buoy moves up and down, the diaphragm drives air in and out of an aperture at the top end of the tube. This flow of air turns a turbine to generate electricity. An underwater cable carries the electricity to shore. This method requires a floating object that can be seen and collided into.

The most publicized system is the Polamis ("sea snake" in Greek),[88] designed to capture wave energy. It looks like a series of giant sea snakes floating in the ocean. Each snake points in the direction of wave motion, perpendicular to the wave crests, and consists of metal cylinders with the size of railway cars hinged to each other so that the snake flexes with the waves. In between the cylinders are air pistons pushing air back and forth with the wave motion. This air drives onboard electric generators. These Polames have been built in several countries, Portugal for one.

Cost and power have been calculated, but none of these ideas has been worked out for impact on the environment, wildlife, and ship traffic. Engineering for 30-year lifetime in the sea may be difficult. The resistance of materials against salt water damage, so important in offshore wind turbines, has not been mentioned, for instance. The power is also not constant, so that some storage mechanism is needed to level it out. It is clear that these entrepreneurial ventures cannot yet be taken seriously.

Biomass

Organic waste from human activities or natural swamps contains energy. Many societies already produce methane from cow dung or even human waste. Low-tech companies have sprouted up to make biofuels from deep-fry oil, left-over beer, or even onions. Almost all of these efforts are to produce fuel for transportation, which has already been treated in this chapter. There is only one application to general energy production. This is to mix biomass with the fuel in a fossil-fuel plant. The same amount of power can be generated with less coal.[89] Small plants burning only biomass would be very inefficient.

Artificial photosynthesis is an interesting development that does not generate energy. Using chlorophyll, plants convert water, carbon dioxide, and sunlight into carbohydrates and oxygen. Daniel Nocera at the Massachusetts Institute of Technology has been able to split the water molecule in the laboratory using special catalysts and energy from solar cells (or the grid). Two electrodes, one of indium-tin oxide and the other of platinum, are immersed in a solution containing cobalt and potassium phosphate.[90] When a voltage is applied, oxygen bubbles came out at one electrode and hydrogen at the other. The catalysts reform themselves. This process does not produce energy; it produces hydrogen, which can store solar energy during the night.

Wild Schemes

The inventiveness of the human mind has spawned a large number of crazy ideas for generating energy or slowing global warming. Some ideas are described in the Solar Power and Geoengineering sections. For instance, there is a plan to put square miles of silicon solar panels into synchronous orbit around the earth, convert the solar power into microwaves, and then beam the microwaves back to earth. Another is to place a huge mesh of wires at the point where the sun's and earth's gravitational fields cancel. The mesh scatters sunlight so that not as much falls onto the earth, thus reducing global warming (and perhaps trigger the next ice age). There are wind scrubbers that catch CO_2 as it comes by in the wind. Dumping huge amounts of iron filings into the ocean would spawn huge blooms of plankton which absorb CO_2. These ideas appear often in the popular literature.[91,92,93] Astute readers will recognize the ridiculous ones and have a good laugh.

Notes

1. This is the same as energy return on energy invested (EROEI). Until renewables are well established, however, all that energy will come from fossil fuels, hence casting a fossil footprint.
2. Nature Conservancy Magazine, Autumn 2009.
3. Boston Globe, September 11, 2009. Original source: *Science*.
4. Audubon Magazine, September–October 2006.
5. *California Guidelines for Reducing Impacts to Birds and Bats from Wind Energy Development*, California Energy Commission CEC-700-2007-008-CMF (October 2007).
6. For instance, http://www.nationalwind.org/workgroups/wildlife/.
7. http://www.wind-watch.org/.
8. National Geographic, August 2005.
9. IEEE Spectrum, July 2009.
10. All these data are from a Wikipedia article.
11. E.ON Netz Wind Report, 2005.
12. http://www.eia.doe.gov/emeu/reps/enduse/er01_us_tab1.html.
13. http://www.spectrum.ieee.org/green-tech/wind, January 2009.
14. Wall Street Journal, September 14, 2009.
15. *Wind Turbine Blade Flow Fields and Prospects for Active Aerodynamic Control*, NREL/CP 500-41606, August 2007.
16. National Wildlife Magazine, December/January 2009.

17. H. Kudsk, Vestas, private communication.
18. Adapted from a presentation by Chris Varrone, Chief Strategist, Technology R&D, Vestas Wind Systems.
19. *Breakthrough in Power Electronics from SiC*, NREL/SR-500-38515 (March 2006).
20. Wind Power Note, No. 16 (December 1997), Danish Wind Turbine Manufacturers Association.
21. Life-cycle assessment of offshore- and onshore-sited wind power plants based on Vestas V90-3.0 MW turbines (June 2006), Vestas Wind Systems A/S, Denmark.
22. Vestas Wind, No. 16, April 2009.
23. Business Week, October 8, 2007.
24. IEEE Spectrum, February 2008.
25. IEEE Spectrum, March 2008.
26. http://www.spectrum.ieee,org/energy/renewables/, June 2008.
27. Physics Today, July 2008.
28. http://www.cdc.gov/HomeandRecreationalSafety/Falls/adultfalls.html. From personal experience, sense of balance gets worse by the age of 80. My wife warns against going on the roof to check the swimming pool solar panels, but not everyone will listen.
29. http://www.cdc.gov/nchs/data/dvs/LCWK10_2006.pdf.
30. http://www.allcountries.org/uscensus/135_deaths_and_death_rates_from_accidents.html.
31. http://www.msha.gov/stats/charts/coalbystate.asp.
32. http://nextbigfuture.com/2009/11/climategate-coal-mine-deaths-air.html.
33. Vacuum Technology and Coating, June 2008.
34. US Energy Information Administration.
35. http://www.azocleantech.com.
36. Washington Times, December 11, 2009.
37. Physics Today, June, 2009.
38. http://www.emcore.com and http://www.spectrolabs.com.
39. Microns are micrometers denoted by μm, where the μ is a Greek mu. We have avoided this symbol for the sake of Grecophobes. A micrometer is an instrument for measuring small thicknesses.
40. Source: Wikipedia.
41. IEEE Spectrum, August 2008.
42. X. Wu et al., National Renewable Energy Laboratory report NREL/CP-520-31025 (2001).
43. Scientific American web article, April 25, 2008.
44. National Geographic TV program *Five Years on Mars* (2008).
45. An electrode is a piece of metal or other conducting material used to collect or transmit electric current. They usually come in pairs, a positive one (an *anode*) and a negative one (a *cathode*). Electrons go to the anode and ions go to the cathode.
46. Y. Yang, University of California, Los Angeles, private communication.
47. This phrase is an Americanism. It really means "…*couldn't* care less…"
48. V. Manousiouthakis, University of California, Los Angeles, private communication.
49. http://www.howstuffworks.com.
50. IEEE Spectrum, March 2005.
51. C. Etievant and P. Millet, private communication.
52. The NiMH battery in the 2009 Prius consists of 168 1.2-V cells giving 201.6-V total. Weighing 44 kg (97 lbs), its capacity is 1.3 kWh. The gasoline engine is 98 HP (73 kW). It works in an efficient "Atkinson cycle." There are two electric motors which start and drive the car and can run in reverse to charge the battery. Together the motors provide 100 kW (134 HP). These numbers (from Wikipedia) have been increased in the 2010 model. A computer controls the gas and electric motors so that they can work individually or together at each instant. To ensure the longevity of the battery, it is kept between 40 and 60% of full charge normally, and is not permitted to exceed the range 20–80%.
53. The drag coefficient C_d is the ratio of the wind force on a car to that on a flat sheet with the same frontal area. A truck or a Hummer has a C_d of about 0.6. A station wagon has a C_d of about 0.38, the same as for the original Volkswagen Beetle. Large modern cars have trouble getting C_d below 0.3, but small sedans can get below that. The latest hybrids can get down to

0.25. Note that C_d depends on streamlining, but the total force can be reduced also by a smaller frontal cross-section.

54. Environmental assessment of plug-in hybrid electric vehicles, EPRI=NRDC, July 2007.
55. The data are from Wikipedia except for the dollar figures, which were added by the author.
56. IEEE Spectrum, March 2009.
57. IEEE Spectrum, January 2010.
58. http://www.powergeneration.siemens.com/press/press-pictures/.
59. L.A. Times, May 9, 2010.
60. http://spectrum.ieee.org/energy/the-smarter-grid/fastcharging-lithium-batteries-disputed (June 2009).
61. IEEE Spectrum, November 2007.
62. B. Dunn, University of California, Los Angeles, private communication.
63. Scientific American, January 2007.
64. National Geographic, October 2007.
65. US Department of Energy, Energy Efficiency and Renewable Energy site.
66. Time Magazine, April 7, 2008.
67. Consumer Reports, October 2006.
68. Audubon Magazine, October 2007.
69. IEEE Spectrum online, January 2010.
70. Los Angeles Times, August 5, 2010.
71. Sierra Club, in http://www.sierraclub.org/policy/conservation/nuc-power.aspx. Admittedly, this policy was formed before global warming was well known.
72. Los Angeles Times, in editorial *No to Nukes*, July 23, 2007.
73. World Nuclear Association, in http://www.world-nuclear.org/info.
74. Wikipedia: Nuclear Power by Country.
75. The terms *fissile* and *fissionable* have slightly different meanings. This does not concern us here.
76. Decay here means beta-decay, which is an emission of an electron and a neutrino from the nucleus, changing a neutron into a proton. Free neutrons will decay into protons in 16 min, but neutrons bound inside a nucleus do not normally decay. However, unstable nuclei can beta-decay into a more stable isotope.
77. I have not seen an explanation of this, but here is a thought. The nucleons in a nucleus are arranged in shells, like the electron shells in the Bohr atom. The shells have even numbers of nucleons. With an odd number, one nucleon has no place to go and sticks out. When another neutron comes along, the nucleus tries to rearrange itself to accommodate it but finds that it is easier to break up into two pieces. It is like adding one more card to a house of cards.
78. Physics World, July 2007.
79. http://www.instablogsimages.com/images/2007/09/21/ausra-solar-farm_5810.jpg.
80. http://thoughtsonglobalwarming.blogspot.com/2008/03/solar-thermal-company-says-it-could.html.
81. Physics World, June 2010.
82. Scientific American, August 2009.
83. Wikipedia: Civilian Nuclear Accidents.
84. This information came from *The Truth About Chernobyl*, by Grigori Medvedev, as told by Garwin and Charpak [41].
85. International Energy Agency, *Key World Energy Statistics* (2008).
86. US Energy Information Agency.
87. http://www.vortexhydroenergy.com.
88. http://www.pelamiswave.com.
89. Energy Efficiency and Renewable Energy Biomass Program, US Department of Energy; http://www.eere.energy.gov/biomass.
90. M.I.T. Technology Review, July 2008.
91. Popular Science, August 2005.
92. Scientific American, September 2006.
93. Physics World, September 2009.

References

1. F. Ardente, G. Beccali, M. Cellura, V. Lo Brano, Renewable Energy **30**, 109 (2005)
2. Y. Lechón, C. de la Rúa, R. Sáez, J. Sol. Energy Eng. **130**, 021012 (2008)
3. C.W. Gellings, K.E. Yeager, Transforming the electric infrastructure, Physics Today, December 2004
4. P.M. Grant, C. Starr, T.J. Overbye, *A Power Grid for the Hydrogen economy* (Scientific American, July 2006); P.M. Grant, *Extreme Energy Makeover* (Physics World, Oct 2009). Chauncey Starr, the founding director of EPRI, wrote one of the first detailed articles on the energy problem (Scientific American, Sept 1971)
5. M.Z. Jacobson, M.A. Delucchi, A path to sustainable energy, Scientific American, New York, Nov 2009
6. K. Zweibel, J. Mason, V. Fthenakis, A solar grand plan, Scientific American, New York, 2008
7. J.A. Mazer, Photovoltaic technology and recent developments, Vacuum Technology and Coating, April 2008
8. M. Powalla, D. Bonnet, *Advances in OptoElectronics* (Hindawi Publishing Corporation, Cairo, 2007), 97545
9. A. Gupta et al., Mat. Res. Soc. Symp. Proc. **668** (2001)
10. M. Raugei, S. Bargigli, S. Ulgiati, Energy **32**, 1310 (2007)
11. V.M. Fthenakis, Energy Policy **28**, 1051 (2000)
12. V.M. Fthenakis et al., Environ. Sci. Technol. **42**, 2168 (2008)
13. V.M. Fthenakis, Renewable Sustainable Energy Rev. **8**, 303 (2004)
14. V.M. Fthenakis, H.C. Kim, Renewable Sustainable Energy Rev. **13**, 1465 (2009)
15. J. Mason et al., Prog. Photovolt. Res. Appl. **16**, 649 (2008)
16. G.W. Crabtree, N.S. Lewis, Physics Today, March 2007
17. E.S. Aydil, Nanotechnol. Law Bus. **4**, 275 (2007)
18. G. Li et al., Nat. Mater. **4**, 864 (2005)
19. H.Y. Chen et al., Nat. Photonics **3**, 649 (2009)
20. J.B. Baxter, E.S. Aydil, Appl. Phys. Lett. **86**, 053114 (2005)
21. P.M. Margin, Vacuum Technology and Coating, March 2008
22. A.J. Nozik, Physica E **14**, 115 (2002)
23. J.B. Baxter et al., Nanotechnology **17**, S304 (2006)
24. K.S. Leschkies et al., Nano Lett. **7**, 1793 (2007) and supporting info
25. D.T. Morelli et al., Phys. Rev. Lett. **101**, 035901 (2008)
26. J.P. Heremans et al., Phys. Rev. Lett. **88**, 216801 (2002) and **91**, 076804 (2003)
27. L.D. Hicks et al., Phys. Rev. B **53**, 10493 (1996)
28. E.M. Choi et al., J. Phys. Conf. Ser. **25**, 1 (2005)
29. S. Satyapal, J. Petrovic, G. Thomas, Scientific American, April 2007
30. V. Ozolins, E.H. Majzoub, C. Wolverton, J. Am. Chem. Soc. **131**, 230 (2009)
31. O.M. Yaghi, Q. Li, Mater. Res. Soc. Bull. **34**, 682 (2009)
32. P. Millet, D. Dragoe, S. Grigoriev, V. Fateev, C. Etievant, Int. J. Hydrogen Energy **34**, 4974 (2009)
33. J. Voelcker, IEEE Spectrum, Sept. 2007, p. 27
34. B. Kang, G. Ceder, Nature **458**, 190 (2009)
35. T. Brezesinski, J. Wang, S.H. Tolbert, B. Dunn, Nat. Mater., Advance Online Publication (2010)
36. D. Pimentel, T.W. Patzek, Nat. Resour. Res. **14**, 65 (2005)
37. M. Wang, in *15th International Symposium on Alcohol Fuels*, Sept. 2005
38. G.W. Huber, B.E. Dale, Scientific American, July 2009
39. S. Atsumi, T. Hanai, J.C. Liao, Nature **45**, 186 (2008)
40. S. Atsumi, W. Higashide, J.C. Liao, Nat. Biotechnol. **27**, 1177 (2009)
41. R.L. Garwin, G. Charpak, *Megawatts and Megatons* (University of Chicago Press, Chicago, IL, 2001). This book contains complete information on nuclear power and weapons, all explained in a readable way

42. F.F. Chen, Double helix: the Dawson separation process, in *"From Fusion to Light Surfing," Lectures on Plasma Physics Honoring John M. Dawson*, ed. by T. Katsouleas (Addison-Wesley, New York, 1991)
43. H.G. Wood, A. Glaser, R.S. Kemp, The gas centrifuge and nuclear weapons proliferation, Physics Today, Sept 2008
44. David Petti (Idaho National Laboratory), *The Next Generation Nuclear Plant: Mission, Design Status and Directions in Technology Development*, Seminar, UCLA, Feb 2008
45. W.M. Stacey, J. Fusion Energy **38**, 328 (2009)
46. W.M. Stacey et al., *Georgia Tech SABR Studies of a Fusion–Fission Hybrid Fast Burner Reactor*, American Nuclear Society Annual Meeting, San Diego, CA, June 2010

Part II
How Fusion Works and What It Can Do

Chapter 4
Fusion: Energy from Seawater*

Fission and Fusion: Vive La Différence!

The energy of the nucleus can be tapped two ways: by splitting large nuclei into smaller ones (fission) or by combining small nuclei into larger ones (fusion). The first yields what we know as atomic or nuclear (fission) energy, together with its dangers and storage problems. The second gives fusion energy, which is basically solar power, since that is the way the sun and stars generate their energies. Fusion is much safer than fission and requires as fuel only a little bit of water (in the form of D_2O instead of H_2O, as will soon be clear). Fission is a well-developed technology, while fusion is still being perfected as an energy source. *The object of this book is to show how far fusion research has gone, how much further there is to go, and what we will gain when we get there.*

Binding Energy

How can we get energy by fusing two nuclei when normally we have to split them? To understand this, we have to remember that atomic nuclei are composed of protons and neutrons, each of which weighs about the same[1] but has a different electric charge: +1 for protons and 0 for neutrons. When these *nucleons* (a general term for protons and neutrons) are assembled into a nucleus, they hold themselves together with a nuclear force measured by the so-called *binding energy*. The size of this binding energy varies from element to element in the periodic table, as shown in Fig. 4.1. There we see that elements near the middle of the periodic table are more tightly bound than those at either end. At the peak of the curve, with the highest binding energy, is iron. It is labeled as Fe^{56}, 56 being its atomic number, meaning that this is the number of nucleons in its nucleus.

Energy is released when elements are transmuted into other elements which have higher binding energy. Starting with a heavy element like uranium, one has to

*Numbers in superscripts indicate Notes and square brackets [] indicate References at the end of this chapter.

F.F. Chen, *An Indispensable Truth: How Fusion Power Can Save the Planet*,
DOI 10.1007/978-1-4419-7820-2_4, © Springer Science+Business Media, LLC 2011

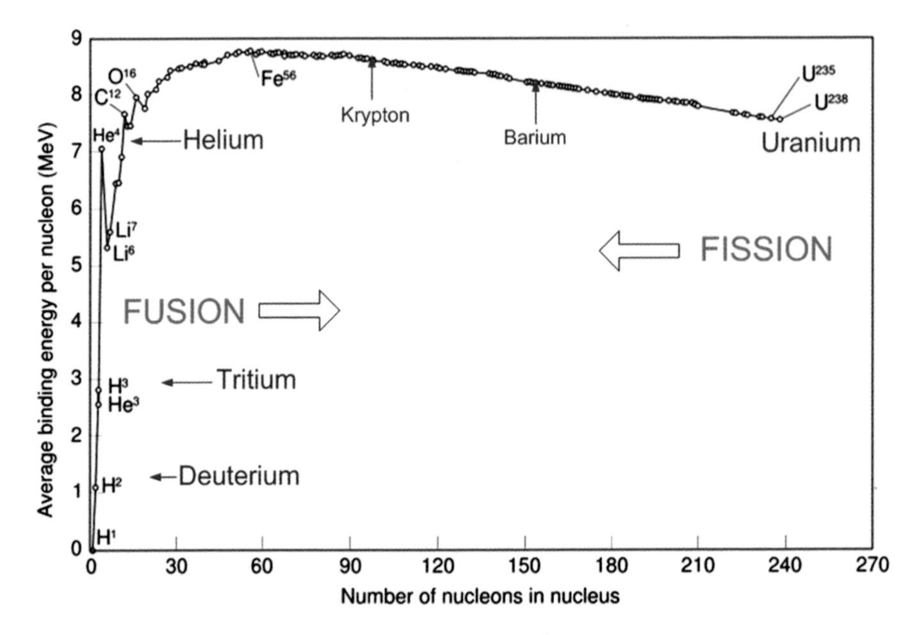

Fig. 4.1 Binding energy vs. atomic number for all elements from hydrogen to uranium (redrawn from Wikipedia.com). The energy units will be explained later

split it to get atoms with lower atomic number. If one starts with a light element like hydrogen, one has to fuse two nuclei together to get higher atomic number and move toward the peak of the curve. As labeled, fission goes from right to left, and fusion goes from left to right.

You may wonder why binding energy is i*ncreased* in both fission and fusion. Would not that require an *input* of energy rather than yield an *output* of energy? Yes, it is confusing; but to move forward without such distractions, the explanation is relegated to Box 4.1. Figure 4.1 would make more sense if we turn it upside down and plot binding energy downwards. This is done in Fig. 4.2. There we see that both fission and fusion go downhill, generating energy in the process.

Fission and Fusion Reactions

Fissionable uranium (U^{235}) cannot break up into two iron atoms because iron has 56 nucleons, and uranium has 235, which is a lot more than two times 56. To break it up into three or four big pieces would be very unlikely. So uranium fissions into two atoms larger than iron: typically, into krypton (Kr^{89}) and barium (Ba^{144}), whose atomic numbers add up to 233. There are then two neutrons left over, and it is these that carry off the generated energy and keep the chain reaction going. The energy released is not maximal since uranium moves only about halfway down the right-hand slope.

Now look at fusion at the extreme left of Fig. 4.2. When heavy hydrogen in the form of deuterium (H^2) and tritium (H^3) combine to form helium (He^4), with one

Box 4.1 What is Binding Energy?

Suppose we have two pitchers, one 30 cm (1 feet) tall and the other 60 cm (2 feet) tall. We then drop a ripe tomato into the short pitcher. The tomato releases some energy by making a *plunk*! sound. It is bound to the pitcher because it takes energy to lift it out. Now we drop another ripe tomato into the tall pitcher. It releases more energy by, perhaps, going *splat*! It is more tightly bound to its pitcher because it takes twice as much energy to lift it out. When each tomato drops down, it loses gravitational potential energy and gains binding energy. Therefore, binding energy is the *negative* of potential energy. That is why Fig. 4.2 makes more sense than Fig. 4.1. Since the sum of potential energy and kinetic (motion) energy remains the same, kinetic energy is increased when potential energy is decreased; or, equivalently, binding energy is increased. In nuclear reactions, the increase in kinetic energy goes mainly to the lightest resultant particles, usually the neutrons. In both fission and D–T fusion, the neutrons are captured and their kinetic energies turned into heat.

Of course, it is impossible to pick a nucleus apart one nucleon at a time to measure how tightly each is bound. Binding energy is actually inferred from the mass difference. Einstein's equation $E=mc^2$ predicates that energy and mass can be converted into each other. The mass of a uranium atom can be measured to be larger than the sum of the masses of the fission products. In splitting uranium, therefore, mass has been lost. This mass has been converted into the energy of the products that fly out of the reaction. Since the velocity of light, c, is a very large number, c^2 is larger yet, and a small *mass defect* leads to a large energy output. Similarly, in combining deuterium with tritium, the masses of the helium and neutron that are produced are smaller than those of the fusing hydrogen nuclei, and therefore mass has been lost and energy gained.

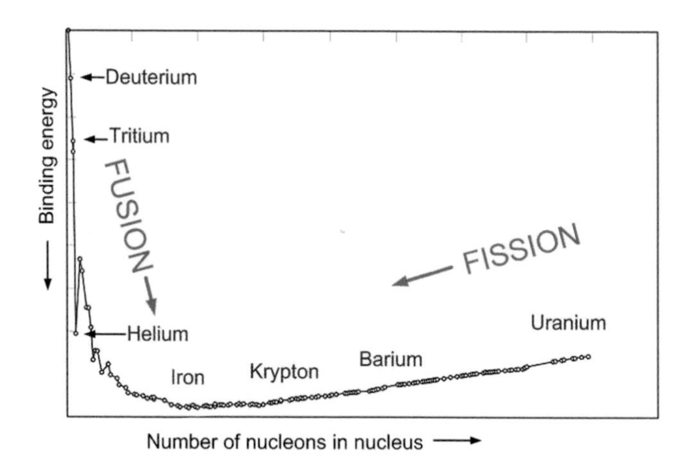

Fig. 4.2 An inverted binding energy diagram showing that going downhill from either side will release energy

neutron left over, there is a very sharp increase in binding energy. The curve is so steep that a lot more energy is released than in fission. However, this is energy *per nucleon*, and uranium has many more nucleons than hydrogen. After this is accounted for, the total energy gained *per reaction* is larger in fission than in fusion. This is not important. The end result is that both processes create large amounts of energy by forming elements closer to the middle of the periodic table.

The materials involved are, however, very different. In fission, uranium has to be mined and transported to huge isotope separation plants. Raw uranium is mostly U^{238}. Only 0.7% of it is U^{235}, the part that is fissionable. The separation plants enrich the mix so that there is a higher percentage of the good stuff. The products of fission are highly radioactive, some for thousands or millions of years. This is a well-known problem with fission.

By contrast, fusion uses only hydrogen, which occurs in three forms. Normal hydrogen, labeled as H^1 in Fig. 4.1, contains only a single proton. Deuterium (H^2) contains one proton and one neutron; it is "heavy hydrogen." Tritium (H^3) is heavier, containing one proton and two neutrons. The sun produces its energy by converting H^1 hydrogen into helium through a sequence of reactions which we cannot duplicate on earth. Here, we cannot do as well and must be content with converting heavy hydrogen, H^2 or H^3, into helium, but the energy gain is still very large. The reaction product is helium, whose nucleus, also called an alpha particle, consists of two protons and two neutrons. It is very tightly bound, so helium is very stable. This stability causes it to be the harmless gas used to fill birthday balloons. Deuterium, which we will call D, occurs naturally in water. In heavy water, D replaces the H in H_2O. There is one part of D_2O for every 6,400 parts of H_2O, and it is easy to separate it out. No mining or large separation plants. However, the other fuel, tritium (H^3 or T for short), does not occur naturally. It is also radioactive and decays in 12.3 years. It has to be bred from lithium in a fusion reactor. You may have noticed that deuterium contains one proton and one neutron, while helium contains two protons and two neutrons. Why not fuse two D's together to get helium? Well, this is hard and will come only in the second generation of fusion reactors. Right now, we are trying to fuse D with T to get helium plus an extra neutron. That neutron carries away most of the energy generated, but it also causes some radioactivity, but much less than in fission. For the future (Chap. 10), there are other advanced reactions involving helium-3 (He^3), lithium, or boron which are completely free of radioactivity. Note that lithium and boron are abundant and safe elements on earth.

How Fusion Differs from Fission

That the binding energy curve peaks in the middle is the reason that both fission and fusion can produce energy, but the way to tap these resources leads to entirely different types of reactors. In a fission reactor, a chain reaction is sustained as neutrons created in one fission move on to split other atoms nearby. The material, uranium or plutonium, is held in tubes which can be moved so that the number of "nearby" atoms can be controlled. If control is lost, the reaction runs away, and there is an accident. In fusion, the hydrogen fuel is heated into a gaseous, electrified state called a *plasma*. Since the plasma is hotter than the interior of the sun, it must be held in place by a magnetic field rather than a walled container. The problem is that this *magnetic bottle* leaks, and the problem is to keep the fire burning. There is certainly no possibility of a runaway reaction in this case. However, plugging these leaks has been a long and difficult journey for fusion researchers, whose story will soon unfold.

People used to confuse astronomy with astrology. With the great success of the Hubble telescope, the difference between science and fortune-telling is now clear in the public's mind. If fusion should succeed, perhaps the difference between fission and fusion would be equally well recognized.

The Size of Energy

Large amounts of energy can be measured in, say, millions of barrels of oil equivalent or kilotons of TNT equivalent. A more familiar household unit is the kilowatt-hour (1,000 Wh), which is used in our electric bills. A 100-W bulb will use 100 W h of electricity every hour. Since there are 3,600 seconds in an hour, a watt-second (which is called a *joule*) is 1/3,600 of a watt-hour, or the energy used by a 1-W cell phone in 1 s. These are units that we use on a human scale. When we talk about atoms, however, we have to use much smaller units because atoms are *very* small. There are some 100,000,000,000,000,000,000,000 atoms in a teaspoon of water. So the energy of an atom would be that much smaller than the energy units, like a watt-second, that we encounter in real life.

First, let's find a way to avoid writing all those zeroes. Scientific notation is an easy shorthand to do this. The large number above has 23 zeroes and is written as 10^{23}, where the superscript, called an exponent, tells how many zeroes follow the 1. A thousand (1,000) would be written as 10^3 and pronounced "ten to the third power" (or ten cubed in this case). Three thousand would be 10^3 multiplied by 3, written as 3×10^3. 3,600 would be written 3.6×10^3, and so forth. This works also for fractions if we use negative exponents. One thousandth (1/1,000) would be 10^{-3}. Two hundredths would be 2×10^{-2}. The only thing to note is that if we write decimals, 1/1000 would be 0.001, and the number of zeroes is one less than the exponent. But you need not worry about that; just remember that 10^{-3} is a thousandth, 10^{-6} is a millionth, 10^{-9} is a billionth, and so forth.

How much energy is released in fusing two hydrogen atoms? It is approximately 3×10^{-18} J. Joules are too large when dealing with atoms. A more convenient unit of energy is in order. The unit used is the electron-volt, or eV, which is more like the size of the energies of atomic particles. One electron-volt is 1.6×10^{-19} J. Now we can use eVs and stop counting zeroes. Since we will be talking about atoms in the next few chapters, we will use eVs and not worry about changing to more familiar units until we have to design reactors.

Let's get an idea of how big 1 eV of energy is. Molecules, CO_2 for instance, are held together with an energy of about 1 eV. An atom is a nucleus surrounded by electrons, equal in number to the protons in the nucleus. The outermost electron in an atom is bound to the nucleus with about 10 eV. A fusion reaction yields about 10 *million* eV or 10 MeV. A fission reaction yields about 100 MeV. The advantage of nuclear power is now obvious. Chemical reactions involve molecules and atoms, as in the burning of gasoline. These reactions yield eVs of energy each, and therefore a large number of molecules (read tankfuls of gasoline) are needed in normal use. Chemical energy is already very efficient. Witness monarch butterflies going 2,000 miles from Canada to Mexico or demoiselle cranes going from Russia to India over the Himalayas with no food or stopping. But chemical energy is infinitesimal compared with nuclear energy. Nuclear reactions yield tens to hundreds of *millions* of eVs each, so that the fuel needed for even a large power plant occupies a relatively small volume. Some think of hydrogen fusion as "burning" water. To do this in a chemical sense means that you first have to separate the hydrogen from H_2O and then ignite the hydrogen. The energy you get is relatively small, since it is a *chemical* reaction. In any case, you can't get any more energy out than it took to separate the hydrogen from the oxygen in the first place. But "burning" the hydrogen in a *nuclear* sense yields many *million* times more energy than in chemical burning.

How Fusion Works

We have shown that transmuting hydrogen into helium would release a large amount of energy. Let's be more specific. The first step is to use the easiest reaction possible, which is the following:

$$D + T \rightarrow \alpha + n + 17.6\ \text{MeV}.$$

Remember that D stands for deuterium, a hydrogen isotope containing one proton and one neutron, and T stands for tritium, containing one proton and two neutrons. Alpha (α) stands for a helium nucleus (He^4), containing two protons and two neutrons. There is one neutron (n) left over, which flies off carrying most of the energy released in the fusion, which is 17.6 million electron volts. This reaction is depicted in Fig. 4.3. The intermediate state shown there with five nucleons in it is not stable and immediately breaks up into an α particle and a neutron. The α particle, being ordinary helium, is very stable; and its energy will be used to keep the reaction going. The neutron carries 80% of the energy released (about 14 MeV), and it has to be captured and its energy transformed into heat, replacing the heat we now get from burning fossil fuels to run a power plant. Although neither reaction product is itself radioactive, the neutron can induce radioactivity in the walls of the reactor, and this material has to be buried. We shall see in Chap. 9 that the amount of long-lived radioactive waste is about 1,000 times smaller than for fission reactors. The D–T reaction is the worst of the fusion reactions in this regard, but it is the easiest to start with. Chapter 10 will show advanced reactions which have less radioactivity or none at all.

Fig. 4.3 The D–T reaction

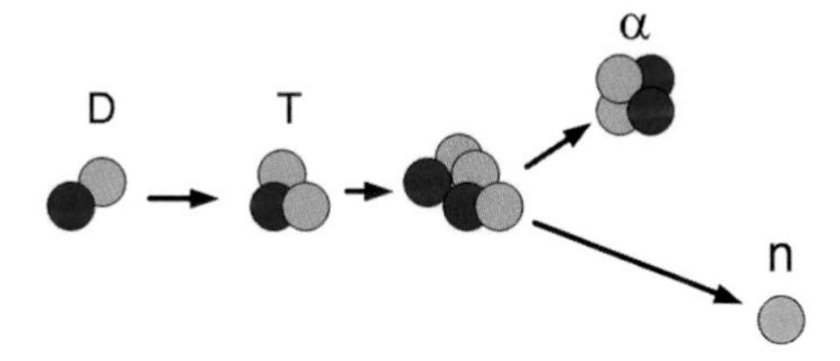

As pointed out before, deuterium is easy to separate out from water, but tritium has to be made in a nuclear reaction. In a fusion reactor, tritium is regenerated in a "blanket" containing lithium. Leaving this Chap. 9 topic aside for the moment, let us see how we can make this reaction go, because it's not easy. Since D and T have one proton apiece, they each have a positive charge. Like charges repel, so if we fire a beam of deuterons into a tritium target, the D's will most likely bounce off the T's without ever getting close enough to combine. Only a head-on collision with energy larger than 280 keV can overcome the electric repulsion (the so-called Coulomb barrier). Once inside this barrier, the nuclear force takes over, and the force becomes attractive instead of repulsive. Most of the time, however, the D's will bounce off without penetrating the barrier and lose most of the energy used to accelerate them. It is possible to use beams of around 60 keV energy and get net energy out, but not enough to justify the large number accelerators needed to make a dent on the power grid. There is a better solution. And that is *not* to use beams of particles at all but to heat a hydrogen gas, half in the form of deuterons and half in the form of tritons (tritium nuclei), to such a high temperature that there are always *some* high-energy collisions that result in fusion. The energy in failed collisions is not lost; it returns to the gas to keep it hot. This hot gas, called a plasma, perks away steadily, releasing enough fusion energy to keep itself hot and generate power besides. That's what happens in the interior of the sun. The fusion power generated comes out as solar radiation, of which the earth receives a small portion.

It is hard for people to understand why a hot plasma is necessary when you can simply shoot a beam of deuterons from a particle accelerator and hit a solid tritium target with enough energy to penetrate the electric barrier and get the D and T close enough to fuse. Or, one might circulate a beam of deuterons in one direction and a beam of tritons in the opposite direction in a round accelerator. Once in a while there will be a head-on collision and a fusion. But not often enough to pay for the energy used in accelerating the beams! Believe me, many proposals for using beams for fusion have been tried and have failed. Here is an analogy to illustrate how *plasma* fusion works. Imagine a friction-free pool table which has no pockets at the edge. However, there are pockets all over the middle of the table, and each pocket is surrounded by a hill, like a deep crater at the center of a volcano. The hills represent Coulomb repulsion. A pool player then adds billiard balls randomly, shooting them with insufficient accuracy and speed to climb the hills and get into the hole. Since there is no friction, the numerous balls keep bouncing around at random until one is lost by chance by jumping off the table, whereupon it is replaced by another shot. Since the balls bounce against one another, once in a while, one will undergo several favorable bounces in a row and end up with more than the average energy. If it has enough energy and is going right toward a crater, it will be able to climb the hill and get into the pocket at the top. This represents a fusion reaction yielding 17.6 MeV of energy. You might have to wait a long time for this to happen, but after the initial energy used to shoot the balls, no more energy is needed other than to replace those lost over the edge. This is the idea of plasma-induced fusion. A small amount of energy is invested in shooting the balls in, and then one waits for a long time before a ball by chance climbs a hill and gets into the pocket. But the payoff in energy is so huge that there is a large energy gain even if it takes many collisions to get one fusion.

Plasma, the Shining Gas

At this point, we should define what "hot" means. When a gas like air or steam has a temperature, it means that the velocities of the molecules are spread out in a particular way, known as a Maxwellian distribution. This is the same bell-shaped curve, called a Gaussian, that teachers use to grade exams. Gaussian and Maxwellian mean the same thing. Physicists tend to use Maxwellian while mathematicians use Gaussian. Figure 4.4 shows such a curve representing the relative number of hydrogen ions having different velocities in a gas at about 10,000 K.[2] When a material is in thermal equilibrium, it has such a "Maxwellian" distribution. The temperature is proportional to the width of this curve, so the velocities are higher at higher temperature. By raising the temperature, we can assure that there will be enough D and T ions in the "tail" of the distribution with enough energy to fuse. The "tail" is either end of the Gaussian curve, far from the center, where there are few particles of very high velocity. The ones that collide without fusing go back into the body of the distribution. In our billiard ball analogy, those would be the balls that go up a hill and come down again without falling into the hole. Multiple collisions

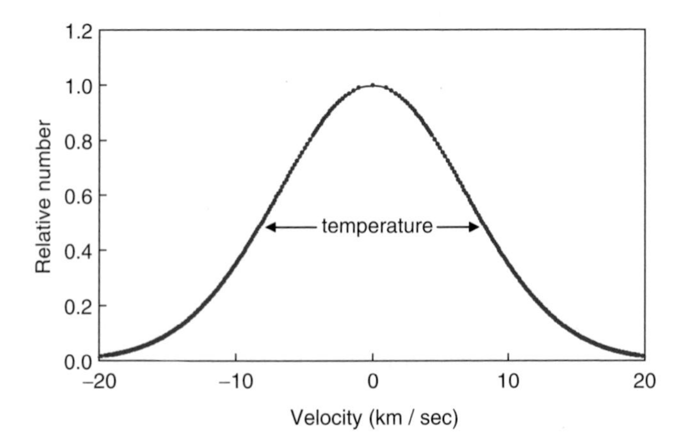

Fig. 4.4 A Maxwellian distribution of velocities

automatically maintain the shape of this most probable distribution. That means that the energetic particles in the tail which are lost in fusion reactions are replenished by successive favorable collisions. The collisions are random, so a particle can gain or lose energy each time. Only a fortuitous sequence of energy-gaining collisions can get a particle to high energy; that is why there are so few of them in the tail. Fusion reactors will require gas temperatures over 100,000,000 K!

At these temperatures, or even at 20,000 K, which is the temperature of the electrons inside a fluorescent light bulb, the gas no longer resembles the gases that we are familiar with, like air, helium, or CO_2. Molecules become dissociated, and atoms become ionized. An oxygen molecule O_2, for instance, first becomes dissociated into two O atoms, and then each O is ionized into an ion (O^+) and an electron (e^-). Normally, an oxygen nucleus with charge +8 is surrounded by eight electrons, so that the atom as a whole is neutral. When one of the electrons is stripped off by colliding with a free electron, it becomes free, and the nucleus is left with an excess charge of +1. The gas is now a gas of ions thoroughly mixed with a gas of electrons, the way NaCl molecules are mixed with H_2O molecules in a saline solution. But there is a big difference: this gas mixture is electrically charged. The ion fluid is positive, and the electron fluid is negative, so there can be electric fields inside the mixture. This type of electrically charged fluid is called a *plasma*. A plasma as a whole is neutral, with the same number of positive and negative charges. It is not *exactly* neutral, however, because there are electric fields inside a plasma. These fields are created by a very small charge imbalance of the order of one part in a million. If these fields were not there, nuclear fusion would not be a problem. So we call these gases "quasineutral" plasmas. Figure 4.5 shows what this new kind of gas is like. The ions are the small (blue) dots. They are given tails to show that they are moving in random directions. The big, fuzzy objects are the electrons, which provide the negative charges to make the plasma quasineutral. They are fuzzy

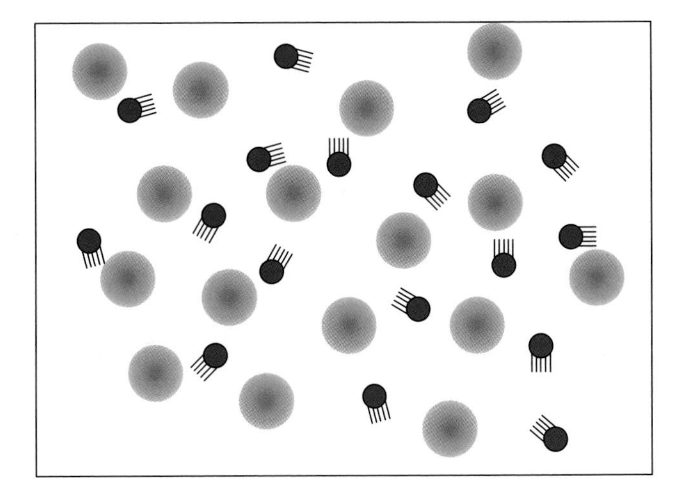

Fig. 4.5 A cartoon of a quasineutral plasma

because you can never tell exactly where a given electron is. These particles move around at their thermal velocities and bump into one another, preserving the Maxwellian distribution. At the temperatures we are dealing with, the electrons and ions move too fast to stick to each other and recombine into an atom.

Often called the fourth state of matter, plasma is what you get when you heat a solid into a liquid, then into a gas, and finally into an ionized gas. Plasma emits light when electrons collide with atoms, kicking one of the orbiting electrons into a higher orbit. Light is emitted when that bound electron goes back to its original orbit. Although 85% of the matter in the universe is believed to be dark matter, the part that we see can be seen because it is in the plasma state. That includes all the stars, galaxies, and nebulae. On earth, plasmas cannot survive in our dense atmosphere, but we can see them in the Aurora Borealis and in fluorescent lights. You may have encountered plasmas without knowing it. Sparks are plasmas at atmospheric pressure. When there is a high voltage, electrons can jump across and make a transient plasma. This happens when you touch a doorknob on a cold day or plug in the power brick of a laptop computer. Lightning is a huge spark between a cloud and the earth or another cloud. These breakdowns are uncontrolled; but the steady, voluminous plasmas that we create on purpose are well behaved. They cannot spark because they are already completely broken down!

Plasma behavior is extremely complicated, and a whole new science of *plasma physics* has grown up from the effort to produce fusion energy. This science has now permeated into other fields. Computer chips cannot be made without plasmas. Plasma TVs are commonplace. Chaos theory and supercomputers were spawned by plasma research. How did we get to this subject? We found that particle beams cannot create fusion with a net energy gain. We had to heat a whole gas up to an extreme temperature so that, in a thermal equilibrium with a Gaussian velocity distribution, there are ions in the tail of the distribution with enough energy to fuse together. This is, then, a thermally generated nuclear reaction or *thermonuclear* reaction. This word has bad connotations and is no longer used by fusion researchers. Nonetheless, this clever method underlies the hydrogen bomb.

It is obvious that no solid material can withstand temperatures of millions of degrees, so we cannot hold the plasma with walls. We can hope to hold it with invisible forces, such as gravity, electricity, or magnetism. The sun produces fusion energy by holding plasma in its core with a large gravitational field. We cannot do this on earth because our gravity is much too weak, and we cannot shape it. That leaves electricity and magnetism. We can make strong electric fields, but these would not do it. The real proof is subtle, but basically you can see that an electric field will pull ions one way and push electrons the other way. The field will pull a plasma apart rather than confine it. That leaves magnetic fields. The name of the game is to make a *magnetic bottle* to hold plasma. That is the main subject of this book. All magnetic bottles leak. It is like holding Jello(R) with rubber bands. A plasma has a mind of its own. Fixing one leak reveals another one that you didn't see before. Nobel Laureate Irving Langmuir chose the unfortunate name *plasma*, which had already been adopted by the blood people. It means something that can be shaped or molded. Nothing can be further from the truth! But the problem has been solved, and the end is in sight.

Designing a Magnetic Bottle

What Is a Magnetic Field?

So we have found that the best way to produce fusion reactions in a continuous manner is to make a very hot plasma, so hot that it cannot be held in place by any material container. We also decided that of all the forces that we can use to make a wall-less container, only the magnetic force would work. What would a magnetic bottle look like? Actually, it looks like a bagel; but before we get to this, we have to review what we know about magnetic fields. Most people know that the earth has a magnetic field, as shown in Fig. 4.6. The lines with arrows show the direction of the field. A compass needle aligns itself with the field line that passes through it on the earth's surface, and therefore points toward the magnetic pole, which is close to the geographic pole. The earth's field is already a magnetic bottle, but an imperfect one. Protons and electrons coming from the sun in the solar wind[3] get trapped in this field because charged particles tend to move along field lines, not across them. But the trap has large leaks at the north and south poles where the field lines run into the ionosphere, bringing the particles with them. When electrons strike oxygen atoms in our atmosphere, visible light is emitted which we call the Aurora Borealis. Since the plasma particles can travel in either direction, the same thing happens in the southern hemisphere. The Aurora Australis is not as well known because few people stay out on a winter's night in Antarctica to watch it, and penguins have other agenda.

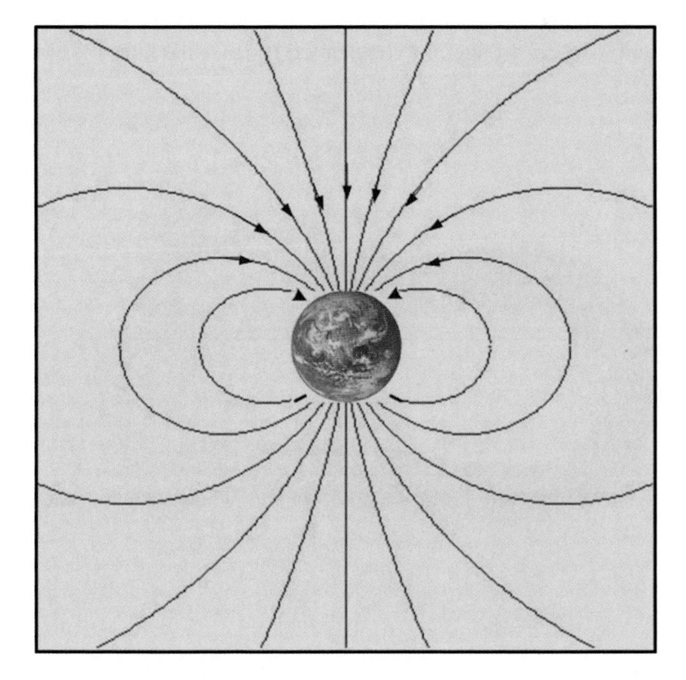

Fig. 4.6 The earth's magnetic field

Magnetic field lines are, of course, only a mathematical construct. Electric or magnetic fields can be detected only by the forces that they exert. It was the great Scottish physicist James Clerk Maxwell[4] who invented the concept of a "field" to describe action at a distance. Once a field at a given position is known, one can calculate the forces which that field would exert on an object there. To depict the shape of a field, one can draw any number of lines. A visual display of magnetic field lines is commonly given in textbooks, where the pattern of iron filings traces the field lines around a horseshoe magnet, as in Fig. 4.7.

Magnetic field lines are sometimes called "lines of force," but this is a misnomer. The magnetic force is actually *perpendicular* to the lines! A compass needle points north–south because, when it is not aligned, the north pole of the needle is pushed one way by the magnetic field of the earth, and the south pole the other way, until the needle is aligned with it. Similarly, each elongated iron filing in the horseshoe demonstration acts like a miniature compass needle and points in the direction of the field at its location. It is important to understand what a field line represents, because how a magnetic bottle works depends critically on how these lines are shaped.

The problem with permanent magnets is that the strongest magnetic field it generates is *inside* the iron of the magnet, where we cannot put any plasma. Fortunately, we can create magnetic fields with electromagnets. In Fig. 4.8a, we show the field around a bar magnet, which is a magnetized iron cylinder; it has basically the same shape as the earth's field. In Fig. 4.8b, we have replaced the iron bar with a glass tube of the same length and diameter, and we have wound many turns of wire around the tube. When we hook the wire up to a DC voltage source, such as a battery, the current in the wire generates a magnetic field *of the same shape* as that of the bar magnet! But now we can put plasma *inside* the glass tube, where the field is much stronger, as you can tell because the lines are closer together.

Fig. 4.7 The magnetic field of a horseshoe magnet as revealed by iron filings

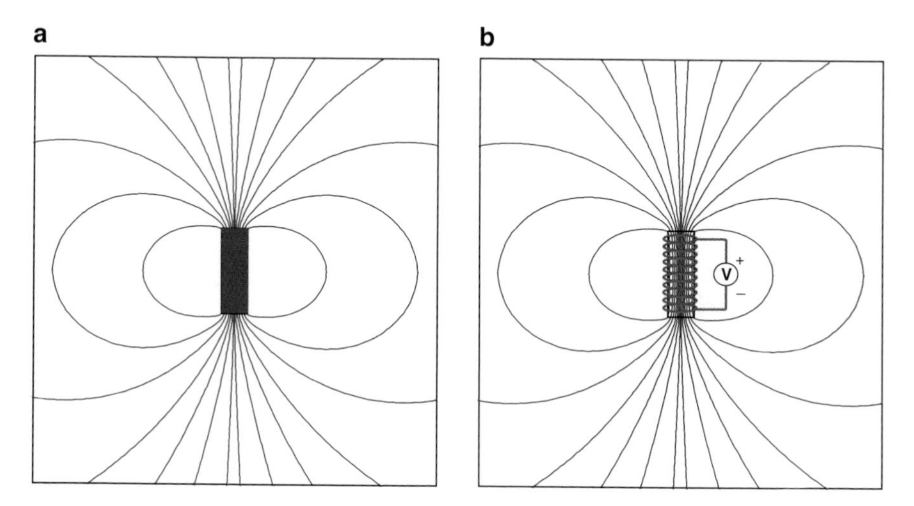

Fig. 4.8 The magnetic field around (**a**) a bar magnet and (**b**) an electromagnet of the same size

Now we can move on to see how to make a leak-proof magnetic bottle for plasma using cleverly shaped wire coils to produce field shapes that will plug all the leaks.

How Can a Magnetic Field Hold a Plasma?

When one puts a note on the refrigerator door with a magnet, one gets the impression that the attractive force is in the direction of the magnetic field. On the other hand, we said that the magnetic force is perpendicular to the field lines. Before we resolve this apparent contradiction, let's see what the magnetic force on a particle (an ion or electron) is supposed to be. The force is called the Lorentz force,[5] and it has five main features. (1) It acts only on particles with an electric charge. (2) It is proportional to the strength of the magnetic field, as one would expect. (3) It does not affect a particle that is stationary nor one that moves only along a field line. Only the perpendicular motion of a particle – that which takes it from one field line to an adjacent one – counts. (4) The force is perpendicular to both the particle velocity and the field line. (5) The force depends on the electric charge on the particle and is in opposite directions for positive and negative charges. This is a mouthful, but here is what it means. If a proton, say, is stationary, it feels no force. If the proton moves strictly along a field line, it also feels no force. If it moves across field lines, however, the magnetic field will push it, not backwards, but in a perpendicular direction. An ion and an electron both have the same charge, but of opposite signs, so the Lorentz force on them is in opposite directions. As we shall see, this will cause the protons and electrons to revolve in small circles around a field line. Refrigerator magnets seem to pull *along* the field, though. This is because permanent magnets are more complicated.[6]

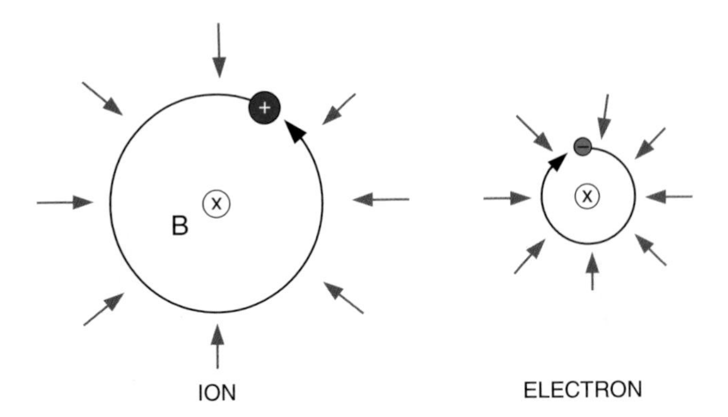

Fig. 4.9 Gyration orbits of an ion and an electron in a magnetic field B pointing into the page. The electron's orbit is greatly enlarged for clarity

A cartoon of the orbits of an ion and an electron in a magnetic field is shown in Fig. 4.9. The X in the center indicates that the magnetic field, labeled **B**, points into the page. The arrows indicate the Lorentz force, which is everywhere perpendicular to both the particle velocity and the magnetic field. If the velocity is constant, the force is inward everywhere with the same strength, so the orbits are circles. Note that the motions are in opposite directions because the charges have opposite signs. Imagine taking a yo-yo, stretching it out, and swinging it with a steady motion in a circle over your head. The string pulls the yo-yo inward with the same force at all times, so the yo-yo moves in a circle. Here, the magnetic field applies a force just like that of the string. This gyration orbit is called a *cyclotron orbit*, since the first cyclotrons used this principle to keep the protons inside a circular chamber. It is also called a Larmor orbit because in science you can get something named after you without paying a huge endowment. The radius of the circle is called its Larmor radius.

Since the magnetic force is always perpendicular to the field's direction, particles move in the parallel direction without being influenced by the magnetic field. A *magnetized* plasma, then, doesn't look like Fig. 4.4, where ions and electrons are free to move in any direction. Instead, it would look like Fig. 4.10, where the charged particles gyrate in their Larmor orbits and move unimpeded in the direction of the magnetic field **B**. Field lines are like invisible railroad tracks that guide the motion of charged particles.

How big is a Larmor orbit? In a cyclotron, the orbit is the size of a large laboratory because the protons have very large energies. In a fusion reactor, a deuteron has a Larmor radius of about 1 cm, when compared with a plasma radius of about a meter. An electron's orbit is much smaller than a deuteron's, even if it has the same energy. This is the result of two effects. With the same energy, an electron would move much faster because it is much less massive than a deuteron. So you would think that its orbit would be larger than a deuteron's. However, the Lorentz force that curves the orbit is stronger with higher velocity. The upshot is

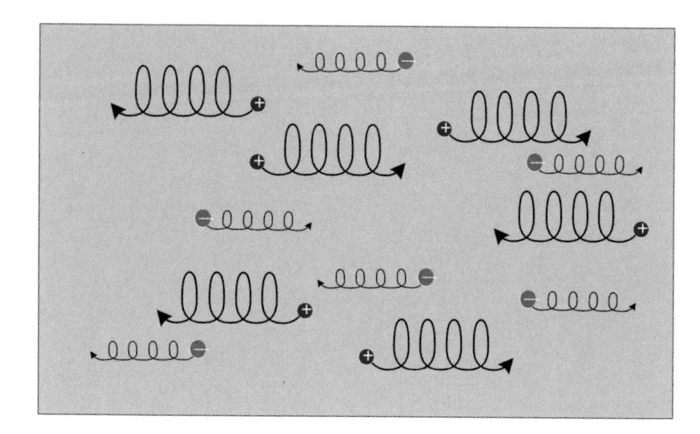

Fig. 4.10 A cartoon of a plasma in a magnetic field. Ions are *blue*, and electrons are *red*

that the electron's orbit is smaller by the square root of the mass ratio, or about 60 in this case. In Fig. 4.10, the electron orbit was greatly enlarged in order to be visible.

Since these gyration orbits are so much smaller than the plasma that they are immersed in, we don't have to track the particle motions in such detail. We only have to track the motion of the centers of the circles, which are called *guiding centers. In the future, when we talk about the motion of plasma particles, we will mean the motion of the guiding centers.*

We can now return to the question, "How can a magnetic field hold a plasma?" We have seen that a magnetic field does not apply a force to a particle that will stop it from following field lines, so field lines that end on a boundary somewhere cannot prevent a plasma from hitting a wall.[7] On the other hand, plasma cannot go across field lines because the magnetic force simply keeps charged particles spinning in small Larmor orbits around the same field line. Obviously, the solution is to make a field with lines that close on themselves and do not end. That's the very first step in designing a magnetic bottle!

The Hole in the Doughnut

Looking at a globe, we see lines that do not end. The latitude lines go in circles around the earth and end on themselves (Fig. 4.11). The longitude lines go north and south until they reach the poles, where they continue over to the other side of the earth (Fig. 4.12). Why can't we make a magnetic bottle shaped like a sphere with magnetic field lines that go either north–south or east–west? Here's why. If we look down at the north pole, say, in Fig. 4.11, we see that the field lines go around in smaller and smaller circles. As one gets closer to the pole, the magnetic field must get weaker and weaker, since the fields on opposite sides of the circle are in opposite directions and tend to cancel each other. Exactly at the pole, the field must be zero, since it cannot be in two directions at the same time. This is called an

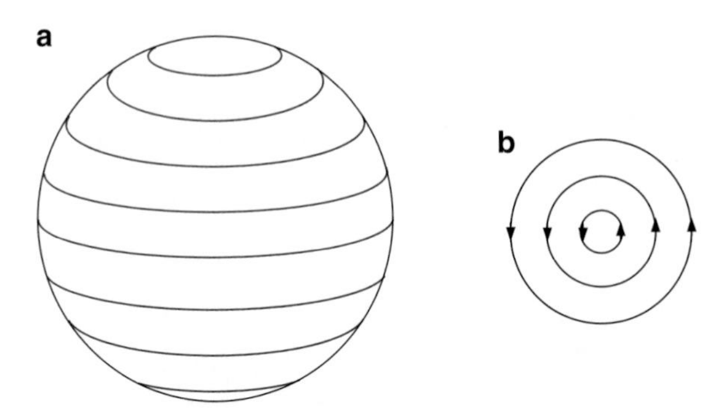

Fig. 4.11 A magnetic field with an O-point

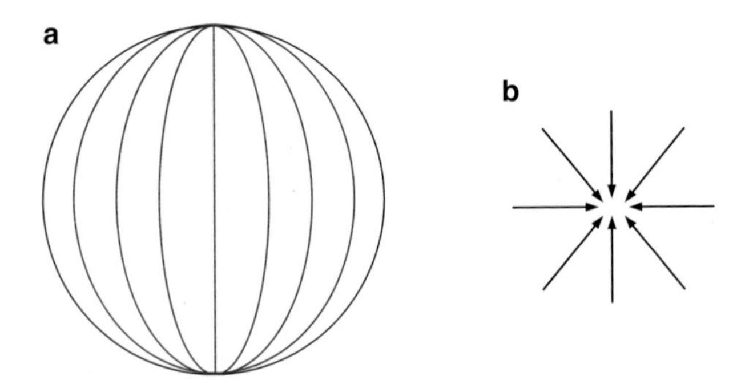

Fig. 4.12 A magnetic field with an X-point

O-type null. The plasma will leak out at the poles, since there is no magnetic field there to confine it. If we now look at a configuration in which the field lines are like longitude lines, Fig. 4.12 shows that the field lines point toward (or away from) one another at the poles, or cross one another at an angle. Again, the field at the pole must be zero, since it cannot be in two directions at once. This is called an X-type null. A simple shape that is topologically equivalent to a sphere cannot be made into a magnetic bottle. It will have a big leak at the poles, where there is no magnetic field to hold the plasma.

The simplest shape that will work is a torus, a three-dimensional volume like a tire or a doughnut, with a hole in it, as shown in Fig. 4.13. Mathematicians would call it a doubly connected space. Field lines that have no ends can be imbedded in such a chamber in such a way that ions and electrons cannot find a way out by moving along the field lines. Such closed field lines are of two types. *Toroidal* field lines, of which one is shown in Fig. 4.13a, go around the torus in the long way, encircling the hole. *Poloidal* field lines, shown in Fig. 4.13b, go around the short way and do not encircle the hole. Remember that field lines are just a graphic way to show the direction of the magnetic field. There is an infinite number of field lines. The torus is entirely filled with magnetic field, so that plasma placed inside will not, in principle, escape. The ion and electron guiding centers simply move along the field lines and never hit the wall, as long as the field lines they're on do not wander out of the torus.

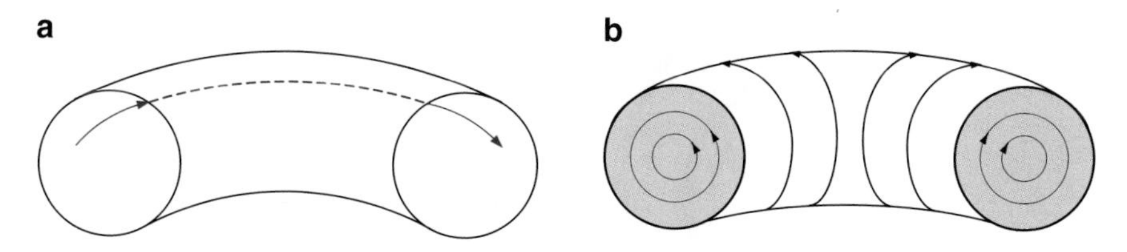

Fig. 4.13 Toroidal (**a**) and poloidal (**b**) closed field lines in a torus

Now imagine combining toroidal and poloidal fields into the same torus. A toroidal field line going around the long way will also bend the short way, like an old-fashioned barber pole or the stripes on a candy cane. The field line will look like a Slinky® toy stretched around a lamppost; it is a helix bent into a circle. The generic toroidal and poloidal types of magnetic field will not work. Combining them into a helix is the beginning of the art of making magnetic bottles. All this is necessary because magnetic fields do not stop particles from moving longitudinally, and therefore they must not end on a material wall.

Why the Field Lines Have to Be Twisted

If we had straight field lines in a cylinder, there would be no problem; but when we bend the cylinder into a torus so that the field lines do not strike a wall, the first of several toroidal effects comes into play. In Fig. 4.8b, we saw that an electromagnet generates a magnetic field by driving electric current in a coiled wire. The field lines are then formed inside the wire coil. When we bend the cylinder into a torus, as in Fig. 4.13a, the wire coil will also have to be bent to surround the torus, resulting in the configuration shown in Fig. 4.14. Each turn of the coil carries current in the direction of the arrows, generating a magnetic field purely in the toroidal direction.

Two of the field lines have been drawn. Notice how the coils crowd together as they go through the hole in the torus. The bunched current then creates a larger field at point A than at point B, which is farther from the doughnut hole. The magnetic field is always larger on the inside of a torus than on the outside. This is a toroidal effect that does not happen in a cylinder. The consequence of this effect is that charged particles no longer gyrate in perfect circles. Let's look at the orbit of an ion in the right-hand cross section of the torus in Fig. 4.15. Normally, it will gyrate clockwise in a circular orbit, but here its orbit has been distorted into a spiral. Remember that it is the Lorentz force that makes the ion gyrate, and this force is proportional to the magnetic field strength. On the left-hand side of the orbit, the ion will feel a stronger force than it does on the right-hand side, where the field is weaker, so it will turn more tightly on the inside. The result is that the ion's guiding center drifts downwards in this diagram. Observe that an electron drifts upwards because it has negative charge, and therefore gyrates in the opposite sense to that of the ion. This drift has been greatly exaggerated here, but nonetheless it has a huge effect on the plasma, collecting the positive charges on the bottom and the

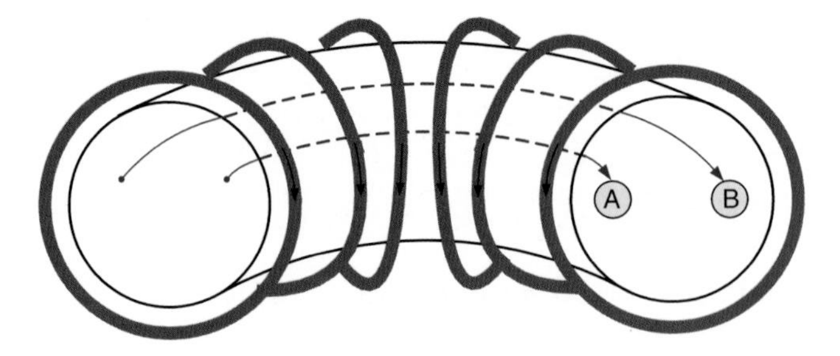

Fig. 4.14 Coils that generate a toroidal field

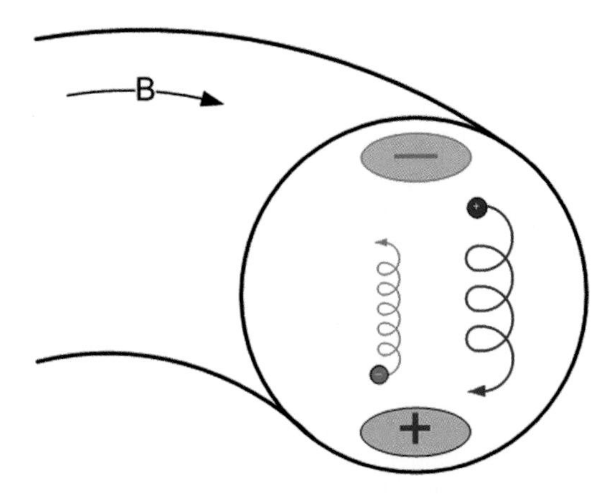

Fig. 4.15 Particle drifts in a torus. Ions are *blue*, the electrons *red*, and the size of the electron's orbit has been exaggerated

negative charges on the top, as shown. These charge bunches will create an electric field going from the positive charge bunch at the bottom to the negative one at the top. Such a vertical electric field, as we shall see, will blow the whole plasma out toward the outer wall. Such a simple magnetic bottle will not work!

This problem was recognized very early in the game. The famous astronomer Lyman Spitzer, Jr., was riding up the long ski lift at Garmisch-Partenkirchen when he thought of a solution. Incidentally, Spitzer was the prime mover in getting the Hubble telescope built. It was not named after him, but after his death a Spitzer telescope was finally put into orbit. Spitzer's solution was to twist the torus into a pretzel shape, as in Fig. 4.16. If you were a particle traveling along the depicted field line starting at B, you would feel a stronger magnetic field on the left than on the right. When you reach A, the strong field is on the right, now that the torus has been twisted. This is different from the circular torus of Fig. 4.14, where the strong field is always on the same side. Let's look at the two cross sections in Fig. 4.16 in more detail. These are shown larger in Fig. 4.17. Cross section A is the same as that in Fig. 4.15, with the magnetic field pointing out of the page and with the ions drifting downwards. In cross section B, on the opposite side, the field *also points out of the page* instead of into the page, as in a circular torus. The fat arrows are

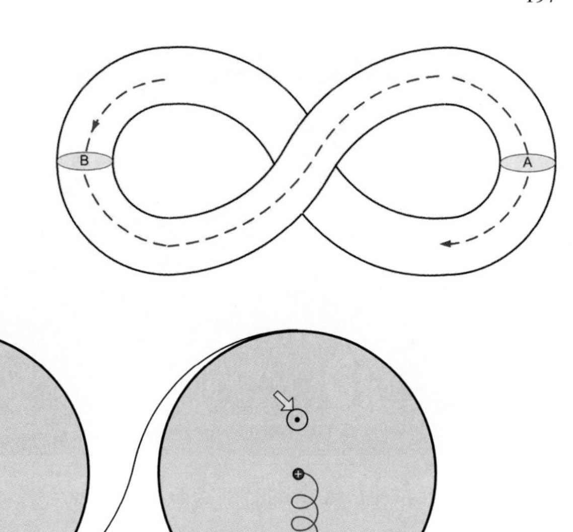

Fig. 4.16 A twisted torus

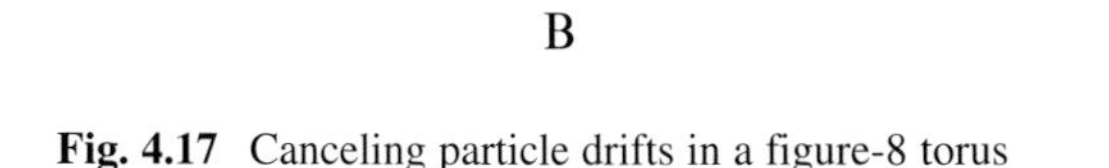

Fig. 4.17 Canceling particle drifts in a figure-8 torus

supposed to show this. With the field in the same direction, the ion gyrates clockwise in B, as it does in A. In B, however, the strong field is on the right side of the orbit, so the ions drift upwards instead of downwards. The vertical drifts then cancel as the ion moves along the figure-8 along a field line, and the catastrophic separation of charges, in principle, does not occur.

This type of magnetic bottle was named a *stellarator* by Spitzer because it was intended to reproduce the conditions inside stars which allow them to generate fusion energy. A series of a half-dozen figure-8 stellarators was built at the Plasma Physics Laboratory in Princeton University in the 1950s to test this confinement idea. A model of a figure-8 stellarator (Fig. 4.18) was shown at the 1958 Atoms for-Peace conference in Geneva, in which thermonuclear fusion was declassified and different nations showed off their inventions. The individual coils carrying the current to generate the magnetic field can clearly be seen in this model. There was also an electron gun inside the chamber that could emit electrons that visibly traced the magnetic field lines. In addition to this model, an entire real, working stellarator was shipped to Geneva and reassembled there in the US exhibit. The Russians proudly displayed their Sputnik satellite, but their fusion exhibit was an unimpressive, unintelligible black hunk of iron called a *tokamak*. It was only many years later that the world realized that *that* was the real star of the show.

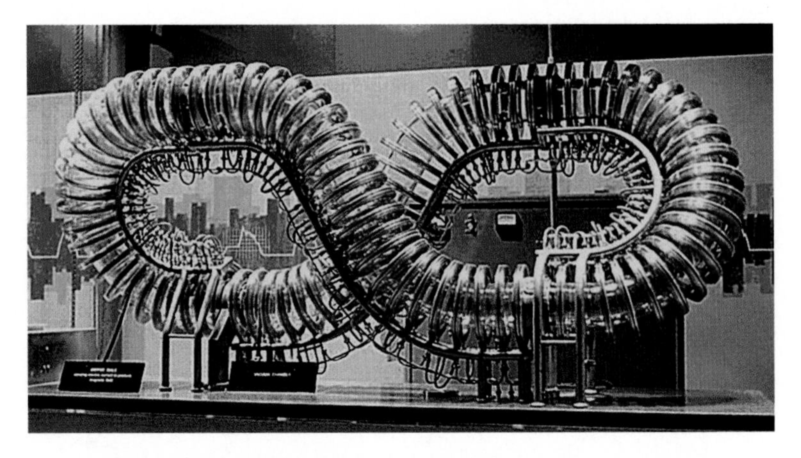

Fig. 4.18 A demonstration model of a figure-8 stellarator

Mappings, Chaos, and Magnetic Surfaces

Figure-8 stellarators are hard to make, especially since the coils have to be accurate enough to keep the field lines from wandering out to the walls.[8] It was soon realized, however, that the necessary twist of the field lines can be produced without twisting the entire torus. We mentioned that the field lines in a toroidal magnetic bottle are twisted like the stripes on a candy cane. The way to produce such helical field lines can be visualized more easily if we decompose them into *toroidal* lines, as in Fig. 4.13a, and *poloidal* lines, as in Fig. 4.13b. Adding these two types of fields together will result in a field with helical field lines. To produce the toroidal part of the field, we can use coils like those in Fig. 4.14. Now we want to add coils that will produce the poloidal field. Figure 4.19 shows how this is done. Let there be a number of toroidal hoops placed all around the torus; two of these are shown in Fig. 4.19. If each hoop carries a current in the toroidal direction, as shown by the horizontal arrows, it will produce a magnetic field around itself in the direction shown by the arrows on the small circles around each hoop. The part of this field that extends into the plasma will be mostly in the poloidal direction. Imagine that there are an infinite number of these hoops covering the surface of the torus. Their fields inside the plasma will add up to give a purely poloidal field, as shown by the dashed arrows.

You have no doubt noticed the complementarity here: poloidal windings create toroidal magnetic fields (Fig. 4.14), and toroidal windings create poloidal fields (Fig. 4.19). In the same way that the toroidal and poloidal fields add up inside the torus to make helical field lines, the poloidal and toroidal windings can be combined into a helical winding! One turn of such a winding is shown in Fig. 4.20. The dotted line is a helical field line. Because it contains both toroidal and poloidal components, it may start near the top and then go to the bottom in another cross section. Now look at what an ion does.[9] On the right, an ion starts drifting

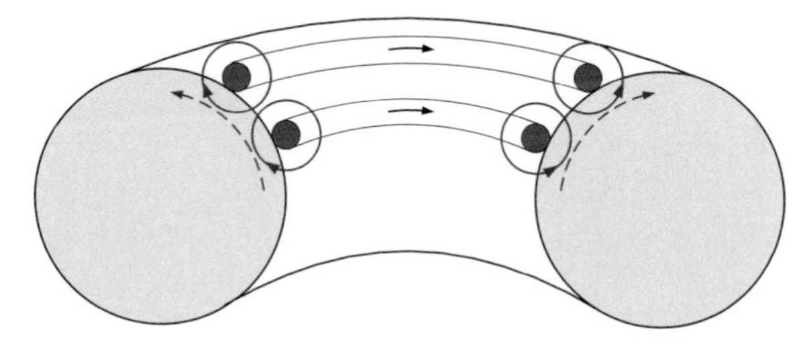

Fig. 4.19 Generation of poloidal fields with coils

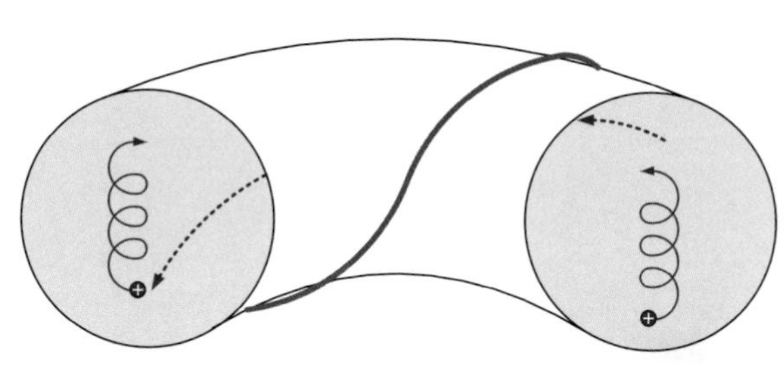

Fig. 4.20 Example of a helical winding and particle drifts in a helical field

upwards – not downwards, as in Fig. 4.17 – because here I have drawn the magnetic field going *into* the page instead of *out of* the page. When the ion reaches the left side, it is still drifting upwards – not downwards as in a figure-8 stellarator – but this is fine, because the ion is now near the bottom, and an upward drift will bring it back away from the wall. So there are two ways to skin the cat. Either a figure-8 stellarator or a stellarator with helical field lines made by helical windings can cancel the dreaded vertical drift of ions and electrons caused by bending a cylinder into a torus.

We started with the concept that field lines have to end on themselves so that particles moving along them will never leave the magnetic trap. Of course, the field lines do not have to meet their own tails exactly. All that is required is that the line never hits the wall. In general, field lines do *not* close on themselves. Rather, they come back to the same cross section in a different position after going around the torus the long way. This is illustrated in Fig. 4.21. An imaginary glass sheet has been cut through the torus so that we can see where the field lines strike this cross section. Let's assume that a field line intersects this cross section at position 1. After going around the torus once, it might intersect at position 2. On successive passes, its position might be 3, 4, 5, 6, etc. On the seventh pass, the field line almost comes back to position 1, but it does not have to. One can define a mapping function such that for every position on that plane, there is a definite position for the next pass. Thus, whenever the line goes through position 2, it will pass near position 3 the next time. The line does not ever have to come back to itself. It can cover the entire cross section randomly, and the plasma will still be confined as long as the line never hits the wall.

At this point, we should define a quantity that will be very useful for understanding twisted magnetic fields: the *rotational transform*. This is the average number of times a field line goes the short way around a cross section for each time

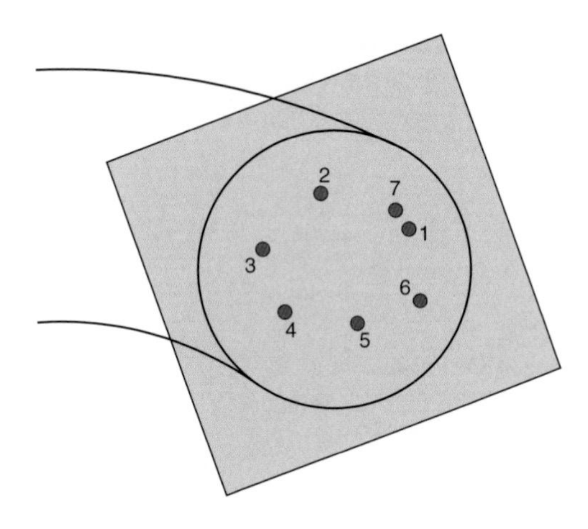

Fig. 4.21 Mapping of a field line

it goes the long way around the whole torus. In Fig. 4.21, suppose pass No. 7 fell exactly on pass No. 1, then it took six trips around the torus for the field line to make one trip around the cross section. The rotational transform is then about one-sixth. The field line does not have to trace a perfect circle in the cross section, and the crossings do not have to be evenly spaced. The rotational transform is an average that more or less measures the amount of twist.

You have no doubt heard of fractals and chaos theory, topics that have been developed since the invention of fast computers. It was the mapping of field lines in magnetic bottles that gave impetus to the development of these concepts. Ideally, with well designed and fabricated windings for creating the magnetic field, the locus of intersection points in a stellarator can be perfect circles, with the field lines coming back to a different angle on the circle each time. With a finite number of turns on the helical coil instead of an infinite number, the circle can be distorted into, say, a triangle; but a field line will come back to the same triangle on each pass. In real life, magnet coils are not made perfectly, and there are small perturbations. These can cause wild behavior in the map, causing strange attractors, where the points tend to clump at a particular place; or magnetic islands, which we will discuss later; or complete chaos in the way the points are distributed. The name of the game in stellarators is to create *nested magnetic surfaces*, in which the magnetic lines always stay on the same surface and intersect each cross section on the same curve. An idealized case is shown in Fig. 4.22. Once created on a magnetic surface, an ion or electron stays on that surface as it goes around the torus thousands or millions of times. The surfaces do not have to be circles, but they never touch or overlap, so the plasma remains trapped by the magnetic field.

A stellarator requires such precision in its manufacture that in the early days they could not hold a plasma very long. In the next chapter, we shall introduce the tokamak. This is a torus, of course, since it has to be doubly connected; but its poloidal field is not generated by external coils but by a current in the plasma itself. This allows it to have self-healing features which can overcome small imperfections in its construction.

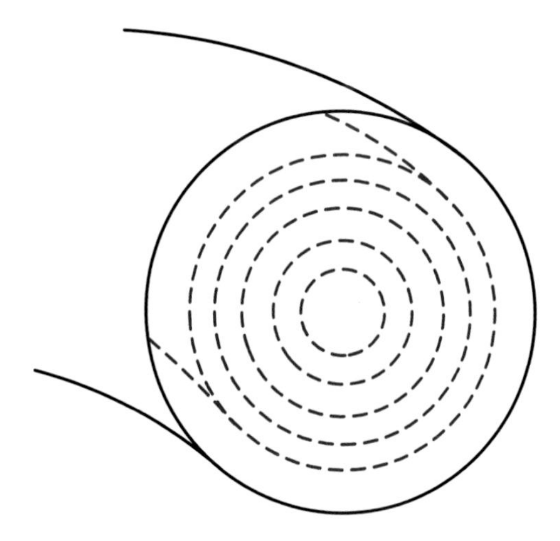

Fig. 4.22 Nested magnetic surfaces. A particle stays on its surface as it goes around and around the torus

Notes

1. The difference between "weight" and "mass" is purposely ignored at this point.
2. 1 K is the same size as 1°C (also called Celsius) and about twice (actually 9/5) as large as 1°F. The only difference between kelvin and degree centigrade is that kelvin is measured from absolute zero (–273°C), while degree centigrade is measured from the freezing point of water. Fahrenheit is measured from an archaic point such that water freezes at 32°F. At –40°, the temperature is the same in both °C and °F. When we are dealing with millions of degrees, the 273° difference between K and °C is totally insignificant, and even the factor of 2 difference between °K and °F can be ignored unless you are a scientist. The average person couldn't care less whether the sun is at 1 million degrees or 2 million degrees; it is just hot beyond comprehension.
3. These are the particles of the van Allen belt.
4. "Clerk" is pronounced "Clark."
5. For those who prefer a formula, the Lorentz force is $\mathbf{F}_L = q(\mathbf{v} \times \mathbf{B})$, where q is the charge and $\mathbf{v} \times \mathbf{B}$ is the cross-product between the vectors for the particle velocity and magnetic field, respectively.
6. Refrigerator magnets seem to have a force in the direction of the magnetic field coming out of them. The reason permanent magnets do not move sideways like a charged particle is that they are macroscopic objects which feel the sum of the forces on all the individual atoms in them. If we cut the magnet in Fig. 4.8a horizontally, we get two bar magnets which attract each other directly. But permanent magnets are made up of small current loops like the large one in Fig. 4.8b. Suppose we divide that large coil into two coils, one on top of the other, with a gap in between, each coil represents one permanent magnet. Now consider the force between the loops just above or below the gap. The electrons inside the wire carry the current in the circular (azimuthal) direction. The magnetic field of the upper loop flares out so that, *at the position of the lower loop*, it is partly in the radial direction. The Lorentz force on the electrons in the lower loop is then perpendicular to both the azimuthal and the radial directions, and it is therefore in the vertical direction. The two coils will then attract each other. Thus, the force *appears* to be in the direction of the field of the entire system.
7. Magnetic mirrors are an exception. They will be described in Chap. 10.

8. Modern stellarators are still a major option being developed. These will be described in Chap. 10.
9. For those who want to follow this more closely, the direction an ion gyrates in its cyclotron orbit is given by the right-hand rule. If the thumb of the right-hand points along the magnetic field, curled fingers will point in the direction an ion gyrates. Electrons will gyrate in the opposite direction, to the delight of lefties. Similarly, if the thumb points in the direction of a current, the magnetic field it generates will point in the direction of the fingers.

Chapter 5
Perfecting the Magnetic Bottle*

Some Very Large Numbers

The last chapter had a lot of information in it, so let us recapitulate. To get energy from the fusion of hydrogen into helium as occurs in the sun and other stars, we have to make a plasma of ionized hydrogen and electrons and hold it in a magnetic bottle, since the plasma will be too much hot to be held by any solid material. The way a magnetic field holds plasma particles is to make them turn in tight circles, called Larmor orbits, so that they cannot move sideways across magnetic field lines. However, the ions and electrons can move along the field lines in their thermal motions without restraint. Consequently, the magnetic container has to be shaped like a doughnut, a torus, so that the field lines can go around and around without ever running into the walls. The field lines also have to be twisted into helices to avoid a vertical drift of the particles that occurs in a torus but not in a straight cylinder. Ideally, each field line will trace out a magnetic surface as it goes around many times without ever coming back exactly on itself. The plasma is then confined on nested magnetic surfaces which never touch the wall. This ideal picture will be modified in this and later chapters as we understand more about the nature of these invisible, nonmaterial containers.

We've gotten an idea of what a magnetic bottle looks like, but how large, how strong, or how precise does it have to be? The sun, after all, has a tremendous gravitational force to hold its plasma together; but we on earth have much more limited resources. The size of a fusion reactor will be large if it is to produce backbone power. The torus itself may be 10 meters in diameter, and the reactor with all its components will fill a large four-story building. A better picture will be given in the engineering section later in this book. For experiments on plasma confinement, however, much smaller machines have been used. The figure-8 stellarators, for instance, were only about 3 meters long. Modern torus experiments are about half or a quarter the size of a reactor.

The temperature of the plasma in the interior of the sun is about 15,000,000 (1.5×10^{7}) degrees, but a fusion reactor will need to be about ten times hotter, or

*Numbers in superscripts indicate Notes and square brackets [] indicate References at the end of this chapter.

F.F. Chen, *An Indispensable Truth: How Fusion Power Can Save the Planet*,
DOI 10.1007/978-1-4419-7820-2_5,

150,000,000 (1.5×10^8) degrees. We can use the electron-volt (eV) to make these numbers easier to deal with. Remember that 1 eV is about the amount of energy that holds a molecule together. Remember also that the temperature of a gas is related to the average energy of the molecules in the gas. It turns out that 1 eV is the average energy of particles in a gas at 11,600 K or roughly 10,000 K. So instead of saying 150,000,000°, we can say that the temperature is 15,000 eV or 15 keV. By that we mean that the particle energies in the gas are of the order of 15 keV. When we say degrees, do we mean Fahrenheit, Centigrade, or Kelvin (absolute)? For this discussion, it doesn't matter, since Fahrenheit and Centigrade degrees differ by less than a factor of 2, and Centigrade and Kelvin differ by only 273°. We do not really care whether the sun is at 10 million or 20 million degrees! It makes a difference to scientists, who use degrees Kelvin, but not for this general overview.

Why do we need a plasma temperature as high as 10 keV? This is because positive ions repel one another with their electric fields, and they must have enough energy to crash through the so-called Coulomb barrier before they can get close enough to fuse together. In Chap. 3, we discussed why a hot plasma is a better solution than the beams of fast ions. Here, we give more details. Figure 5.1 shows a graph of the probability of deuterium–tritium fusion plotted against the temperature of the ions in keV.[1] Note that the probability peaks at around 60 keV, but the ion temperature does not have to be that high because the ions have a Gaussian distribution of energies. When the ions are at 10 keV, there are enough ions in the tail of the distribution (Fig. 3.3), near 40 keV, which fuse rapidly enough. Note that at the sun's 2 keV, the reactivity is very low; so low that ions stick around for millions of years before they undergo fusion. But on earth we do not have that kind of time!

Exactly how much time do we have? A magnetic bottle cannot hold a plasma forever because a plasma will always find a way to escape. From Fig. 5.1, we see that the lower the temperature, the slower is the fusion rate, so the plasma containment time has to be longer. The relation among plasma density (n for short), ion temperature (T_i for short), and confinement time t was originally worked out by J.D. Lawson and is commonly known as the *Lawson Criterion*. A modified form of this

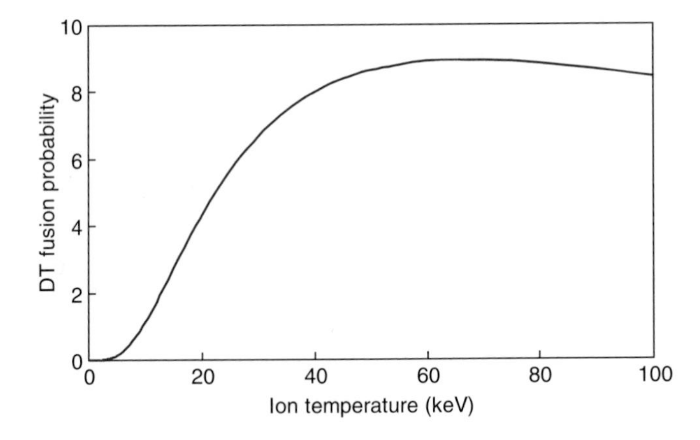

Fig. 5.1 Probability of DT fusion vs. ion temperature

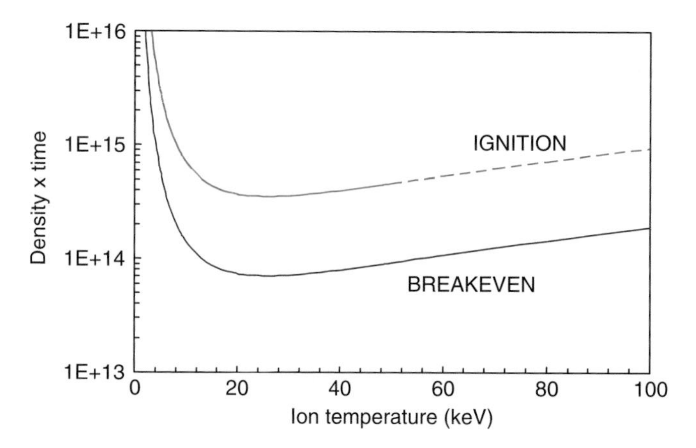

Fig. 5.2 Lawson criterion for DT fusion (For those already familiar with fusion, the ordinate is actually $n\tau_E$, the *energy* confinement time in units of s/cm^3. The curves were recomputed using the modern data of Bosch and Hale [1] and assuming a thermal conversion efficiency of 30%. The time τ_E is more honest than the *particle* confinement time, called *t* here, because it includes losses in the form of electromagnetic radiation.)

is shown in Fig. 5.2. The criterion says that the product of density and confinement time – that is, $n \times t$ – has to be higher than a value that varies with T_i. There are two curves. The lower one, marked BREAKEVEN, stands for *scientific* breakeven, in which the fusion energy just balances the energy needed to create the plasma. Real breakeven would include all the power needed to run the rest of the plant, requiring higher *nt*. The upper curve, labeled IGNITION, is the *nt* required for a self-sustaining plasma, in which the plasma heats itself without additional energy. That happens because one of the products of a DT reaction (see Fig. 3.2) is a charged α-particle (a helium nucleus), which is trapped by the magnetic field and stays in the plasma keeping the D's and T's hot with its share of the fusion energy. Clearly, the goal of fusion research is to reach ignition, and present plans are to build an experiment that can generate enough α-particles to see how they thermalize.

Now we can answer the question as to how long we must hold the plasma. The breakeven curve in Fig. 5.2 says that *nt* must be *at least* 10^{14} sec/cm^3 (marked as 1E+14 on the graph). A reasonable value for the plasma density *n* is 10^{14}/cm^3 (100 trillion ion–electron pairs per cubic centimeter). Therefore, *t* is of the order of 1 sec. We have to hold the plasma energy in a magnetic bottle for at least 1 sec, not a million years, as in the sun. This has already been achieved, albeit not at such a high density. The progress in fusion can be appreciated when one recalls that the confinement in figure-8 stellarators was about 1 *micro*second. Our work has paid off a million-fold.

To confine the plasma in a stellarator, the magnetic field has to be carefully made. Figure 5.3 shows the average distance an ion travels, in kilometers, before it makes a fusion collision.[2] The curve is lowest at ion energies of around 60 keV, since the fusion probability in Fig. 5.1 peaks there. At the more normal energy of 40 keV, as explained above, an ion covers about the circumference of the earth as it goes around and around a torus! One might think that a magnetic bottle cannot be made this accurately, but it turns out that confining *single particles* is not a problem.

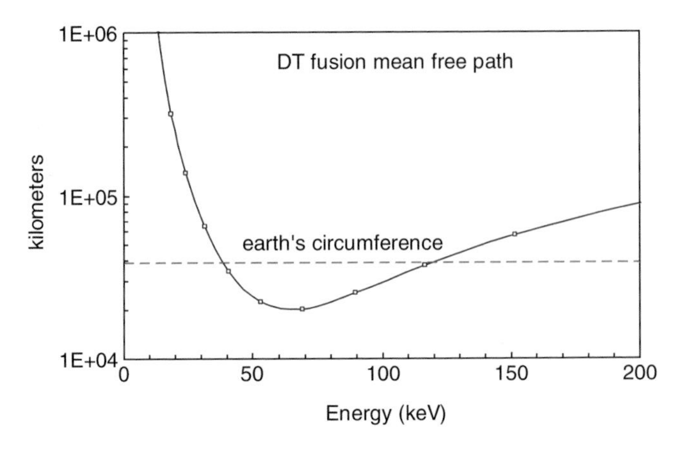

Fig. 5.3 Ion mean free path for fusion vs. energy, at a density of 10^{14}/cm^3

After all, storage rings in atom smashers can hold protons for hours or even days. Toroidal fusion experiments do not use focusing magnets as in particle accelerators, but electrons have been shown to be confined for millions of turns even in a primitive stellarator [3]. To hold a *plasma* is much harder because the ions and electrons can cooperate with one another to form their own escape paths. The accuracy of the magnetic field is not the problem.

So far we have considered the shape of the magnetic field but not its strength. A hot gas like a plasma exerts a lot of pressure, and the magnetic bottle has to be strong enough to hold this pressure. How does this pressure compare with our everyday experience? Pressure is density times temperature. Let's first talk about temperature. Room temperature is about 300 K. Expressed in electron-volts, this is 300/11,600 = 0.026 eV. A fusion plasma has a temperature of, say, 15 keV, about 600,000 times higher. Fortunately, the density is much lower. Atmospheric density is about 3×10^{19} molecules/cm^3, while a fusion plasma has about 2×10^{14} particles/cm^3, about 150,000 times fewer. So the net result is that the magnetic field has to hold a pressure about 600,000/150,000 times higher than normal: roughly 4 atmospheres (atm). This is the pressure at which water comes out of a faucet or that felt by a diver at 40 m depth, as one can figure out from the well-known fact that atmospheric pressure is about 1 kg/cm^2 or 15 lbs./sq.in.^{2}. Four atmospheres is not a huge number, but the pressure has to be exerted by a massless magnetic field! The strength of a magnetic field is measured in Teslas (T), each Tesla being 10,000 gauss (G), which may be a more familiar unit to old-timers. A magnetic field can exert a pressure of about 4 atm/T. Thus, the field strength required to hold a fusion plasma is about 1 T (10,000 G). This is a conservative number, because actual machines go up above 3 T. This is to be compared with the earth's magnetic field, which is about 0.5 G, or with the strength of a memo-holding refrigerator magnet, about 40 G. However, MRI (Magnetic Resonance Imaging) machines need about 1 T. You can hear the field during an MRI exam because the field has to be oscillated, causing parts of the machine to rattle and hum. To create a 1-T field requires large, heavy "coils" consisting of copper windings or superconductors imbedded in a solid material to hold them in place. This is not a problem and is routinely done in fusion experiments. Though it is the magnetic field that applies pressure to the plasma, the field is held in place by the current-carrying coils; and it is the coils that ultimately bear the pressure. This is also not a problem because the coils have to be made quite sturdily in any case.

Instabilities: The Fly in the Ointment

So far, we have encountered no insuperable problems. We can build a torus with a helical magnetic field and nested magnetic surfaces which should contain a plasma. We know how to make coils that will generate the 1-T magnetic field to hold the plasma pressure. Even if the plasma pressure is higher than 3–4 atm, a mere doubling of the field strength to 2 T will hold *four times* as much plasma because the field pressure increases as the *square* of the field strength. That toroidal fields which hold single particles for millions of traverses around a torus was shown very early in the game [3]. As we shall see, the Lawson criterion on the *nt* product would be easy to attain if a plasma behaved like a normal gas. The problem is that plasma is a special kind of gas, and an ornery one at that.

We said before that "plasma" is a misnomer because a plasma is not easily shaped or formed. Nature abhors a vacuum. A magnetic field is a vacuum. Plasma will try to cross the field and expand to fill the material container. Although the magnetic field keeps each ion and electron spiraling in a Larmor orbit so that each particle *by itself* cannot cross the field lines, the ensemble of particles can form ways to escape. This is because the particles are charged and can clump together to create electric fields, and these electric fields can take plasma across field lines. The plasma behaves more like a fluid (air or water, for instance) than like a collection of particles, each acting by itself. Since the particles are charged, a plasma can pull Houdini tricks that air or water cannot. Like an ant colony, the community can accomplish more than the individuals. Metaphors aside, these escape mechanisms are called *instabilities*, which are responsible for the slow progress in fusion up to now, and which are the subject of most of the technical literature on plasma physics. Before we can describe instabilities, we have to tell more about how a plasma behaves.

Hot Plasma as a Superconductor

Ions and electrons will collide with one another, but not like billiard balls because they have electrical charges. Like charges repel, so an ion approaching another ion will feel the repulsion well before they come close and will veer off. There is an occasional head-on fusion collision, of course, but these are very infrequent. The result of the more distant collisions is to form the most probable distribution of velocities; namely, the Maxwellian distribution shown in Fig. 3.3. Electrons will do the same thing, only faster because they are lighter and move faster at the same energy. So their velocities will also fall into a Maxwellian distribution. However, it does not have to be at the same temperature as the ion distribution. The way we heat a plasma usually heats one species preferentially. For instance, driving a current through a plasma will preferentially heat the electrons, so that the electron temperature, called T_e, will be higher than the ion temperature, T_i. A plasma can have two different temperatures, T_e and T_i, at the same time, or even more if there are other species in the plasma. It may seem unusual that a plasma can have two temperatures at the same time, but imagine turning on the heat in a cold room. The air will get hot first, while the furniture stays cold. It will take some time for everything to come to the same temperature. Though ions and electrons in a plasma are intermixed, they exchange their heat comparatively slowly because they collide infrequently and have vastly different masses. Plasma particles are always being regenerated as they leave the container, and usually they leave before they can come into equilibrium with other species, so it is normal for T_i to be different from T_e.

When an electron collides with an ion, their opposite charges attract, and the electron will orbit the ion the way a comet orbits the sun. These collisions will tend to equalize T_e and T_i, but it takes much longer because an electron is so much lighter than an ion that very little energy is exchanged at each collision. Generally, particles do not stay in the plasma long enough for T_e and T_i to equalize, so the temperatures are usually different.

What do we mean by a collision when particles do not actually touch? The magnitude, so to speak, of a collision depends on how much the particles' paths have been deflected or how much their energies have changed. In this type of collision at a distance, each particle feels the electric field of the other particle during the time when they are close. This time becomes very short when the particles are moving fast. An electron with 10 keV of energy, for instance, will go past an ion so fast that there is hardly any time for the ion's electric field to deflect the electron or change its energy. It makes sense, therefore, that a hot plasma, whose particles have large velocities, hardly makes any collisions at all; in other words, it is a superconductor. Even plasmas with only 100-eV temperature can act like superconductors. We call these *collisionless plasmas*. Being able to neglect collisions makes theory much simpler, and most of the early work concerned collisionless plasmas. In most cases, this was a good approximation, since an electron can travel around a torus many times before it makes an effective collision. Later in the development of magnetic confinement, people finally realized that these weak collisions cannot be neglected after all.

How Plasma Moves in Electric Fields

In Chap. 4, we saw that the guiding centers of ions and electrons gyrating in a toroidal magnetic field have vertical drifts because the field is nonuniform; that is, it varies horizontally. The reason is that the particle feels a different magnetic field on

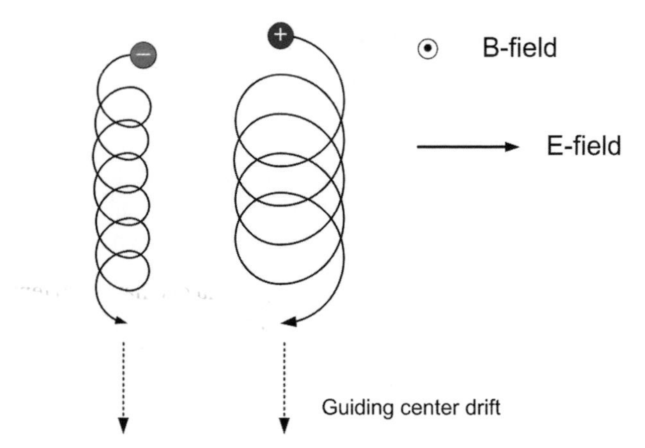

Fig. 5.4 Illustrating the drift of an electron (*left*) and an ion (*right*) in crossed electric and magnetic fields

each side of its Larmor orbit. A similar effect occurs in the presence of an *electric* field. This is shown in Fig. 5.4. There, the magnetic field (B-field) is coming out of the paper, and the electric field (E-field) points from left to right. Consider first the positive ion. It tries to follow its usual circular path, but it is pushed to the right by the E-field. Having higher energy, its orbit becomes larger. As it cycles back to the left, it moves against the E-field and is slowed up, so its orbit is smaller on the left side. This clearly causes the center of the orbit, the guiding center, to drift downwards. Now consider the electron on the left. Since it has opposite charge, it gyrates counterclockwise instead of clockwise, and is pushed to the left instead of to the right by the E-field. The result is that it also drifts downwards. Furthermore, since it is lighter and moves faster than the ion, it executes more orbits in the same time interval, ending up with exactly the same downward drift! The result is that particles have an $\mathbf{E}\times\mathbf{B}$ (E-cross-B) drift that is *perpendicular to both the B-field and the E-field, and which has the same speed and same direction for ions and electrons regardless of their energie*s.

It may seem strange that when you push in one direction, the particle goes in a perpendicular direction, but this effect is the same as that in a toy gyroscope. When the gyroscope tips down from vertical, gravity pulls it downwards, but the gyroscope precesses horizontally. If you follow a point on the rotating ring, you will see that under gravitational pull the whole ring will move sideways, just as do the orbits in Fig. 5.4. The same effect causes a rolling hoop to go a long way before falling over. When the hoop starts to lean over to the left, say, gravity will pull the hoop downwards, and the gyroscopic effect will turn the hoop to the left, so that it travels in a direction that will straighten it up. The front wheel of a bicycle also benefits from this effect, but only in a small way. There are stronger stabilizing forces in a bicycle.

The Rayleigh–Taylor Instability

When you turn a bottle of mineral water upside down, the water falls out even though the atmospheric pressure of 15 lbs./sq.in is certainly strong enough to support the water. This happens because of an instability called the Rayleigh–Taylor instability, which is illustrated in Fig. 5.5. If the bottom surface of the water remained perfectly flat, it would be held by the atmospheric pressure. However, if there is a small ripple on the surface, there is slightly less water pressing on the top of the ripple than elsewhere, and the balance between the weight of the water above the ripple and the atmospheric pressure is upset. The larger the ripple grows, the greater is the unbalance, and the ripple grows faster. Eventually, it grows into a large bubble which rises to the top, allowing water to flow out under it. If you hold the end of a straw filled with water, the water does not fall out because surface tension prevents the interface from deforming like that. A similar instability occurs in a plasma held by *magnetic* pressure, as we'll soon see.

Instabilities occur because of positive feedback. There are many examples of this in real life. Microphone screech occurs because the loudspeaker feeds into the microphone the tone to which it is most sensitive. The audio system amplifies that tone, making it louder in the speaker, which then drives the microphone harder. Forest fires are instabilities. A small fire dries the wood around it so that it catches fire more readily. The larger fire then dries a larger amount of wood near it, which then starts to burn, and the instability spreads like… a wildfire. Stock market instabilities can go both ways, as a rise or fall in the market induces more people to buy or sell. A more subtle instability creates snow cups when a field of snow melts or sublimes, as can be seen in Fig. 5.6.

If the sun shines evenly on a perfectly flat surface of snow, it should melt evenly, retaining a smooth surface. It never does, because there are ripples in the snow. A depression in the snow will cause some sunlight to scatter onto its walls, heating them before reflecting out into space. The deeper the hole is, the more light will deposit energy into it to hasten the melting. A snow cup can be started by a twig or pebble, which, being dark, will absorb more heat. But instabilities will always start and grow because there is always some imperfection or noise in the system. It just takes longer if the system starts out being almost perfect.

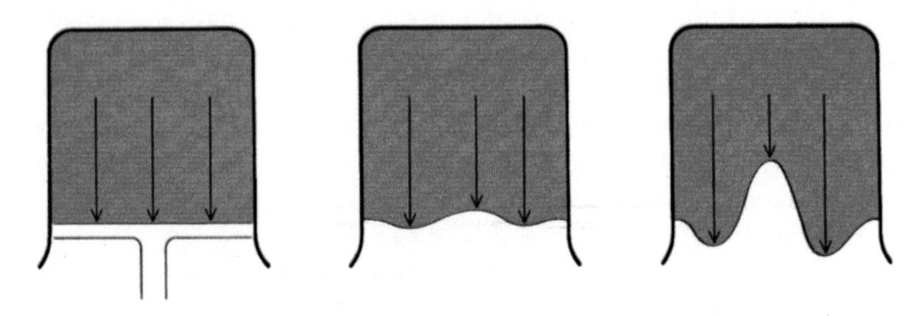

Fig. 5.5 Development of a Rayleigh–Taylor instability

Fig. 5.6 Snow cups: an instability in melting snow

The main obstacle to making a leak-proof magnetic bottle is instability. There are many instabilities, and the first step is to know your enemy. This first instability, however, was known from the beginning because it is similar to the well-known Rayleigh–Taylor instability in hydrodynamics. A plasma weighs almost nothing, so the instability is driven not by gravity but by pressure. To see how this works, we have to extend the concept of $\mathbf{E}\times\mathbf{B}$ drifts to drifts caused by other forces. Or, you can skip the next two diagrams and move on to see how this instability is stabilized.

Fig. 5.7a is the same as Fig. 5.4 except that the small gyrations have been suppressed, and only the guiding center drift due to an electric field is shown. In Parts (b) and (c), the E-field has been turned to different directions, and the drifts have rotated correspondingly. In Part (c), the E-field applies a downward force on the ions. If we apply to the ions another type of downward force, such as a pressure force, the ions will also drift to the left, as shown in Fig. 5.7d. Note that the electrons and ions now drift in opposite directions. The reason that the electric-field drifts are the same for both species is that both the electric force and the Lorentz force of the magnetic field depend on the sign of the charge, and these two dependences cancel each other. The pressure force, however, is in the same direction regardless of charge, so this cancelation does not take place, and the pressure drift depends on the sign of the charge.

Figure 5.8a shows a part of the plasma boundary when it is perfectly smooth, like the first drawing of a water bottle in Fig. 5.5. The upper part is plasma, and the lower part is vacuum, containing only the magnetic field. The plasma pressure is held back by the magnetic pressure, just as the water in Fig. 5.5 is supported by the atmospheric pressure. The force that now tries to push the denser fluid into the less dense fluid is now the plasma pressure rather than gravity. The pressure force, according to Fig. 5.7, causes ions to drift to the left and electrons to the right. As long as the plasma surface is straight and smooth, these drifts are perfectly harmless, and the magnetic field prevents the plasma from leaking out. Now suppose there is a small ripple in the surface, like the one in the Rayleigh–Taylor instability

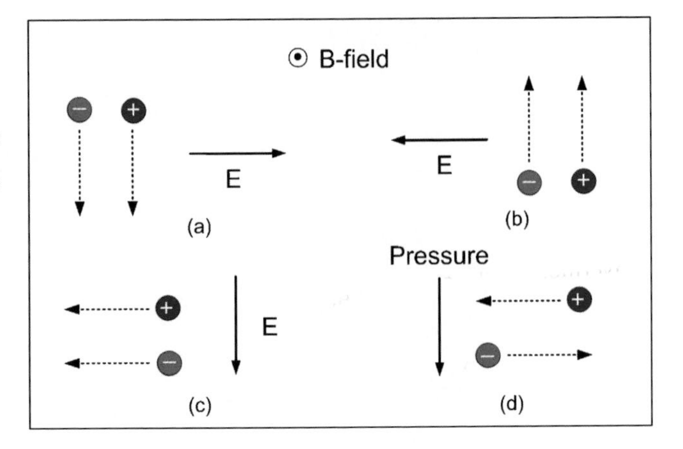

Fig. 5.7 Guiding center drifts caused by electric fields (*top* and *bottom left*) and by pressure forces (*bottom right*). In all cases, the magnetic field is out of the page

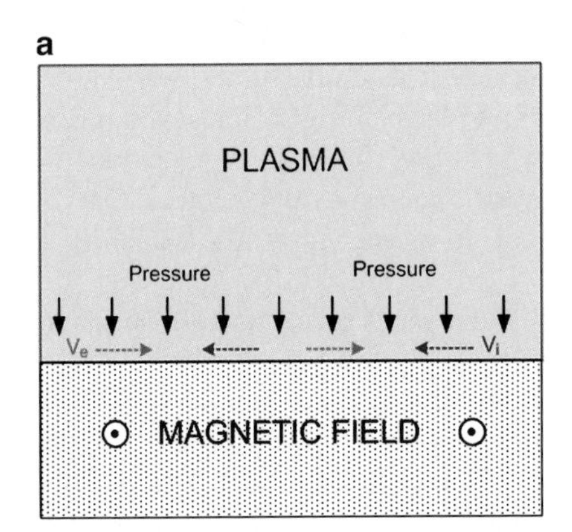

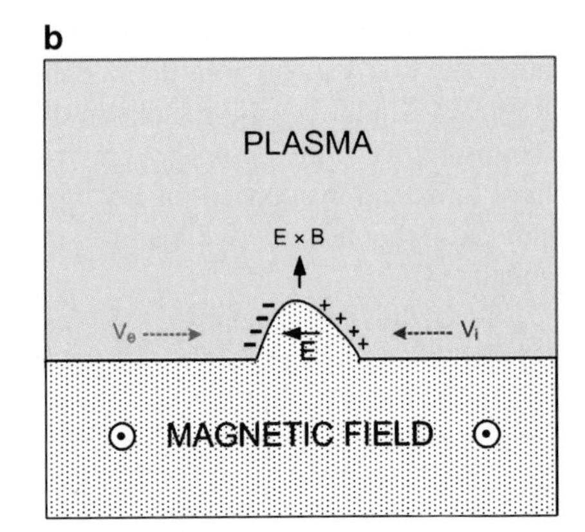

Fig. 5.8 Development of a Rayleigh–Taylor instability in a plasma

for water. What happens is shown in Fig. 5.8b. The ions, drifting to the left, pile up on the right side of the ripple, and the electrons, drifting right, pile up on the left side. These charges create an E-field pointing to the left, as shown in Fig. 5.8b. From Fig. 5.7b, we see that this E-field causes both ions and electrons to drift upwards, thus enhancing the ripple. The ripple or bubble then grows unstably, with the magnetic field forcing its way into the plasma, ejecting the plasma outwards in a way reminiscent of Fig. 5.5. The plasma escapes from the magnetic trap by organizing itself to create electric fields which can push it out! Since in the long run the magnetic field has basically changed its place with the plasma, with the field on the inside and the plasma on the outside, this instability is also called an *interchange instability*.

Stabilization by Sheared Fields

The Princeton Gun Club was a small shack on the side of the runway of the Princeton airport and was purportedly used for skeet shooting at one time. It was an ideal location for a classified meeting of Project Sherwood in 1955. The Robin Hood connection came from one of the participants, James Tuck (Friar Tuck) of Los Alamos. Representatives of the four US laboratories working on fusion (Livermore, Oak Ridge, Los Alamos, and Princeton) fit into the small room. Edward Teller was there. After hearing about our trying to hold a plasma with a magnetic field, he exclaimed, "It's like holding jello with rubber bands!" Indeed, the jello would squeeze out between rubber bands, exchanging places with an equal volume of rubber, so that the rubber bands were on the inside and the jello on the outside.

A solution to the basic interchange instability was formulated: weave the rubber bands into a mesh. In a toroidal magnetic field, this is done by magnetic shear. Figure 5.9 shows several magnetic surfaces in a torus, each containing magnetic field lines that are twisted. The twist angle, however, changes from surface to surface, so if a ripple starts on one surface and is aligned with the field lines there, as in Fig. 5.8, it finds itself misaligned with the field on the next surface. The difference in pitch angle from one surface to another has been greatly exaggerated. It does not take a very fine mesh of field lines to kill the interchange instability; in fact, we will see later that the amount of twist is limited by another instability.

A graphic picture of how shear stabilization works was provided by an experiment by Mosher and Chen [4]. The plasma in Fig. 5.10 was in a straight cylinder with a magnetic field up out of the page. The shaded circle in the center represents a thick rod inside the plasma carrying a current into the page and creating a "poloidal" magnetic field that gives the field lines a helical twist. At the left, a bump on a magnetic surface is shown which might represent an instability getting started.[3] In successive views to the right, the current in the rod is increased, twisting

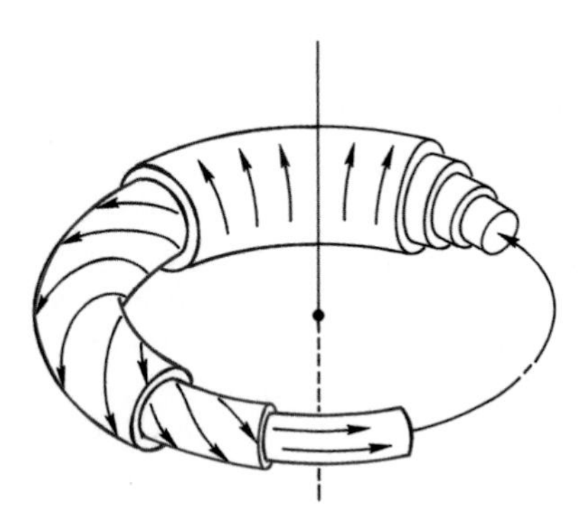

Fig. 5.9 A torus with a sheared helical field

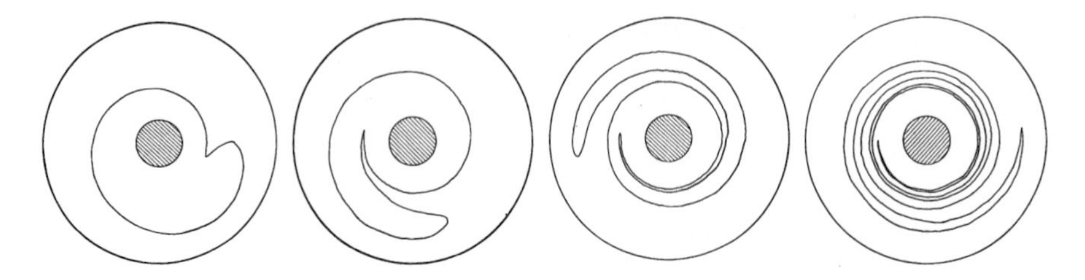

Fig. 5.10 Effect of shear on a bump in a plasma

the field lines more and more. Finally, at the right, the measurements show that the bump has been twisted into a thin spiral, so thin that the charges that create the electric field in the Rayleigh–Taylor instability (Fig. 5.8) can leak across the spiral, short-circuiting the electric field and killing the instability. In addition, short-circuiting by electrons moving in the toroidal direction (perpendicular to the page) also happens, and in fact is the main stabilizing effect of shear on the interchange instability.

To summarize this first of many instabilities, we saw that a plasma cannot fall out of its magnetic container the way water falls out of a bottle because a plasma weighs practically nothing. Its gas pressure, however, is always pushing against the magnetic field. With the slightest perturbation, the plasma organizes itself to create an electric field that causes a tongue of plasma to leak out and a bubble of field to leak in. Being wise to the plasma's tricks, we can thwart the plasma's moves by short-circuiting its self-generated electric fields with magnetic shear.

Plasma Heating and "Classical" Leak Rates

You are probably wondering how we can heat a plasma to 100 million degrees (10 keV). We can do that because a plasma is not a collisionless superconductor after all! Although much of the theory of instabilities is done with the approximation of collisionlessness, we now have to take into account collisions between electrons and ions, infrequent though they are. First of all, plasmas can be made only inside a vacuum chamber because its heat would be snuffed out by air. Vacuum pumps create a high vacuum inside the torus. Then a gas such as hydrogen, deuterium, or helium is bled in up to a pressure that is only three parts in a million (3×10^{-6}) times as high as atmospheric pressure. These atoms are then ionized into ions and electrons by applying an electric field, as we will soon show. Although the plasma is heated to millions of degrees, it is so tenuous that it does not take a lot of energy to heat the plasma particles to a million degrees (100 eV) or even 100 million degrees (10 keV). This is the reason a fluorescent tube is cool enough to touch even though the electrons in it are at 20,000°. The density of electrons inside is much, much lower than that of air.

Once we have the desired gas pressure in the torus, we can apply an electric field in the toroidal direction with a transformer (this will be explained later). There are always a few free electrons around due to cosmic rays, and these are accelerated by the E-field so that they strip the electrons off gas atoms that they crash into, freeing more electrons. These then ionize more atoms, and so on, until there is an avalanche, like a lightning strike, which ionizes enough atoms to form a plasma. This takes only a millisecond or so. The E-field then causes the electrons to accelerate in the toroidal direction, making a current that goes around the torus the long way. The ions move in the opposite direction, but they are heavy and move so slowly that we can assume that they stay put in this discussion. If the plasma were really collisionless, the electrons would "run away" and gain more and more energy while leaving the ions cold. However, there *are* collisions, and this is the mechanism that heats up the whole plasma.

Running an electric current through a wire heats it because the electrons in the wire collide with the ions, transferring to them the energy gained from the applied voltage. According to Ohm's law, the amount of heating is proportional to the wire's resistivity and to the square of the electric current. In toasters, a high-resistance wire is used to create a lot of heat. High resistance is hard to get in a plasma because it is almost a superconductor. The number of ions that electrons collide with may be 10 orders of magnitude (10^{10} or 10 billion times) smaller than in a solid wire. Nonetheless, heating according to Ohm's Law ("ohmic heating") is effective because very large currents can be driven in a plasma, currents above 100,000 A (100 kA), and even many megamperes (MA). This is the most convenient way to heat a plasma in a torus, but when the resistance gets really low at fusion temperatures, other methods are available.

Calculating the resistance of a plasma is not easy because the collisions are not billiard-ball collisions. The transfer of energy between electrons and ions occurs through many glancing collisions as they pass by at a distance, pushing one another with their electric fields. This problem was first solved by Spitzer and Härm [5], and their formula for plasma resistivity ("Spitzer resistivity") allows us to compute exactly how to raise a plasma's temperature by ohmic heating.

This resistivity formula allows us to calculate something of even more interest; namely, the rate at which plasma collisions can move plasma across magnetic field lines. Every time an electron collides with an ion, both their guiding centers shift more or less in the same direction, so both of them move across the field lines. The plasma, then, spreads out (diffuses) across the magnetic field the way an ink drop diffuses in a glass of water until the ink reaches the wall. This is a slow process, but nonetheless it limits how long a magnetic bottle can hold a plasma. There is, however, a big difference between ordinary diffusion and plasma diffusion in a magnetic field. In ordinary diffusion, collisions slow up the diffusion rate by making the ink molecules, for instance, undergo a random walk. The more the collisions, the slower the diffusion. A magnetically confined plasma, on the other hand, does not diffuse at all *unless* there are collisions. Without collisions, the particles would just stay on the same field line, as in Fig. 4.5. Collisions cause them to random walk across the B-field, and the collision rate actually speeds up the diffusion. Since a

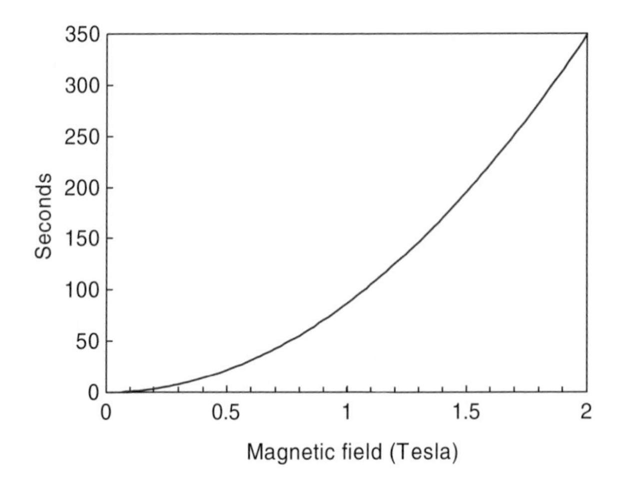

Fig. 5.11 "Classical" confinement time of a fusion plasma

hot plasma makes very few collisions, being almost a superconductor, this "classical" diffusion rate is very slow. This is called "classical" diffusion because it is the rate predicted by standard, well-established theory and applies to normal, "dumb" gases. Unfortunately, plasma can diffuse out rapidly by generating its own electric fields; and it leaks out much faster than at the classical rate.

Figure 5.11 shows the classical confinement time of a hot plasma as a function of magnetic field. We have assumed fusion-like electron and ion temperatures of 10 keV and a plasma diameter of 1 m – a large machine, but smaller than a full reactor. What is shown is the time for the plasma density to drop to about one-third of its initial value. This is similar to the "half-life" of a radioisotope used in medicine, a concept most people are familiar with. We see that at a field of 1 T (10,000 G), which we found before to be necessary to balance the plasma pressure, the time is about 90 secs – a minute and a half. This is much longer than what the Lawson criterion requires, which, we recall, is about 1 sec. It was this prediction of very good confinement that gave early fusion researchers the optimistic view that controlling the fusion reaction was a piece of cake. It did not happen, of course. Numerous unanticipated instabilities caused the confinement times to be thousands of times shorter than classical, and it is the understanding and control of these instabilities that has taken the last five decades to solve.

Notes

1. The data are from Bosch and Hale [1]. The vertical axis is actually reactivity in units of 10^{16} reactions/cm^3/sec.
2. Such data were originally given by Post [2] and have been recomputed using more current data.

3. What is actually shown here is an equipotential of the electric field, which is the path followed by the guiding centers in an $\mathbf{E}\times\mathbf{B}$ drift. The short-circuiting occurs when the spacing becomes smaller than the ion Larmor radius, so that the ions can move across the field lines to go from the positive to the negative regions on either side of the equipotential. The curves are measured, not computed.

References

1. H.S. Bosch, G.M. Hale, Nucl. Fusion **32**, 611 (1992)
2. R.F. Post, Rev. Mod. Phys. **28**, 338 (1956)
3. F.F. Chen, *Observations of X-rays from the Stellarator* (USAEC, Washington, DC, 1955) Tech Rept. 289, pp. 297–302
4. D. Mosher, F.F. Chen, *Convective losses in a thermionic plasma with shear*. Phys. Fluids **13**, 1328 (1970)
5. L. Spitzer, R. Härm, Phys. Rev. **89**, 977 (1953), summarized in *Physics of Fully Ionized Gases*, 2nd edn, by L. Spitzer, Jr. (Wiley-Interscience, New York, 1962)

Chapter 6
The Remarkable Tokamak*

A Special Kind of Torus

The name *tokamak* comes from the Russian words ***to**roidalnaya* ***ka**mera* ***ma**gnitnaya* ***k**atushka* meaning "toroidal chamber magnetic coils," though it might have been appropriate to name it after the Russian word *tok*, meaning current. As mentioned in Chap. 4, this device was unveiled at the 1958 Geneva Conference. In those days, the Russians had the lead in space satellites, but their fusion research was done with poor equipment and considered primitive. The Americans and Britons, by contrast, had shiny, expensive, and well-engineered machines which they proudly displayed. The tokamak, however, turned out to be the one that worked the best and is the leading type of magnetic plasma container today. It was developed by a team led by Academician Lev Artsimovich on an idea of Andrei Sakharov and Igor Tamm and has been adopted by all nations working on magnetic fusion energy.

In Chap. 5, we showed that a magnetic bottle had to be a topological torus and that it had to have helically twisting magnetic field lines in order to compensate for the vertical particle drifts caused by the toroidal shape. The field lines also had to be sheared to stabilize the Rayleigh–Taylor interchange instability. In a stellarator, the proper magnetic field shape can be created with external helical windings carrying current. In a tokamak, this is simplified by driving a large amount of current through the plasma itself. The current flows in the toroidal direction (the long way around the torus), and it generates a poloidal magnetic field (the short way around the cross section). When this poloidal field is added to the main toroidal field from the large outside coils, the magnetic field inside the plasma is twisted into helices. Moreover, since the poloidal field is not the same on every magnetic surface, the helical field also has shear. This is illustrated in Fig. 6.1. A strong field in the toroidal direction is created by external coils, of which only three are shown for clarity. Inside the plasma, one magnetic surface is shown. A toroidal current is driven through the plasma inside this surface, and this creates a poloidal field, which adds to the toroidal field to form a twisted helical field. Depending on how much current

*Numbers in superscripts indicate Notes and square brackets [] indicate References at the end of this chapter.

F.F. Chen, *An Indispensable Truth: How Fusion Power Can Save the Planet*,
DOI 10.1007/978-1-4419-7820-2_6, © Springer Science+Business Media, LLC 2011

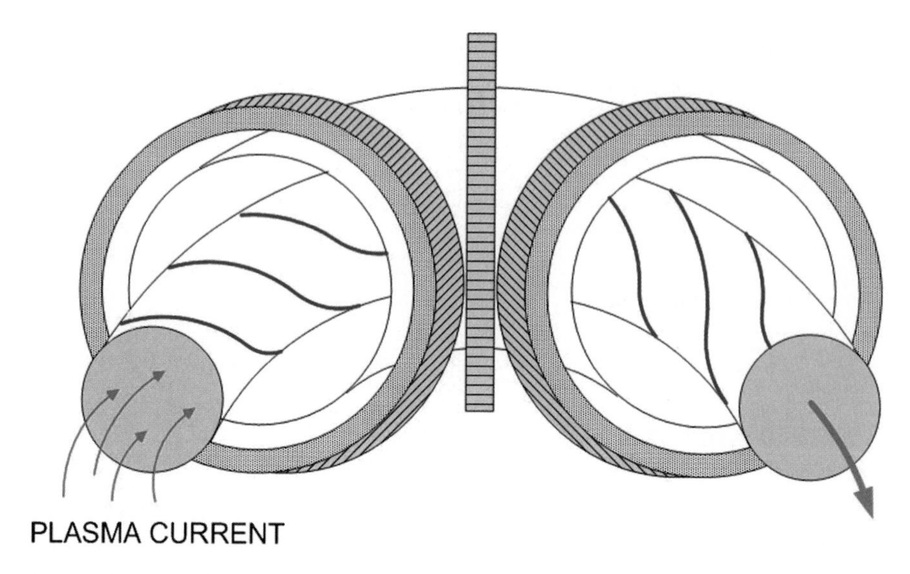

Fig. 6.1 Helical field lines created by external coils and a plasma current

there is inside each magnetic surface, the amount of twist differs from one surface to the next, and so the field is also sheared to prevent instabilities.

Using the plasma itself as a current-carrying coil to generate the twisting field would seem to be a great simplification, but we have not yet shown the hardware needed to drive this current. The advantage of the tokamak is more subtle. The current path for the poloidal field is not fixed by an external coil but can be varied by the plasma; and, fortuitously, the plasma has a self-curing property that distributes the current in a beneficial way. We explain this more fully later on.

Kink Instability and the Kruskal Limit

A toroidal plasma current serves two purposes: it generates the necessary twist in the magnetic field, and it can also raise the plasma temperature by ohmic heating. However, there is a limit to how much current can be driven because of yet another instability: the kink instability. Figure 6.2 shows an initially straight current path in the plasma that has bent itself into a kink. The circles show the field lines of the poloidal field that the current generates (the toroidal field is from left to right). Note that the lines are closer together on the inside of the kink than on the outside, indicating that the field is stronger on the inside. The magnetic pressure, therefore, is stronger at the bottom of this picture than at the top, and the kink is pushed further out. The bigger the kink, the larger the pressure difference; and the instability grows rapidly and disrupts the current. Remember that the poloidal field shown here is *not* the main (toroidal) field that supports the plasma pressure; it is the relatively small field that provides the twist. The toroidal field has a stabilizing influence, since it resists being pushed around by the plasma current. The onset of instability, therefore, depends on how strong the toroidal field is relative to the current. Conversely, onset of instability depends on how much current there is for a given toroidal field strength.

The limiting current for stable operation is called the Kruskal–Shafranov limit, and it is conveniently expressed in terms of the rotational transform, which is the number of times a field line goes around a torus the short way for each time it goes around the long way (Chap. 4). The critical rotational transform is exactly ONE! The critical current is that which creates a poloidal field large enough to twist the field lines just enough to give unity rotational transform, taking into account the strength of the main toroidal field. Transforms larger than 1 are unstable to kinks; transforms smaller than 1 are stable. The criterion for kink stability is actually quite complicated, since it depends on how the current varies across the plasma, but we can give a rough picture of why a rotational transform of 1 is a magic number.

The kink shown in Fig. 6.2 is in a straight plasma, but the current channel actually flows around the torus and joins back on itself. Figure 6.3 shows the largest unstable kink, which is actually an off-center displacement of the plasma. The plasma has been made unrealistically thin in order to have room to show the effect. In the top view (a), the dashed lines indicate the cross sections viewed in panel (b). Let us assume that the rotational transform is exactly 1. On the right-hand side of either view, the plasma has been displaced toward the outer wall. On the left-hand side, half-way around the torus, the field lines have rotated half-way around the cross section, so the plasma is now close to the inside wall. If the transform is

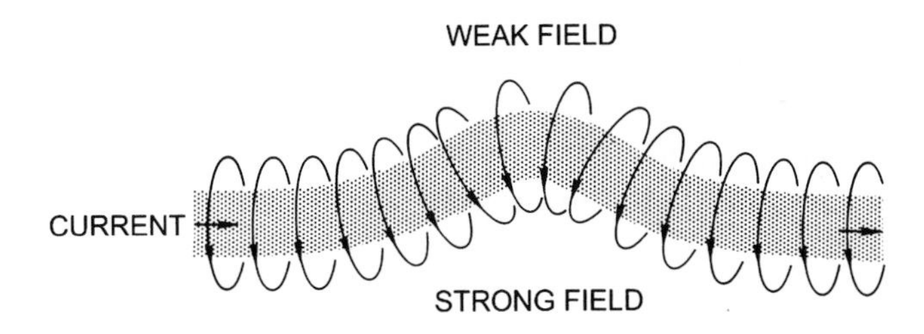

Fig. 6.2 A kink instability

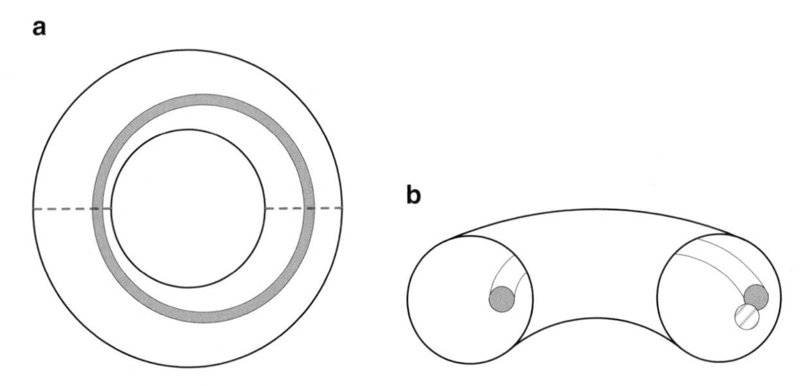

Fig. 6.3 A large kink distortion of the plasma in a torus: (**a**) top view and (**b**) cross-sectional view

exactly 1, when the field lines come back to the right-hand side, they will be in the same place where they started, so the current can flow in a closed path. Remember that the plasma is almost a superconductor; so, without collisions, the electrons carrying the current must stay on the same field line. Now let us assume that the rotational transform is less than 1. Then, upon coming back to the right-hand cross section, the current channel is in the position shown by the cross-hatched circle, which does not match up with its initial position. Since current must flow in a continuous path, this distortion of the current channel is not possible, and this kink cannot form. The plasma is stable for rotational transforms less than 1. In this simple picture, the plasma would also be stable if the transform is greater than 1, as long as it is not exactly 1. However, in that case, the current is strong enough to drive other shapes of kinks, and the plasma is kink-unstable in a way that is not easy to explain.

Since small rotational transform is good while large transform is bad, the *reciprocal* of the transform is used in tokamak lore. This is the *quality factor q* ("little q"), which is high when the plasma is kink-stable and low when it is kink-unstable. If the rotational transform is larger than 1, q is less than 1, and the plasma iskink-unstable. If the rotational transform is smaller than 1, q is larger than 1, and the plasma is kink-stable. What if q is a rational fraction so that the current channel joins up to itself after several trips around the torus? Then very interesting things happen, which we will get to.

Mirrors, Bananas, and Neoclassicism

Walking past Harold Furth's office one day, I saw this huge Chiquita Banana balloon hanging down from the ceiling. "What's going on?" I asked. "Welcome to banana theory," he replied, "the *fruitful* approach to fusion!" This was the beginning of a new understanding of how particles move in a torus. We knew that bending a cylinder into a torus would induce vertical drifts, and we knew how to counteract those by twisting the field lines into helices. But there were more subtle toroidal effects that we did not know about for the first 15 years. To explain banana orbits, we first have to describe magnetic mirrors.

If a magnetic field is not uniform – that is, if its strength changes as you move along a field line – it can reflect a charged particle and cause it to go backwards. This is the same effect that makes two permanent magnets repel each other when you turn one around so that their polarities don't match. There are toys that use this repulsion effect to suspend a magnetic object in midair. In Fig. 4.3b in Chap. 4, we showed that an electromagnet can create a magnetic field with coils of wire carrying a current. The ions and electrons gyrating in their circular orbits in a magnetic field are like electromagnets, since they are like one-turn coils carrying a current, even if the current is lumped into one charged particle. Figure 6.4 shows the field of a gyrating ion immersed in the nonuniform field of a normal electromagnet. The ion's magnetic field is always in the opposite direction to that of the field it's immersed in. Why? Because a physical system always tries to fall into the

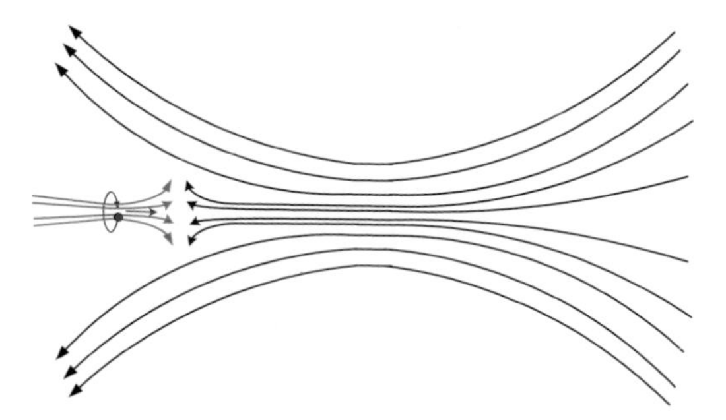

Fig. 6.4 Reflection of an ion heading into a stronger magnetic field. The field generated by the ion's gyration is shown in *red*

lowest energy state. By canceling part of the background magnetic field, the ions can lower the total magnetic energy. Electrons will do the same even though they have negative charge. They rotate in the opposite direction, but being negative, they carry current in the same direction as the ion do.

In Fig. 6.4, an ion, carrying the magnetic field that it generates, moves to the right. The field lines on the right are of a background magnetic field generated by large coils outside the plasma. The field lines generated by the current of the gyrating ion are shown in red. The opposing fields push the ion backwards, like two permanent magnets with opposite polarity. The ion's motion to the right is slowed up. The ion is moving into a stronger field, since the black lines are getting closer together. When the external field gets too strong, the ion cannot go any farther and is reflected back. How far the ion goes depends on how fast it was moving from left to right. However, not all ions will get reflected because the background field has a maximum strength. If the ion comes in with enough energy to go through the maximum, it gets slowed up there, but it is able to go through and regain its velocity on the other side. A converging magnetic field is a magnetic mirror that can reflect all but the fastest ions. This mechanism of magnetic mirroring was used by Enrico Fermi to explain the origin of cosmic rays. There, the interstellar magnetic fields are moving very rapidly, and they can push ions up to very high energies. Why can't we use magnetic mirrors to trap and hold a plasma? Indeed, we can, but magnetic mirror systems have not worked out as well as tokamaks. Mirrors will be described in Chap. 10.

Now we can get to the bananas. Tokamaks also have magnetic mirrors, but they hinder rather than help the confinement. Recall from Fig. 4.14 in Chap. 4 that the magnetic field is always stronger on the inside of a torus, near the hole, than on the outside because the coils are closer together in the hole, and therefore the field near one coil also gets contributions from the neighboring coils. That means that there is a nonuniform magnetic field, and particles going from a weak field to a strong field might get reflected. Ideally, particles travel along helical field lines on a magnetic surface and never leave it. However, magnetic mirroring prevents this, as shown in Fig. 6.5.

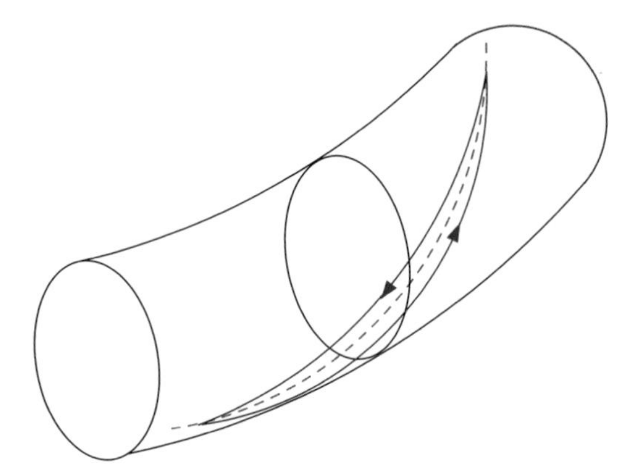

Fig. 6.5 A banana orbit in a tokamak. In reality, this orbit drifts around the torus

In this figure, the dashed line is a helical field line. An ion does not actually follow this line exactly unless its Larmor radius is zero. When it gyrates in a finite-sized circle, it will drift slowly from one line to another, as shown in Fig. 4.10, if the magnetic field strength is not the same on every side of its orbit.[1] The helical twisting cancels out the vertical drift on the average, but the averaging is disrupted by the mirror effect. The actual ion orbit is like the one shown by the solid line in Fig. 6.5. This ion starts out on the outside of the torus, where the field is weak, and it loops around toward the inside, where the field is strong. If it is not moving fast enough, it will be reflected by the magnetic mirror effect and come back on a slightly different path. Only ions with enough energy parallel to the field line will make it around to the inside of the torus and sample all parts of a magnetic surface as we envisioned in our earlier naïve picture of magnetic bottles. If we project the path of the ion in Fig. 6.5 onto the cross section of the torus shown there, it will look something like Fig. 6.6.

These are the so-called banana orbits. In each case, the outside of the torus is on the right side of the cross section, and the strong field near the hole in the doughnut is on the left. The small banana in panel (a) is for a particle with small velocity parallel to the magnetic field; it gets reflected before it gets very far toward the inside. The *dashed* line is the path of a passing particle, one that gets through the mirror and can come all the way around. In panel (b), the particle has larger parallel velocity and goes farther to the left, describing a larger banana. The limiting case is shown in panel (c), where the particle nearly makes it through the mirror. Tom Stix whimsically dubbed this the WFB, the *World's Fattest Banana*.

Banana orbits were discovered theoretically. They have never been seen in experiment because it would be very hard to track the path of a single ion or electron in a plasma with more than a trillion particles per cubic centimeter. However, theory predicts the consequences of banana orbits, and these unfavorable effects are well established by experiment. It's easy to understand why these bananas bear

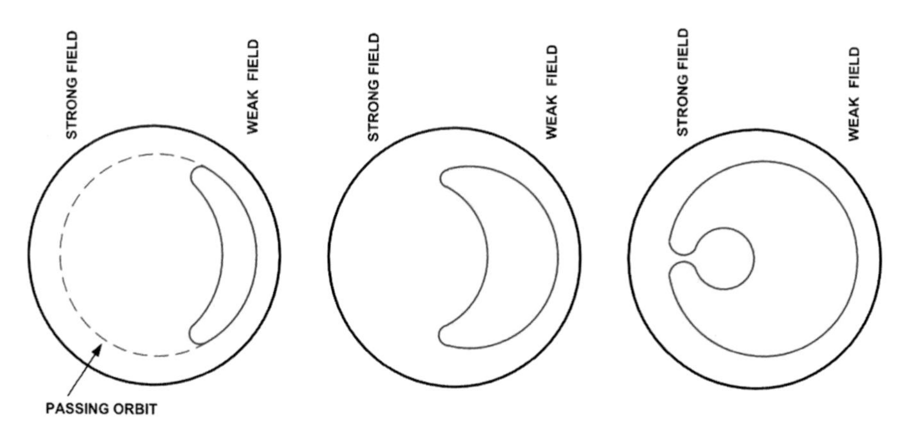

Fig. 6.6 Banana orbits of particles with increasing parallel velocities

bitter fruit. When an ion makes a collision, instead of jumping from one Larmor orbit to an adjacent one, it jumps from one banana orbit to the next; and banana orbits are much wider.[2] Instead of the very slow rate of "classical" diffusion that we described in Chap. 5, the rate of plasma transport across the magnetic field is much faster in a torus than in a straight cylinder. The rate of banana diffusion can be calculated easily and is called *neoclassical diffusion*. It is a characteristic of toruses that was not initially foreseen. The good news is that it is still a classical effect; that is, it can be calculated using a known theory. Figure 6.7 shows how banana diffusion differs from classical diffusion. At the left-hand side, the collision rate between ions and electrons is very small, so small that an ion can traverse one or many banana orbits before making a collision. In the middle, flat part of the curve, the trapped ions (those making banana orbits) make collisions during a banana orbit, but the passing particles, being faster, do not. In the right-hand part, the collision rate is high enough that all particles make collisions in traversing the torus. Under fusion conditions, the plasma is so hot and so nearly collisionless that it is well into the banana regime, at the extreme left of the graph. Therefore, it is clear that the banana diffusion rate is much higher than the classical one, shown by the straight line at the bottom.

One might think that the closer a torus is to a cylinder, the smaller the banana effects will be. The aspect ratio A of a torus is the major radius R divided by the minor radius a, as shown in Fig. 6.8. A fat torus would have small A and a skinny one, large A. One would think that large A would have smaller banana diffusion, but this is not always true. It depends on many subtle effects which can cancel one another. The Kruskal–Shafranov limit states that q (the inverse rotational transform) has to be larger than 1. For a given value of q, banana diffusion is actually larger for large A. This is primarily because the ion has to go a long way around the torus before it turns around, and it is drifting vertically the whole time.

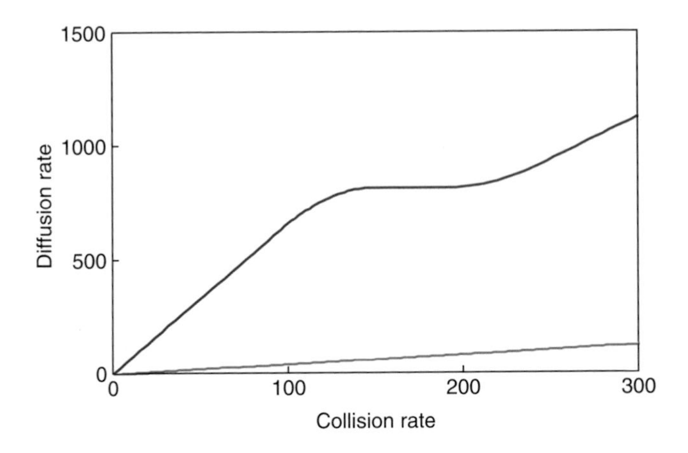

Fig. 6.7 Neoclassical (*top curve*) and classical (*bottom curve*) diffusion rate for ions as a function of collision frequency[3]

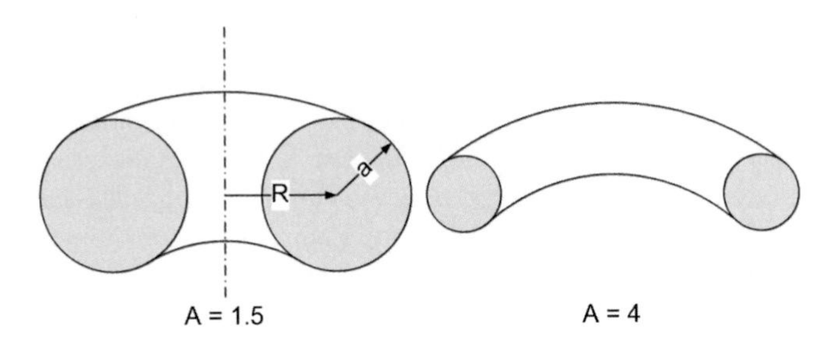

Fig. 6.8 Toruses with small and large aspect ratios

An even stranger, counter-intuitive effect has to do with the width of a banana orbit. It turns out that this width depends only on the strength of the poloidal field B_p and not on the toroidal field B_t. Remember that B_p is only the small field generated by the plasma current that gives the field lines a small twist. The banana width is approximately the Larmor radius of an ion calculated with B_p instead of B_t. This is much larger than the real Larmor radius, calculated with B_t. Since banana diffusion goes by steps of the size of a banana width, which depends only on the relatively weak B_p, does this mean that the much stronger toroidal field is useless? No! The toroidal field is needed to make the real Larmor radius small so that we can consider only the movement of the guiding centers, not the actual particles. If the toroidal field were eliminated,[4] the gyration orbits would be so large that magnetic confinement would be no good at all, and furthermore there would be nothing to hold the plasma pressure.[5]

Turbulence and Bohm Diffusion

A picture of David Bohm was taped to the wall of Bob Motley's office, and our group of experimentalists at Princeton's Plasma Physics Laboratory took turns throwing darts at it. The frustration came from an unexplained phenomenon called "Bohm diffusion," which caused plasmas in toruses to escape much faster than any classical or neoclassical theory would predict.[6] In spite of all efforts to suppress the known instabilities, the plasma was always unstable, vibrating, rippling, and spitting itself out, like the foam on violently breaking surf. In Chap. 5, classical diffusion was described. This is a process in which collisions between ions and electrons cause them to jump from one field line to another one about one Larmor radius away. The classical confinement time is long, of the order of minutes. In this chapter, we described neoclassical diffusion, in which particles jump from one banana orbit to the next. The neoclassical confinement time is still of the order of seconds, longer than needed. Bohm diffusion caused the plasma to be lost in milliseconds. Major instabilities like the Rayleigh–Taylor or kink were no longer there, else the confinement time would have been microseconds. There were obviously some other instabilities that the theorists had not foreseen.

Bohm diffusion was first reported by physicist David Bohm when he was working on the Manhattan project and, in particular, on a plasma device for separating uranium isotopes. From measurements of the plasma's escape rate, he formulated a scaling law for this new kind of diffusion. It reads as follows. The diffusion rate across the magnetic field, given by the coefficient $D_\perp$ (pronounced *D*-perp), is proportional to 1/16 of the electron temperature divided by the magnetic field:

$$D_\perp \propto \frac{1}{16}\frac{T_e}{B}.$$

The 1/16 makes no sense here because I have not said what units T_e and B are in, but that number has a historical significance. Whenever Bohm diffusion is observed, there are always randomly fluctuating electric fields in the plasma. Regardless of what is causing these fluctuations, the plasma particles will respond by drifting with their $\mathbf{E}\times\mathbf{B}$ drifts (Chap. 5). Since the size of the noise is related to T_e, which supplies the energy for it, and the drift speed is inversely proportional to B, it is not hard to show that the T_e/B part is to be expected [1]. But how did Bohm come up with the number 1/16? Bohm had disappeared from sight after he was exiled to Brazil for un-American activities. In the 1960s, Lyman Spitzer tracked him down and asked him where the 1/16 came from. He didn't remember! So we'll never know. It turns out that the Bohm coefficient depends on the size and type of turbulence and can have different values, but always in the same ballpark.

Plasma turbulence is the operative term here. Any time there was unexplained noise, it was called "turbulence." Doctors do the same thing with "syndrome" or "dermatitis." Figure 6.9 is an example of turbulence; it is simply a wave breaking on a beach. As the wave approaches the beach, it has a regular, predictable up and down motion. But as the water gets shallower, the wave breaks and even foams. The

Fig. 6.9 Turbulence at the beach

motion of the water is no longer predictable, and every case is different. That's the turbulent part. The regular part is called the *linear* regime; this is a scientific term that has to do with the equations that govern a physical system's behavior. Linear equations can be solved, so the linear behavior is predictable. The turbulent part, in the *nonlinear* regime, can be treated only in a statistical sense, since each case is different. Nonlinear generally means that the output is not proportional to the input. For instance, taxes are not proportional to income, since the rate changes with income. Compound interest is not proportional to the initial investment, even if the interest rate does not change, so the value increases nonlinearly. Population growth is nonlinear even with constant birth rate, in exact analogy with compound interest. Waves, when they are small and linear, will have sizes proportional to the force that drives them. But they cannot grow indefinitely; they will saturate and take on different forms when the driving force is too large. What a wave will look like after it reaches saturation can be predicted with computers, but the detailed shape will be different each time because of small differences in the conditions. Then you have turbulence. The smoke rising from a cigarette in still air will always start the same way, but after a few feet each case will look different.

The turbulence in every fusion device in early experiments was always fully developed; we could never see the linear part, so we could not tell what caused the fluctuations to start in the first place. An example of plasma turbulence in a stellarator is shown in Fig. 6.10. This is what "foam" looks like in a plasma. These are fluctuations in electric field inside the plasma. These noisy fields make the particles do a random walk, reaching the wall faster than classical diffusion would take them.

Turbulence is well understood in hydrodynamics. If you try to push water through a pipe too fast, the flow breaks up into swirling eddies, slowing down the flow. Hydrodynamicist A.N. Kolmogoroff once gave an elegant proof, using only dimensional analysis, that the sizes of eddies generally follows a certain law; namely, that the number of eddies of a given size is proportional to the power 5/3 of the size.[7] Attempts to do this for plasmas yielded a power of 5 rather than 5/3 [1],

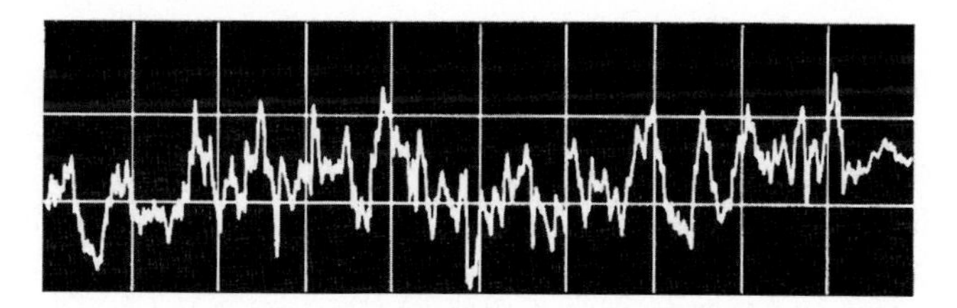

Fig. 6.10 Fluctuations in a toroidal plasma

and this has been observed in several experiments. However, plasmas are so complex (because they are charged) that no such simple relation holds in all cases.

The importance of turbulence and Bohm diffusion is not only that it is much faster than classical diffusion, but also that it depends on $1/B$ instead of $1/B^2$. In classical diffusion, doubling the magnetic field B would slow the diffusion down by a factor of 2^2 or 4. In Bohm diffusion, it would take *four* times larger B to get the same reduction in loss rate. It was this unforeseen problem of "anomalous diffusion" that held up progress in fusion for at least two decades. Only through the persistence of the community of dedicated plasma physicists, was the understanding and control of anomalous diffusion achieved. Modern tokamaks have confinement times approaching those required for a D–T reactor.

The Culprit: Microinstabilities

If the plasma in a torus always thrashes around violently, there must be an energy source that drives the thrashing. An obvious source is the electric field applied to drive the current in ohmic heating. In the 1960s, a new method was devised for heating without a large DC current in the plasma. This was Ion Cyclotron Resonance Heating or ICRH. A radiofrequency (RF) power generator was hooked up to an antenna around the plasma, the way an FM station is hooked up to its antenna on a tower. The frequency was tuned to the gyration frequency of the ions in their cyclotron orbits. As the ions moved in circles, the RF field would change its direction so as to be pushing the ions all the time, just as in a real cyclotron. This could heat up the plasma without having to drive a DC current in it.[8] Would this kill the turbulence and make the plasma nice and quiet, without Bohm diffusion? A case of champagne was bet on it. It didn't work. The thrashing was as bad as ever. The darts in Bohm's picture stayed there.

The problem was a failure of magnetohydrodynamics, MHD for short. MHD theory treats a plasma as a pure superconductor, with zero resistivity, and neglects the cyclotron orbits of the particles, treating them as points moving at the speed of their guiding centers. Though this simplified theory served us well in the design of toroidal confinement devices and in the suppression of the gravitational and kink instabilities, it did not treat a plasma in sufficient detail. First of all, there have to be *some* collisions in a fusion plasma or else there wouldn't be any fusion at all! These infrequent collisions cause the plasma's resistivity to be not exactly zero, and that has dire consequences on stability. The fact that the Larmor orbits of the ions are not mathematical points gives rise to the finite Larmor radius (FLR) effect. In some cases, even the very small inertia of the electrons has to be taken into account. Finally, instabilities could even be caused by distortions of the particles' velocity distributions away from a pure Maxwellian, an effect called Landau Damping. These small deviations from ideal MHD turned out to be important, making the theorists' task much more difficult.

The first inkling of what can happen was presented by Furth, Killeen, and Rosenbluth in their classic paper on the tearing mode [2]. If a current is driven along the field lines in a plasma with nonzero resistivity, the current will break up into filaments; and the initial smooth plasma will tear itself up into pieces! So "tearing" rhymes with "bearing," and not "fearing," though the latter interpretation may have been more appropriate. The tearing mode is too complicated to explain here, but we describe other instabilities which caused even more tears.

One of the tenets of ideal MHD is that plasma particles are "frozen" to the field lines, as shown in Fig. 4.10. Without collisions or one of the other microeffects named above, ions and electrons would always gyrate around the same field line, even if the field line moved. Bill Newcomb once proved a neat theorem about this [3], saying that *plasma cannot move from one field line to another as long as* $E_{\|}$ *(E-parallel) is equal to zero*. Here, $E_{\|}$ is the electric field along a magnetic field line, and it has to be zero in a superconductor, since in the absence of resistance even an infinitesimal voltage can drive an infinite current. But if there are collisions, the resistivity is not zero, $E_{\|}$ can exist, and plasma is freed from one of its constraints.

So it was back to the drawing board. While the theorists enjoyed a new challenge and a new reason for their employment, the experimentalists pondered what to do. In previous chapters, we showed that (1) a magnetic bottle had to be shaped like a torus, (2) bending a cylinder into a torus caused vertical drifts of ions and electrons, (3) these drifts could be canceled by twisting the field lines into helices, (4) this twist could be produced by driving a current in the plasma, and (5) this current could cause other instabilities, even in ideal MHD, but that those could be controlled by obeying the Kruskal–Shafranov limit. In spite of these precautions, the plasma is always turbulent, even when the current is removed by using a stellarator rather than a tokamak. How can we get a plasma so smooth and quiet that we can see a wave grow bigger and bigger until it breaks into turbulence, as in Fig. 6.9? Obviously, if one could straighten the torus back into a cylinder, much of the original cause of all the trouble would be removed. But how can one hold the plasma long enough just to do an experiment? The plasma will simply flow along the straight magnetic field into the endplates that seal off the cylinder so that it can hold a vacuum. The solution came with the invention of the Q-machine (Q for Quiescent). Developed by Nathan Rynn [4] and Motley [5], this is a plasma created in a straight cylinder with a straight magnetic field. Inside each end of the vacuum chamber is a circular tungsten plate heated to a red-hot temperature. A beam of cesium, potassium, or lithium atoms is aimed at each plate. It turns out that the outermost electron in these atoms is so loosely bound that it gets sucked into the tungsten plate upon contact. The electron is then lost, and the atom comes off as a positively charged ion.

Fig. 6.11 Example of a Q-machine

Of course, a plasma has to be quasineutral, so the tungsten has to be hot enough to emit electrons thermionically, the way the filament in light bulb does. So both ions and electrons are emitted from the tungsten plates to form a neutral plasma. *No electric field has to be applied!* Only tungsten or molybdenum, in combination with the three elements above, can perform this kind of thermal ionization. In this clever device, all sources of energy to drive an instability have been removed or so we thought. Figure 6.11 shows a typical Q-machine, covered with the coils that create the steady, straight, and uniform magnetic field.

The plasma in a Q-machine *has to be quiescent*, right? To everyone's surprise, it was still turbulent! The trace shown in Fig. 6.10 actually came from a Q-machine. Fortunately, it was possible to stabilize the plasma by applying shear, as shown in Fig. 5.9, or by applying a small voltage to the radial boundary of the plasma. A quiescent plasma in a magnetic field was finally achieved. Then, by adjusting the voltage, one could see a small, sinusoidal wave start to grow in the plasma; and, with further adjustment, one could see the wave get bigger and bigger until it broke into the turbulence seen in Fig. 6.10. With a regular, repetitive wave like a wave in open water, one could measure its frequency, its velocity, its direction, and how it changed with magnetic field strength. These were enough clues to figure out what kind of wave it was, what caused it to be unstable, and, eventually, to give it a name: a *resistive drift wave*.

As its name implies, the wave depends on the finite resistivity of the plasma. It also depends on microeffects: the finite size of the ion Larmor orbits. Before showing how a drift instability grows, let's find the source of energy that drives it. In a Q-machine, we have eliminated all toroidal effects and all electric fields normally needed to ionize and heat the plasma. In fact, the plasma is quite cold, as plasmas go. It is the same temperature as the hot tungsten plates, about 2,300 K, so that the plasma temperature is only about 0.2 eV. You can heat a kiln up to that temperature, and it would stay perfectly quiescent. A magnetically confined plasma, however, has one subtle source of energy: its pressure gradient. When everything is at the same temperature and there are no energy sources such as currents, voltages, or drifts, there is still one source of energy when the plasma is *confined*. And confinement is the name of the game. Since ions and electrons recombine into neutral atoms when they strike the wall, plasma is lost at the walls. The plasma will be denser at the center than at the outside, and this causes a pressure that pushes against the magnetic field. By Newcomb's theorem, the plasma would remain attached to the field lines, and nothing can happen; but once there is resistivity, all bets are off. The plasma is then able to set up electric fields that allow it to move across the magnetic field in the direction that the pressure pushes it. Even if there are no collisions, other microeffects like electron inertia or Landau damping can cause the drift instability to grow. For this reason, the resistive drift instability and others in the same family are called *universal instabilities*. They are fortunately weak instabilities because the energy source is weak, and they can be stabilized with the proper precautions.

The Drift Instability Mechanism

There are many microinstabilities, but they all share the same types of plasma motion. As an example, we shall try to explain how a resistive drift wave goes unstable. This instability has stood the test of time as other theoretical predictions have come and gone. In general, it is easier to derive an instability mathematically than to figure out exactly what the plasma is doing. If this part is difficult to follow, you can skip to the next section without losing essential information. Let's start with a plasma in a straight cylinder with a straight magnetic field, as shown in Fig. 6.12. The plasma is necessarily denser at the center than on the outside. The white arrows show a density ripple, like a wave, propagating in the azimuthal direction. We shall focus on the plasma's behavior inside the small rectangle at the bottom. This rectangle is shown enlarged in Fig. 6.13. On the left, we see Larmor orbits of ions whose guiding centers may be outside the rectangle. If the magnetic field is

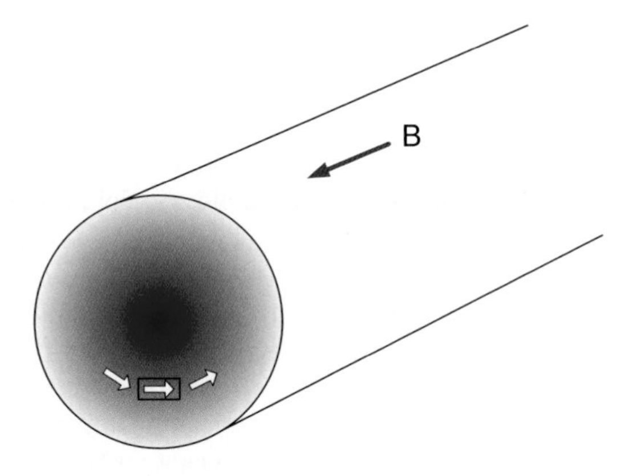

Fig. 6.12 A drift wave in an inhomogeneous plasma

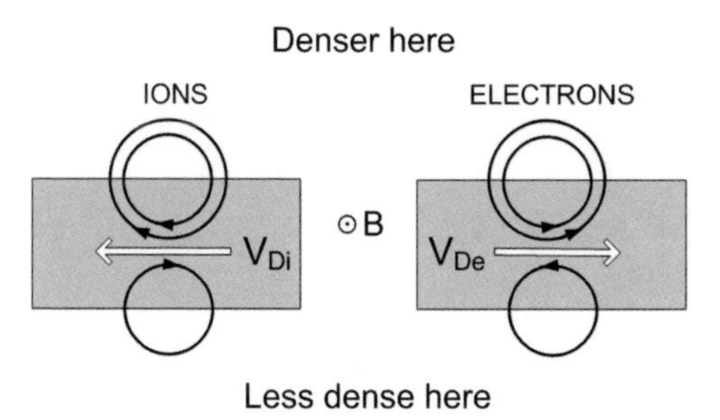

Fig. 6.13 Definition of the diamagnetic drift. The electron orbits are actually much smaller than those of the ions

out of the page, as shown, the ions will be rotating clockwise. Remember that the plasma density is higher at the top than at the bottom because the top is closer to the center of the plasma. To show this, we have drawn two orbits at the top and only one at the bottom. There are obviously more ions going left than going right. The ion fluid in this small volume therefore has an average flow toward the left. This effect is called the *ion diamagnetic drift*, and the drift velocity is called v_{Di}. Note that this drift is perpendicular both to the magnetic field and to the direction in which the density is changing. The diagram on the right is the same thing for electrons. With their negative charge, electrons gyrate counterclockwise. Their diamagnetic drift velocity, v_{De}, is therefore in the opposite direction, to the right. This motion of the ions and electrons, considered as fluids occupying the same space, is there even if the guiding centers are not moving. The existence of the diamagnetic drift depends on the gradient in density and would be zero if the density were uniform everywhere. If you have a problem with two fluids occupying the same space, just think of the vermouth and vodka in a martini.

Now we can proceed with the wave. Our little rectangle is shown three times in Fig. 6.14. At the bottom of the first diagram, (a), a density ripple is shown. A slice of the rectangle near the peak of the wave, where the density is high, is shown in a darker shade. The *background* density is high at the top and low at the bottom, as seen in Fig. 6.12. The diamagnetic drift of the ions in the *background* density gradient is to the left for ions and to the right for electrons, as shown in Fig. 6.13. Because the *wave density* is high near its peak, the diamagnetic drifts bring an excess of positive charge to the left side of the small slice and an excess of negative charge to the right side. These electric charges create the electric field **E** shown in panel (b). Recall from Chap. 5, Fig. 5.4, that an electric field causes an **E** × **B** drift, v_E, perpendicular to both **E** and **B**. In this case, the drift is downwards, as shown in panel (c). Since the *background density* is high at the top, v_E brings more density into the slice, and the wave gets more density where the *wave density* was already high. Therefore, the wave grows; it is unstable. Figure 6.15 shows what happens at a wave trough. There, the density is less than average, so the diamagnetic drifts bring *less* density to the edges of the slice, causing the buildup of charges of the opposite sign. The resulting electric field, shown in panel (b), is in the opposite direction from before. This causes the **E** × **B** drift in panel (c) to be upwards instead

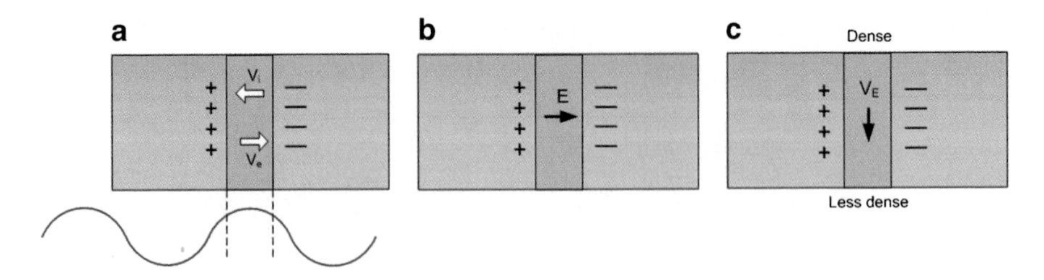

Fig. 6.14 The charges, fields, and velocities at the peak of a drift wave

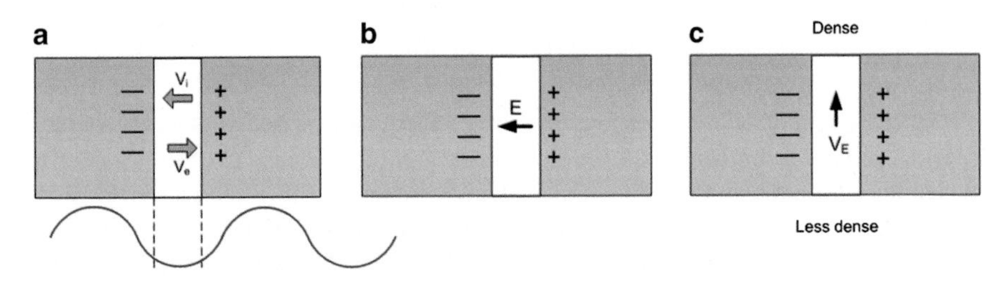

Fig. 6.15 The charges, fields, and velocities in the trough of a drift wave

of downwards. But an upward motion brings lower *background* density into the slice where the *wave* density is already low. This adds to the growth of the wave. We can now give it its rightful name: a *drift wave*. If we average over the cycles of the drift wave, more density is moved downwards at the peaks of the wave than is lost at the troughs, and consequently the wave causes plasma to move outwards, away from the center, toward the wall. Another insidious, cunning way the plasma finds to escape from its magnetic trap.

However, we are not quite finished; there is a three-dimensional part, shown in Fig. 6.16. The rectangular slices at the peaks and troughs of the wave in the last two figures are shown together at two cross sections of the cylinder, now considered as part of a torus. There are four slices: peak, trough, peak, trough. Between the slices are the electric charges shown in Figs. 6.14 and 6.15. Recall that a toroidal confinement requires a poloidal field to twist the magnetic field. This twist causes the field line going through a positive charge to connect to a negative charge in a cross section further downstream. Electrons, being very light and mobile, almost instantaneously move along the field line to cancel the charges. The electric field of the drift wave is nullified, and the wave can never grow. Ah, but if there are collisions, the electrons are slowed down, and they cannot cancel the charges fast enough. This is another example of Newcomb's theorem: if $E_{\|}$ is not zero, all bets are off! The growth of the drift instability depends on the existence of resistivity, one of our microeffects. Even without collisions, electron inertia or Landau damping can slow down the electrons and allow the instability to grow. Hence, it is a universal instability which can occur whenever there is a density gradient in a magnetic confined plasma.

Fig. 6.16 A drift wave in three dimensions

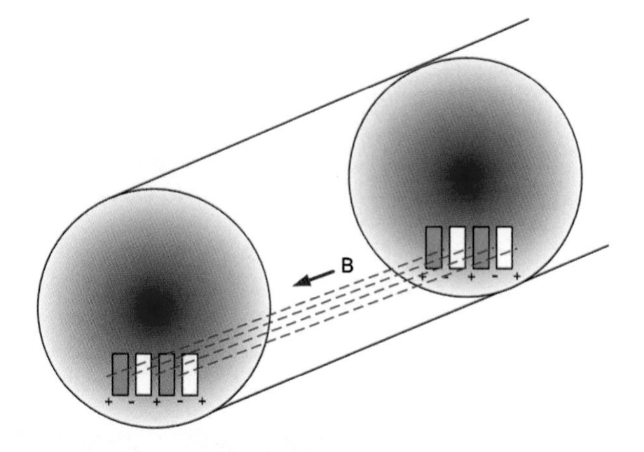

The obvious question is, "What if the plasma density is uniform all the way out to the wall?" That can't happen, since the density has to be essentially zero at a cold wall. If the density gradient occurs in a thin layer near the wall, the sharp gradient there will make the instability grow even faster. It then eats away the plasma so that the thickness of the gradient layer gets larger and larger. Drift instabilities can be stabilized by shortening the *connection length* between the cross sections shown in Fig. 6.16, so that the electrons can move between them fast enough. This requires a larger helical twist of the magnetic field. Fortunately, this can be done without violating the Kruskal–Shafranov limit.

There are many other possible microinstabilities. The ion-temperature-gradient instability is another one that is worrisome. This example of the resistive drift instability serves to give an idea of how complicated plasma behavior is and how Bohm diffusion was solved. What happens when an unstable wave breaks and becomes turbulent? It is no longer possible to identify which instability started the turbulence, but one can apply known stabilization methods to see if the fluctuations can be suppressed. There are turbulence theories that purport to predict how the turbulence will look and how much anomalous diffusion it will lead to. A powerful modern method is to do a computer simulation. A computer does not care whether an equation is nonlinear or not. It does not even need to solve an equation; it just follows the particles around to see where they go. There will be some examples later; it's not as simple as it sounds. Or, one can use physical intuition to make a guess. Figure 6.17 shows a guess on what a resistive drift instability might become when it goes nonlinear. The waves break up into blobs of density which are drifted out to the boundary by their internal electric fields. Thus, plasma is lost in bunches. This guess was made in 1967, before diagnostic techniques were available to detect such blobs. However, in 2003, physicists at M.I.T. (Massachusetts Institute of Technology) developed a special technique which allowed them to take pictures of blobs as they carried plasma radially outward. One such picture of two simultaneous blobs is shown in Fig. 6.18 [6]. This behavior is not accidental, since it was observed also in several other tokamak machines. However, this is just a example

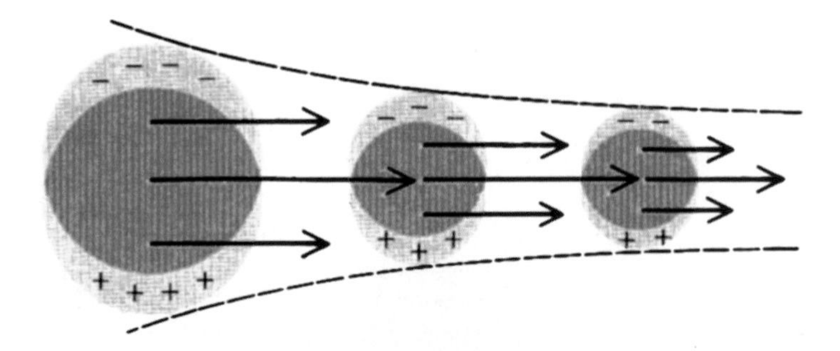

Fig. 6.17 Anomalous transport of plasma in blobs (adopted from Chen [7]). These are not spheres but long tubes of plasma curving with the magnetic field lines

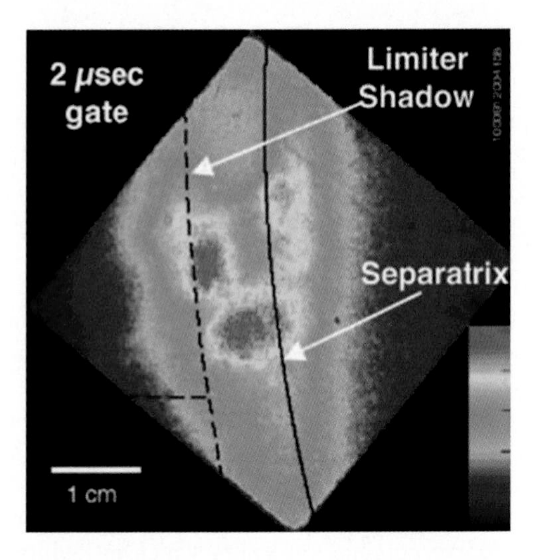

Fig. 6.18 A picture of blobs leaving a tokamak, taken at a shutter speed of 2 millionths of a second. The outside of the torus is on the *left* [6]

of how an instability, starting as a simple wave, can grow and carry plasma outwards. Other instabilities have been found to develop into other shapes as they do their dirty work.

In Chap. 4, we showed why a torus was chosen as a possible shape of a magnetic bottle used to hold a plasma hot enough to produce fusion power. In Chap. 5, we discussed the general features that had to be built into toruses in order to hold plasmas. In this chapter, we described the unexpected difficulties that were encountered in tokamaks and how these were overcome. These are the concepts which guided our work in the early days of fusion. In the four decades since that time, experiments on dozens, or even hundreds, of tokamaks, stellarators, and other magnetic devices throughout the world have led to improvements in design and advances in theoretical understanding. Tokamaks no longer look like simple circular toruses. The next chapter will tell why.

Vertical Fields

Before leaving this basic description of a tokamak, there is one more essential part that needs to be described: the vertical field. A ring of hot plasma will try to expand. Its internal pressure will push outwards so as to make the cross section fatter, and we have countered this force with a strong toroidal magnetic field. However, the plasma pressure will also tend to make the entire ring expand in radius, as shown in Fig. 6.19. The toroidal field is not good at restraining this motion because it is weaker on the outside of the ring than on the inside. Furthermore, the toroidal current in a tokamak creates a hoop force that also pushes on the ring to expand its major radius. This force arises from the magnetic field that the plasma current generates. This field is also stronger inside the hole of the torus than outside, so that its magnetic pressure is outward. Fortunately, these hoop forces are easily balanced by applying a small magnetic field in the vertical direction. Remember that in a tokamak there is always a current in the toroidal direction to give the field lines a twist. This current is mostly carried by the electrons. The Lorentz force on a moving charge, described in Chap. 4, is perpendicular to both the velocity of the particle and the magnetic field. By superposing a magnetic field in the vertical direction, either up or down, depending on the direction of the current, the tokamak current creates a Lorentz force that pushes the plasma ring inwards, toward the center of the torus.

Note that this effect is different from all the plasma drifts that we discussed previously. Those concerned individual particle motions; it did not matter how many particles there were. Here, we are considering the immense pressure of a hot gas.

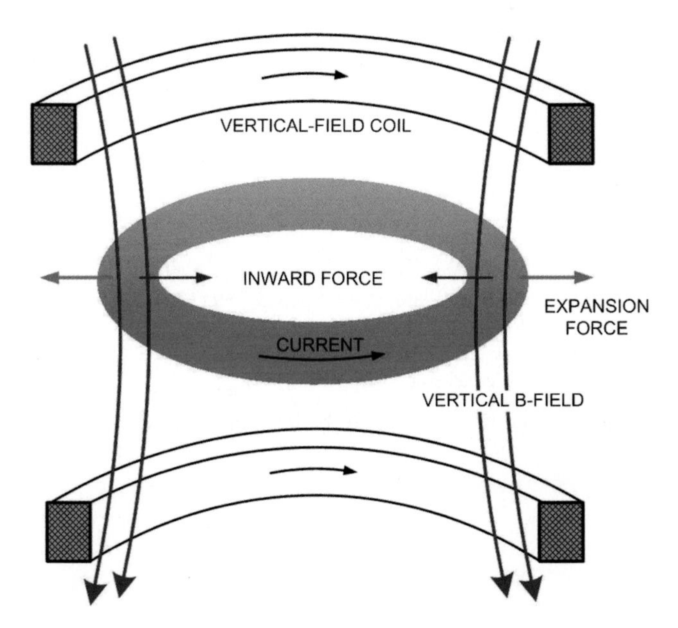

Fig. 6.19 Application of a vertical B-field to keep a ring of plasma from expanding

Thus, there are three main types of fields in a tokamak: the toroidal field generated by poloidal coils; the poloidal field generated by a plasma current; and a vertical field generated by large toroidal coils above and below the torus. These vertical-field coils can be combined with the coils that drive the plasma current, as will be described in the next chapter, so they do not always appear as a separate set.

Notes

1. In addition to the vertical drift due to the gradient of the toroidal field, there is also a smaller vertical drift due to the centrifugal force of particles whizzing around the torus the long way.
2. More likely, a collision takes a particle from a banana orbit to a passing orbit, and a second collision takes it from the passing orbit into another banana orbit.
3. (What is plotted here is perpendicular diffusion coefficient in m^2/s against Spitzer collision frequency in kHz. We have assumed 10 keV ions, 1 T magnetic field, aspect ratio $A=2.5$, quality factor $q=2$, and major radius $R=1$ m. The densities required to trace this whole curve would be unreasonably high. Fusion conditions have the very low diffusion rates in the extreme lower left corner.)
4. There are other devices, called reversed-field pinches, that have a very large toroidal current and only a small toroidal B-field. These depend on other stabilization mechanisms such as wall currents. But we are concentrating on tokamaks here because their development is further along.
5. The fact that banana diffusion does not depend on B_t comes from a cancelation between the vertical drift velocity, which varies as $1/B_t$, and the time a particle spends drifting in one direction, which varies as B_t. This is because increasing B_t for fixed B_p decreases the twist of the field lines.
6. For historical accuracy, neoclassical diffusion was discovered *after* Bohm diffusion was.
7. This holds only for an intermediate range of sizes.
8. This was not in a tokamak but in a stellarator. In tokamaks, a DC current is needed to create the rotational transform; in stellarators external coils are used to do this.

References

1. F.F. Chen, *Spectrum of low-β plasma turbulence*. Phys. Rev. Lett. **15**, 381 (1965)
2. H.P. Furth, J. Kileen, M.N. Rosenbluth, Phys. Fluids **6**, 459 (1963)
3. W.A. Newcomb, Ann. Phys. **3**, 347 (1958)
4. N. Rynn, N. D'Angelo, Rev. Sci. Instrum. **31**, 1326 (1960)
5. R.W. Motley, *Q-Machines* (Academic, New York, 1975)
6. S.J. Zweben et al., Phys. Plasmas **9**, 1981 (2002)
7. F.F. Chen, *The leakage problem in fusion reactors*. Sci. Am. **217**, 76 (1967)

Chapter 7
Evolution and Physics of the Tokamak*

In the exploration of space, the launching of *Sputnik* proved the possibility of sending an object into orbit around the earth. Subsequent development of spacecraft led to the landing of man on the moon with Apollo 11, followed by construction of the space station, serviced by shuttles that could re-enter the atmosphere repeatedly. In the development of fusion reactors, the success of early experimental tokamaks is analogous to the success of *Sputnik*. The simple drawings of tokamaks in previous chapters resemble modern tokamaks no more than *Sputnik* resembles Apollo 11. A lot of development has occurred, and a lot more has been learned about tokamaks. There have been both pleasant and unpleasant surprises. There is, however, a big difference between the two programs. In space science, the basic physics – namely, Newton's laws of motion – were already known; while in fusion, the physics of plasmas and of toroidal confinement had to be worked out first. After the initial successes, there was much more to learn about spacecraft, such as their interaction with the plasmas, solar wind, and magnetic fields in the solar system. This chapter describes what we have learned about tokamaks, once they were up and running.

Magnetic Islands

Figure 6.1 of the last chapter showed how a plasma current circulating around a tokamak generates a poloidal magnetic field to give a twist to the field lines. This twist, or helicity, is necessary to average out the vertical drifts that the particles have in a torus. These drifts arise when a straight cylinder is bent into a circle to form a torus. We then defined a quantity q, the *quality factor*, which tells how much twist there is; actually, how *little* twist there is. Large q means the twist is gentle, and small q means that the twist is tight. It is called the quality factor because the plasma is stable if q is larger than 1 and unstable if q is smaller than 1, so larger q gives better stability. You may recall that the culprit was the kink instability, and the boundary at $q=1$ was called the Kruskal–Shafranov limit. If $q=1$, a field line goes around the torus the short

*Numbers in superscripts indicate Notes and square brackets [] indicate References at the end of this chapter.

F.F. Chen, *An Indispensable Truth: How Fusion Power Can Save the Planet*,
DOI 10.1007/978-1-4419-7820-2_7, © Springer Science+Business Media, LLC 2011

way (the poloidal direction) exactly once after it goes once around the long way (the toroidal direction). It then joins on to its own tail. If $q=2$, the twist is smaller, so it takes two trips the long way before the field line joins onto itself, and so on.

In general, q is not a rational number like 1, 2, 3, 3/2, 4/3, and so forth. Except in such cases, a field line never comes back to itself; rather, after numerous turns, it traces out a magnetic surface. The field lines of neighboring surfaces cannot be parallel to one another either, because the magnetic field has to be sheared. Shear has a stabilizing effect on almost all instabilities. That means that q has to vary with radius within a cross section of the torus, so that the amount of twist is different on each magnetic surface. Scientifically, we say that q is a function of minor radius r, written as $q(r)$. By now you may have guessed that something special happens when q is a rational number, like 2. At the radius where $q(r)=2$, a field line joins onto itself after traversing the torus twice the long way. Remember that the tokamak current (the one that creates the helicity) is driven by an electric field (E-field). How this is done is shown later in this chapter. It is easier for the E-field to drive a current if the field lines are closed, since the electrons can then run around and around on the same field line. The current can break up into filaments. Each filament acts like its own little tokamak with its own magnetic surfaces, and the magnetic surface at $q=2$ breaks up into two magnetic islands. Other chains of islands could form at the $q=3$ surface, and so on. Between rational surfaces, the filamentation does not occur, and there are no islands. Figure 7.1 shows a computed picture of islands at the $q=3/2$ surface.[1] Since the rotational transform is $1/q$, it has the value of 2/3 here. That means that a field line inside the top island, after going around the whole torus once, will end up in the island at the lower right, say, two-thirds of the way around the cross section. After the next revolution, it will go another two-thirds of the way around, ending up in the island at the lower left. After the third traversal, it will be back in the top island, but not exactly where it started. It will be on the same small

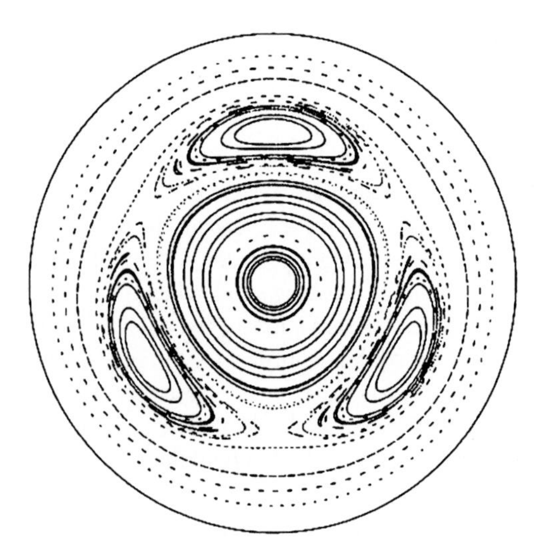

Fig. 7.1 Magnetic islands in a tokamak at the $q=3/2$ surface

magnetic surface *inside the island*, but at a different point. It is only after many, many traversals that the island is traced out. Our previous naïve picture of nested magnetic surfaces has taken on a fantastic character!

Ions and electrons can cross an island in between collisions; and since the island width is much larger than a Larmor diameter, the escape rate is faster than classical just as in banana diffusion. Fortunately, not all island chains are large, and higher fractions like 5/6 would not yield noticeable islands at all.

Exactly where these island chains lie depends on how much current there is at each radius. The amount of current depends not only on the strength of the E-field, but also on the temperature of the electrons. The higher the temperature, the lower the resistivity, and the higher the current. Since the plasma tends to be hotter at the center, the plasma current generally has a peak at the center. Figure 7.2 gives an example of where island chains can occur, in principle. The curve shows how q typically varies with distance from the center of the plasma's cross section. In this case, the rational surfaces $q=1$, 2, and 3 occur at radii of about 3, 7, ad 9 cm, respectively, and there are no places where q is 4 or higher. There is a special region where q is less than 1.

The shape of the curve $q(r)$ is determined by the distribution of the plasma current. Figure 7.3 gives examples of different current profiles $J(r)$ and the $q(r)$ curves that they produce. The uppermost curve, corresponding to the most peaked current, would have more rational q surfaces. Tokamak operators have some control over $J(r)$, since there are various ways to heat the plasma. If the electron temperature changes, however, $J(r)$ will change, and so will the magnetic topology. Where the q curves cross the line $q=1$ is of utmost importance, as will be explained next.

Islands were first observed experimentally by Sauthoff et al. [1] in the famous "sombrero hat" experiment. Electrons emit a small X-ray signal when they collide with ions. By collecting these signals with detectors surrounding the plasma, the plasma density distribution can be reconstructed by computer the same way as in a medical CAT scan. Figure 7.4 shows a typical result at one instant of time.

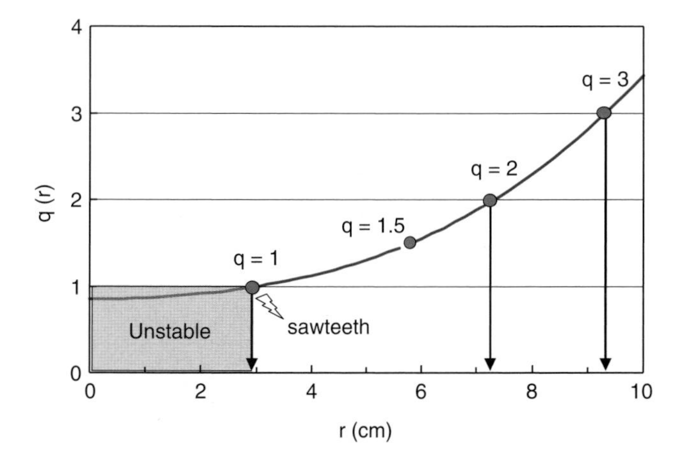

Fig. 7.2 Example of a $q(r)$ curve, showing the radii at which islands could occur

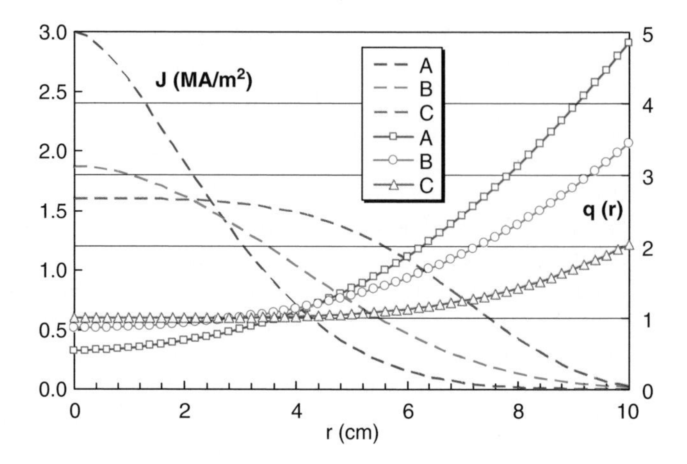

Fig. 7.3 Three different current density profiles $J(r)$ (*dashes*) and the $q(r)$ profiles (*curves with points*) that they produce. Each case (A, B, or C) has the same color. The scale for $J(r)$ is on the *left* and has units of megamperes per square meter. The scale for $q(r)$ is on the *right*. Islands can occur where the $q(r)$ *curves* cross the *horizontal lines*, and this depends on how the tokamak current is distributed

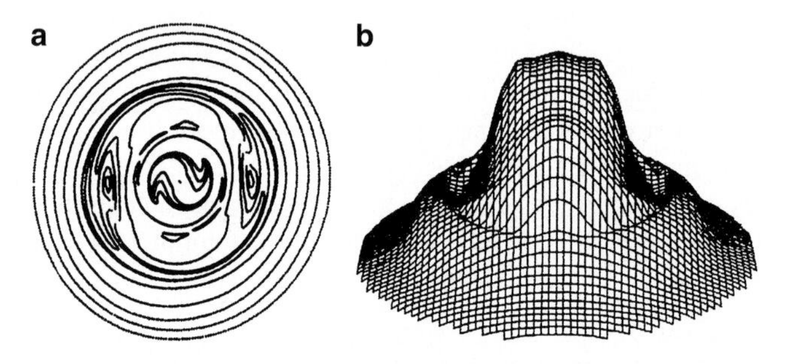

Fig. 7.4 (**a**) Density distribution in a tokamak cross section modified by islands. (**b**) Measured density contours showing island structure [1]

The contours of constant density in Fig. 7.4a shows a $q=2$ island structure. A 3D plot of this in Fig. 7.4b resembles a sombrero.

Sawtooth Oscillations

In every tokamak discharge, there is a magnetic surface where $q=1$. Inside that surface, where q is less than 1, the plasma is unstable to kinks, according to the Kruskal–Shafranov limit. Therefore, it is turbulent and a jumble of oscillations, and

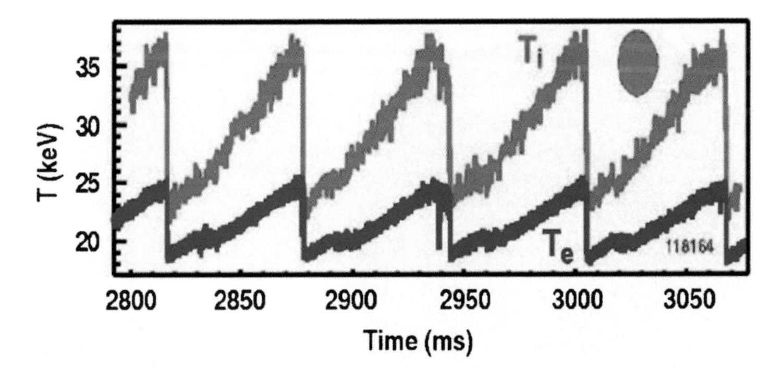

Fig. 7.5 Sawtooth oscillations of both ion and electron temperatures at the $q=1$ surface of a tokamak [2]

there is no magnetic confinement. Only when plasma gets outside the $q=1$ surface and enters the nested magnetic surface and island structure, does it get restrained by the magnetic field and diffuse slowly to the wall.

Very early in tokamak research, experimenters using a synchrotron-radiation method to detect changes in electron temperature observed regular oscillations near the $q=1$ surface. These were observed in all tokamaks and always had a sawtooth shape, rising slowly and falling sharply each time, as seen in Fig. 7.5. Since the current is largest inside the $q=1$ surface, near the center, the plasma gets hotter there. Higher temperature means less resistivity, and that makes the current even larger and more peaked. When the shape of the current profile changes, so does the whole island structure, as seen in Fig. 7.3. Finally, the magnetic structure is so disturbed that the steady state can no longer be maintained, and the plasma has to change. What the tokamak does is to eject the overly hot plasma in outward bursts, thus cooling the center back to normal. This explanation was for a long time only a conjecture, but recent advances in instrumentation have enabled actual movies of these sawtooth bursts to be taken in real time. These movies show that the temperature actually oscillates several times before the big crash, when hot plasma is shot out and replaced by cooler plasma. Still frames from the movie by H.K. Park of the Princeton Plasma Physics Laboratory are shown in Fig. 7.6, but they do not do justice to the actual product [3].

Diagnostics

The figure showing a sawtooth crash was generated with modern measurement techniques and brings up the question, "How does one measure anything inside a fusion plasma?" At temperatures over a million degrees, nothing put into the plasma will survive. Plasma diagnostics is a whole field in itself, but here is a brief summary. There have to be enough windows, or "ports" to get light or other beams

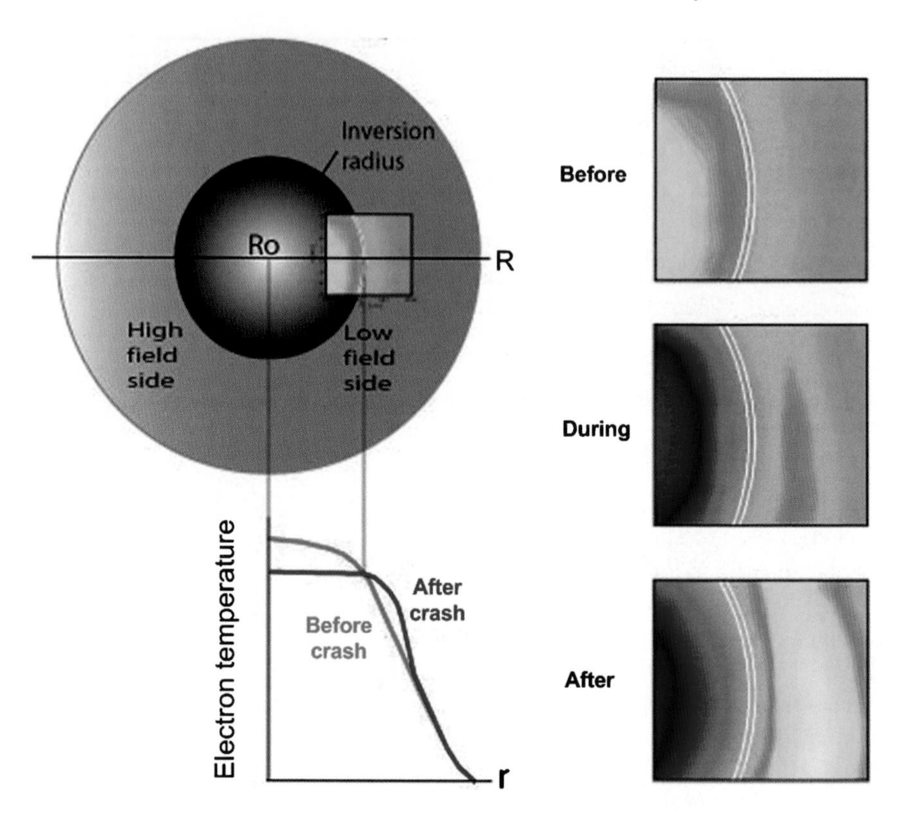

Fig. 7.6 The temperature distribution before, during, and after a sawtooth crash (*top right* and *colored squares*). *Yellow* or *light color* is hot, and *blue* or *dark color* is cold. The pictured region at the $q=1$ surface of the tokamak cross section is shown at the *left*. The *small graph* at the *bottom* shows the temperature distribution before and after the crash, showing that hot plasma has been moved out of the $q=1$ surface [3]

into and out of the plasma. The scattering of a laser beam can give information about the electron temperature and density of the plasma. Crossed laser beams can be used to measure the ion temperature. The transmission or emission of electromagnetic radiation in the microwave, X-ray, or infrared range can show oscillations inside the plasma. Sawteeth were first observed from the fluctuations in the soft X-ray radiation emitted by electrons at twice their cyclotron frequencies. Since the frequency depends on the magnetic field, and the field varies with position, this also tells where the radiation is coming from. Diagnostic beams of neutral atoms or heavy ions can be injected into the plasma since they can penetrate the magnetic field. In beam emission spectroscopy (BES), a beam of neutral hydrogen atoms is injected and reaches the plasma interior. There, it collides with the ions and is ionized by the electrons. In the process, it emits light whose spectrum can be analyzed by computer. This light carries information about the density and velocities of the ions and even the strength and direction of the local magnetic field. Heavy ion beam probes (HIBPs) work even better and can even measure the internal electric fields. These are the main ways, developed over many years, to probe inside a fusion plasma and get the knowledge we now have on how a magnetized plasma behaves.

Self-Organization

Sawteeth are an essential feature of tokamak discharges and are important because they show that a tokamak is *self-healing*. Toruses such as stellarators do not have such a feature because the magnetic structure is fixed by magnetic coils outside the plasma. A stellarator plasma cannot adjust its own magnetic topology by sawtooth-shaped hiccups. This brings up the general subject of self-organization. Many physical systems have been found which are self-organized. It may seem inconceivable that an insentient object can organize itself, but there are many examples in real life. Snowflakes are self-organized. No one had to program a computer to make these beautiful, symmetric art pieces (Fig. 7.7).

Our own bodies are self-organized. Complicated organs such as the eye, with its cornea, iris, lens, retina, and macula; and the ear, with its ossicles, cochlea, hair cells, and stereocilia, are self-organized, though some programming had to be done with the DNA. In the new field of nanotechnology, the objects are so small that they are difficult to make; and people are hoping that self-organization will help. In magnetic fusion, tokamaks have taken the lead partly because of their ability to heal themselves. There are magnetic bottles other than the standard tokamak emphasized here that depend even more on self-organization (Chap. 10).

Fig. 7.7 A snowflake is a self-organized object

Magnetic Wells and Shapely Curves

Up to now, we have suppressed plasma instabilities by applying magnetic shear, creating a mesh of field lines that plasma cannot easily penetrate. There is another good way to eliminate instabilities, and that is to create a magnetic well. This is a magnetic bottle that surrounds the plasma with a stronger field on every side. The plasma then does not have enough energy to climb out of the hole that it is in. It is not possible to make such a container without a leak, which is why tokamaks do not depend on this effect as much as some other confinement concepts do. However, understanding the magnetic-well effect will help in the design of better tokamak shapes.

A simple magnetic well can be made with four infinitely long rods with opposite current in neighboring rods, as shown in Fig. 7.8. The magnetic field lines are the circles, and their spacing shows that the field gets stronger as one approaches each rod. A plasma trapped in the center would see the field increasing in every direction and would be held stably. However, there are leaks at each of the four cusps, where the field lines meet. An ion or electron following a field line toward one of the four cusps, where the field is strongest, would be reflected by the magnetic mirror effect described in Chap. 6. Unfortunately, that effect depends on the transverse momentum of the particles – the momentum that makes the particles gyrate in Larmor orbits. Those particles that have their velocities almost parallel to the field lines would not be reflected and would go right out at the cusps. There are enough of those particles to bring the confinement time of the plasma well below the many

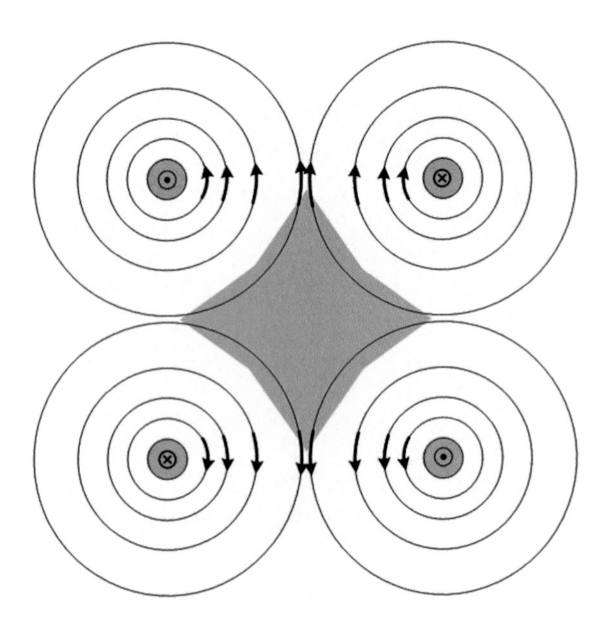

Fig. 7.8 Plasma in a magnetic well

Fig. 7.9 A "picket fence" confinement scheme

seconds required in a fusion reactor.[2] In the early days of fusion, one of the fanciful magnetic buckets that were proposed was the Picket Fence, a veritable Great Wall of China, as shown in Fig. 7.9. But if one had done his homework, he would have found that the leak at even one of the many cusps would have been insufferable.

Why does the magnetic field in cusp geometry look so different from the tokamak fields we have seen so far? It is because the field bulges out *towards* the plasma instead of away from it. In a magnetic well, the field lines are convex as seen by the plasma, not concave. This generally means that the field is stronger on the outside than on the inside, and such field lines are said to have *good curvature*. Conversely, field lines that bulge outwards have *bad curvature*. This concept is much more general than its use in magnetic confinement. In Fig. 7.10, we see that a board which is bent upwards will support more weight than one which is level or sags downwards. Roman arches and those highly arched wooden bridges in Japanese gardens have good curvature.

Tokamaks have mostly bad curvature, but they can be designed, as we shall see, to minimize that effect. A true magnetic well is called a *minimum-B* device, where the plasma is in a magnetic field minimum. The twisting field lines in a torus can go through regions of both good and bad curvature. In that case, what matters is how much there is of each kind. If an electron sampling all regions of a magnetic surface sees mostly good curvature, it would be in an *average-minimum-B* device. It is hard to do this in a tokamak, but other toroidal systems which cannot be described here can be designed to be average-minimum-B. The idea is to minimize the time a particle spends in a region where the field is sharply bent in the bad direction. When an instability is concentrated in a region of bad curvature, it is

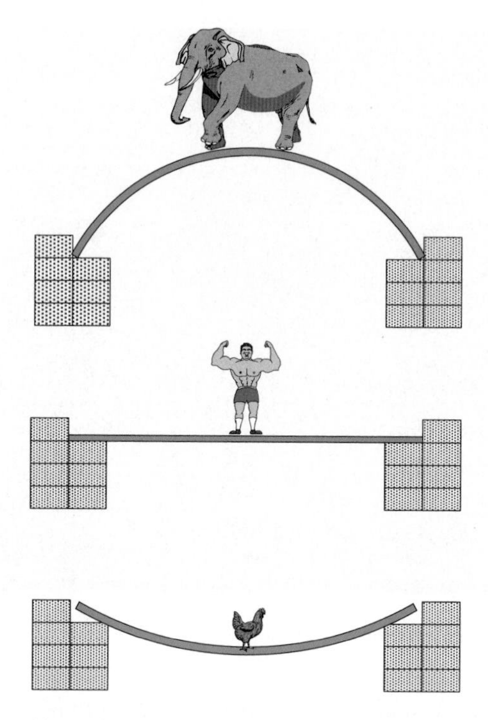

Fig. 7.10 A structure with good curvature will support more weight than one with bad curvature

called a *ballooning mode*. The plasma escaping is such a region pulls the field lines with it, further weakening the field. This could be called a plasma hernia, but ballooning is a more dignified term!

Evolution of the D-Shape

The reason that a toroidal magnetic bottle has to have twisted field lines is that ions and electrons drift vertically in opposite directions, as explained in Chap. 4. This drift arises from the fact that the magnetic field is necessarily weaker on the outside of the torus than on the inside, near the hole in the doughnut. An obvious idea to get a larger volume of plasma without changing the drifts is simply to make the tokamak taller, without changing its radius. This is shown in Fig. 7.11a. The sharp corners have very bad curvature, so they have to be rounded off. A machine built at General Atomics in San Diego, the Doublet, is shown in Fig. 7.11b. This looks like two merged tokamaks, one on top of the other, connected by a region with good curvature. The bean-shaped cross section studied at Princeton University has good curvature on the inside of the torus and shown in Fig. 7.11c [4]. It turns out that it is not necessary to curve the inside surface; keeping it straight is almost as

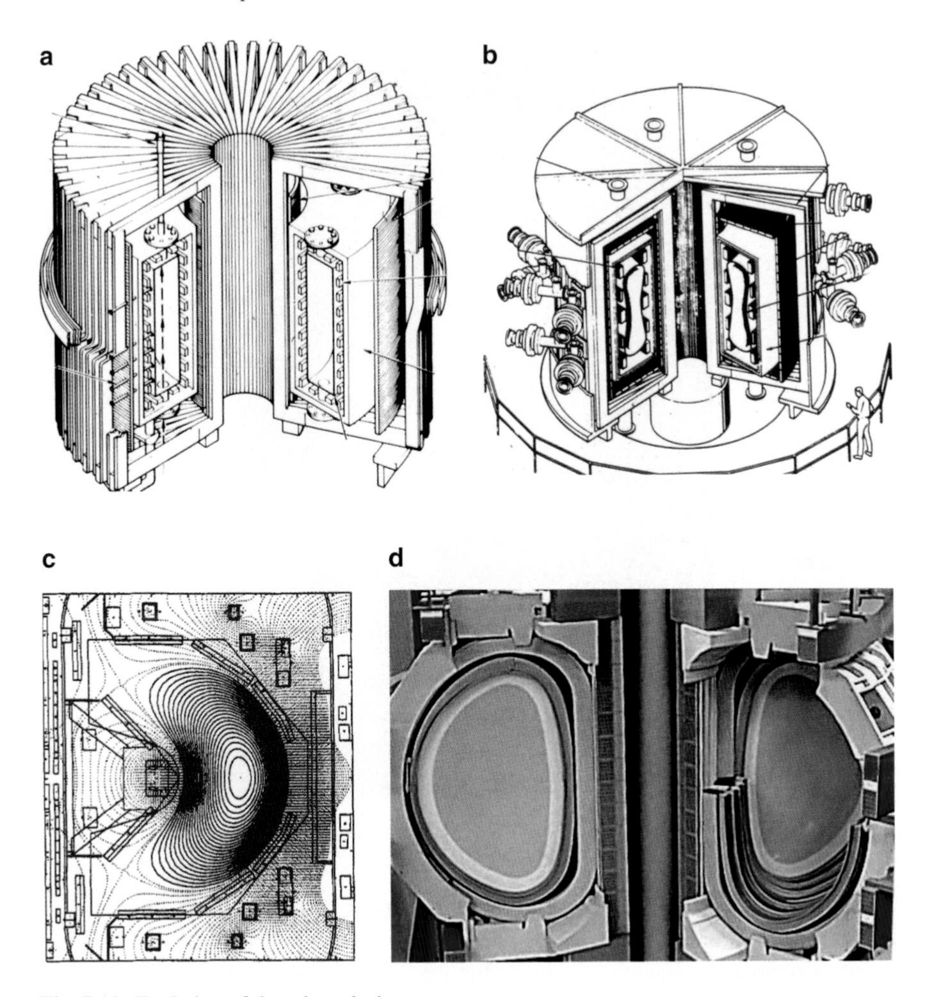

Fig. 7.11 Evolution of the tokamak shape

good, since the magnetic field naturally gets stronger as the plasma tries to escape toward the inside of the doughnut. We then have the D-shape shown in Fig. 7.11d. The outside of the D still has bad curvature, but it curves more gently than in a circular tokamak because of the elongation. Figure 7.12 is a D-shaped toroidal-field coil shown during the construction of the ASDEX tokamak in Germany. This was one of the first large tokamaks of the time (ca. 1980) but is small compared with those operating today.

The D-shape is not all gravy: the bad curvature at the corners of the D is very sharp, but at least it occurs in only a small part of the total surface. Actually, this part of the D can be used for a necessary function – that of plasma exhaust. A product of D–T fusion is helium (alpha particles). This "ash" has to be taken out since confining it would use up the magnetic confinement capability reserved for the DT. Furthermore, the normal escape of DT plasma, though slow, still carries out

Fig. 7.12 A D-shaped ASDEX coil

more heat than the walls of the chamber can stand. By channeling the escaping plasma into the corners of the D, special devices called *divertors* can be placed there to handle the heavy heat load. Figure 7.13 shows a diagram of the cross section of a D-shaped tokamak with divertors. The last closed magnetic surface is changed with locally placed coils so that the field lines leave the surface and lead outwards into the divertor. Plasma diffusing to that surface then enters the divertor, where it is captured by high-temperature, rapidly cooled materials.

How to Heat a Plasma to Unearthly Temperatures

We saw in Chap. 5 that the plasma in a fusion reactor has to have a temperature of at least 10 keV (about 100 million degrees), but most of our deliberations have been about the problem of keeping a plasma from leaking out of its magnetic container. Isn't heating to 50 times the temperature of the sun a bigger problem? The problem is nontrivial, but there have been no unexpected effects comparable to, say, microinstabilities. The simplest way to heat a plasma is to drive a current through it.

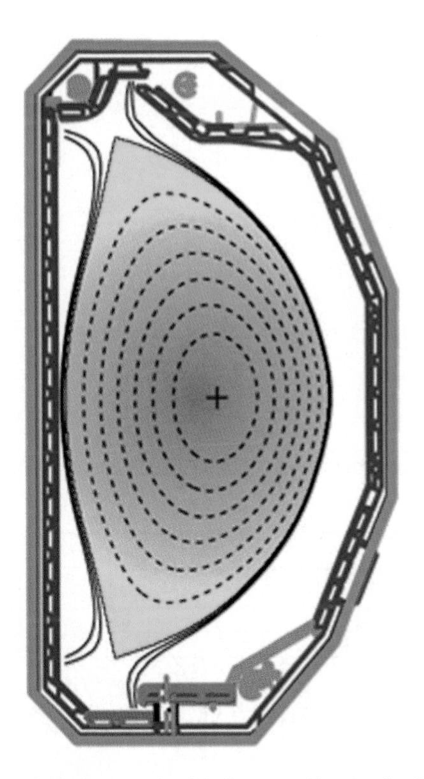

Fig. 7.13 Diagram of a D-shaped tokamak with divertors (drawing by Tony Taylor of the DIII-D tokamak configuration at General Atomics, San Diego, California)

A current is needed in a tokamak anyway to produce the poloidal field. This is *ohmic heating*, which happens whenever there is resistance in a wire carrying a current, such as in a toaster. The plasma in a tokamak can be considered as a one-turn wire loop, even though it is a gaseous one. It has a resistivity due to electron–ion collisions. When a voltage is applied around the loop, the electrons carry the current; and when they collide with ions, their velocities get randomized into a bell-shaped distribution, raising the temperature. The usual way to apply an electric field to loops of wire is to use a transformer, a common household device. It is the heavy piece of iron found in fluorescent lights and in the power bricks of electronic devices like cell phone chargers. Very large transformers are used to convert the high voltage of the power line (as much as 10,000 V) down to the household 115 V AC that we use in the USA or the 230 V in Europe. We know about these because they sometimes blow up, causing a power outage.

The first tokamaks used transformers for ohmic heating, as illustrated in Fig. 7.14. A pulse of current in the primary winding (shown as the three turns on the outer legs) drives a larger current in the plasma, which forms a one-turn secondary winding. This method was OK for small research machines, but the transformer would be too large in a large machine. Instead, one can use an air-core

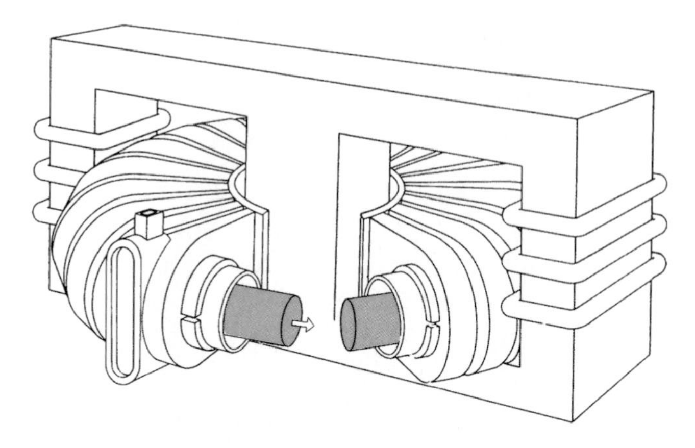

Fig. 7.14 Use of an iron-core transformer to drive ohmic heating current

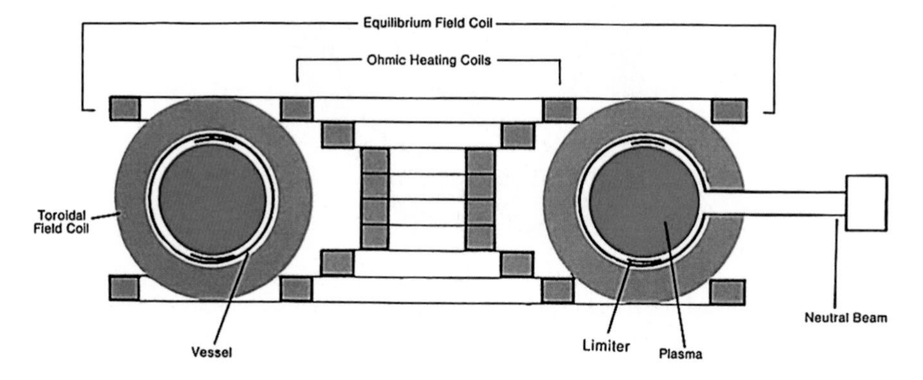

Fig. 7.15 Use of an air-core transformer to drive ohmic heating current

transformer without the iron, as shown in Fig. 7.15. What are shown are toroidal coils, known as OH (ohmic heating) coils, which go around the torus the long way. A pulsed current in the OH coils induces a current in the opposite direction in the plasma. This is inefficient compared to an iron transformer, but it is easier to drive a large current in the OH coils than to create the space for a large iron transformer. The "Equilibrium Field Coil" in that figure generates the vertical field described at the end of Chap. 6. Note that Fig. 7.15 is intended only to show the principle; actual "poloidal-field coils" are numerous toroidal coils located mostly on the outside of the torus and combine the currents necessary for equilibrium, ohmic heating, and shaping of the plasma.

At this point, the words *poloidal* and *toroidal* have been used so often that it may be well to review what these terms mean to avoid any further confusion. A *toroidal* line goes along a doughnut, or even a pretzel, the long way, tracing out a circle in the case of a doughnut and a figure-8 in the case of a pretzel. A *poloidal* line goes the short way around the cross section of a doughnut, encircling the dough but not the hole. What is confusing is that magnetic and electric fields are generated differently by currents flowing in coils. For magnetic fields, a toroidal field is generated by poloidal coils which pass through the hole and encircle the plasma. Thus, the main toroidal magnetic field of a tokamak is generated by *poloidal* coils called *toroidal-field* coils! These are the blue coils seen in Fig. 7.15. A toroidal *coil* generates a magnetic field passing *through* the coil. Thus, the largest red coils in Fig. 7.15 generate a more or less vertical magnetic field, which is *poloidal* even though it does not actually encircle the plasma the short way. For electric fields, the opposite is true: a toroidal coil will generate a toroidal current. Thus, the smaller toroidal red coils in Fig. 7.15 are used to induce toroidal currents in the plasma. These are the OH coils. It is not necessary to understand this. Creating the fields we need is straight electrical engineering, and there are no unexpected plasma instabilities!

Ohmic heating cannot be the primary heating method in a fusion reactor for two reasons. First, OH cannot raise the plasma temperature high enough for fusion because, as explained in Chap. 5, the plasma is almost a superconductor at those temperatures. Collisions are so rare that the plasma's resistance is almost zero, and resistive heating becomes very slow. Second, transformers work only on AC, whereas a fusion reactor must be on all the time in a DC fashion. The current induced in the secondary depends on an increasing current in the primary, and that current cannot increase forever. That is why tokamaks up to now have been pulsed, though very long pulses, of the order of minutes, are now possible. Other heating methods are used which can operate in steady state. Remember, however, that aside from ohmic heating, a current is necessary in a tokamak for producing a rotational transform – the twisting of the field lines. Fortunately, there are other ways to generate DC current for that purpose. One way is to launch a wave in the plasma that can push electrons along the magnetic field. Another is the "bootstrap current," a naturally occurring phenomenon that we describe in the "Mother Nature Lends a Hand" section. Stellarators are toroidal machines that do not need a current, since the rotational transform is generated by twists in the external coils. Hence, stellarators avoid the problem of current drive. They may ultimately be the way fusion reactors are constructed, but up to now we have had much more experimental experience with tokamaks.

Another way to heat a fusion plasma to the required millions of degrees is *Neutral Beam Injection* or NBI to those who like acronyms. This is now the preferred method, and it works as follows. Neutral atoms of deuterium with high energy (between 100 and 1,000 keV) are injected into the plasma. Being neutral, these atoms can cross the magnetic field. Once inside the plasma, the atoms are rapidly ionized into ions and electrons, producing beams of energetic deuterium ions. The velocity of the neutral atoms is adjusted so that they can go far into the plasma before they are ionized. Once ionized, the beam becomes a beam of fast deuterium ions, and these give their energy to electrons by "electron drag" and to the plasma ions by colliding with them, raising their temperature. Neutral atoms cannot be accelerated by an electric field because they have no charge, so to make a neutral beam one must start with charged particles. One can start with a positive ion, accelerate it, and then add an electron to make it neutral; or one can start with

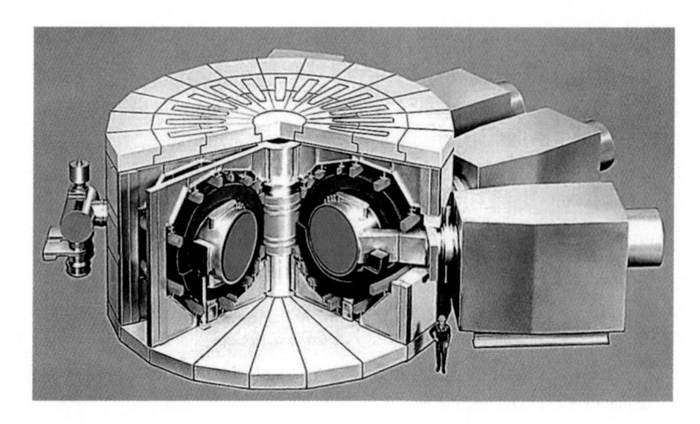

Fig. 7.16 Neutral beam injectors on a tokamak

a negative ion and then strip its extra electron to make it neutral. It is easier to do the latter. Hydrogen has an affinity for electrons, so negative deuterium (D^-) ions are not hard to make. They are then accelerated in a relatively simple accelerator. The extra electron in D^- is loosely bound, so it is easily stripped off when the beam passes through a little bit of gas; and a fast neutral is formed. Neutral beam injectors are very large and tend to take up more space than the tokamak itself. Figure 7.16 shows what a tokamak looks like when surrounded by neutral beam injectors. These beams can be aimed in different directions to give momentum to the plasma. Normally, it is best to use co-injection; that is, injection in the same direction as the tokamak current. This method of heating is powerful and necessarily changes the conditions of the plasma from what simple theory would predict. On the other hand, adjusting the beam affords another way to control the plasma.

There are three other major methods for heating worth noting: ion cyclotron resonance heating (ICRH), electron cyclotron resonance heating (ECRH), and lower-hybrid heating (LHH). In cyclotron heating, a high-frequency electric field is launched into the plasma, and its frequency is adjusted to match the gyration frequency of the particles as they rotate around in the magnetic field. These circular Larmor orbits were shown in Fig. 4.9. The electric field changes its direction at the cyclotron frequency, so that as the particle moves in a circle and changes its direction, the electric field follows it so that it is always pushing the particle. Those particles that start out out-of-phase are decelerated by the field but then get into phase and are pushed. They collide with one another to thermalize, thus raising the temperature of the whole gas. This works for both ions and electrons, but the technology is entirely different.

ICRH requires power generators with frequency in the tens of MHz (million cycles per second). This is in the radiofrequency range, between the bands used by AM and FM radios. Therefore, the generators are like those used by radio stations, only more powerful. The antenna, however, is not mounted on tall towers. It is a series of coils inside the vacuum chamber of a tokamak but outside the plasma, so that it does not get damaged.

ECRH requires generators of the much higher cyclotron frequency of electrons, around 50 GHz (billion cycles per second). This is in the microwave range. Microwave ovens and some telephones operate at the standard frequency of 2.4 GHz, some 20 times lower. The magnetron that is used in microwave ovens puts out about a kilowatt of power. In fusion, special gyrotrons have been developed which can produce tens of megawatts continuously. As in a microwave oven, ECRH does not need an antenna; the waves go through a hole. A very useful feature of cyclotron heating is that it is localized. Cyclotrons work because the frequency does not change with particle energy (until it goes beyond an MeV), but the frequency *does* change with magnetic field. Since the magnetic field in a tokamak is not the same everywhere, this means that only the plasma located at the right magnetic field gets heated, and this position can be changed by changing the frequency. We have seen how the profile of the tokamak current can change the magnetic topology and the q-value of the rotational transform. Localized heating can change all this, giving operators a way to control the stability of the tokamak.

Heating can also be accomplished by launching waves into the plasma using different frequencies and different types of antennas. These waves bear names like lower-hybrid wave or fast Alfvén wave and belong to a large array of waves that can exist in a magnetized plasma. By contrast, the unmagnetized, un-ionized air that we breathe can support only two kinds of waves, light and sound. It remains to be seen whether wave heating will be practical in a real fusion reactor.

Mother Nature Lends a Hand

Many a frustrated physicist has complained that Mother Nature is a bitch. After the instability problems we described in previous chapters, fusion physicists would have agreed had the problems not been so challenging but soluble. There have even been several pleasant surprises when unexpected benefits were found that could not have been foreseen when fusion reactors were first envisioned. Some of these effects are now well documented; others still cannot be explained. The most remarkable of these surprises is the H-mode, a high-confinement mode on which present designs depend. It is so important that it deserves its own section, which follows this one at the next major heading.

Bootstrap Current

Since tokamaks depend on an internal plasma current to produce the required twist of the magnetic field lines, the current has to be produced even if it is not needed for ohmic heating. Fortunately, the plasma automatically generates such a current, figuratively "pulling itself up by its own bootstraps." This comes about as follows. Since the plasma is not perfectly confined but gradually diffuses to the wall, there will be a density *gradient*, with the density high in the center and low near the walls, where the plasma can leave quickly. Think of a packed football or soccer stadium where, at the end of a game, the crowds storm onto the field in spite of the guards. The density of people is high at the top, but the crowd disperses on the field, where the density is low, forming a density gradient. It is this density gradient in a tokamak that causes the bootstrap current. Technically, it is the *pressure gradient*, where the pressure is density times temperature. Consider a tokamak with its helical magnetic lines, as shown in Fig. 7.17. The twist in the lines of the magnetic field is created by a toroidal current J, which generates the poloidal component, B_p of the field. It is this poloidal part of the field which is important here.

Figure 7.18 is a closer look at the minor cross section of the plasma showing the same tokamak current J seen in Fig. 7.17. The black arrows show the force on the electrons exerted by the plasma pressure pushing outwards. We can neglect the ions because they move so slowly that they cannot carry much current. The electrons gyrate in small circles, so we need only to consider the drift of their guiding centers. In Chap. 5, we showed the gyroscopic effect on guiding centers, in which a force moves the guiding center in a direction perpendicular to that force. The relevant part of Fig. 5.7 is reproduced in Fig. 7.18a, showing that the B-field, pressure force, and electron velocity are mutually perpendicular. Note that the *current* is opposite to the electron velocity because of their negative charge. In Fig. 7.18b, the force is in the radial direction, pushing outwards, while the poloidal field B_p is in the azimuthal direction, going around in the circular direction. The electrons therefore drift in the direction perpendicular to both, which is the toroidal direction, the same as that of J. This toroidal electron drift constitutes the bootstrap current. It turns out that *this current is always in the same direction as J*, so that it adds to the total current. Once a *seed current* is induced in the torus so that the field lines twist enough to confine the plasma, the bootstrap current can then take over most of the work. There is, of course, also a pressure-caused drift perpendicular to the main *toroidal* component of the B-field, but this drift is in circles in the poloidal direction and does not contribute to the main tokamak current J.

Mother Nature made it hard for us to confine a plasma in a torus by requiring that the field lines be helical, but she then provided the benefit of bootstrap current so that this helicity can be mostly self-generated. It does not matter which direction the toroidal field is in, or which direction the toroidal current is in; the bootstrap current will always add to the toroidal current. In present experiments, the bootstrap current has been observed to contribute more than half the total current. In planned experiments, the bootstrap fraction will be more than 70%, and in fusion reactors more than 90%.

Detailed calculations[3] of bootstrap current can be made using the neoclassical banana theory described in Chap. 6. Although collisions between passing particles and those in banana orbits cause the major part of bootstrap current, the final answer does not depend on knowing the collision rate. Collisions cause the pressure gradient, but it is only the resulting pressure gradient that matters. Going back to the stadium full of fans, we see that a density gradient of fans will occur regardless of whether they bump and shove one another or whether they do not touch.

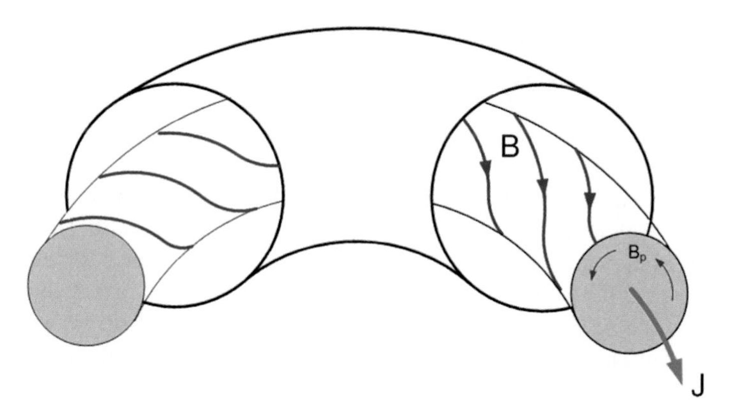

Fig. 7.17 A tokamak with a toroidal current J, which generates a poloidal field B_p, giving a twist to the magnetic field

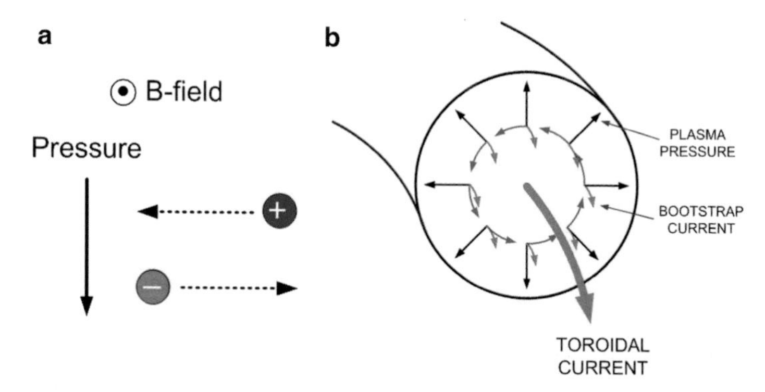

Fig. 7.18 Generation of the bootstrap current perpendicular to both the pressure gradient and the poloidal field

In designing tokamaks, the shape of the bootstrap current depends on the shape of the magnetic field, which itself depends on the bootstrap current, so a delicate optimization problem has to be solved. The so-called Advanced Tokamak designs with high bootstrap fraction have hollow current profiles with larger current at the edge than in the center.

The Isotope Effect

This is a beneficial but baffling effect that is still unexplained. In comparing the confinement times of tokamak discharges using hydrogen, deuterium, and helium, it has been carefully documented [5] that the confinement time *increases* with the mass of the ion, contrary to all neoclassical and instability theories. Heavier ions have larger Larmor radii, so their step size in diffusion across the magnetic field should be larger, leading to shorter rather than longer confinement times. If ions cross the field not by collisions but by instability, most theories predict one of two scalings with atomic number A. (Here A is not the aspect ratio that we used before but the more familiar A used in chemistry. A is 1 for hydrogen, 2 for deuterium, 3 for tritium, and 4 for helium.) The crudest estimate is Bohm diffusion, which we discussed in Chap. 6. That diffusion rate is independent of A. More refined theories predict gyro-Bohm scaling, which takes into account the Larmor radii of ions, which vary as the square root of A or $A^{1/2}$. That leads to confinement times that vary as $1/A^{1/2}$, shorter for heavier ions. What is observed, however, is that confinement varies more like $A^{1/2}$, improving by a factor between 1.4 and 2 between hydrogen and deuterium. This means that confinement will be even better with tritium, which is not normally used in small experiments because it is radioactive.

The isotope effect seems to be universal, occurring in many different types of tokamak discharges. At first it was proposed that it is caused by impurities in the gas, but very clean discharges also exhibit this effect. There have been several theories on specific instabilities that could have nonlinear behavior that depends on A in this fashion, but so far these have not been confirmed in tokamak experiments. A factor of 1.5 or 2 may be trivial in an experiment but would have great commercial benefits in a power plant.

The Ware Pinch

The first attempts at fusion were carried out with a simple device called a "pinch," which we will describe first. This was a tube filled with a low-pressure gas in which a large pulsed current was driven by a voltage applied to electrodes at either end. As shown in Fig. 7.19, the current first ionizes the gas into a plasma and then generates a magnetic field surrounding the plasma. If the cylinder were turned into a torus, the current would be in the *toroidal* direction, and the field in the *poloidal* direction. This is like the current in a tokamak, but in a pinch there is no toroidal

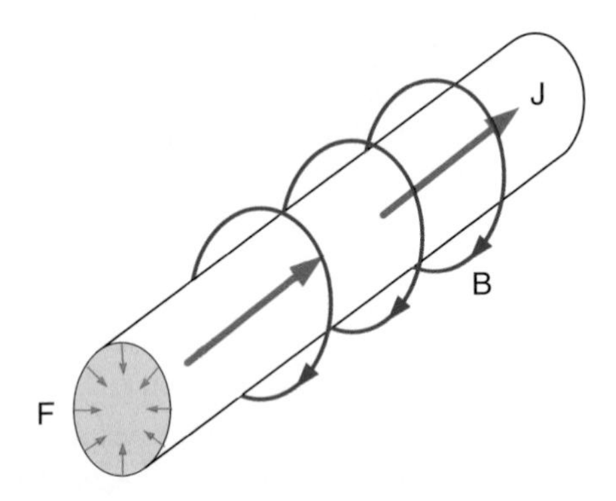

Fig. 7.19 A linear pinch carries a large current J, which creates an external magnetic field B. This field pushes the plasma inward with a force F, thus "pinching" it

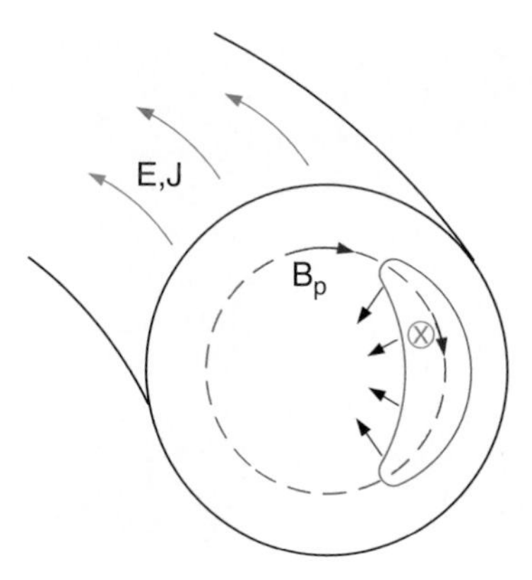

Fig. 7.20 In a Ware pinch, particles in banana orbits are pushed inwards by the **E**×**B** force of the toroidal electric field and the poloidal magnetic field

field from external coils. The magnetic pressure of the "poloidal" field in Fig. 7.19 then compresses the plasma to a smaller diameter, whereupon the magnetic field gets stronger, compressing the plasma even more. Since compression heats the plasma, the hope was that the heating would reach fusion temperatures. Of course, the system suffered from the kink instability described in Chap. 6, and the kinks drove the plasma into the walls.

The Ware pinch [6] is a more subtle effect occurring in tokamaks and affecting mostly particles which move in banana orbits. The mechanism is illustrated in Fig. 7.20. In tokamaks in which the toroidal current J is driven by a toroidal electric field E, the poloidal-field component B_p which J produces is in the direction shown in diagram. This is the field that gives the necessary twist to the magnetic lines. Crossed electric and magnetic fields give rise to a perpendicular **E** × **B** drift of the guiding centers, as shown in Fig. 5.4. This drift is always toward the center of the cross section regardless of where the particle is in its banana orbit, and the drift has the same direction and magnitude for both ions and directions. Note that B_p is small compared to the toroidal component B_t, but B_t is parallel to E, not crossed with it, so it does not give an **E** × **B** drift. Thus, the principal fields in a tokamak generate a drift that counteracts the outward diffusion of the plasma, at least for particles trapped in bananas. The Ware pinch effect was invented to explain observations of oscillations occurring when the pinching reached its limits and started over again. This effect has been observed in other tokamaks and is not an artifact of neoclassical theory. It is another of Mother Nature's gifts.

Zonal Flows

The major instabilities encountered in early toroidal confinement research have been controlled. The remaining microinstabilities are of the drift-wave type, which we described in some detail in Chap. 6. They differ only in the energy source that

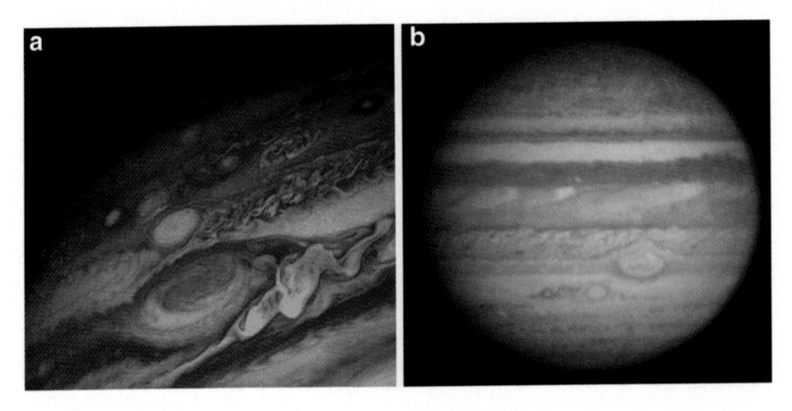

Fig. 7.21 (**a**) Turbulent eddies in the clouds of Jupiter. (**b**) Zonal flows in Jupiter's atmosphere

feeds them and in the collisional process that allows guiding centers to be unglued from magnetic lines. The effect of these instabilities on how long a plasma can remain trapped depends on the type of turbulence that the waves grow into – their nonlinear behavior, as physicists would say. In fluids like water or air, turbulence can the form of swirling eddies. For instance, in the picture of the surface of Jupiter shown in Fig. 7.21a, turbulence driven by winds is visible in the cloud patterns, including the largest eddy, the famous Great Red Spot. In water or air, flows are driven by pressure differences. In a magnetized plasma, flows across the magnetic field are driven instead by electric fields (the aforementioned **E** × **B** drift), but can also give rise to turbulent eddies. However, in a tokamak, Mother Nature reveals another of her helpful tricks: these eddies are self-limiting in their sizes! This means that large eddies like the Great Red Spot cannot occur – eddies that could otherwise bring plasma toward the wall rapidly across their diameters.

Referring back to Fig. 6.17, we see that drift waves create poloidal electric fields by the bunching of alternately positive and negative charges. These E-fields cause inward or outward flow of plasma in the radial direction, and the net loss of plasma comes about because the E-fields are phased so that the drift is always outward where the density is high and inward where the density is low. A better picture of these eddies is shown in Fig. 7.22. The distribution of "+"and "–" charges is, as shown in Fig. 6.17, generating the alternating electric field shown by the short red arrows. Together with the toroidal magnetic field, this E-field causes an **E** × **B** drift of the plasma in the closed loops or eddies, also called *convective cells*. The density pattern of the drift wave is displaced with respect to these eddies in such a way that the density is higher (blue) where the drift is outward and lower (red) where the drift is inward. Thus, the net motion of the plasma is outward. The danger is that these convective cells could be long "streamers" in the radial direction, as drawn here, so that the plasma can move a long way toward the wall in each cycle of the wave.

Fortunately, this does not happen because the turbulence takes on a different form as it grows. Alternating drifts in each radial layer automatically arise,

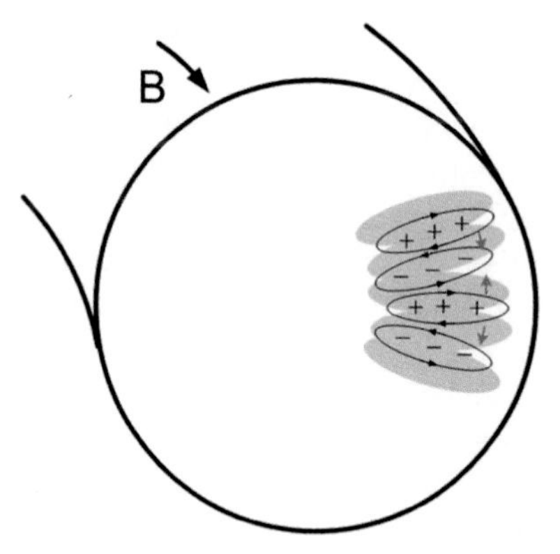

Fig. 7.22 Cross-sectional view of eddies caused by microinstabilities in the outer part of a torus. The electric charges and fields and the resulting drifts are shown, as well as the density fluctuation

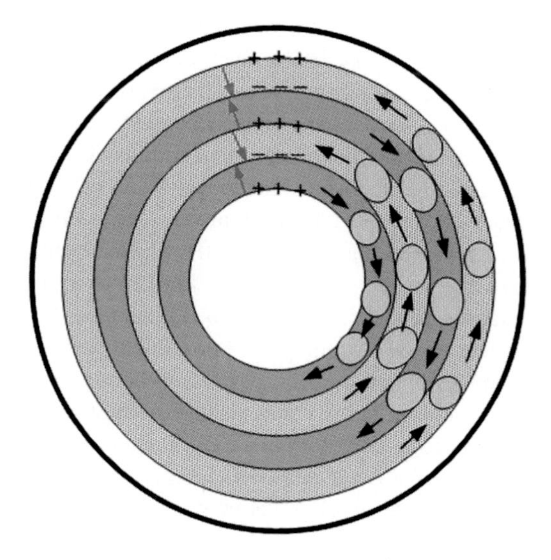

Fig. 7.23 Turbulent flows in the poloidal direction break up the eddies into smaller ones. This pattern oscillates in time but also has a steady-state component. The flows are **E**×**B** drifts in the electric fields (*red arrows*) of the + and – charges shown

as shown in Fig. 7.23. These are the zonal flows. The flows are **E** × **B** drifts driven by "+"and "–" charges on the boundaries of each zone. They break up the large convective cells into small ones, only about a centimeter wide, the size of an ion Larmor radius, so that the rapid convection in each cell can move the plasma only a short distance. The flows themselves cannot remove plasma, since they are parallel to the wall. In the picture of Jupiter taken by the Hubble Space Telescope shown in Fig. 7.21b, one can see zonal flows clearly in the upper half of the picture. In those stripes, the wind blows in alternating directions. The shear in the wind speed at the zone boundaries causes the turbulence seen more clearly in the bottom half of the picture. The zonal flows in a toroidal plasma, however, are fundamentally different. In the plasma, the zonal flows do not create the turbulence. *It is the turbulence that creates the zonal flows!* In other words, a zonal flow is an instability that is driven by another instability! Since a zonal flow is the same all around the torus in both the toroidal and the poloidal directions, it takes little energy to set it into motion. There is no need to add angular momentum to spin the flows around the poloidal direction, since the flows are in opposite directions in adjacent layers, so that the net angular momentum is zero. Microinstabilities in a torus develop into a turbulent state that incorporates zonal flows, a type of turbulence that is *self-limiting* in its eddy sizes. In principle, this should cause anomalous diffusion to be slower than theoretically expected, though this has not yet been shown experimentally.

Zonal flows were seen in many computer simulations of the nonlinear state of microinstabilities and have received extensive treatment by theorists [7]. The tool of computer simulation has greatly advanced progress in fusion in the past decade; this subject will be described shortly. Theories have been proposed on many details of zonal flows, including how a drift-wave instability can drive zonal flows through what is called a *modulational* instability. Such details have not been verified by experiment, but the existence of plasma flows that do not vary in either the poloidal or toroidal direction *has* been established experimentally [8]. In two Japanese laboratories, one with a tokamak and the other with a compact helical system (a type of stellarator), a sophisticated diagnostic called a heavy ion beam probe was used for this purpose. A beam of ions, usually cesium (Cs^+), is accelerated to such high energy that it has a Larmor radius larger than the plasma radius, and so it can be aimed at any part of the plasma. When it gets ionized to a doubly charged state (Cs^{2+}), its Larmor radius gets smaller and its orbit changes. By catching the Cs^{2+} ions at a particular part of the periphery, it is possible to tell the exact spot inside the plasma where this re-ionization occurred. Then the number and energy of the Cs^{2+} ions can tell the electron density and electric field at that spot, even if these are fluctuating at a high frequency. With this tool, fluctuations matching the characteristics of zonal flows have been detected. However, the predicted connection between the existence of zonal flows and an improvement in confinement time has yet to be quantified in the laboratory.

Time Scales

At this point, you may wonder how the complicated picture of banana orbits and magnetic islands jibes with the seemingly unrelated picture of convective cells and zonal flows in turbulence. These phenomena have different time scales. Hot electrons move almost at the speed of light, which is about one foot per nanoscecond. One trip around a large tokamak may be 20 feet, taking 20 ns. If it takes 100 trips to describe a banana orbit, that amounts to 2,000 ns or 2 μs. These individual particle motions therefore occur on microsecond time scales. Microinstabilities, on the other hand, have typical frequencies of 10 kHz, corresponding to a wave period of 100 μs. Growing into turbulence takes several periods, so the time scale is of the order of 1 ms. On this time scale, the plasma can be described as a fluid, but the fluid is not like water or air, in which the particles move randomly. *The fluid that participates in microinstabilities and turbulence in a tokamak consists of particles moving in the very peculiar orbits existing in toruses.*

There are two longer time scales. With a steady level of turbulence, the plasma settles into a steady state, arranging the distribution of toroidal current to give a stable q profile, possibly with magnetic islands. The radial distributions of density and of ion and electron temperatures arrange themselves so that everything is consistent. If these profiles become untenable, there are sawtooth crashes once in a while to rearrange them. All this happens in many milliseconds. Meanwhile, the plasma and its energy are leaking out slowly at rates described by the particle and energy *confinement times*. As discussed in Chap. 5, this time scale is of the order of seconds, possibly longer in a reactor-size machine.

To keep a discharge going in a reactor, DT fuel in the form of pellets is injected, and the helium "ash" is removed by the divertors. Before it is removed, the helium deposits its energy in the plasma to keep it hot. The length of a pulse in current tokamaks is determined by the transformer action needed to drive the toroidal current, since transformers cannot run DC-wise. Pulse lengths of the order of an hour are already possible. Reactors have to operate continuously, so the part of their current not generated by "bootstrap" will have to be driven "noninductively," by waves, for instance. Or else, reactors will have to be stellarators, which do not need a current to produce the twist of the field lines. A practical power plant will have to be designed to run continuously for months or years between maintenance shutdowns. *That* is the longest time scale.

High-Confinement Modes

The H-Mode

When neutral-beam heating was installed and turned on in the ASDEX tokamak in Garching, Germany [9] in 1982, Mother Nature came up with a major surprise that no one could have predicted. When the heating power was increased slightly from 1.6 to 1.9 MW, the plasma snapped into a new mode. Its temperature went up; its density went up; and the confinement times of both the plasma energy and the plasma particles went up, as dramatically shown by a sudden drop in the measured flux of escaping ions. It was as if a wall or dam, called a transport barrier, had formed, as depicted in the cartoon of Fig. 7.24a. The plasma would diffuse as it normally does up to this barrier, and then it would be held up by the barrier and leak out slowly in small bursts. This high-confinement mode, called the H-mode, came about from two innovations: the increase in heating power possible using neutral beams and the use of a single divertor of the type shown in Fig. 7.13. When the

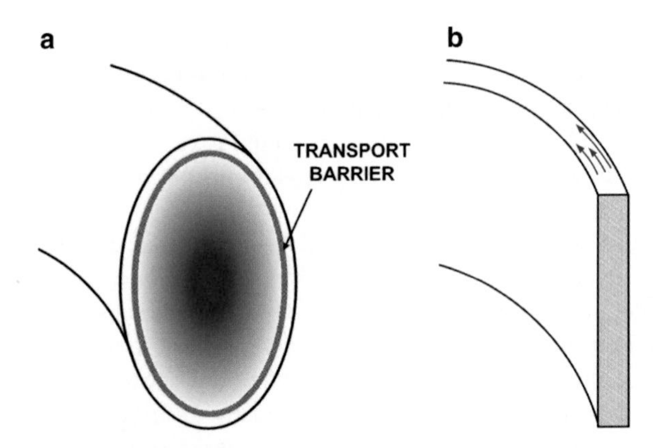

Fig. 7.24 (**a**) Location of the H-mode transport barrier in a tokamak. (**b**) Illustration of the sheared **E**×**B** drifts inside the thin barrier

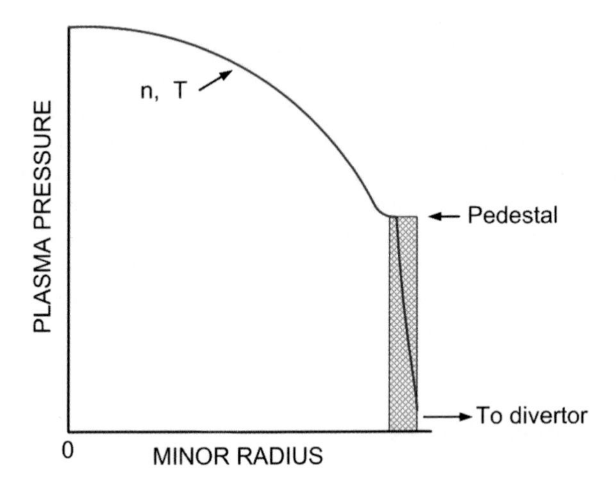

Fig. 7.25 Behavior of the plasma pressure when it encounters the H-mode pedestal

neutral beams are turned on below 1.6 MW, the confinement time actually gets a little worse because the beam disturbs the plasma equilibrium that was set up by ohmic heating. This is called the low-confinement L-mode. Once the power is increased beyond the H-mode threshold, the L- to H-transition occurs and a pressure *pedestal* forms.

Figure 7.25 shows what is meant by the pedestal. This is a graph of the plasma pressure as it varies across the minor radius; that is, from the center of the tokamak's cross section to the outside. Up to the pedestal, the plasma density and temperature (whose product is the pressure) fall gently from their maxima as in normal diffusion; but they do not fall all the way to zero. They hang up at a high value, so that the average pressure inside is higher than in the L-mode. At the pedestal, the pressure falls rapidly to nearly zero as the plasma is drained off to the divertor, where it recombines into gas and is pumped out. What happens inside the barrier is illustrated in Fig. 7.24b. Large electric fields in the direction of the minor radius are set up, and these cause perpendicular **E** × **B** drifts in the toroidal direction, as shown in Fig. 5.6. These drifts are not uniform but are highly *sheared.* Apparently, this sheared motion stabilizes the microinstabilities and slows down the diffusion from the instability-controlled diffusion in the interior. Note that this is *electric* shear stabilization, as opposed to the *magnetic* shear stabilization used in elementary forms of toroidal confinement devices.

The H-mode barrier layer is very thin, about 1–2 cm in a large tokamak with meter-sized cross sections. The H-mode is not a peculiarity of the tokamak, since it has been seen in stellarators and other toroidal devices. It is also not a phenomenon of neutral beam heating. It seems to have only two requirements: (1) that the input power be high enough and (2) that the plasma be led out by a divertor into an external chamber rather than be allowed to strike the wall. The latter requirement is due to the fact that impurity atoms or neutral atoms prevent the pedestal from forming. In the H-mode, the confinement time improves by about a factor of 2 (see Fig. 7.26a), and the plasma pressure by about 60%. A factor of 2 does not seem a lot, considering that confinement times have increased a million-fold since fusion research began; but we are now talking about a machine that is almost ready to be designed into a reactor. A factor of 2 can turn a 1-GW reactor into a 2-GW reactor, serving 1,000,000 homes instead of 500,000. All current designs for fusion reactors assume H-mode operation. The power produced by a reactor depends critically on the density and temperature of the pedestal.

How can we understand this freak of nature that we have stumbled on? There are two main problems: (1) How do the sheared fields in the barrier layer reduce the diffusion rate and (2) what causes this layer to form and how can we control that?

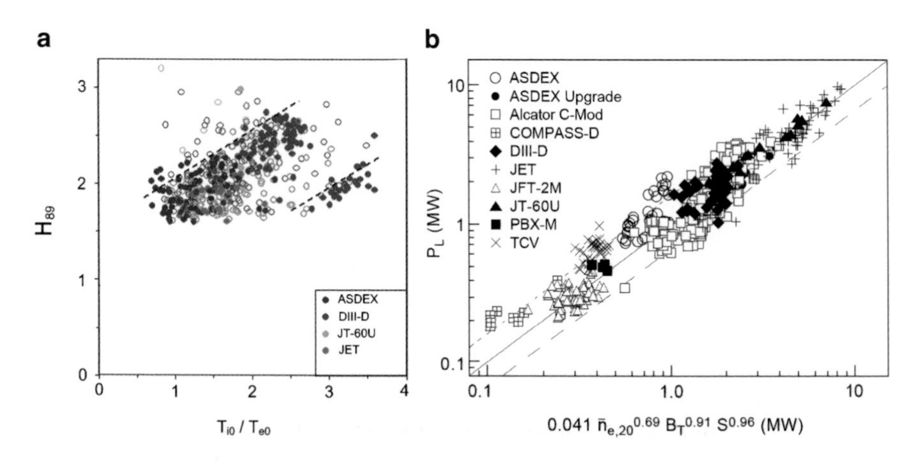

Fig. 7.26 (**a**) The H-mode confinement enhancement factor vs. ion–electron temperature ratio, as measured in four large tokamaks (adapted from A.C.C. Sips, Paper IT/P3-36, 20th IAEA Fusion Energy Conference, Vilamoura, Portugal, 2004). (**b**) Scaling law for H-mode threshold power vs. plasma density, toroidal magnetic field, and plasma surface area [10]

These have occupied the thoughts of a large fraction of fusion physicists for over two decades. One annual conference devoted to this topic has been going on for over 20 years. Sheared flows have good and bad effects. On the one hand, they can *cause* an instability, called the Kelvin–Helmholtz instability, which is well known in hydrodynamics. It is the instability that causes wind to ripple the surface of water. On the other hand, shear can quench an instability or at least limit its growth. In hydrodynamics, there is a simple theorem that tells what shape of shear is stable or unstable. In plasma physics, no such simple result is possible because so many kinds of waves can exist in a magnetized plasma. It is also difficult to make measurements in such a thin layer. The physics of the transport barrier – "edge physics" – is an ongoing study. The transport task force, a conference devoted to this topic, has been meeting yearly since 1988. More important, however, is to know how to turn on the H-mode. The threshold power depends on magnetic field, plasma density, and machine size. Since the H-mode threshold has been observed in so many machines, it was possible to formulate a scaling law that tells how the threshold depends on these various parameters. This is shown in Fig. 7.26b.

The H-mode has benefited not only our ability to confine plasma, but it has also improved our knowledge of plasma physics. Even the way in which the plasma's energy escapes from the barrier has turned out to be a considerable problem. It escapes by means of yet another instability, call an ELM. This is described in the next chapter.

Reversed Shear

The $q(r)$ profile of a tokamak discharge is perhaps its most important characteristic. It controls the stability of the plasma, where the magnetic islands form, and other essential features. An example of how the quality factor q varies with minor radius is shown in Fig. 7.2. It typically increases from 1 at the core to some number between 3 and 9 at the edge. Remember that q is the *reciprocal* of the rotational transform, so the twist of the magnetic field lines decreases gradually from the center to the edge of the plasma's cross section. The changing degree of twist provides the shear stabilization of instabilities. To increase the amount of shear would require $q(r)$ to change over a wider range than 1–9. However, q cannot be too large, because then the twist would be too weak to cancel the particles' vertical drifts; and q cannot be smaller than the Kruskal–Shafranov limit of 1, because, as we saw in Chap. 6, kink instabilities would occur. An obvious solution to this dilemma would be to make the twist change its angle several times, which would increase the shear without exceeding the bounds on q. This idea was never taken seriously earlier because there was no way to produce tokamak currents that would have to vary with radius in a screwy way. But now, all the large tokamaks have been able to produce "hollow" current profiles that are not peaked at the center but at some radius halfway out, as shown in Fig. 7.27. This generates a $q(r)$ that is large at the center, falls to a minimum somewhere inside, and then rises again to a normal value at the edge.

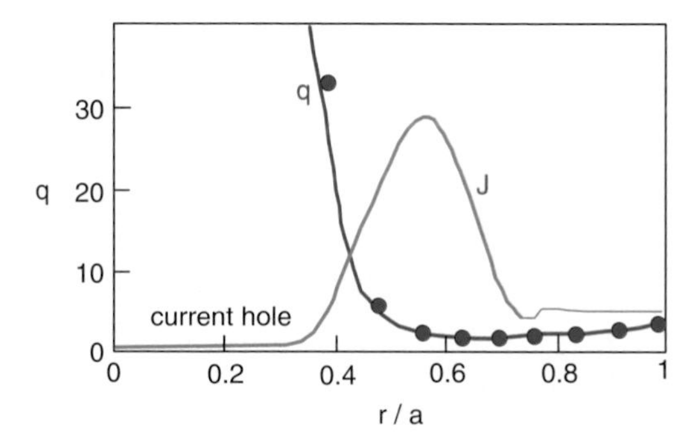

Fig. 7.27 A current distribution J with a hole at the center, creating a reversed-shear q profile. This is from an experiment on the JT-60 tokamak in Japan (adapted from T. Fujita and the JT-60 Team, Nucl. Fusion 43, 1527 (2003))

Physically, the twist of the magnetic lines is very small near the center, gets tighter halfway out, and then gets gentle again near the edge. The twist angle changes more rapidly with radius, thus increasing the shear. A lower turbulence level is observed as well as a corresponding increase in confinement time.

Initially, hollow current profiles were produced transiently by a combination of ramped neutral beam heating (increasing the power in a prescribed way) and auxiliary heating. This would not work for a reactor, which has to run in steady state; but by a fortuitous circumstance, bootstrap current can create hollow current profiles. This is yet another of Mother Nature's gifts. With the large bootstrap fractions in reactor-level machines, it is theoretically possible to design an "Advanced Tokamak" scenario in which the pressure profile leads to a bootstrap current profile that produces reversed shear, and the resulting reduced diffusion rate is consistent with the pressure profile! This sounds like a pipedream, but, as we shall see, much of this has already been accomplished in experiment.

Internal Transport Barriers

The achievement of reversed shear led to an even more important discovery: internal transport barriers or ITBs (another acronym that I shall avoid). These are like the H-mode pedestal but can be created inside the plasma, away from the walls. They effectively stop the fast transport of plasma to the walls caused by instabilities and turbulence. At the radius where the q profile is at a minimum, the shear in both the poloidal magnetic field and the poloidal $\mathbf{E} \times \mathbf{B}$ drift is so strong that most instabilities are quenched, and anomalous diffusion comes to a stop, as if there were a wall in the middle of the plasma. This is another unexpected benefit discovered only by painstaking experiment on large tokamaks. Figure 7.28 shows how an internal transport barrier should be designed. If it is placed close to the axis (dashed line), the hot, dense plasma will be limited to the small volume inside the barrier. Furthermore,

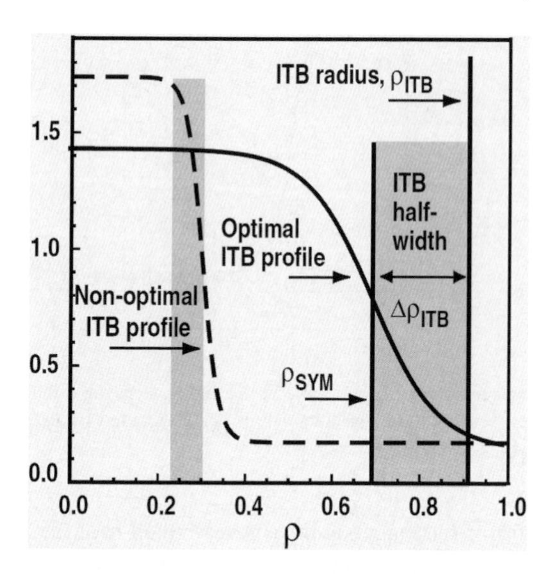

Fig. 7.28 Schematic of internal transport barrier profiles. The abscissa ρ (rho) is the fractional minor radius *r*/*a*. The *curves* are generic and can represent density or ion or electron temperature. The *dashed curve* shows a barrier that is too narrow and too close to the axis. The *solid curve* is an optimal profile, centered at ρ_{SYM} and spread out to a width $\Delta\rho_{ITB}$ on either side of its center [11]

it turns out that a barrier placed further out, as shown by the solid lines, is more consistent with the current profiles achievable with bootstrap current. The width of the barrier also makes a difference. It has to match the size of the turbulent eddies to be suppressed. Since the large eddies are more dangerous, the barrier should not be too narrow.

To create a good internal barrier, the current profile has to be manipulated so that it does not peak at the center, as it tends to do. This is done by adjusting the current in the ohmic heating coils and using waves to drive additional currents (noninductive current drive). The wave used is primarily the so-called lower-hybrid wave, but electron cyclotron waves are also used. The radial location of the currents driven by waves can be adjusted by changing their frequencies. In the most intense discharges produced to date, the bootstrap current can make a significant contribution. Internal transport barriers have been produced in all four of the largest tokamaks in operation: the ASDEX Upgrade in Germany; the DIII-D in General Atomics of San Diego, California; the JT-60U in Japan; and JET, the European tokamak in England. A fifth large machine, the TFTR in Princeton, New Jersey, has been decommissioned and scrapped as a result of budget cuts by the US Congress. The example shown here is from the DIII-D [11].

In the following example, a double barrier was actually achieved, consisting of an H-mode edge barrier in addition to an internal barrier. The *q* profiles are shown in Fig. 7.29, one with the internal barrier alone and one with a double barrier. In neither case does the *q*-value drop below the Kruskal–Shafranov limit of 1.

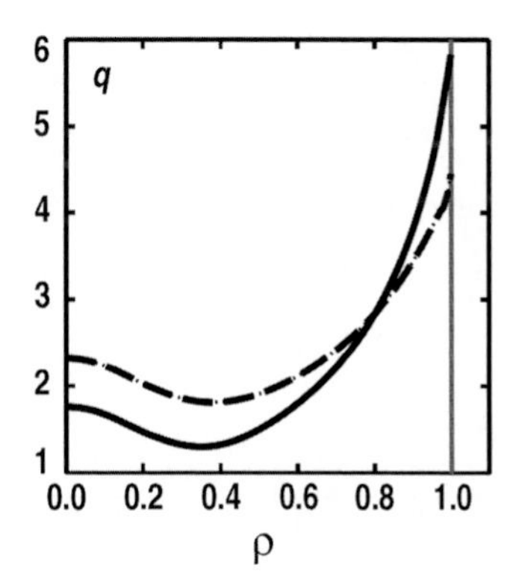

Fig. 7.29 The q profile of a discharge with an internal barrier (*dashed line*) and with a double barrier (*solid line*) [11]

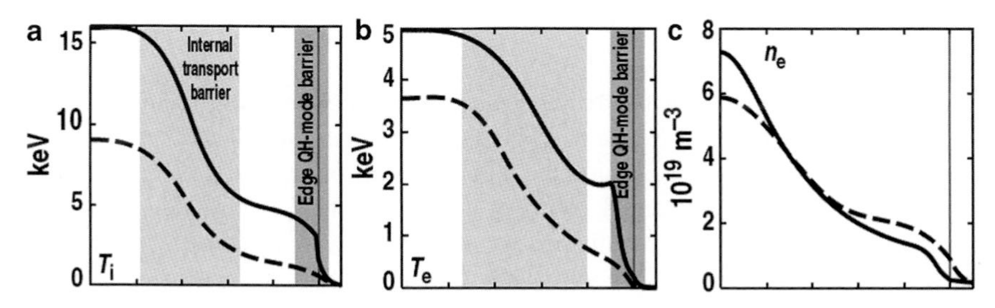

Fig. 7.30 Radial profiles of ion temperature (**a**), electron temperature (**b**), and plasma density (**c**) in single internal barrier (*dashed line*) and a double barrier (*solid line*) discharges [11]

The effect of the barriers on the plasma is shown in Fig. 7.30. Both curves show the high temperatures and density inside the internal barrier, and the solid curve shows the large increase in temperature when the edge barrier is added.

That the transport barriers dramatically reduce the losses and increase the energy containment in a tokamak is shown in Fig. 7.31. What is shown is the radial variation of the thermal diffusivity χ (chi) of ions and electrons; that is, the rate at which their energies are being transported outwards at each radius in the discharge. A low value is good; a high value is bad. The dotted curves show χ when there are no transport barriers. As before, the dashed curves show the case with an internal barrier alone and the solid one with both barriers. These dip well below the barrier-less curve inside the barriers. At the bottom of the χ_i plot, are thin lines showing what χ_i should be according to neoclassical theory; that is, if there were no instabilities. Normally, the turbulence level from microinstabilities makes χ_i much larger than the ideal theoretical value. Here, we see that the internal barrier has brought χ_i down to the ideal level for the first time, at least in the inside part of the plasma.

These results were obtained in a powerful tokamak, with 1.3 MA (megamperes) of toroidal current, a toroidal field of 2 T (20,000 G), and a bootstrap-current fraction of 45%. For best barrier formation, it was found that it was better to heat the

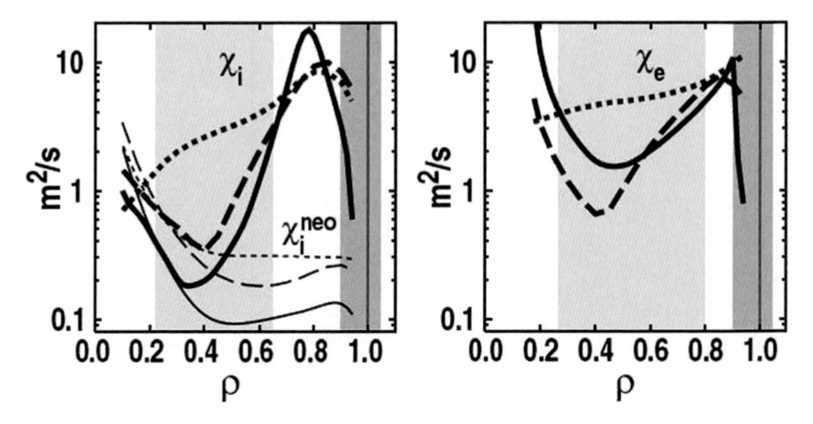

Fig. 7.31 Radial thermal diffusivity profiles of ions (*left*) and electrons (*right*) for double barrier (*solid curves*), single barrier (*dashed curves*), and no barrier (*dotted curves*) discharges in DIII-D, when compared with theoretical values (*thin curves*, *left*) [11]

plasma with neutral beams injected opposite to the direction of the tokamak current than along it, as in the usual case. This could be done without moving the large beam injectors by simply reversing the polarity of the current in the ohmic heating coils. In the larger JET tokamak, running with DT instead of pure deuterium, ion temperatures up to 40 keV, maintained for almost 1 s, were achieved with an internal transport barrier. The magnetic field there was 3.8 T, and the plasma current was 3.4 MA [12]. Taken together, the data from the large tokamaks, especially those with large bootstrap fraction, give credence to the hope that the Advanced Tokamak scenarios can be used in the design of a practical reactor.

Although the possibility that reversed shear and internal transport barriers could reduce the plasma loss rate was predicted theoretically [13, 14], turning the idea into reality depended on the availability of machines large enough to produce this effect and on hands-on twiddling of these machines to attain the right conditions. The diagnostics needed to quantify these results with detailed measurements inside the plasma also required major equipment and advanced technique. For instance, to get the $q(r)$ profiles a sophisticated method called Motional Stark Effect was used which actually measures the pitch of the field lines at every radius.

Fusion has suffered from the reputation that it is always promised to be available in 25 years. This was because the difficulties were not initially known. They have been overcome, but it took time and funding to build the necessarily large research machines, to train a generation of plasma physicists, and to develop the diagnostic tools to be able to see what we are doing. The underlying physics is now understood well enough that more accurate estimates of what it takes to make magnetic fusion work can be made. Thousands of dedicated physicists and engineers labored for decades to bring fusion within the foreseeable future. There are still a few physics problems to be solved, as described in the next chapter. Engineering is another matter. What has to be done to make fusion reactors practical is the subject of Chap. 9.

Notes

1. Courtesy of Roscoe White.
2. An acute reader would ask, "Why don't we just let those non-gyrating particles go and confine the rest?" The reason is that those particles which have leaked out would be quickly regenerated by the plasma in what is called a velocity-space instability. It is another of a plasma's tricks to bring itself to thermal equilibrium without waiting for collisions to do so.
3. A nice treatment of this is given by Jeff Freidberg, in *Plasma Physics and Fusion Energy*, Cambridge University Press, 2007.

References

1. N.R. Sauthoff, S. Von Goeler, W. Stodiek, Nucl. Fusion **18**, 1445 (1978)
2. E.A. Lazarus et al., Phys. Plasmas **14**, 055701 (2007)
3. H.K. Park, in *13th International Conference on Plasma Physics*, Kiev, Ukraine, 2006
4. R.E. Bell et al., Phys. Fluids B **2**, 1271 (1990)
5. M. Bessenrodt-Weberpals et al., Nucl. Fusion **33**, 1205 (1993)
6. A.A. Ware, Phys. Rev. Lett. **25**, 15 (1970)
7. P.H. Diamond et al., Plasma Phys. Control. Fusion **47**, R35 (2005)
8. A. Fujisawa et al., Phys. Rev. Lett. **98**, 165001 (2007)
9. F. Wagner et al., Phys. Rev. Lett. **49**, 1408 (1982)
10. ITER Physics Basis, Nucl. Fusion **39**, 2196 (1999). Chap. 2
11. E.J. Doyle et al., Nucl. Fusion **42**, 333 (2002)
12. C. Gormezano et al., Phys. Rev. Lett. **80**, 5544 (1998)
13. C. Kessel et al., Phys. Rev. Lett. **72**, 1212 (1994)
14. J.F. Drake et al., Phys. Rev. Lett. **77**, 494 (1996)

Chapter 8
A Half-Century of Progress*

What Have We Accomplished?

A controlled fusion reaction requires holding together for a long enough time a plasma that is hot enough and dense enough. These critical conditions can be quantified by the triple product $Tn\tau$, a modification of the Lawson criterion explained in Chap. 5. Here, T is the temperature of the ions, the reacting species; n is the density of either the ions or the electrons, since the plasma is quasineutral; and τ (tau) is the energy confinement time, a measure of how fast (or slowly) energy must be applied to keep T constant. Over the years, over 200 tokamaks have been built, and the value of $Tn\tau$ achieved in each has been calculated. Some of these are plotted in Fig. 8.1 as a function of time. This measure of success has increased over 100,000 times in four decades, recently doubling every two years.

Most of this increase has come from the confinement time. The first experimental machines suffered from hydromagnetic instabilities such as the Rayleigh–Taylor and the kink instabilities described in Chap. 5. These can take the plasma to the wall at the speed of a field line wiggle called an "Alfvén wave," which limits the confinement time τ to microseconds. Once these were controlled, τ increased a thousand-fold to several milliseconds, at which point microinstabilities were the limiting factor. After years of understanding banana orbits, magnetic islands, ballooning modes, and connection lengths, these instabilities were minimized; and τ increased another thousand times to the present value of several seconds.

The rate of progress in fusion can be compared with that in the development of computer chips, the famous Moore's Law. Gordon Moore had predicted that the number of transistors on a chip would double every two years, an unbelievable rate which was actually followed almost exactly. Figure 8.2 shows how this growth compares with a range of doubling times. The fusion figure of merit in Fig. 8.1 keeps pace with Moore's law, now also doubling every two years. Both of these outstrip Livingston's law for particle accelerators; where the energy doubling time is three years.

---

*Numbers in superscripts indicate Notes and square brackets [] indicate References at the end of this chapter.

F.F. Chen, *An Indispensable Truth: How Fusion Power Can Save the Planet*,
DOI 10.1007/978-1-4419-7820-2_8, © Springer Science+Business Media, LLC 2011

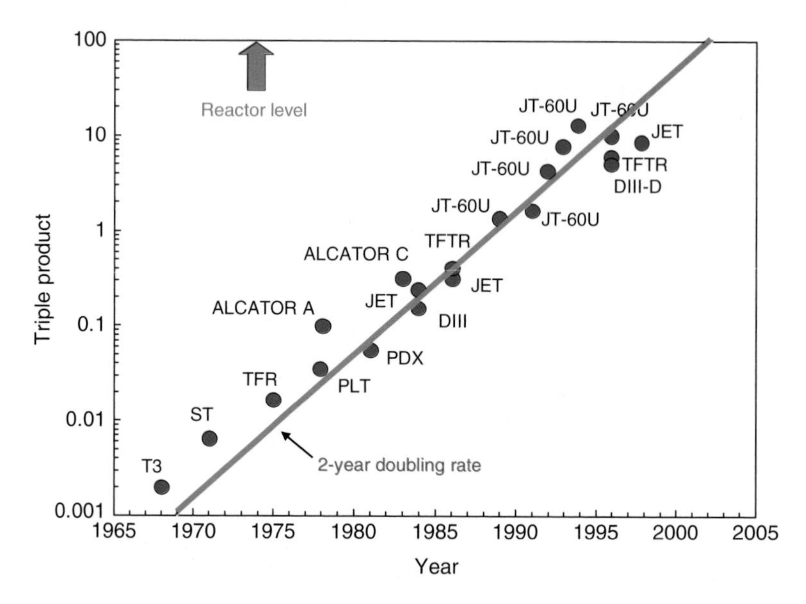

Fig. 8.1 Increase of the triple product $Tn\tau$ with year. The points are labeled with the names of the tokamaks (Data from http://www.efda.org_fusion_energy/fusion_research_today.htm. The units for $Tn\tau$ are 10^{20} m^{-3} keV s)

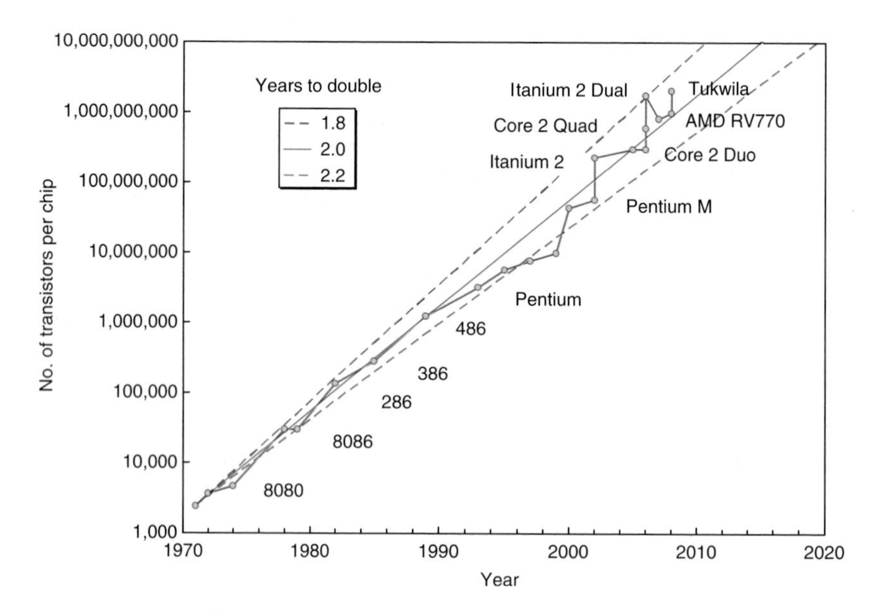

Fig. 8.2 Moore's Law for semiconductors compared with doubling rates

Here are pictures of the four large tokamaks which provided the points at the top of these graphs (Figs. 8.3–8.6).[1]

As you can see, or cannot see, the tokamak itself is hidden behind a jumble of equipment which includes the neutral-beam injectors, power feeds to the coils, the support structure, and diagnostic instrumentation. To show the size of these machines, Fig. 8.7 is an inside view of the vacuum chamber of DIII-D when it is opened up to air.

Fig. 8.3 TFTR: Tokamak Fusion Test Reactor at Princeton, NJ

Fig. 8.4 JET: Joint European Torus at Abingdon, UK

Fig. 8.5 DIII-D: Doublet III at General Atomic, LaJolla, CA

Fig. 8.6 JT-60U: Japan Torus at Ibaraki, Japan

Fits, Starts, and Milestones

How did we get to this point? The scatter in the points in Fig. 8.1 tells a story. In the short term, progress has been sporadic, with fits and starts caused not only by problems of physics, but also by problems of funding and politics. Glimpses of the history of fusion research can be found in popular books by physicists Amasa Bishop [1], Hans Wilhelmsson [2], McCracken and Stott [3], and Ken Fowler [4].

Fig. 8.7 Inside the vacuum chamber of DIII-D when it is opened up to air

Less technical coverage of people and politics is given in books by journalists Joan Lisa Bromberg [5] and Robin Herman [6], and in an article by Gary Weisel [7]. Here is a nutshell account.

In the USA, three groups started research on controlled fusion in 1951–1952: one at Livermore, California, headed by Richard F. Post; one at Los Alamos, New Mexico, headed by James Tuck, and one at Princeton, New Jersey, headed by Lyman Spitzer, Jr. It was obvious that the hydrogen bomb reaction was a source of a huge amount of energy, if only it could be released slowly in a controlled way. It was not obvious how to do it. All agreed that trapping and holding a hot plasma would be necessary. Dick Post proposed to use magnetic mirrors, which we shall describe in Chap. 10. Jim Tuck proposed to use pinches (Chap. 7), in which the entire magnetic field is generated by plasma currents. These devices suffered, of course, from the kink instability, which was not known at that time. Tuck had the foresight to name his machine the Perhapsatron. At Princeton, Lyman Spitzer, an astronomer, designed the figure-8 torus, which he named, of course, a Stellarator. A little later, a fourth program started at Oak Ridge, Tennessee, based on another mirror machine, the DCX. This group emphasized experiments which ran continuously (hence DC) rather than in pulses, and eventually included the curiously named ELMO Bumpy Torus. In England, the initial efforts concentrated on pinches, particularly the toroidal pinch, which is a torus like a tokamak, but with a poloidal confining field produced by a large toroidal current. In Russia, research began at the Kurchatov Institute in Moscow with a small torus which they named the Tokamak, invented by Igor Tamm and Andrei Sakharov. Other nations did not join in until after the first milestone, the Geneva conference of 1958, when these secret programs were declassified and revealed.

In the years before that, the US program grew rapidly with the enthusiastic support of Atomic Energy Commission chairman Lewis L. Strauss. The program was named Project Sherwood after the name of James Tuck, reminiscent of Friar Tuck of Sherwood Forest. Strauss kept the program classified and well funded with the aim of beating out the UK and the USSR in achieving fusion. Sherwood conferences were held yearly, and there were some memorable occasions. In 1956, the meeting was hosted by Oak Ridge at Gatlinburg, Tennessee, and most attendees found out for the first time the meaning of "dry town." Even without lubrication, Lyman Spitzer regaled the group with his rendition of songs by Gilbert and Sullivan, which he sang from memory. In 1957, the meeting was in Berkeley, California, and a movie theater had to be taken over in the day time and secured for the classified meeting. By sheer coincidence, the movie that was playing that week was "Top Secret." From 1952 to 1954 James van Allen, who discovered his famous radiation belts, built the B-1 stellarator at Princeton, a machine which the newly hired young experimentalists inherited in 1954.

Meanwhile, Spitzer had assembled a strong theoretical group, whose *magnum opus* was the elegant paper *An energy principle for hydromagnetic stability problems*, published in 1958 [8]. This paper by Bernstein, Frieman, Kruskal, and Kulsrud did more than anything else to establish plasma physics as a respectable new field in the eyes of all physicists. A calculational method based on minimization of energy was given that could predict the boundaries of stable MHD operation even in toroidal machines with complicated magnetic geometries. This tool allowed experimentalists to build machines that were stable against the Rayleigh–Taylor and kink instabilities, among others, that were discussed in Chaps. 5 and 6.

The 1958 Atoms for Peace conference was organized by the IAEA (International Atomic Energy Agency), formed in 1957 by the United Nations. Based in Vienna, Austria, the IAEA has sponsored the plasma physics and controlled fusion conference every two years since then. A large contingent from Project Sherwood was sent to Geneva, flying across the Atlantic on propeller planes. Preceding the team were tons of display equipment managed by the Oak Ridge experts. Not only were there models such as the figure-8 stellarator shown in Fig. 4.18, but actual operating machines were also transported, including the power supplies and control equipment needed to make them work. No expense was spared. England also put on a large and splendid exhibit, featuring their toroidal pinch, the Zeta. Meanwhile, the USSR exhibit featured the Sputnik, which they had just launched to open the space age. Their fusion machine, the tokamak, was secondary. The tokamak on exhibit looked like a formless, dark, unrecognizable piece of iron and was not made to work. This was how the tokamak age began. But the gauntlet was thrown by the USA, the UK, and the USSR; and the race was on.

At the Geneva conference, the British team announced that neutrons characteristic of fusion reactions had been observed in Zeta. This would have been the first demonstration of fusion created by hot plasma. Unfortunately, it was found that these neutrons came from energetic ions striking the wall, not from the thermal ions in the body of the plasma. As explained in Chap. 3, ion beams cannot produce net energy gain; that requires a *thermo*nuclear reaction. The Brits had been careless and had stumbled. It was an embarrassing moment for their leaders, Peter Thonemann and Sebastian "Bas" Pease, two gentlemen who were the best friends one could have. The idea of a toroidal z-pinch (zed-pinch to Englishmen) has survived, however, as a possible advanced alternative to the tokamak, aided by a brilliant theory by their countryman, Bryan Taylor.

The 1960s saw progress on many fronts. The most important was the announcement in 1968 by Lev Artsimovich, the driving force of the Russian effort, that the confinement time was 30 times longer than the Bohm time and record-breaking electron temperatures had been achieved in their T-3 tokamak. Recall that Bohm diffusion, caused by microinstabilities, was limiting confinement times to the millisecond regime, so this was important progress if it could be believed. The scientific community was skeptical, since Russian instruments were comparatively primitive. In 1969, an English team headed by Derek Robinson flew to Kurchatov with a laser diagnostic tool that the Russians did not have. They measured the plasma in the T-3 and found that the Russian claims were correct. The tokamak had to be taken seriously. Soon thereafter, research tokamaks began appearing at General Atomics and several universities in the USA, as well as in many locations in Western Europe and Japan. Even the venerable Model C stellarator at Princeton was converted to a tokamak in 1970. In retrospect, the invention of the tokamak was a lucky break. Its self-curing feature of sawtooth oscillations was not foreseen, nor were the gifts from Mother Nature listed in Chap. 7. The cures for Bohm diffusion could have been laboriously found in any of a number of magnetic bottles, some of which may turn out to be more suitable for a reactor than a tokamak. It was concentrating on a single concept, the first promising one, that advanced the tokamak to its present status.

Throughout the 1960s, the Princeton group whittled away at the Bohm diffusion problem, clarifying the microinstabilities responsible for that enhanced loss rate. Much of this work was basic experimentation done in linear machines, which did not suffer from the complicated field lines of stellarators and tokamaks. In the USSR, Mikhail Ioffe at his institute in St. Petersburg invented the "Ioffe bars." These were four bars carrying current to form a magnetic well ("minimum-B") configuration in a mirror machine, thus stabilizing the most troublesome instability in those confinement devices. Though mirror confinement is outside our scope here, the minimum-B concept is also used in tokamak configurations. These results, as well as the ones from the T-3 tokamak, were presented in the memorable IAEA meeting of 1968. After the technical sessions in Moscow, Artsimovich led the entire conference to a big party in Novosibirsk, the science city deep in Siberia. The party was held at a large artificial lake made by cutting down trees and covering the stumps with water. Long picnic tables were set up on the shores and food served with Russian hospitality. It seemed that the tables for 60-second chess games must have stretched for 100 yards. Here, plasma physicists from many countries got acquainted on a personal level. It was the beginning of international cooperation and competition.

Another milestone was announced at the Novosibirsk meeting when the General Atomics group showed the picture of Fig. 8.8, which completely surprised the Russians. Had the Americans trumped them with the resources to build a torus large enough to hold a person standing up? Actually, it was not a tokamak or stellarator

Fig. 8.8 Inside the toroidal octopole at General Atomics (courtesy of Tihiro Ohkawa and published in Chen [10])

but an "octopole," spelled "octupole" when another one was built at the University of Wisconsin by Don Kerst. It had four current-carrying rings suspended by thin wires within the plasma, creating a magnetic well. The plasma was absolutely stable in such a magnetic field, and the classical diffusion rate, caused by collisions alone, was observed for the first time [9]. Being a pure physics experiment, the octopole did not require a large, expensive magnetic field, and it was not the advanced fusion machine that the Russians had feared. Internal conductors would not be practical in a real reactor.

The 1970s was a period of euphoria, with Artsimovich predicting scientific breakeven by 1978, and Bob Hirsch, then head of fusion research in the Atomic Energy Commission, pushing for an even earlier date. The prospect of an infinite energy source evoked such lyrical epithets as "Prometheus Unbound!". With the difficulty of magnetic confinement recognized, the importance of controlling fusion was compared with that of inventing fire. Funding started to increase when James R. Schlesinger became AEC chairman on the way to the CIA and Defense. Support for fusion energy was further escalated by the oil crisis of 1973, when a speed limit of 55 miles per hour was mandated throughout the USA. The dramatic increase in the fusion budget is shown in Fig. 8.9, reaching a peak of almost $900M annually in 2008 dollars. Championed by Representative Mike McCormack (D-WA), Congress passed the Magnetic Fusion Engineering Act of 1980, which laid out the plans and the budget needed to build a demonstration reactor DEMO by the year 2000. The Act was never funded as passed. Tired of promises that fusion would be achieved in 25 years regardless of when the question was asked, Congress began cutting the fusion budget. Ed Kintner took over the fusion office from Hirsch in 1976 and had to reorganize priorities to fit available funds. Many alternative

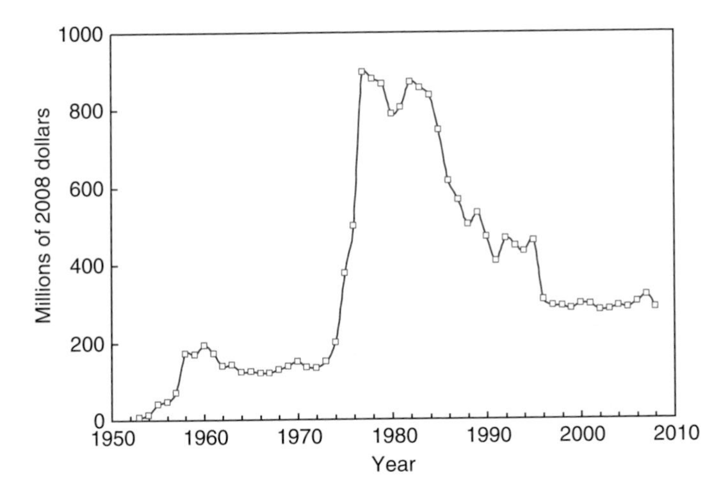

Fig. 8.9 US fusion research budget in 2008 dollars (adapted from data from Fusion Power Associates, Gaithersburg, VA)

approaches to magnetic confinement still existed at that time,[2] and they should be explored while keeping the tokamak as the flagship, while critical engineering tests are made. Nonetheless, several large projects ultimately had to be canceled, including the Fusion Materials Test Facility and MFTF-B, the world's largest superconducting magnet built for mirror fusion. That fusion would always be 25 years in the future was made a self-fulfilling prophecy by the decrease in funding.

Curiously enough, the peak in funding in Fig. 8.9 follows a similar graph of the price of oil at the time.[3] Unfortunately, this did not happen in the oil crisis of 2008, since other energy alternatives such as solar and wind power were available, and the USA was at war in Iraq. The dissolution of the Soviet Union in 1991 had a major effect on the willingness of Congress to support fusion. The threat of being outdone by the Russians was no longer there, and the attitude was to let the friendly nations which are more dependent on foreign oil bear the main expense. As a result, the USA, which had been the world leader in fusion development, slowly lost its preeminent position to the UK and Japan.

The peak funding levels of the 1970s nonetheless enabled the start of the billion-dollar machines that set milestones two decades later. The TFTR at Princeton[4] began construction in 1976 and ran from 1982 to 1997. This was a big step because it was the first machine made to run with DT rather than helium or deuterium. Once tritium is introduced, the DT reaction would produce 14-MeV neutrons, which would activate the stainless steel walls. Massive shielding would be required, and maintenance could be done only by remote control. By 1986, TFTR had set records in ion temperature (50 keV or 510,000,000°C), plasma density (10^{14} cm^{-3}), and confinement time (0.21 s), but of course not all at the same time. In 1994, a 50–50% DT mixture was heated to produce 10.7 MW of fusion power. This is only about 1% of what a power plant would give and occurred only in a pulse, but it was the first demonstration of palpable power output. Before it was decommissioned, TFTR also demonstrated bootstrap current and reversed shear, effects described in Chap. 7.

Close on the heels of the TFTR, western Europe built an even larger machine, the Joint European Torus, JET, also capable of using DT fuel. Designed in 1973–1975 and constructed in 1979, it has operated from 1983 until now. It was funded by the countries of Euratom and is now operated under the European Fusion Development Agreement, with participation of over 20 countries.[5] Currently, the world's largest tokamak with a major radius of 3 m, it is also powered impressively with a magnetic field of 3.45 T (34.5 kG), total heating power of 46 MW, and a toroidal current of 7 MA. It set a record with a pulse of 2 MA that lasted 60 s. In 1997, JET announced a new world record with DT fuel, producing 16 MW of fusion power and keeping 4 MW going for 4 s. JET is being modified for experiments in support of ITER, the large international project described at the end of this chapter.

The third large tokamak of this era is Japan's JT-60, which started operating in 1985. It plays a leading role in researching the effects on the forefront of tokamak science, such as reversed shear, H-modes, and bootstrap current. Much of this is too technical for this book, but JT-60 has set some world records which are easy to understand. In 1996, it achieved the highest fusion triple product. Recall that the triple product is, more exactly,

$$\text{Triple product} = nT_{\text{i}}\tau_{\text{E}},$$

where τ_{E} is the *energy* confinement time. The value achieved was 1.5×10^{21} keV s/m^3, close to the value needed for energy breakeven, and only about seven times less than that required for a reactor. Of course, this was in a pulse and not in steady state. In 1998, JT-60 set a record for Q, the ratio of fusion energy to plasma heating energy, at $Q = 1.25$. However, since JT-60 was not designed to handle tritium, the experiment was done in deuterium and the result extrapolated to DT. The highest ion temperature of 49 keV was also reported in JT-60. The machine excelled in long pulses, running steadily for as long as 15 s, or for 7.4 s while the bootstrap fraction was 75%. Perhaps most impressive was the production in 2000 of a plasma with zero current over 40% of the minor radius. The current in an outer shell held the plasma even though there was no confinement in the current hole. This is exactly the profile that is suitable for operation with a large bootstrap current fraction.

By focusing on these three machines, we have had to omit the great contributions of other large machines such as DIII-D and ASDEX, as well as those of hundreds of smaller tokamaks built to study particular effects. Though not tokamaks, there are also large machines of the stellarator type, such as Wendelstein 7 in Germany and the Large Helical Device in Japan. No large tokamaks had been built since the turn of the century until two Asian machines went online in 2007: the KSTAR in Daejeon, Korea and the EAST (Experimental Advanced Superconducting Tokamak) in Hefei, China. You can guess what KSTAR stands for. Both of these machines use superconducting coils cooled by liquid helium, requiring a second vacuum system to keep the coils cold. The development of large superconductors is an important step toward a fusion reactor.

As can be seen in Fig. 8.9, the US fusion budget steadily declined in the 1980s and 1990s. Construction of large machines had been completed; there was no oil crisis or competition from the USSR; and people were disillusioned about the prospect of ever achieving fusion. In particular, members of Congress were reluctant to support a project that could not be completed in their terms of office. Major sources of funding shifted to countries which have very limited fossil fuel reserves, and the USA slowly lost its lead at the forefront of fusion research. In 1995, a Fusion Review Panel headed by John P. Holdren and Robert W. Conn submitted a report[6] to President Clinton's Commission of Advisors on Science and Technology on a requested evaluation of the fusion situation. The Panel estimated that progress to a demonstration reactor by 2025 would require annual funding levels averaging $645M between 1995 and 2005, with at peak of $860M in 2002. Should budgetary constraints not permit this level, alternate scenarios were also given. At a realistic level of $320M/year, the best that could be done was to maintain the expert community in plasma science and fusion technology while expanding international participation. With this devaluation, the Magnetic Fusion *Energy* Program was changed to the Fusion Energy *Sciences* Program. The restructured program was presented to the DOE Office of Energy Research by the Fusion Energy Advisory Committee, chaired by Conn, in 1996 [13]. As seen in Fig. 8.9, the budget has been maintained the $300M level since that time, partly through the efforts of Undersecretary for Science Raymond Orbach under President Bush. With DIII-D, the largest tokamak extant in the USA, the level of fusion science and innovation nonetheless leapt forward with many intermediate-sized devices in universities and with advances in computation and theory.

It was in this period that *burning plasma* became the catchword, and planning for a large international tokamak to achieve this, the ITER, began. The success story of the negotiations deserves its own section. This is presently our best chance to move forward in making our own sun. Meanwhile, we need another scientific interlude to clarify the uncertainties that still exist in fusion science.

Computer Simulation

Before describing some effects that are not yet completely understood, we should mention the basis for believing that these problems are not insoluble. That's the important subject of computer simulation. In the 1970s and 1980s, when unanticipated difficulties with instabilities arose, computers were still in their infancy. To the dismay of both fusion scientists and congressmen, the date for the first demonstration reactor kept being pushed forward by decades. The great progress seen in Fig. 8.1 since the 1980s was in large part aided by the advances in computers, as seen in Fig. 8.2. In a sense, advances in fusion science had to wait for the development of computer science; then the two fields progressed dramatically together. Nowadays, a $300 personal computer has more capability than a room-size computer had 50 years ago when the first principles of magnetic confinement were being formulated.

Fig. 8.10 Hokusai's painting of the Big Wave

Computer simulation was spearheaded by the late John Dawson, who worked out the first principles and trained a whole cadre of students who have developed the science to its present advanced level. A computer can be programmed to solve an equation, but equations usually cannot even be written to describe something as complicated as a plasma in a torus. What, for instance, does wavebreaking mean? In Hokusai's famous painting in Fig. 8.10, we see that the breaking wave doubles over on itself. In mathematical terms, the wave amplitude is double-valued. Ignoring the fractals that Hokusai also put into the picture, we see that the height of the wave after breaking has two values, one at the bottom and one at the top. Equations cannot handle this; Dawson's first paper showed how to handle this on a computer.

So the idea is to ask the computer to track where each plasma particle goes without using equations. For each particle, the computer has to memorize the x, y, z coordinates of its position as well as its three velocity components. Summing over the particles would give the electrical charge at each place, and that leads to the electric fields that the particles generate. Summing over their velocities gives the currents generated, and these specify the magnetic fields generated by the plasma motions. The problem is this. There are as many as 10^{14} ions and electrons per cubic centimeter in a plasma. That's 200,000,000,000,000 particles. No computer in the foreseeable future can handle all that data! Dawson decided that particles near one another will move together, since they will feel about the same electric and magnetic fields at that position. He divided the particles into bunches, so that only, say, 40,000 of these superparticles have to be followed. This is done time step by time step. Depending on the problem, these time steps can be as short as a nanosecond. At each time step, the superparticle positions and velocities are used to solve for the E- and B-fields at each position. These fields then tell how each particle moves and where they will be at the beginning of the next time step. The process is repeated over and over again until the behavior is clear (or the project runs out of money). A major problem is how to treat collisions between superparticles, since, with their large charges, the collisions would be more violent than in reality. How to overcome this is one of the principles worked out by Dawson.

Fig. 8.11 Electric field pattern in a turbulent plasma (from ITER Physics Basis 2007 [26], quoted from [14]. The plot is of electric potential contours of electron-temperature-gradient turbulence in a torus)

Before computers, scientists' bugaboo was nonlinearity. This is nonproportionality, like income taxes, which go up faster than your income. Linear equations could be solved, but nonlinear equations could not, except in special cases. A computer does not care whether a system behaves linearly or not; it just chugs along, time step by time step. A typical result is shown in Fig. 8.11. This shows the pattern of the electric fields generated by an instability that starts as a coherent wave but then goes nonlinear and takes on an irregular form. This turbulent state, however, has a structure that could not have been predicted without computation; namely, there are long "fingers" or "streamers" stretching in the radial direction (left to right). These are the dangerous perturbations that are broken up by the zonal flows of Chap. 7.

The simulation techniques developed in fusion research are also useful in other disciplines, like predicting climate change. There is a big difference, however, between 2D and 3D computations. A cylinder is a 2D object, with radial and azimuthal directions and an ignorable axial direction, along which everything stays the same. When you bend a cylinder into a torus, it turns into a 3D object, and a computer has to be much larger to handle that. For many years, theory could explain experimental data after the fact, but it could not predict the plasma behavior. When computers capable of 2D calculations came along, the nonlinear behavior of plasmas could be studied. Computers are now fast enough to do 3D calculations in a tokamak, greatly expanding theorists' predictive capability. Here is an example of a 3D computation (Fig. 8.12). The lines follow the electric field of an unstable perturbation called an ion-temperature-gradient mode. These lines pretty much follow the magnetic field lines. On the two cross sections, however, you can see the how these lines move in time. The intersections trace small eddies, unlike those in the previous illustration. It is this capability to predict how the plasma will move under complex forces in a complicated geometry that gives confidence that the days of conjectural design of magnetic bottles are over.

The science of computer simulation has matured so that it has its own philosophy and terminology, as explained by Martin Greenwald [15]. In the days of Aristotle, physical models were based on indisputable axioms, using pure logic with no input from human senses. In modern times, models are based on empiricism and must agree with observations. However, both the models and the observations are inexact. Measurements always have errors, and models can keep only the essential elements. This is particularly true for plasmas, where one cannot keep track of

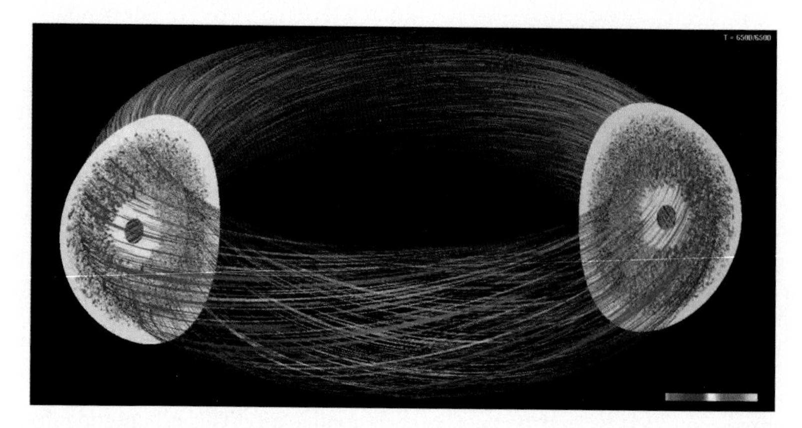

Fig. 8.12 A 3D computer simulation of turbulence in a D-shaped tokamak (courtesy of W.W. Lee, Princeton Plasma Laboratory)

every single particle. The problem is to know what elements are essential and which are not. Computing introduces an important intermediate step between theory (models) and experiment. Computers can only give exact solutions to inexact equations or approximate solutions to more exact (and complicated) equations. Computer models (codes) have to be introduced. For instance, a plasma can be represented as particles moving in a space divided into cells, or as a continuous fluid with no individual particles. *Benchmarking* is checking agreement between different codes to solve the same problem. *Verification* is checking that the computed results agree with the physical model; that is, that the code solves the equations correctly. *Validation* is checking that the results agree with experiment; that is, that the equations are the right ones to solve. Plasma physics is more complicated than, say, accelerator physics, where only a few particles have to be treated at a time. Because even the models (equations) describing a plasma cannot be exact, the development of fusion could not proceed until the science of computer simulation had been developed.

Unfinished Physics

Edge-Localized Modes

In fusion, ELMs are not trees but edge-localized modes. The name itself suggests that they are not understood, not unlike the term assigned to the Irritable Bowel Syndrome. The name has even spawned an adjective, ELMy, and a participle, ELMing, which should give philologists conniptions. ELMs occur at the pedestal in H-mode plasmas (Chap. 7). Recall that in this high-confinement mode, a transport barrier, shown earlier in Fig. 7.25, is formed at the edge of the plasma.

This thin layer holds back the plasma because it quenches all instabilities with strong electric field shear. But it can't do that forever. If the plasma escaped at the classical diffusion rate due to collisions alone, the plasma pressure in the interior would rise so high that the barrier would break down. This breakdown occurs in short bursts, called ELMs, so that there is a steady release of plasma to the outside. Actually, this is a good thing because the "ash" of the DT reaction has to be taken out. This ash is the cleanest ash ever – pure helium – but it has to be removed because otherwise the expensive magnetic field would be used up in confining the ash rather than the fuel.

The H-mode occurs only when the heating power exceeds a certain threshold value. ELMs occur when the power is just above this threshold and are really localized near the plasma edge. Recall that the "edge" of the plasma is defined by the divertor, like the one at the bottom of Fig. 8.13. The plasma edge is defined by the last closed magnetic surface, the one at the X made by the field lines just above the divertor. Plasma venturing beyond that is led into the divertor, where it strikes high-temperature materials with heroic cooling to dissipate the heat. Also shown in the figure is the layer where the H-mode barrier exists and, inside that, the core plasma. The problem with ELMs is that the heat comes in short bursts – less than 1 ms – occurring a few times a second, and divertors cannot handle a heat flow that is not steady. A single ELM, while it lasts, can carry 20 GW of power, an energy flow

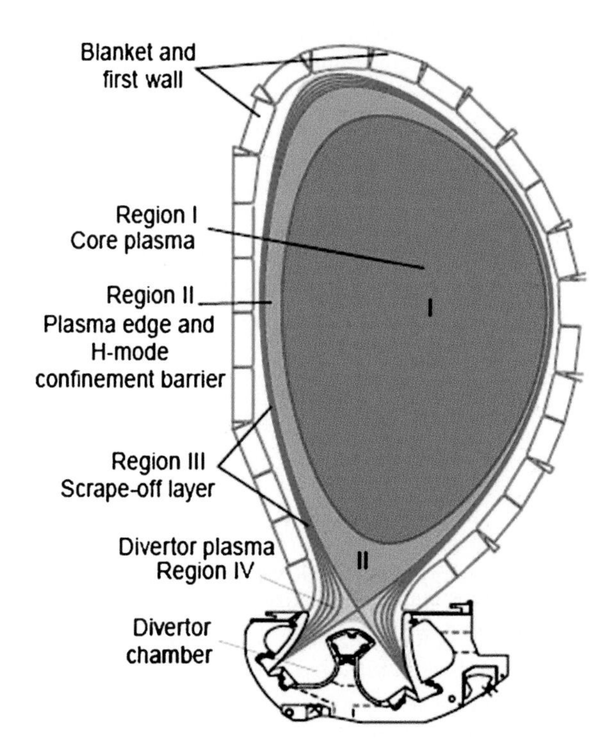

Fig. 8.13 Cross-section of a tokamak with a single-null divertor, showing the scrape-off layer [16]

comparable to that of the Three Gorges Dam in China [17]. There are thus three tasks: measuring what ELMs do, explaining what causes them, and devising a way to suppress them.

It's hard to measure what goes on inside the thin barrier layer during the unpredictable time when a burst occurs, but there is a large data base on the different types of ELMs and the conditions before and after they occur [18]. Three types of ELMs have been observed. As the heating power is increased past the H-mode threshold, Type 3 ELMs first occur. These occur rapidly, each with a small energy release. They come after a magnetic precursor signal can be detected. As the power is raised, the ELM frequency decreases until there are no ELMs at all. Then Type 2 ELMs, called "grassy" ELMs, occur; they are very small, rapid bursts whose time traces resemble grass. Further increase in power produces Type 1 ELMs. These occur in most H-mode tokamaks and release energy in rather regular bursts. Each pulse occurs when the density and temperature at the top of the pedestal reach critical values, and these drop when an ELM occurs. Density and temperature then recover slowly until the next burst is triggered. Although ELM-free discharges can be produced, they cause the temperature and density at the top of the pedestal to be rather low, and these control the quality of the fusion plasma in the main volume. It is found that the best fusion conditions can be produced by ELMy H-mode plasmas, in which the plasma is allowed to escape in regular Type 1 ELMs.

Many theorists [19] have worked on the ELM problem, and the consensus is that ELMs are a magnetic instability called a "peeling–ballooning" instability. Computations can predict the temperature and density values in the pedestal that can trigger an ELM, but they are far from explaining all the features that have been observed. And, as usual, there is no guarantee that another theory can't also explain the ELM threshold. There is, however, good news. The DIII-D team at General Atomics have figured out a way to suppress ELMs *without* degrading the quality of the core plasma [20]. They apply "resonant magnetic perturbations" with an array of small coils just outside the plasma edge. These produce small magnetic islands in the edge region which work some kind of magic. Experimental results are promising enough that such coils are being considered and designed to be added to ITER.[5]

Fishbones

The colorful language of plasma physics cannot compete with the charmed and colored quarks of high-energy theory, but we have so far had bananas, sawteeth, and ELMs. We now have fishbones. These arise from their oscilloscope traces, not from the hunger for better funding. Fishbones were first seen in the PDX tokamak at Princeton during neutral-beam injection [21]. Recall that the most powerful way to heat a plasma is to inject beams of high-energy deuterium atoms. Since the atoms are not charged, they can penetrate the magnetic field and get inside the plasma. Once there, they are rapidly ionized by the electrons and become a beam of deuterium ions of 50-keV energy. Oscillations in the plasma could be seen with several different

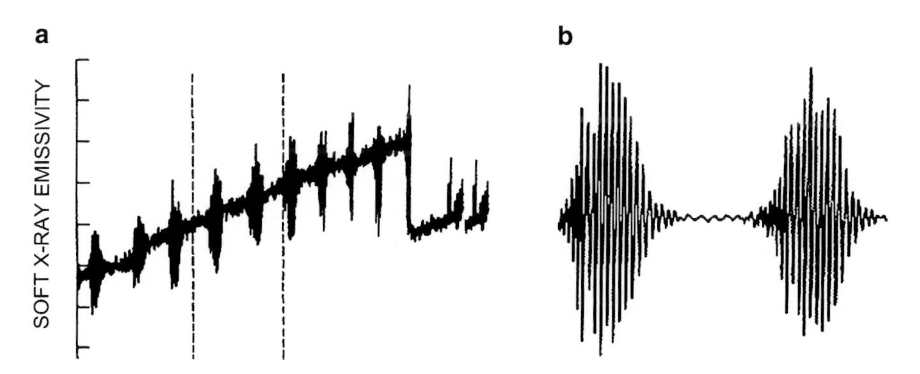

Fig. 8.14 (**a**) Fishbone oscillations on a sawtooth. (**b**) An expanded view reveals the origin on the name [21]

diagnostics, and they look like those in Fig. 8.14. Fishbones often occur on the $q=1$ surface where the sawtooth oscillations (Chap. 7) occur, and sometimes they can excite the sawteeth and appear simultaneously with them. The bad news is that fishbones cause injected ions to be lost before they have transferred their energy to the plasma. As much as 20–40% of the energy can be lost, greatly reducing the efficiency of this primary heating method.

Beams are notorious in exciting plasma instabilities. As usual, the plasma finds a way to come to thermal equilibrium rapidly by generating an instability. Theorists had no problem in finding a suitable instability for this. Initially, there were two somewhat different theories [22, 23], each having to do with an internal kink mode. In Chap. 6, we described the kink instability that occurs to the whole plasma when too large current is driven through it. A localized current can also drive a kink inside a plasma, and this is what happens in the sawtooth region in the presence of a current of fast injected deuterium ions.

The theories could predict the frequency of the oscillations and the conditions when they would occur. Computations of the nonlinear behavior gave traces very much like the experimental ones in Fig. 8.14b. Subsequent work has cleaned up many of the details of the fishbone instability.

The fact that fast ions can be lost via instability is worrisome not only because of the loss of heating power, but even more so because of the fast helium ions (the "ash") that are generated in fusion. The helium has to remain in the plasma long enough to give up their energy to keep the plasma "burning." Fortunately, the theorists can tell us not to worry. Roscoe White et al. [24] have found that there is a regime in a fusion-quality plasma in which neither sawteeth nor fishbones will occur, and this parameter regime is actually larger at higher temperatures and with more fast particles. This has yet to be tested, but there is another mitigating factor. In the next generation of tokamaks, starting with ITER, the plasma will be much larger than the widths of the banana orbits. Since the fast ions are lost with a step size of the order of the banana width, it will take many steps for them to reach the wall. Though not finished, the physics of fishbone instabilities is far enough advanced to tell us that this is not a big problem.

Disruptions

No picturesque name here, because this is a really serious problem. Tokamak discharges are known to disrupt themselves, suddenly stopping and releasing all the energy put into them into the containment chamber. Unless we can stop disruptions from occurring, the entire structure of the tokamak, especially the divertors, would have to be beefed up to absorb all that energy. This is not the kind of accident that can happen in *fission*, because in fusion no energy is released that has not already been put in; it is just that we do not want it to come out all at once and melt or otherwise harm the tokamak structure. The problem is so serious that a large experimental data base has been accumulated on numerous tokamaks, even in the interim between the two ITER planning documents, the ITER Physics Bases of 1999 [25] and 2007 [26].

To get a DT plasma to fuse, we need to heat it to temperatures of the order of a half-billion degrees. The amount of heat in a large experiment like ITER will be about 400 MJ, the energy of 100 pounds of TNT. The poloidal magnetic field created by the tokamak current will hold another 400 MJ of energy. Fortunately, the *toroidal* magnetic field energy, which is much larger, is not released in a disruption unless the toroidal field coils are damaged. Normally, the plasma energy escapes slowly into the divertors, which are designed to handle that heat load; and when the plasma is turned off, the current decays slowly, and the poloidal field energy goes back into the coils that drove the current. In a disruption, all this energy sprays out in a matter of 10 milliseconds and is hard to handle. What happens to the plasma in a disruption has been caught by the M.I.T.[7] group working with the intermediate-size Alcator-C tokamak. In a typical elongated D-shaped tokamak, the plasma has to be kept from drifting up or down with specially shaped coils. When an instability causes a disruption, the plasma moves vertically, as shown in Fig. 8.15, shrinking as it loses its energy and current. In this case, it moves downward toward the divertor, but it could as well move upwards. The time scale shows that the whole event took less than 4 ms.

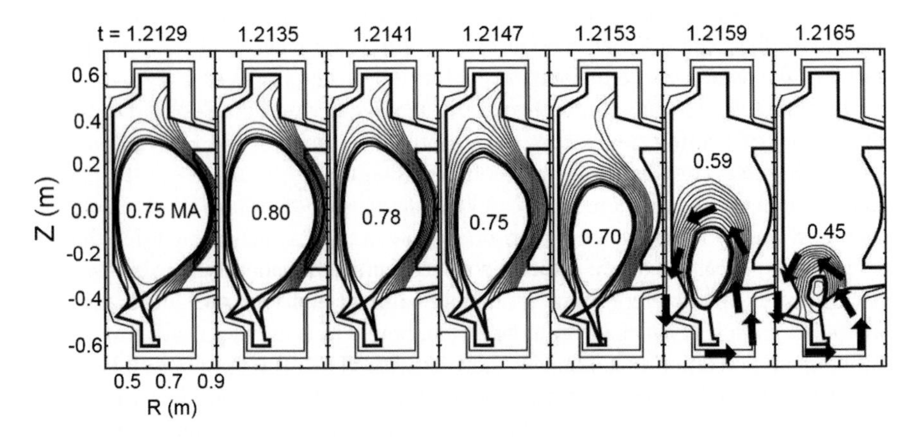

Fig. 8.15 Vertical motion of the plasma in a disruption [27]

The damage caused by a disruption can be divided into three parts: thermal quench, current quench, and runaway electrons. In thermal quench, the plasma's heat is deposited in the walls, vaporizing them in spots. This influx of impure gas raises the resistivity of the plasma, and the tokamak current decays. Even if most of the plasma outflow is channeled into the divertor, there is no time for the heat to be conducted away, and the refractory materials in the divertor – tungsten and carbon – will be vaporized also. In current quench, the fast decrease of the toroidal current will drive a counter-current, by transformer action, in the conducting parts of the confining vessel. Since this counter-current is located inside the strong DC toroidal magnetic field, it will exert a tremendous force on the vessel, moving or deforming it unless it is made sturdy enough. As plasma shrinks toward the divertor, it will drive a "halo current," shown by the dark arrows in Fig. 8.15, flowing through the conducting parts of that structure. The halo current can be as much as 25% of the original tokamak current; and since that current was flowing along helical field lines, the halo current will try to find a helical path through the conducting parts around the divertor.

The third deleterious effect of disruptions is the generation of "runaway" electrons. In Chap. 5, we showed that a hot plasma is almost a superconductor because fast electrons do not make many collisions. The faster the electron, the farther it will go before it collides with an ion. This distance is its free path. If there is a large electric field pushing the electron, its free path can increase faster than the electron is going, and it never makes a collision! It is a runaway and can get up to MeVs of energy before it loses confinement. Of course, this depends on the number of scattering centers; namely, on the plasma density. Normally, runaway electrons occur during the startup of the plasma. If the electric field is turned up too high before the density is high, runaways can occur. Machine operators know how to prevent this. In a disruption, however, there is no control. If the density falls below a critical value while a strong toroidal electric field is still on, a horde of runaway electrons will be created, amounting to 50–70% of the original tokamak current. When these hit the wall, they will certainly cause damage. In ITER, the tokamak current will be 15,000,000 A. By comparison, household circuits carry only 15–20 A.

The obvious questions are then: What causes disruptions? How often do they occur? Can they be eliminated? It turns out that disruptions mostly occur when we try to push the envelope. There are known limits to the plasmas that a tokamak can confine. There is a density limit, called the Greenwald density, which we will describe shortly. There is a pressure limit called the Troyon limit. And there has to be enough shear stabilization, as specified by the quality factor q, which has to be above 2 at the edge. When the plasma is pushed too close to one of these limits, a disruption is likely to occur. Exactly how it occurs is not entirely clear. Sometimes two island chains with different numbers of islands can lock onto each other and merge. If there is a detected precursor, this locking can be avoided by setting the plasma into rotation. Sometimes this change in magnetic geometry brings a bubble of cold gas in from the periphery, disrupting the whole plasma. When the density or pressure limits are approached, known instabilities can occur. These are the ideal MHD instability, called the Rayleigh–Taylor instability in Chap. 5, and the neoclassical

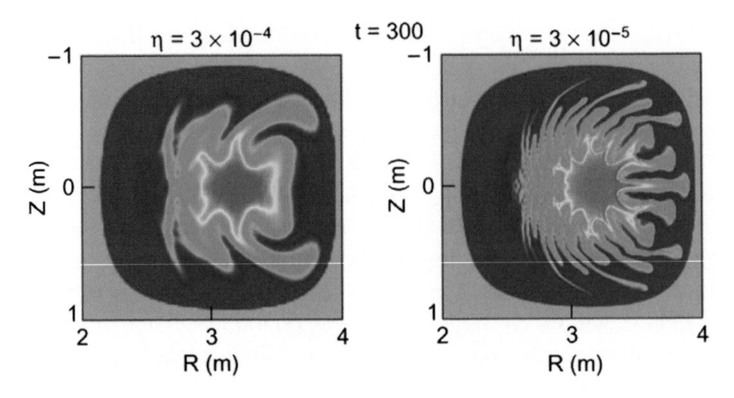

Fig. 8.16 Computer simulation of a disruption [26]

tearing mode, which is triggered by finite resistivity, as described in Chap. 6. Here, "ideal" means that no resistivity has to be considered for the instability to occur, and "neoclassical" means that banana orbits are considered in the calculation. Figure 8.16 shows a computer simulation of how an instability can bring cold plasma in from the edge, thus cooling the core.

Up to now, tokamak discharges have been pulsed and not run continuously as in an eventual reactor. An average over all tokamaks shows that 13% of these pulses have suffered a disruption. This would be an unacceptable rate, but these are experiments meant to probe the stability of a plasma. In long pulses, lasting many seconds in the large tokamaks such as TFTR and JET, the disruption rate is less than 1% because the machine is run conservatively. In the experimental stage, much depends on the experience of the machine operator. He learns the settings on various controls that will produce a stable discharge. For instance, the currents on the various magnetic coils have to be turned on at the right time and increased at the right rate, and the heating power from various sources have to come on at the right time. Operator experience is valuable in the use of almost any machine; snow plows, cranes, and ordinary cars, for instance. Even in the use of a toaster, one sets the darkness level intuitively depending on the dryness of the bread. Nonetheless, in a reactor even one disruption would be disastrous, and methods must be found to eliminate them.

This task is being tackled on three fronts: avoidance, prediction, and amelioration. As already shown in experiment, disruptions can be avoided if the plasma parameters are not pushed close to the instability limits. As shown in Fig. 8.17, these limits have been extensively tested, and the occurrence of disruptions from this cause is predictable. The quantity β_N is a measure of the plasma pressure, and stable discharges are all below the theoretical limit, with disruptions occurring when the limit is exceeded. Prediction of imminent disruption can be obtained from many sensors, for instance of magnetic precursor signals; and neural networks have been successfully used to integrate these signals to give a definite warning of an oncoming disruption. After many trials, these networks can be

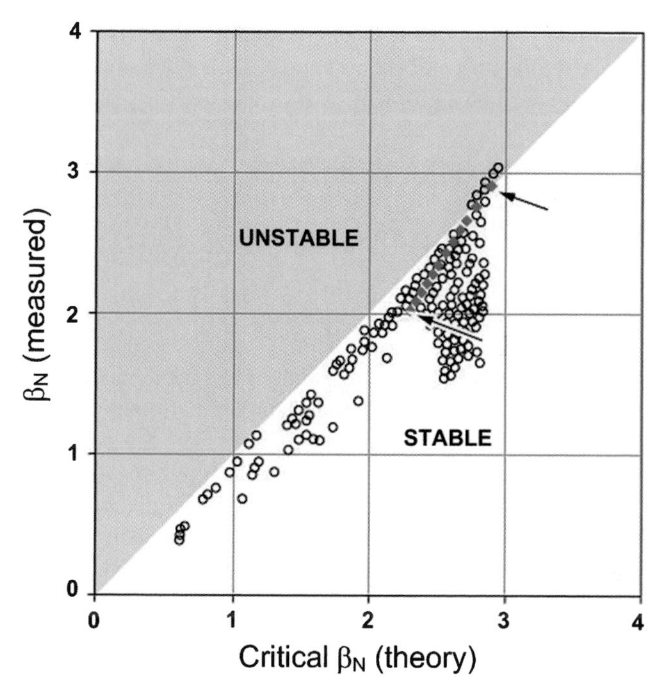

Fig. 8.17 Data from the TFTR tokamak showing the accuracy of theoretical prediction of instability and disruption [25]

trained to suppress false positives. To stop a disruption from occurring, automatic controls can change such parameters as the plasma density, the toroidal current, or the plasma elongation; but this response may be too slow. A faster method would be to drive electron current with electron cyclotron waves in order to change the current profile, and thus the q profile, to a more stable shape. Once an unavoidable disruption starts, there are still ways to ameliorate the damage. For instance, a massive injection of a gas such as neon or argon can reduce the halo currents by 50% and the electromagnetic forces by 75% [26]. Raising the plasma density by about two orders of magnitude this way would also suppress runaway electrons. As tokamaks get larger, the damage from disruptions can be expected to get worse, because the energy released varies as the cube of the radius (i.e., the volume), whereas the energy has to be absorbed by the surface area, which varies only as the square of the radius. On the other hand, the disruptions will evolve more slowly, giving more time to control them.

For tokamaks, the problem of disruptions is receiving a great deal of attention because of its importance. However, tokamaks may not be the machines ultimately chosen for fusion reactors. Stellarators, which do not need large currents, do not suffer from disruptions. The reason that tokamaks are now prevalent is that they gave the best initial results, and there has not been enough money to study other toruses to the same extent. The next generation of tokamaks – the ITER – will allow us to study a burning plasma, one in which the helium products can be used to keep the plasma hot. After that, we still have a choice; we are not stuck with the tokamak if disruptions continue to be a problem.

The Tokamak's Limits

The Greenwald Limit

Ever since the early days of tokamak research, it has been noticed that the plasma density could never be raised above a certain limit. Sometimes this limit was blamed on a loss of confinement via an unspecified instability, sometimes on excessive energy loss by radiation, and sometimes the plasma suffered a disruption. In 1988, Greenwald et al. [28] put together the data from different machines to see what the density limit depended on. They came up with a surprisingly simple answer: roughly speaking, the density limit depended only on the tokamak current per unit area! For those who would rather have a formula, the one for the Greenwald density n_G2 is given in Note 8 hrs.[8] This limit has been found to be obeyed in all tokamaks regardless of what mechanism causes the problem at high densities. No one has yet found a theory that explains this; the Greenwald limit is purely empirical. Figure 8.18 shows how well the Greenwald limit is obeyed in two large tokamaks. In almost all shots, the measured density cannot be raised above the straight line,

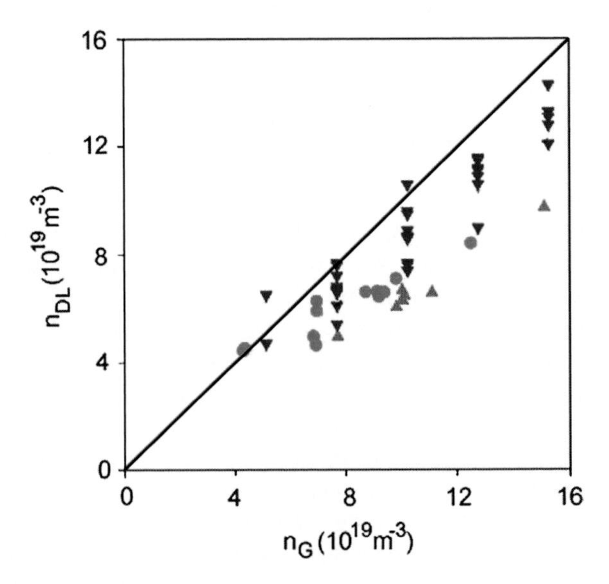

Fig. 8.18 Measured density limit n_{DL} vs. density n_G calculated from the Greenwald formula (modified from a figure in ITER Physics Basis 2007, Chap. 2)

which is the Greenwald limit. This unexplained law is so universal that it is used in the design of future machines. The design would be to achieve, say, 85% of n_G, or 95%, depending on how adventurous one wants to be.

The Troyon Limit

This is a limit on the plasma pressure that a tokamak's magnetic field can hold. Unlike the Greenwald limit, this criterion is rigorously calculated from ideal MHD (MagnetoHydroDynamics) theory. The quantity that measures the balance between the pressure and magnetic forces is called β (beta). Since β is used in many scientific disciplines, especially in medicine, I had refrained from defining it until it was necessary. It is now necessary. Beta is the ratio between plasma pressure and magnetic pressure:

$$\beta = \frac{\text{Plasma pressure}}{\text{Magnetic pressure}}.$$

The plasma's pressure is the product of its density and its temperature, and the magnetic pressure is proportional to the square of the field strength B. These quantities are not constant over a cross section of the plasma, so a reasonable definition would be to take the average pressure and divide it by the average magnetic field before the plasma is created. The last proviso is needed because the plasma is *diamagnetic*, so its very presence decreases the B-field inside it. Since the B-field is the most expensive component, β is a measure of the cost effectiveness of a tokamak. It has a value below 10%, typically 4–5%.

The value of β has been shown to depend on the toroidal current I divided by the plasma radius a and the magnetic field strength B. Figure 8.19 shows how data from different tokamaks all fall on the same line if plotted against I / aB. It is convenient, then, to introduce a *normalized* β, called β_N, which would apply to all tokamaks, regardless of their values of I, a, and B:

$$\beta_N \equiv \frac{\beta \times a \times B}{I}.$$

The Troyon limit (Troyon et al. [30])[9] is when β_N is about 3.5. A numerical formula is given in footnote 10. Figure 8.17 shows how well the experiments in different tokamaks obey the Troyon limit, above which disruptions are likely to occur.

Big Q and Little q

As we now turn our attention from fusion physics to fusion energy, we have to introduce Big Q, as distinct from little q. Little q, as you remember, is the "quality"

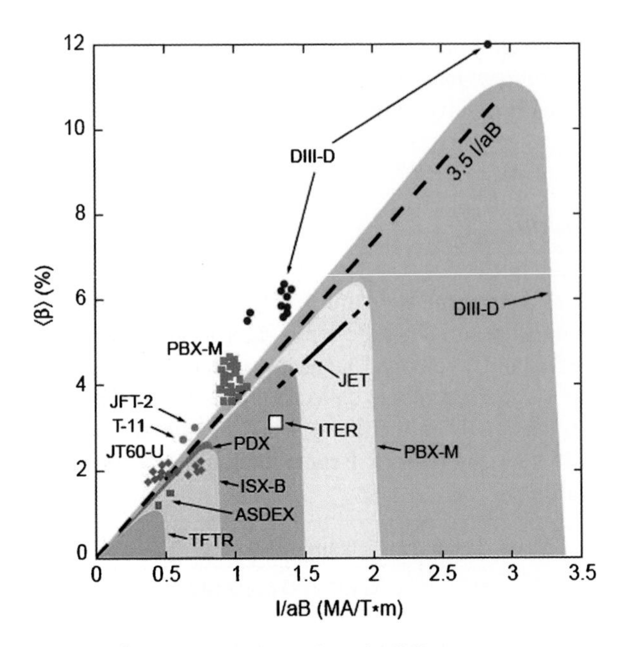

Fig. 8.19 Dependence of β on I/aB in various tokamaks [25]

factor in toruses like tokamaks and stellarators. It is the reciprocal of the rotational transform, which is the number of times a helical field line encircles the minor axis each time it goes around the whole torus. The variation of q with radius r, or $q(r)$, is perhaps the most important feature in the design of toroidal magnetic bottles. Big Q, on the other hand, has to do with how much energy a fusion reactor will produce. It is the ratio of the fusion energy produced to the energy required to make the plasma:

$$Q = \frac{\text{Fusion energy}}{\text{Input energy}}.$$

In Chap. 3, we showed this equation for the DT reaction:

$$D + T \rightarrow \alpha + n + 17.6\ \text{MeV},$$

where α is an alpha particle (a helium nucleus) and n is a neutron. Most of the 17.6 MeV of energy released is carried by a 14.1 MeV neutron, and the other 3.5 MeV is carried by the alpha particle.[11] The neutron energy is the part used to produce the electrical output of the power plant, and the alpha energy is used to keep the plasma hot. Since the α's are charged, they are confined by the magnetic field, and the hope is to hold them long enough that they can transfer their energies to the DT plasma, keeping it at a steady temperature. But since the α's have only one-fifth of the fusion energy, Q has to be at least 5 for this to happen. This is called *ignition*. The plasma is "burning" by itself. The reaction cannot run away as in fission because some instability will quench the plasma as soon as the operational limits are exceeded.

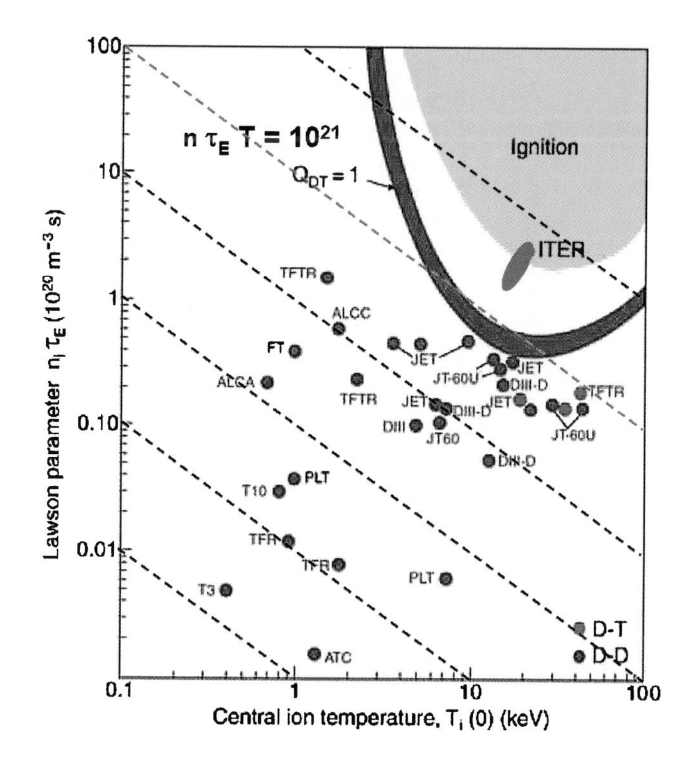

Fig. 8.20 Lawson diagram showing progress toward breakeven and ignition [31]

The first milestone is to achieve $Q=1$, which is called *scientific breakeven*, which assumes that the whole 17.6 MeV is equal to the input energy. The next milestone is to get to ignition at $Q=5$. To produce net energy, you have to count also the energy needed to make the magnetic fields and the plasma currents, as well as all the electricity needed to run the power plant (even the lights!) and the energy used to transmit the power to where it is used. This means that Q has to be at least 10. Figure 8.20 is a Lawson diagram (Chap. 5) plotting $n\tau_E$ vs. T_i and showing what different tokamaks have achieved in DD and DT plasmas. The heavy curve is for $Q=1$ in DT, and we see that this has been reached in JET. The yellow region is ignition at Q greater than 5. The diagonal dashed lines are for constant values of the triple product. The obvious next significant step is to get to ignition, and that is the story of ITER.

The Confinement Scaling Law

The triple product plotted in Fig. 8.20 contains the energy confinement time τ_E, which is how long each amount of energy used to heat the plasma stays in there before it has to be renewed. The plasma energy is lost through three main channels: radiation, mostly in the form of X-rays, and escape of ions and electrons to the wall, carrying their heat with them. The first two of these, radiation and ion loss, follow theory and can be predicted, but electrons escape faster than can be explained. The energy loss by electrons can be measured, but it cannot be predicted. It would be

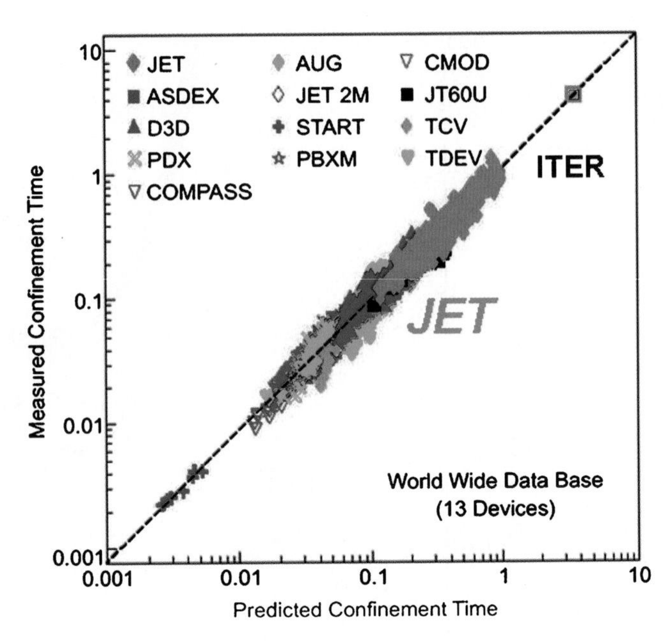

Fig. 8.21 Data from 13 tokamaks showing that the energy confinement time as measured follows an empirical scaling law[12]

impossible to design a new machine accurately without knowing what τ_E would be, but fortunately the over 200 tokamaks that have been built were found to follow an empirical scaling law. This formula[12] gives the value of τ_E in terms of the size and shape of the tokamak, the magnetic field, the plasma current, and other such factors. The result is shown in Fig. 8.21.

This empirical scaling law is the basis on which new tokamaks are designed. It cannot be derived theoretically, but it is followed in a massive database from a variety of tokamaks. This "law" is given in mathematical form in footnote 12. Most of the dependences are consistent with our understanding of the physics. For instance, τ_E increases with the square of the machine size. The strength of the toroidal field does not matter much because the size of the banana orbits depends on the poloidal field. The poloidal field indeed enters in the linear dependence on plasma current. The wonder is that only eight parameters are needed to make all tokamaks fall into line. As seen in Fig. 8.21, the data cover over a factor of 100 in τ_E. To design ITER, the scaling had to be extrapolated by another factor of 4.

ITER: Seven Nations Forge Ahead

The light at the end of the tunnel may be located at the spot marked A in southern France on the map of Fig. 8.22. It is here, in a town called Cadarache near Aix-en-Provence that ITER is being built. Magnetic confinement of plasma gets better with

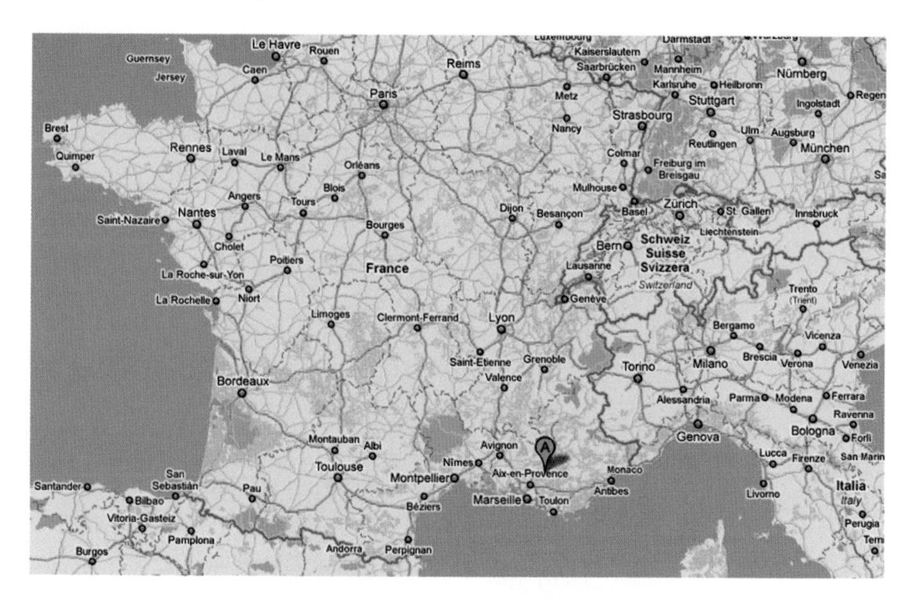

Fig. 8.22 Map of France, showing the location of Cadarache

size, and it has long been clear that a much larger machine has to be built to achieve ignition, a machine so large that no single country can bear the whole cost. Thus was born the international thermonuclear experimental reactor, now known only by its initials, ITER. Coincidentally, ITER in Latin means a path, a journey. It may indeed be the best way to get there.

The reason for the large size is that the amount of power generated is proportional to the volume of the plasma, which increases with the cube of its radius, while the losses are proportional to the surface area of the plasma, which increases only as the square of its radius. To take the next step beyond the four machines shown above, therefore, requires a much larger machine, one so large that its cost has to be shared among many countries. The idea of an international project to achieve fusion energy was born in the 1985 Geneva Superpower Summit, where President Mikhail Gorbachev of the USSR and President Ronald Reagan of the USA, with advice from President François Mitterand of France, agreed to initiate a project involving the USSR, the USA, the European Union, and Japan. (It probably helped that Gorbachev's advisers were Evgeniy Velikov and Roald Sagdeev, both plasma physicists.) More on what ensued afterwards will come later, but first let's see what kind of machine ITER is.

Figure 8.23 is the diagram of the machine being built. Its size is indicated by the small figure at the bottom, representing a standard 2-meter person. The plasma chamber has the standard D-shape, 1.7 times as high as it is wide. The width is 4 m at its widest part, and the major radius (the distance between the center of the chamber and the axis of the whole machine) is 6.2 m. The D-shaped coils that produce the main magnetic field can be seen, but all the other equipment is shown simplified; otherwise, the vacuum chamber would not be visible at all!

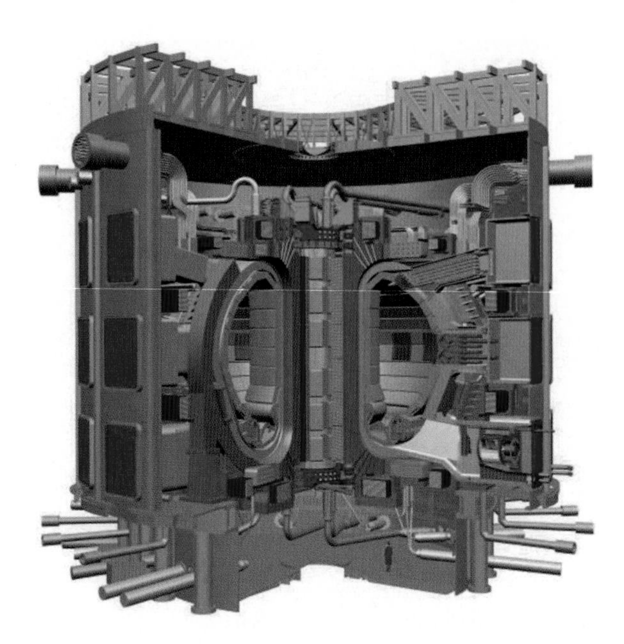

Fig. 8.23 Diagram of ITER (http://www.iter.org)

That includes all the other coils for shaping the plasma, the neutral-beam injectors for heating, the neutron-absorbing blanket, the divertors for catching the plasma, pellet injectors for fueling, and a host of measurement devices. How much bigger ITER is compared with the current champion, JET, is shown in Fig. 8.24. The clutter surrounding a real machine can be seen in the pictures of existing large tokamaks in Figs. 8.3–8.6.

What is ITER designed to do? The primary goal is to produce, for the first time, a "burning" plasma. That is, a plasma that will keep itself hot once it has been heated to several hundred million degrees. Remember that 80% of the fusion energy from DT fuel is in the form of neutrons, and only 20% is in alpha particles (helium ions) which can give energy to the plasma because they are magnetically confined. Therefore, a Q value of at least 5 is needed for burning or ignition. To get a safety margin, ITER is designed to produce a Q of 10, where Q is the ratio of energy out of the plasma to the energy put into the plasma from external sources. $Q=1$ is scientific breakeven (energy in equals total energy out), but most of that energy is in the form of neutrons, which produce the power plant energy but cannot heat the plasma. The best that JET could do was $Q=0.65$, below scientific breakeven. The large step from $Q=0.65$ to $Q=10$ is the reason that ITER has to be so big. The step is not trivial also from a physics point of view. The 3.5-MeV alphas may cause an instability that drives them out of the plasma. Although the stability conditions have been calculated, they have never been tested. The experiment will be considered a success if enough self-heating occurs for these conditions to be established, even if $Q=10$ is not achieved. The self-heating mechanism which powers the sun has never been seen on earth outside of a bomb, and plasma experts are eagerly

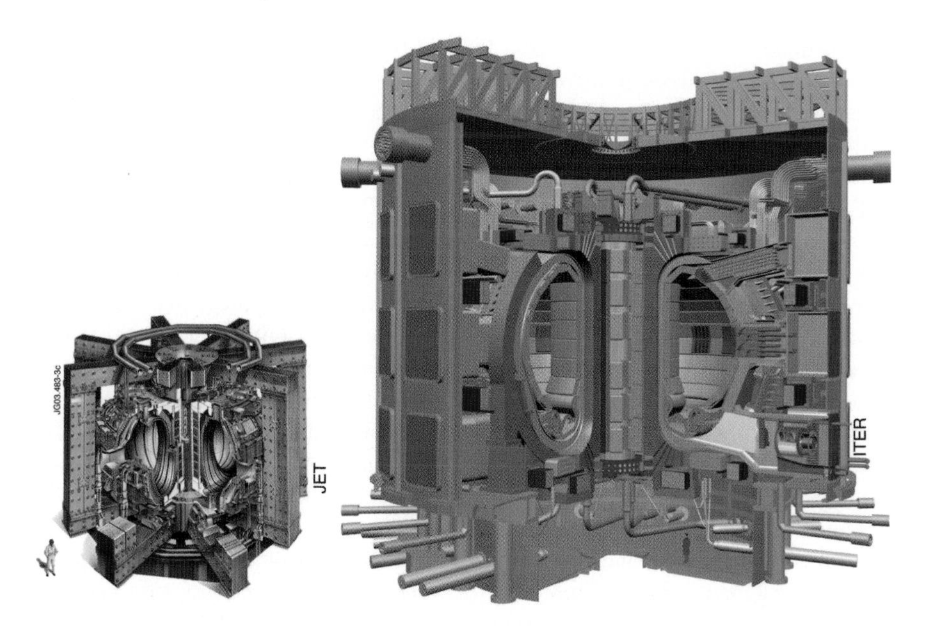

Fig. 8.24 Comparison of ITER with JET (http://www.iter.org)

anticipating this critical test. The term "ignition" may invoke fear that the reaction will run away and cause an explosion. This cannot happen in a fusion reactor because if the density or temperature gets too high, the plasma will disrupt and fizzle out. This may cause melting of parts of the tokamak, but it would be no worse than leaving a pot on a stove after the water has boiled out. The "pot" here would be an expensive one, though!

There are other objectives for ITER besides achieving $Q=10$. It will produce 500 MW of power, about one-sixth that of a full-size reactor. Many large key components of a fusion reactor have to be designed, manufactured, and tested in operation. This includes superconducting magnet coils, wall materials and divertors that can withstand the heat and neutron bombardment, tritium handling, and remote control and maintenance after the walls become radioactive and cannot be approached by personnel. Instability control has to keep the plasma confined steadily for as long as 8 min, using a large amount of bootstrap current and generating 500 MW of power. There will be a first test of a neutron-absorbing "blanket" that can breed tritium. Tritium does not occur naturally. Most of the time, ITER will use tritium coming from fission reactors, of which it is a byproduct; but in a fusion power plant the tritium has to be made internally. This is done in a blanket that captures the 14-MeV neutrons from the reaction, slows them down, and generates heat to run a steam plant. A part of this blanket can be used to breed tritium from lithium, which is an abundant element on earth.

ITER is the logical next step toward fusion power, but it is still primarily a physics experiment. It will lead to DEMO, a demonstration power plant that will run without breakdown and produce a usable amount of power. However, many believe that an intermediate step between ITER and DEMO is necessary to develop engineering concepts that will work in a real reactor. Some of the difficult problems are, for instance, (1) the material to be used in the plasma-facing components (the "first wall"), (2) the handing and breeding of tritium, (3) continuous operation for long periods, (4) maintenance procedures, and (5) plasma exhaust and waste treatment. ITER can provide only a first try on such topics. Engineering will be the topic of the next chapter; this is only an introduction. As an example, the first-wall material has to take the heat of facing a 100,000,000-degree plasma, and it has to allow a large flux of neutrons to pass through without causing such damage that it has to be replaced often. It also cannot contaminate the plasma with impurities of high atomic number, which would cool the plasma. Tests of suitable materials can be done without a tokamak; a fission source of neutrons would do. In fact, most of these engineering tests can be done on a much smaller, cheaper machine than ITER, and such a machine can be built and operated simultaneously with ITER to save time. Most large laboratories have proposed such a machine. For instance, the Fusion Development Facility proposed by General Atomics is a tokamak using normal-conducting coils and producing only 100–250 MW of power at Q less than 5. But it is designed to run continuously for weeks at a time over 30% of a year and breed up to 1.3 kg of tritium per year. Such machines and DEMO are still in the talking stage, but the ITER project is up and running.

As can be imagined, a cooperative project among seven nations is an administrative nightmare. It took over 20 years to get to the present stage. After the initial Gorbachev–Reagan agreement, the four partner nations managed to agree to start Conceptual Design Activities in 1988, and the design was finished in 1990. The resulting tokamak was much larger than the present design. In 1992, an agreement was made to start more serious Engineering Design Activities. Each country had its own home team, and a Joint Central Team was stationed in La Jolla, California. The directors of ITER for this study was at first Paul-Henri Rebut and later Robert Aymar, both of France. After six years of work, it was decided that the tokamak was too large and too expensive, and the activity was extended to 2001. The final design, finished in 2001, is half the price but achieves almost the same objectives. The physics basis for ITER, which we discussed in Chap. 7, was worked out in this period and contributed to the efficiency of the new design. Some $650M was expended to design ITER, with the original agreement that the European Union and Japan would each bear one-third of the cost, while the USSR and the USA shared the other third. To everyone's chagrin, the USA withdrew from the project in 1999, not to return until 2003. The project continued without funding from the US Congress.

Meanwhile, in 1991, the USSR collapsed and was replaced by the Russian Federation. In 2003, the Peoples' Republic of China and South Korea joined ITER. India joined in 2005, raising the number of partners to seven. Canada was temporarily involved but dropped out when its proposed site was turned down. With an area larger than that of Western Europe, Kazakhstan has been considering joining in spite of the fact that it has large fossil reserves. The seven current nations supporting ITER are shown in Fig. 8.25. These countries represent more than half the world's population. Without public support, the USA has been a

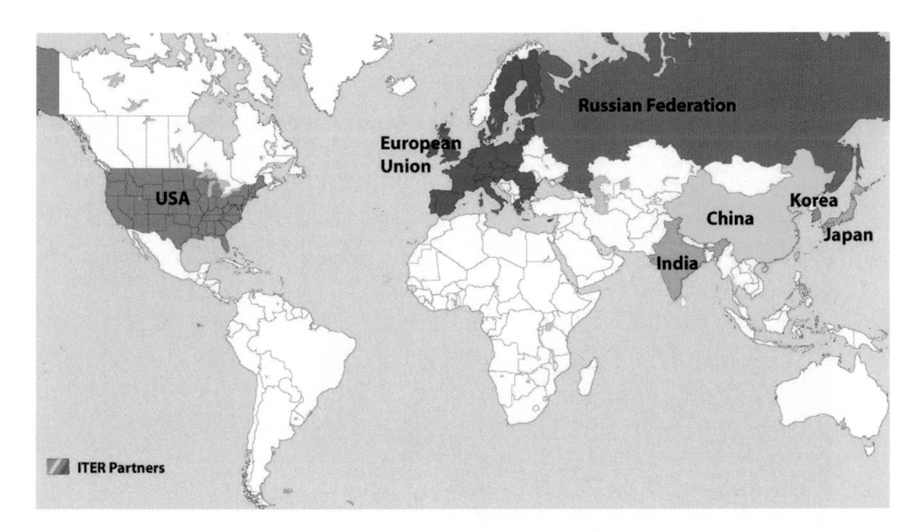

Fig. 8.25 The seven nations in the ITER organization

lukewarm partner in this path-breaking enterprise, and again failed to contribute its financial share in 2008.

By 2003, ITER's design had been agreed upon, and the project was ready to move ahead. The estimated cost was calculated to be five billion euros (about $7B) for ten years of construction, and another 5B euros for 20 years of operation.[13] Then came a totally unexpected delay. There was a deadlock on the site for ITER. The site had to have sufficient power and accessibility for such a large machine. The finalists were a site in Japan and a site in Europe, at first in Spain, but finally in France. The EU, China, and Russia voted for France; and Japan, Korea, and the USA voted for Japan. India had not yet joined. The impasse lasted for two years. Finally, in 2005, the deadlock was broken, and France was chosen. As compensation, Japan was to supply 20% of the staff and had the right to choose the Director. Furthermore, the EU was required to purchase 20% of its ITER material from Japan. As host, the EU has to bear 5/11ths of the cost of ITER, and the other six countries 1/11th each. Kaname Ikeda was chosen to be Director. The 45% contribution by the EU will stimulate its economy.

Once a Joint Implementation Agreement was signed in November 2006 by the seven parties, the ITER Organization sprang into action. Hundreds of scientists, engineers, and administrators began to migrate to Cadarache, settling into temporary offices. Bulldozers began to move two million cubic meters of soil to prepare the flat site for ITER, shown in Fig. 8.26. This amount of dirt would fill the Cheops pyramid, and the area is that of 57 soccer fields. The roads had to be widened to accommodate nine-meter wide truck convoys which will carry the major components of the tokamak. Even traffic circles (roundabouts) like the one at the upper left of Fig. 8.26 had to be enlarged. Those parts manufactured outside Europe would be shipped to the Mediterranean port of Fos-sur-Mer and then barged and trucked to Cadarache. A three-story office building was built in 2008 to house 300

Fig. 8.26 Preparation of the ITER site in 2008

employees, but this was still temporary and off-site. To accommodate their families, a multilingual school was established in Manosque; by 2009 it had 212 students from 21 nations and 80 teachers. In 2010, the school will have its own building and include a nursery school and a junior high. The first ITER baby was delivered in 2008. A weekly bulletin[14] covers not only technical and personnel news but also includes cultural events and introduces the entire international community to the history and traditions of this region in southern France.

ITER is truly an international project. For instance, the vacuum vessel will be made by Europe and Korea, with other parts from Russia and India. The largest components, the magnet coils, will weigh 8,700 tons and will be made of Nb_3Sn and NbTi superconductors. Many different types of magnet coils and their feed-ins are required, and the manufacture of the superconductor material and their formation into coils are shared among most of the ITER partners. The USA will supply 40 tons of expensive Nb_3Sn conductors for the toroidal field, and those for the poloidal field will be shared among China, Russia, and Europe. Superconductor wire is very complicated, wound in many strands and cooled with liquid helium. That these actually work in large coils has been tested in the LHD stellarator in Japan and will be further tested in the new superconducting tokamaks in China, Korea, and Japan.

Domestic Agencies have been established in each country to organize the manufacture of its in-kind contributions to ITER by local industries. Through these agencies, Procurement Agreements have to be drawn up and signed by each member country. As of 2010, 28 PAs have been signed. The site in Fig. 8.25 has been completely leveled, and the construction of 38 buildings on it has begun. The first of these is a six-story 253-m long building for winding the poloidal field coils, which are too large to be shipped, and the superconductor cable is all in one piece.

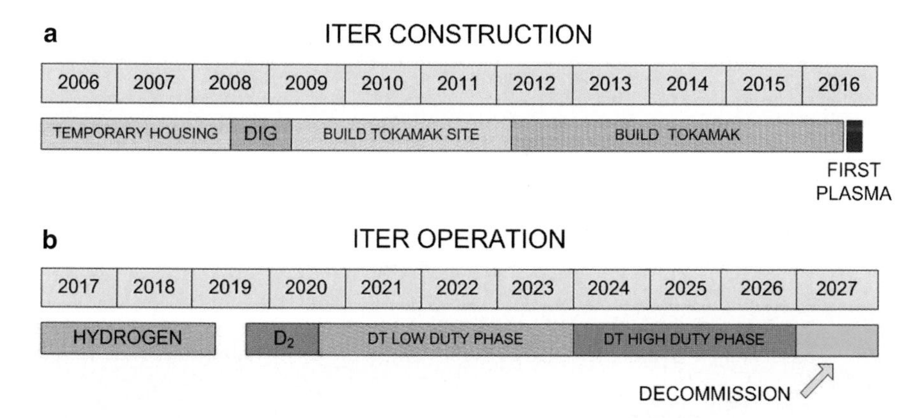

Fig. 8.27 The original ITER timeline

New office buildings will replace the temporary ones. Off-site in Manosque, a new school, will be built for the community.

It is clear that the ITER project is in for the long pull. Figure 8.27 shows the originally agreed schedule for the construction and operation of ITER. The site preparation will not be finished until 2012, but meanwhile the components are being designed, fabricated, and tested in various countries. It will take four years to get all the parts delivered and the tokamak assembled. The first plasma is scheduled to be made near the end of 2016. At first, experiments will be done with hydrogen, which is not radioactive. Remote handling will then be implemented so that deuterium can be used; the D–D reaction creates some neutrons, but not as many as does DT. In 2020, operation with DT will start, first in pulsed (low-duty) operation, to achieve the designed Q value. In the later stages, emphasis will be on quasi-steady state operation (high-duty) to test whether bootstrap current and non-inductive (no transformer) drive can sustain the plasma. At the end of 2026, a decision will be made whether to decommission the machine or to continue it with modifications. De-activating, decommissioning, and disposing of the machine is expected to take another 11 years. The ITER machine will have 30,000 components in ten million pieces. To get these to be delivered on time and fit together requires numerous groups and oversight committees. Their acronyms are overwhelming, but that's the price you pay for organizational efficiency.

At this time, the goal of achieving first plasma in 2016 seems a long way off, but the worldwide economic downturn in 2008–2009 has made it even worse. Both the budget and the schedule had to be revised in 2010. The project will be delayed two years or more by economic constraints. The new construction schedule will look something like Fig. 8.28. DT plasmas will not be attempted before 2027.

These estimates notwithstanding, the project is proceeding nicely under new Director Osamu Motojima. The digging and flattening of the ITER site has been finished and is shown in Fig. 8.29. Parts of the machine are coming in from different countries. Figure 8.30 shows the buildings planned for the site. These will be earth-quake-proof, and some will have containment for radioactivity. The long coil-winding

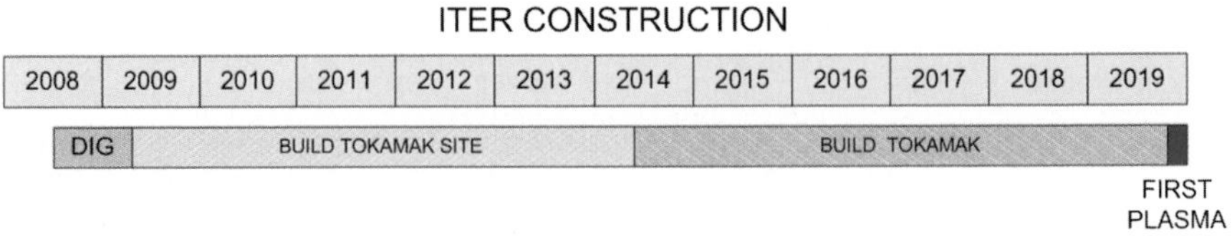

ITER CONSTRUCTION

2008	2009	2010	2011	2012	2013	2014	2015	2016	2017	2018	2019

DIG | BUILD TOKAMAK SITE | BUILD TOKAMAK

FIRST PLASMA

Fig. 8.28 The revised ITER timeline [32]

Fig. 8.29 The ITER site in June, 2010[14]

Fig. 8.30 Planned buildings for the ITER site [32]

building mentioned above can be seen at the top for scale. It is exciting to see international teamwork functioning so well.

Contrary to popular perception, fusion is no longer in a guessing stage. The timeline for its development has been set. Each country has its own ITER organization and its own specialized manufacturing capabilities to contribute to the project.

At the current level of funding, it will take until 2026 to get the information from this experiment. Concurrently, materials testing facilities can be built and run to support DEMO. Design, construction, and operation of DEMO will take until 2050; and, if it is successful, commercial reactors can follow soon thereafter. The present plan is to achieve fusion power by 2050, in time for the present generation of children to enjoy. However, with increased international ambition, the time can be shortened.

There may be some confusion in the public's mind between ITER and another large experiment, the Large Hadron Collider, or LHC, at CERN near Geneva. Geneva can be seen in Fig. 8.21 north of Cadarache. It is quite a coincidence that the two largest physics experiments in the world should be located only a few hundred kilometers from each other. The LHC is a particle accelerator 27 km (17 miles) in circumference, buried in a circular tunnel under France and Switzerland. It is similar to ITER in internationality, cost (6.3B euros), and the extensive use of superconductors; but it is entirely different in technology and purpose. The LHC is a basic physics experiment to explore the subatomic structure of matter and energy: quarks, Higgs bosons, dark matter, and so forth. Protons and antiprotons are accelerated to multi-TeV (trillions of eV) energies and hurled against one another to break them up, one particle at a time. ITER, on the other hand, deals with a gas of multi-billions of particles at KeV (thousands of eV) energies. In the LHC, large magnetic fields are used to bend the protons into circular orbits, their Larmor radius being measured in kilometers. In ITER, large magnetic fields are used to hold a plasma, which exerts a large pressure not because the particles are so energetic but because there are so many of them.

The LHC and its predecessors were inspired by man's urge to understand his place in the universe, not by any practical need. ITER, on the other hand, is being built to develop an energy source that will save mankind, and, if done soon enough may also solve current problems in climate change and fossil fuel depletion. We are living in a golden age in which civilization has advanced to such a point that we can afford to reach for lofty goals. Let us hope that our reach does not exceed our grasp.

Notes

1. http://www.toodlepip.com/tokamak/gallery-ext.htm.
2. Alternate concepts have been described by Bishop [1] and Chen [11, 12].
3. Dale Meade, Astronomy 225 seminar notes, Princeton University, 2005.
4. http://www.pppl.gov/projects/pages/tftr.html.
5. http://www.jet.efda.org/pages/multimedia/brochures.html.
6. PCAST report, 1995: http://www.ostp.gov/pdf/fusion1995.pdf.
7. Massachusetts Institute of Technology.
8. $n_G\,(10^{20}\ \mathrm{m}^{-3}) = I_p/\pi a^2\ (\mathrm{MA/m^2})$, where I_p is the toroidal current and a is the minor radius. There are recent attempts to explain the limit theoretically [29].
9. This original reference does not give the formula that is now used.
10. $\beta_N = \beta(\%)\dfrac{(a(\mathrm{m})B(\mathrm{T}))}{I(\mathrm{MA})} = 3.5$, where the units are meters, Tesla, and megamps, and

$\beta = \frac{\langle n(KT_i + KT_e)\rangle}{B_0^2 / 2\mu_0}$. This will give the value of β (in percent) for each machine when its I, a, and B values are inserted.

11. This is just a 20–80% division of the energy because the alpha weighs four times more than the neutron, and they both have the same momentum.
12. The scaling law is $\tau_E = 0.0562 \times I^{0.93} B^{0.15} P_{loss}^{-0.69} n_{e,19}^{0.41} M^{0.19} R^{1.97} \varepsilon^{0.58} \kappa^{0.78}$, where τ_E is energy confinement time (s), I is the plasma current (MA), B is the toroidal magnetic field (T), P_{loss} is the power to divertor (MW), $n_{e,19}$ is the electron density (10^{19} m^{-3}), M is the average atomic number, R is the major radius (m), ε is the inverse aspect ratio, and κ is the elongation.
13. The latest increases are given in Chap. 11.
14. http://www.iter.org/newsline.

References

1. A.S. Bishop, *Project Sherwood* (Addison-Wesley, Reading, 1958)
2. H. Wilhelmsson, *Fusion, a Voyage Through the Plasma Universe* (Institute of Physics Publishing, Bristol, 2000)
3. G. McCracken, P. Stott, *Fusion, the Energy of the Universe* (Elsevier, Amsterdam, 2005)
4. T.K. Fowler, *The Fusion Quest* (Johns Hopkins Univ. Press, Baltimore, 1997)
5. J.L. Bromberg, *Fusion: Science, Politics, and the Invention of a New Energy Source* (MIT Press, Cambridge, 1982)
6. R. Herman, *Fusion, the Search for Endless Energy* (Cambridge Univ. Press, Cambridge, 1990)
7. G.J. Weisel, *Properties and phenomena: basic plasma physics and fusion research in postwar America*. Phys. Perspect. **10**, 1 (2008)
8. I.B. Bernstein, E.A. Frieman, M.D. Kruskal, R.M. Kulsrud, Proc. Roy. Soc. **A244**, 17 (1958)
9. T. Ohkawa et al., Phys. Fluids **11**, 2265 (1968)
10. F.F. Chen, *Intro. to Plasma Physics*, 1st edn. (Plenum, New York, 1974)
11. F.F. Chen, *Alternate concepts in magnetic fusion*. Phys. Today **32**(5), 36 (1979)
12. F.F. Chen, *Alternate Concepts in Controlled Fusion: Summaries of Four Workshops*, Electric Power Research Institute Rept. ER-429-SR (1977)
13. M.A. Krebs et al., *A restructured fusion energy sciences program: advisory report*, J. Fusion Energy **15**, 183 (1996)
14. F. Jenko et al., Phys. Plasmas **7**, 1904 (2000)
15. M. Greenwald, *Verification and Validation for Magnetic Fusion: Moving Toward Predictive Capability*, Annual Meeting, Div. of Plasma Physics, Amer. Phys. Soc., Atlanta, GA, November 2009
16. ITER physics basis 1999, chap. 1. Nuclear Fusion **39**, 2137 (1999)
17. http://en.wikipedia.org/wiki/Three_Gorges_Dam Nature, **452** (March 6, 2008)
18. H. Zohm, Plasma Phys. Control. Fusion **38**, 105 (1996)
19. P.B. Snyder et al., Nuclear Fusion **44**, 320 (2004)
20. T.E. Evans et al., Nuclear Fusion **45**, 595 (2005)
21. K. McGuire et al., Phys. Rev. Lett. **50**, 891 (1983)
22. L. Chen, R.B. White, Phys. Rev. Lett. **52**, 1122 (1984)
23. B. Coppi, F. Porcelli, Phys. Rev. Lett. **57**, 2272 (1986)
24. R.B. White, M.N. Bussac, F. Romanelli, Phys. Rev. Lett. **62**, 539 (1989)
25. ITER physics basis 1999, chap. 3. Nuclear Fusion **39**, 2321 (1999)
26. ITER physics basis 2007, chap. 3. Nuclear Fusion **47**, S161 (2007)
27. R.S. Granetz et al., Nuclear Fusion **36**, 545 (1996)

28. M. Greenwald et al., Nuclear Fusion **28**, 2199 (1988)
29. M.Z. Tokar, Phys. Plasmas **16**, 020704 (2009)
30. F. Troyon et al., Plasma Phys. Control. Fusion **26**, 209 (1984)
31. J.M. Noterdaeme, 12th International Conference on Emerging Nuclear Energy Systems (ICENES), Brussels, Belgium (2005)
32. G. Janeschitz, *The Physics and Technology Basis of ITER and Its Mission on the Path to DEMO*, Symposium on Fusion Engineering, San Diego, CA, June 2009

Chapter 9
Engineering: The Big Challenge*

Introduction

With the information they have gathered from the public media, most people who have heard of fusion consider fusion energy to be a pipedream. Their information is out of date. As we have shown in the last two chapters, great advances have been made in fusion physics, and our knowledge of plasma behavior in toroidal magnetic bottles is good enough for us to push on to the next step. This does not mean, however, that fusion is *not* a pipedream. There is a large chasm between the understanding of the physics and the engineering of a working reactor. There are problems in the technology of fusion so serious that we do not know if they can be solved. But the payoff is so great that we have to try.

The situation can be compared – or contrasted – with that of the Apollo program to put a man on the moon. In that program, the physics was already known: Newton's laws of motion covered all the physics that was needed. In the case of fusion, it took over 50 years to establish the science of plasma physics, to develop fast computers, and to understand the physics of magnetic confinement; but we have done it. In the Apollo case, there were engineering problems whose solutions could not be fully tested. Could the nose cone material stand up to the heat of reentry? Can humans survive long periods without gravity and then the stress of reentry? Will micrometeorites puncture the space suits of the astronauts? It was a dangerous experiment, but President Kennedy pushed ahead, and it succeeded marvelously. In the case of fusion, we do not know yet how to build each part of a reactor, but the only way to get this ideal source of energy is to push on ahead. The expense will be comparable to Apollo's, but at least no human lives are endangered.

The path to a commercial fusion reactor has been studied intensely in the past decade. There are three or four steps: (1) the ITER experiment now being built, (2) one or more large machines for solving engineering problems, (3) DEMO, a prototype reactor built to run like a real reactor but not producing full power, and (4) FPP, fusion power plant, a full-size reactor built and operated by the utilities industry.

*Numbers in superscripts indicate Notes and square brackets [] indicate References at the end of this chapter.

F.F. Chen, *An Indispensable Truth: How Fusion Power Can Save the Planet*,
DOI 10.1007/978-1-4419-7820-2_9,

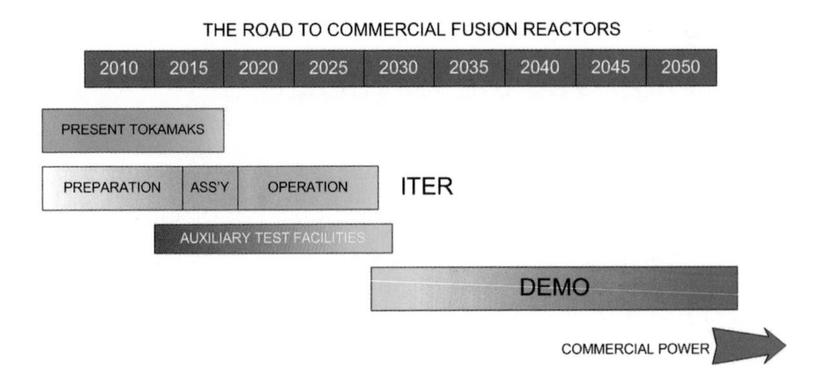

Fig. 9.1 A possible schedule for developing fusion power (Data from G. Janeschitz, *The physics and technology basis of ITER and its mission on the path to DEMO*, Symposium on Fusion Energy, San Diego, California, June 2009)

Step 2 is being hotly debated. Some think that experiments on ITER will give enough information to design DEMO. Others propose intermediate machines designed to solve specific problems such as the tokamak wall material or the breeding of tritium. These problems are described in the main part of this chapter. The time it will take to reach the FPP stage might look something like this (Fig. 9.1). Any additional machines for engineering testing before designing DEMO are shown in Fig. 9.1, although they may not be necessary. Although this timeline is called the "fast track" to fusion, it still will take until 2050 before fusion power becomes a reality. The economic downturn at the turn of this decade has already delayed the construction of ITER. Shortening this timeline can be done only with greatly increased funding. In the meantime, expansion of the other renewable energy sources listed in Chap. 3 is still necessary.

The two toughest engineering problems are the material of the "first wall" and the breeding of tritium. These will be discussed in detail. We also mentioned some physics problems that are not completely solved. One concerns "disruptions" which kill the plasma and must be avoided in a reactor. The best known way to avoid them is to operate safely below the tokamak's limits, and this means less output power. Otherwise, injection of a large puff of gas can stop an incipient disruption; this is a crude solution. A second problem concerns the edge-localized modes (ELMs), instabilities that dump plasma energy into places not designed to absorb it. Currently, internal correction coils are to be inserted inside the plasma chamber to suppress ELMs as well as resistive wall modes (RWMs). This is another crude solution which would not be suitable in a reactor. A third problem concerns the alpha particles (the helium nuclei) which are the products of the D–T fusion reaction. These fast ions can, in theory, excite Alfvén waves, and these electromagnetic waves could disrupt plasma confinement. This instability cannot be studied until we can ignite a plasma to produce these alpha particles.

Although these seem to be formidable problems, there will be a learning curve when ITER and DEMO are built. Once industry gets serious about fusion, progress will be rapid. We will go from Model-T Fords to Mercedes-Benzes. We will go from

DC-3s to Airbus A380s. We may even get lucky with more help from Mother Nature and find that fast alpha particles are stabilizing. Where there's a will, there's a way. With a positive attitude, the fusion community can continue to achieve and live up to its track record of the last 50 years. Further in the future, in the second half of this century, a second generation of fusion reactors will look quite different from the tokamak as described here. There are other magnetic configurations, simpler than the tokamak, that have not been fully developed for lack of funding. These are described in Chap. 10. Better yet, there are fuel cycles that do not require tritium, thus avoiding almost all of the fuel breeding and radioactivity problems of the first generation of fusion reactors. These advanced fuel cycles can run only with hotter and denser plasmas than we can now produce, but which may be possible once we have learned how to control plasma better. Advanced fuels are also presented in Chap. 10. The engineering problems described here are not the end of the story.

The First Wall and Other Materials

The First Wall

Figure 9.2 is a more realistic drawing of the ITER machine than shown in Chap. 8. The plasma will occupy the D-shaped vacuum space surrounded by tiles. These tiles are the plasma facing components (PFCs), commonly called the "first wall." They have to withstand a tremendous amount of heat from the plasma and yet must not contaminate the plasma and be compatible with the fusion products that impinge on them. Early tokamaks used stainless steel, but clearly this is not a high-temperature material. Current tokamaks use carbon fiber composites (CFCs), a light, strong, high-temperature material that is used in bicycles, racing cars, and space shuttles. Just as rebars are used to strengthen concrete, carbon fibers are used to strengthen graphite. However, carbon cannot be used in a reactor because it absorbs tritium, which would not only deplete this scarce fuel but also weaken the CFC. After all, hydrocarbons like methane and propane are very common, stable compounds; and tritium is just another form of hydrogen and can be captured by the carbon to form hydrocarbons.

Tungsten is a refractory metal, but it is high-Z; that is, it has a high atomic number and therefore has so many electrons that it cannot be completely ionized. The remaining electrons radiate energy away, cooling the plasma. The good thing about hydrogen and its isotopes is that they have only one electron, and once that electron is stripped free of the nucleus by ionization, the atom can no longer emit light. Beryllium is a suitable low-Z material, but it has a low melting point, and so has to be cooled aggressively. In preparation for ITER, the European tokamak JET is being upgraded with a beryllium first wall. In short, the first-wall material must not absorb tritium and must have a low atomic number, take high temperatures, and be resistant to erosion, sputtering, and neutron damage.

ITER, of course, is only the first step. There are large steps between ITER and DEMO and between DEMO and a full reactor. Some large numbers on the first wall

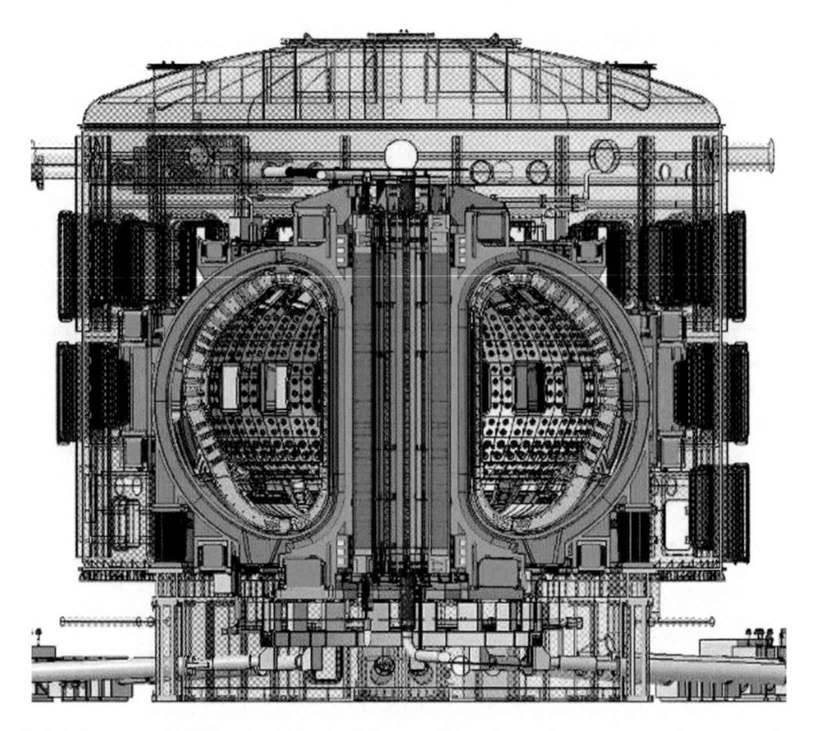

Fig. 9.2 Diagram of ITER, showing the "first wall" and openings (ports) where experimental modules can be inserted for testing [29]

Table 9.1 Loads on the first wall

	ITER	DEMO	Reactor	Units
Fusion power	0.5	2.5	5	GW
Heat flux	0.3	0.5	0.5	MW/m^2
Neutron load	0.78	<2	2	MW/m^2
Neutron load in life	0.07	8	15	MW-years/m^2
Neutron damage	<3	80	150	dpa

are given in Table 9.1. We see that the step between ITER and DEMO is much larger than between DEMO and Reactor. Hence the call for a materials-testing facility intermediate between ITER and DEMO.

The fusion power is given in gigawatts. A typical power plant generates 1 GW of electricity; and perhaps 5 GW of fusion power is needed to give that, since the tokamak needs power to run, and there is still a heat cycle in a steam plant to produce electricity. The heat flux impinging on the first wall is about 0.5 MW/m^2. This translates to 50 W/cm^2 or about 300 W/sq. in. This is not much more than the surface of an electric iron, though the total heat is considerable. The real problem is in the divertor, which has to handle most of the heat from the plasma. Divertors will be covered later.

The neutron load is the energy of the 14-MeV neutrons from the D–T reaction which pass through the first wall. This energy is not deposited in the first wall, but the neutrons damage the wall. The neutron load summed over the life of the wall is what matters. This is much larger for a reactor than for ITER, since ITER is just an experiment, while a reactor should last about 15 years before it has to be revamped. The neutron damage is measured in displacements per atom (dpa). The longer the material is exposed to a neutron flux, the more times one of its atoms will be knocked out of place by a neutron. After many dpa's, the material will swell or shrink and become so brittles as to be useless.

Beryllium melts so easily that it cannot be used in a reactor. Boron coating has been tried successfully, but also cannot take high temperatures. Tungsten seems to be the best available wall material because it does not erode or sputter easily and has a high melting point of 3,410°C. However, it is a high-Z material and also cannot be machined easily. A liquid lithium first wall has been considered, but it is no longer proposed.[1] Silicon carbide (SiC) is a promising material that has been studied extensively in the laboratory but does not have a known method for manufacturing in large quantities [1]. How SiC compares with other materials in operating temperature is shown in Fig. 9.3. These temperature ranges are for irradiated materials so that the swelling and fracture effects caused by neutrons are included. Carbon fiber-reinforced graphite (C/C) can take high temperatures, but carbon cannot be used because of tritium retention. Tungsten and molybdenum are classical refractory metals but will cool the plasma if they sputter into it. Silicon carbide reinforced with layers of SiC fibers (SiC/SiC) seems to be the ideal material for the first wall if it can be made without impurities. It takes high temperature, is quite strong, and is resistant to radiation damage. It can last for the 15-year life of the

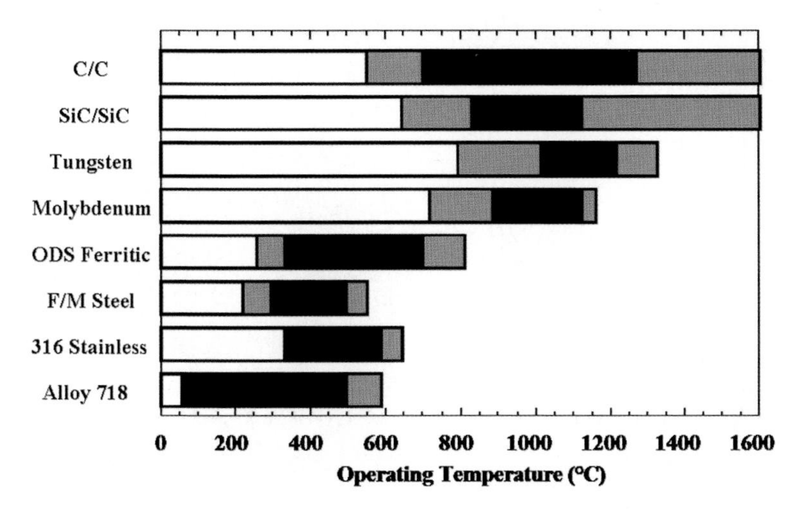

Fig. 9.3 Temperature range of various wall materials under irradiation [1]. The *top* four are refractory materials, and the *bottom* four are structural steels. The *dark center* of each bar is a reasonable operating range; the total bar is an extended range which is possible but not proven

reactor. Its properties have been measured in fission reactors [2]. The main drawback is a thermal conductivity lower than for other materials.

The latest high-tech material is a SiC matrix/graphite fiber composite [1], which has increased thermal conductivity in addition to the other good properties. These advanced materials cannot be designed with existing computer programs, which are applied only to metals. Some reactor studies assume that SiC first-wall material will be available. Though SiC composites have tremendous potential, much research and testing remain to be done before they become a reality.

The Divertor

Sixty percent of the plasma exhaust is designed to go into the "divertor," thus sparing the first wall from the major part of the heat load. Materials and cooling methods can be used in the divertor that cannot be used for the first wall. Figure 9.4 shows how this is done. Special coils located at the bottom of the chamber bend the outermost field lines so that they leave the main volume and enter the divertor. Plasma tends to follow the field lines, so that most of it leaves the chamber by striking the surfaces of the divertor rather than the first wall. Only

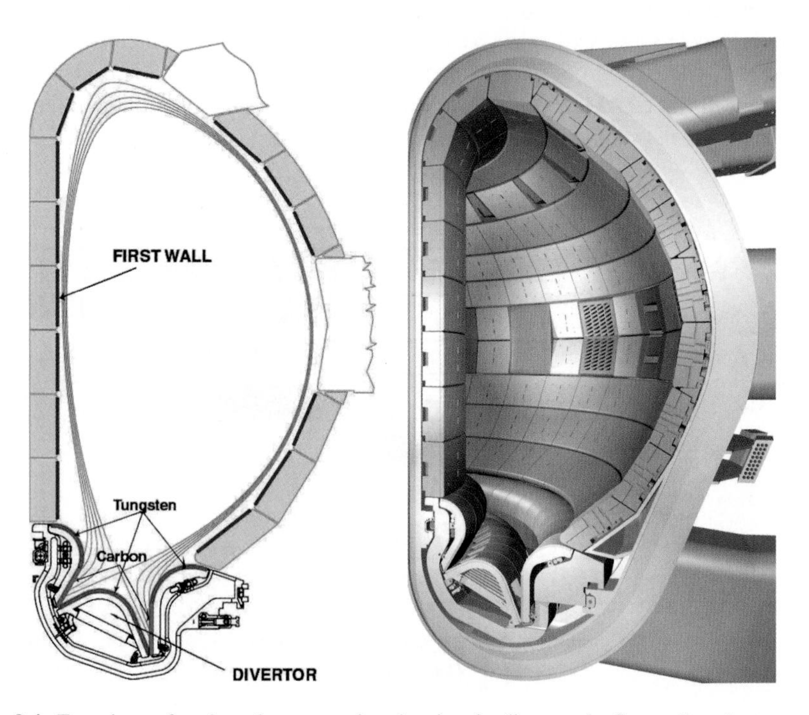

Fig. 9.4 Two views of a tokamak cross section showing the divertor, the first wall, and some ports for heating and diagnostics equipment or for test modules [30, 31]. In the *left diagram*, the outermost magnetic field lines are drawn, showing how they lead the plasma into the divertor. The closed magnetic surfaces in the interior have been omitted for clarity

the plasma that migrates *across* the magnetic field hits the first wall. The heat load on the first wall can be larger than average when there is an instability such as an ELM or a disruption that takes plasma across the field lines suddenly. The first wall of ITER will have to withstand such heat pulses, but DEMO must be built to avoid such catastrophes.

As can be seen in the diagram, the boundary layer of diverted field lines is very thin, only about 6 cm in ITER. In the divertor, these field lines are spread out over a larger area, and the surfaces which the plasma strikes are inclined almost parallel to the field lines so that the heat is deposited over as large a surface as possible. Tungsten can be used for these surfaces, and even carbon compounds can be used in spite of their tritium retention. The divertor parts are easier to replace than the first wall, so the tritium can be removed periodically. The heat load on the divertor surfaces is huge, some 20 MW/m^2, so the cooling system is an important part of the design. Water cooling is possible in ITER, but helium cooling at higher temperatures would have to be used in DEMO and FPPs. The conditions inside a divertor are so intense that they are hard to imagine. Ions with tens of keV energy stream in along the field lines, accompanied by electrons that neutralize their charges. When the ions meet a solid surface, they recombine with electrons to form neutral atoms. There is a dense mixture of plasma with neutral gas made of deuterium, tritium, helium, and impurities, which later have to be separated out in an exhaust processing unit. The neutral gas has to be pumped away fast by vacuum pumps before it flows back into the main chamber and gets ionized again into ions and electrons. To trap the neutrals inside the divertor, a dome-shaped structure has to be added. Figure 9.5 shows the main parts of a divertor designed for ITER. The plasma impinges at a glancing angle onto the high-temperature surfaces made of tungsten and CFC. A heat-sink material, CuCrZr, transfers the heat to water-cooled surfaces.

Water cooling, which is limited to about 170°C, would be insufficient for DEMO and FPP, and cooling by helium gas would have to be used. The helium

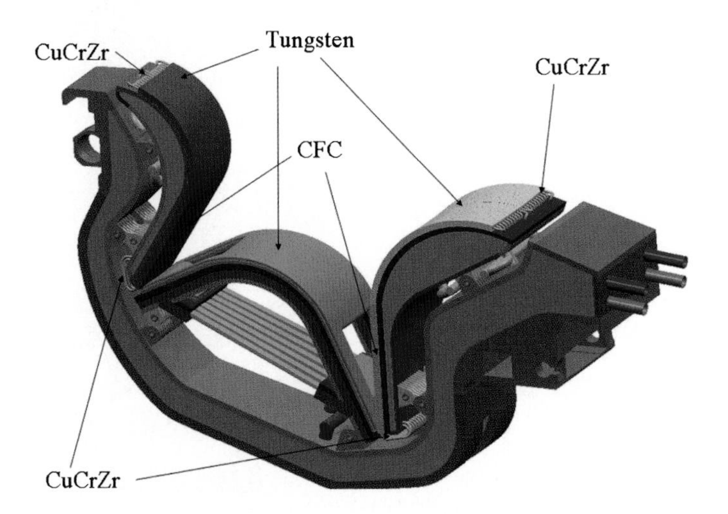

Fig. 9.5 Conceptual diagram of a water-cooled divertor [31]

would be injected at 540°C and be heated to 720°C, while the tungsten and CFC tiles would get to 2,500°C [3]. The coolant would be injected under pressure to cool a small dome as illustrated in Fig. 9.6. These domes are then assembled into nine-finger units, and these units then form a uniformly cooled surface.

Divertor technology is in better shape than other problem areas because divertors are small, and they have already been extensively tested. For instance, meter-sized tungsten and CFC divertor segments (Fig. 9.7) have been tested in Karlsruhe, Germany, up to heat fluxes of 20 MW/m^2. In that large laboratory, divertor materials

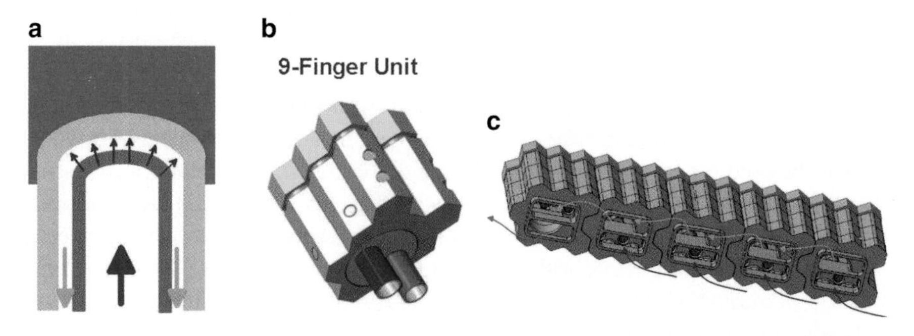

Fig. 9.6 Possible design of a helium cooling system for a divertor [31]. Helium cools a dome-shaped "finger" (**a**), and nine of these are assembled into one unit (**b**). A number of these together then form a cooled surface (**c**)

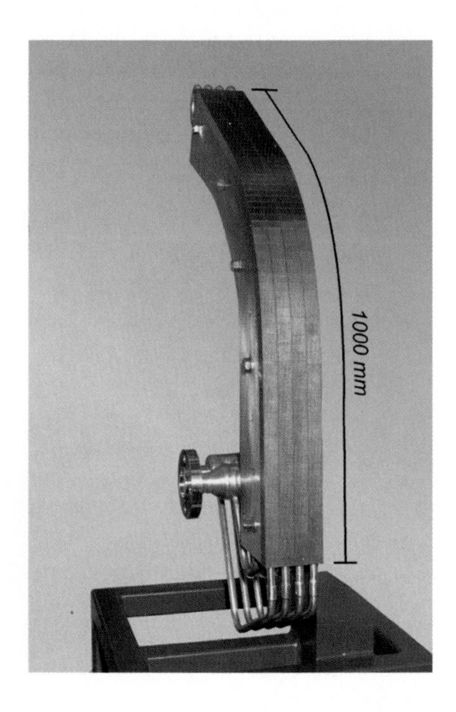

Fig. 9.7 A water-cooled divertor test surface [31]

have been neutron-irradiated, and their manufacturing and assembly techniques have been worked out. Even remote handing techniques for replacing divertors have been tested. It seems possible to design water-cooled divertors for heat fluxes up to 20 MW/m^2 and helium-cooled divertors up to 15 MW/m^2 [31].

Structural Materials

Aside from materials exposed to plasma and large heat fluxes, structural materials have to be chosen to support the huge weight of the reactor elements – the vacuum chamber, magnetic coils, breeding blankets, and so forth. Normally one would use steel; but for fusion, the type of steel has to be carefully designed. The neutrons bombarding the structure will make it radioactive. Only the following elements can be used: iron, vanadium, chromium, yttrium, silicon, carbon, tantalum, and tungsten. Elements like manganese, titanium, and niobium used in other steels would result in long-lived radioactive isotopes. Two *Reduced Activation Ferritic Martensitic Steels* have been designed: Eurofer (in Europe) and F82H (in Japan). These have the following additives to iron [4]:

	Chromium (%)	Tungsten (%)	Vanadium (%)	Tantalum (%)	Carbon (%)
Eurofer	7.7	2	0.2	0.04	0.09
F82H	8.9	1	0.2	0.14	0.12

These steels have only short-lived radioactivity and, unlike fission products, are nonvolatile and can be re-used after storage for 50–100 years. The amount of swelling under neutron bombardment is much smaller than for ordinary stainless steel. Swelling and embrittlement come from helium and hydrogen bubbles trapped in the steel. There are experimental oxide dispersion strengthened (ODS) steels which have nanoparticles of Y_2O_3 that can trap helium and hydrogen, strengthen the material, and reduce creep. Though much has to be done to manufacture these materials with low impurity levels, to study their welding properties, and to test their limits in temperature and radiation resistance in full-time operation, structural materials are not one of the worrisome problems in fusion technology.

Figure 9.8 shows the predicted radioactivity of Eurofer and SiC in a fusion reactor after 25 years of full-power operation. Note that the scales are logarithmic, so that each vertical division represents a factor of 10, and each horizontal division a factor of 100. After 100 years, the radioactivity has decayed by a factor of almost 1,000,000. This material is solid and will not leak out of its containers. The main danger from radioactivity comes from tritium, which decays in 12 years and will be considered in detail later. Note that even this small amount of radioactivity compared with fission is caused by the fact that the D–T reaction emits energetic neutrons. In second-generation fusion reactors using advanced fuels there will be almost no radioactivity.

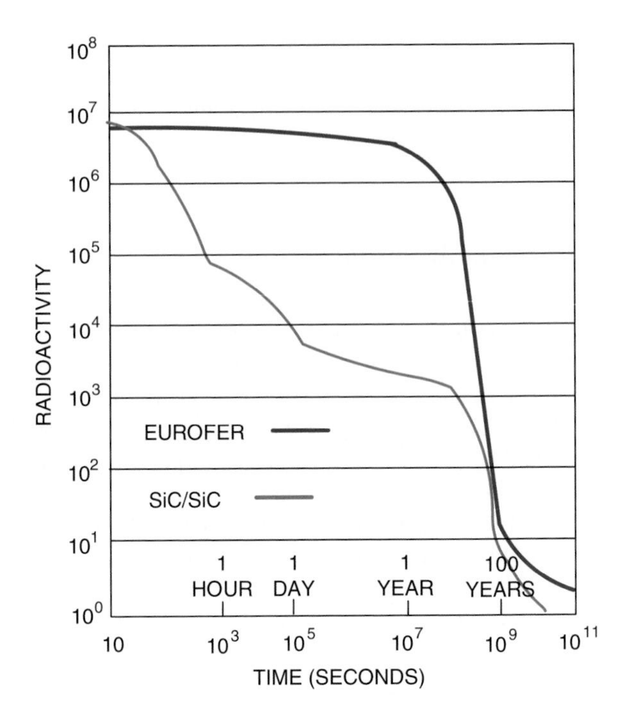

Fig. 9.8 Predicted radioactivity of Eurofer steel and silicon carbide composites versus time. The vertical axis is in units of Becquerels per kilogram. Data from [4]

Blankets and Tritium Breeding

What Is a Blanket?

It is certainly not a thin, soft cover to keep the plasma warm. It is a thick, massive, complex structure that serves three major purposes: (1) capture the neutrons generated by fusion and convert their energy into heat, (2) produce the tritium to fuel the DT reaction, and (3) shield the superconducting magnets from the neutrons. The blanket is divided into modules for easier replacement. Figure 9.9 shows where the blanket is located inside a tokamak. In Fig. 9.9a, we see that the plasma first strikes the first wall (FW), which is also the front surface of the blanket. Then, the neutrons go into the blanket, where their energy is captured, and where the tritium breeding takes place. The heat is taken away by hot gas or liquid coolants to heat exchangers outside. Shielding material protects the vacuum walls and superconducting magnets from the heat and the neutrons. Figure 9.9b gives an idea of how the blanket surrounds the plasma and lies inside the vacuum. Outside the vacuum vessel are the magnetic coils. The Central Solenoid coil is critical, since there is not much room in the hole of the torus to fit this coil into. The symmetry axis of the torus is at the left. The entire machine fits inside a cryostat which insulates the magnet coils from the outside world, keeping them at superconducting temperatures.

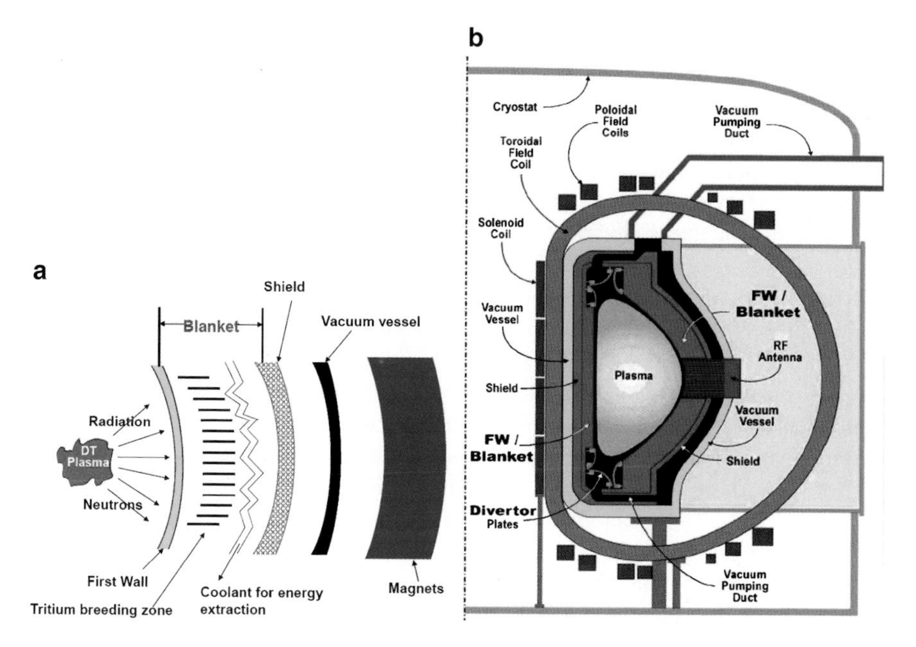

Fig. 9.9 (**a**) The order of the main layers in a tokamak, showing that the entire blanket must be inside the vacuum chamber. (**b**) General scheme of a tokamak's components, showing that the entire machine is inside a cryostat to keep the superconducting magnets cold [32]

In a reactor there could be hundreds of blanket modules, each weighing a ton. There are many ideas for blanket design, and ITER will have three ports available for test blanket modules (TBMs). There are six TBM proposals competing for these three spots [5].

The Role of Lithium

Deuterium and tritium are not the only fuels in fusion; lithium is needed for breeding tritium, which occurs only in minute amounts in nature. Lithium is an abundant element on earth, occurring in two isotopes, 92.6% Li^7 and 7.4% Li^6. (The superscript is the atomic weight, the total number of protons and neutrons in the nucleus.) Lithium-6 is the more useful one and can easily be enriched to 30–90% for use in a blanket. A 1,000-MW fusion plant will consume 50–150 kg of tritium a year, much more than can be supplied by other sources, such as fission reactors. To generate this amount of tritium in blankets, less than 300 kg of Li^6 will be needed by each reactor per year. About 10^{11} kg of lithium is available on land, and 10^{14} kg in the oceans. If all the world's energy is generated by fusion, the lithium will last 30 million years [6]. Deuterium will last even longer. There are 5×10^{16} kg of deuterium in the oceans, and at the rate of 100 kg per reactor per year, that will last 30 *billion* years! That's what we mean when we say that fusion is an infinite power source.

The way tritium is made from lithium-6 is shown in Fig. 9.10. The neutron, which started out at 14-MeV energy, has been slowed down by collisions with a moderator material and collides with a lithium nucleus, breaking it into an alpha (α) particle (helium nucleus) and a triton (tritium nucleus). Together, these carry the 4.8 MeVs of energy which is gained in splitting the lithium nucleus. This energy, as well as the neutron's energy, is transferred to a liquid or gas coolant and eventually transferred to steam for generating electricity. The n-Li^7 reaction is the same, except that a slow neutron is left over which can undergo another tritium-producing reaction. The n-Li^7 reaction works only with fast incoming neutrons, however.

The problem with this scheme is that not enough tritium is produced, since only 20–40% of the neutrons actually react with the lithium [3]. Some of the neutrons are lost through gaps in the blanket needed for plasma heating and measuring equipment. Some are lost by striking support structures instead of the lithium-bearing material, and a few are lost by passing through the whole blanket. To make up for this, there are fortunately neutron multipliers, mainly lead (Pb^{208}) and beryllium (Be^9). These can yield two neutrons for each incoming one. The reaction for beryllium is shown in Fig. 9.11.

Blankets will contain lithium, lead, beryllium, and a structural material; but the main problem is to cool them to take out all the heat that is the power output of the reactor. Blanket designs differ in the way they are cooled and in the form of lithium that is used. To show what is involved, we shall describe three of the leading proposals that have been worked out in detail.

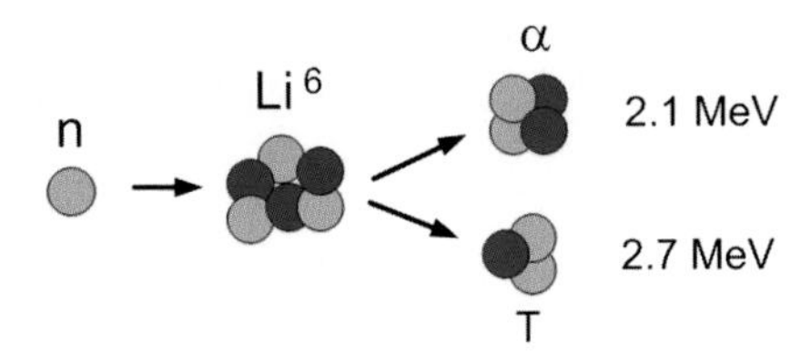

Fig. 9.10 The n-Li^6 breeding reaction, in which a neutron breaks a lithium-6 nucleus into an alpha particle (helium nucleus) and a triton (tritium nucleus). Protons are *blue* and neutrons are *gray*

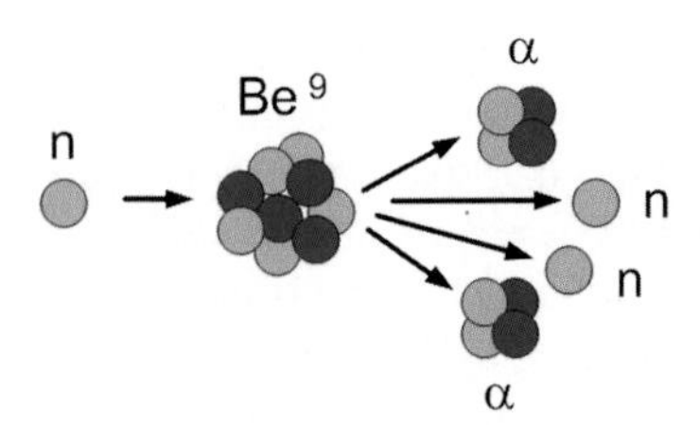

Fig. 9.11 Beryllium acts as a neutron multiplier, breaking up into two helium nuclei and two neutrons when joined by a neutron

Blanket Designs

The main coolants available are pressurized water, liquid metals, and helium. Water can be used only for near-term experiments. Reactors will probably need helium gas at a high temperature. The structural materials would be the same as those considered for the first wall: ferritic steels, vanadium alloys, or silicon carbide composites. The lithium can be in the form of solid pebbles of lithium ceramic, a liquid mixture of lead and lithium, or a molten salt called FLiBe [3]. Figure 9.12 shows how a TBM will be inserted into one of the ports in ITER.

The *helium-cooled ceramic breeder* (HCCB) uses solid material, with the beryllium multiplier and the lithium breeder in separate compartments. Figure 9.13 shows

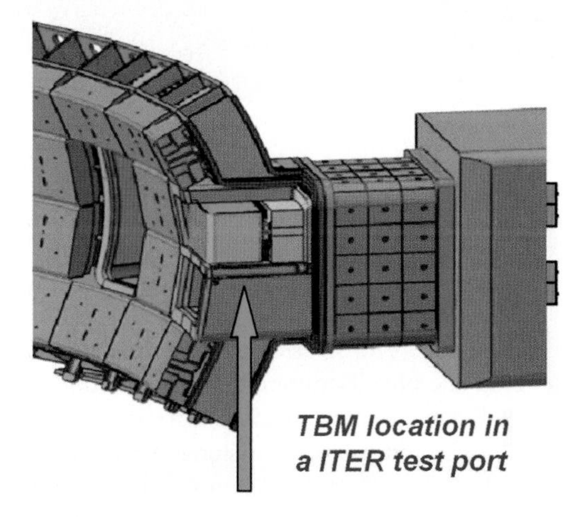

Fig. 9.12 Provision for insertion of test blanket modules in ITER, replacing part of the first wall [33]

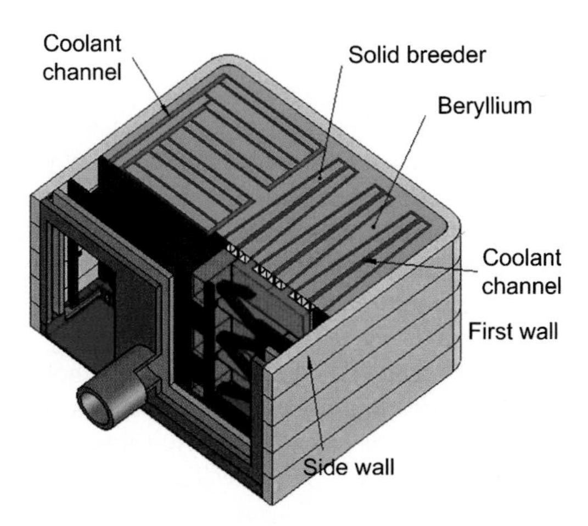

Fig. 9.13 Schematic of a helium-cooled ceramic breeder module [33]

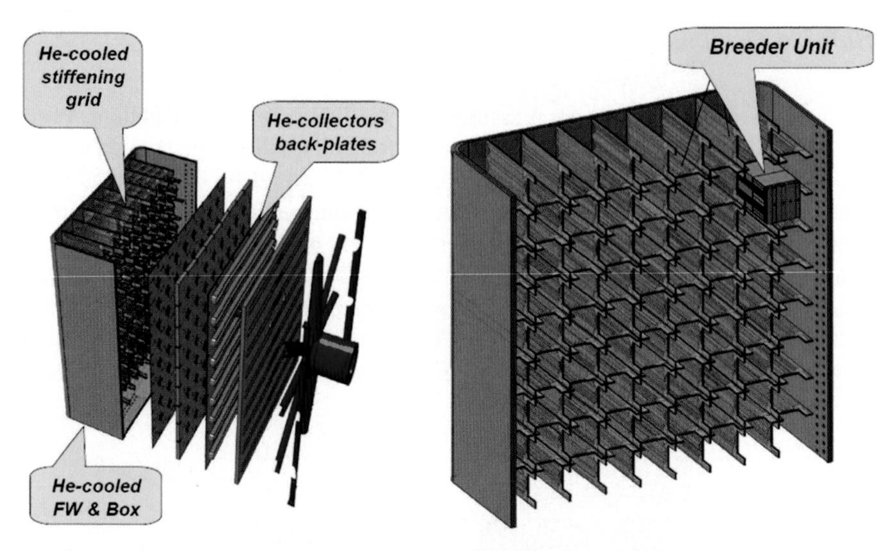

Fig. 9.14 Schematic of a large blanket module. The exploded view at the *left* shows several layers of supporting grids and coolant pipes which have been slid out of the box for clarity. The first wall (FW) is at the *left*. The view at the *right* shows the slots into which the submodules will be placed [3]

the parts of an HCCB module. The slabs containing the beryllium and the lithium ceramic are shown in red and blue. Between the slabs are cooling channels through which helium is pumped under 80 atmospheres of pressure [3]. The temperature of the helium can reach 500°C, and the breeder material can reach 900°C. Note that the front of the blanket is part of the first wall. In a reactor, a blanket module can be assembled from submodules, as shown in Fig. 9.14. The thickness of the blanket is about 50 cm and its width about 3 m.

The solid breeding material consists of ceramic pebbles of lithium orthosilicate (Li_4SiO_4), lithium metatitanate (Li_2TiO_3), or other similar materials. Techniques have been developed to manufacture identical spherical pebbles which can distribute themselves uniformly. The size should be small, less than 1 mm in diameter, to minimize the temperature difference across the radius so that the brittle spheres do not crack [7]. To extract the tritium, a flow of helium containing some deuterium (D_2) or hydrogen (H_2) is passed through the pebble bed, and the tritium (T_2) is carried out in the flow. The gases are then frozen and separated by distillation, since each has a different boiling point. The important thermal properties of a pebble bed have been measured [8].

A *helium-cooled lithium lead* (HCLL) blanket uses a molten alloy of lithium and lead called a eutectic. Meaning *easily melted* in Greek, a eutectic melts at a lower temperature than its constituents. The preferred eutectic is *Pb-17Li*, containing 17% lithium enriched to 90% Li^6. This melts at 234°C, compared with 328°C for lead and 181°C for lithium. In a blanket, the eutectic is heated from 400 to 660°C by the neutrons [3]. Since lead is a neutron multiplier like beryllium (Fig. 9.11), the multiplying and breeding are done in the same liquid. The submodules

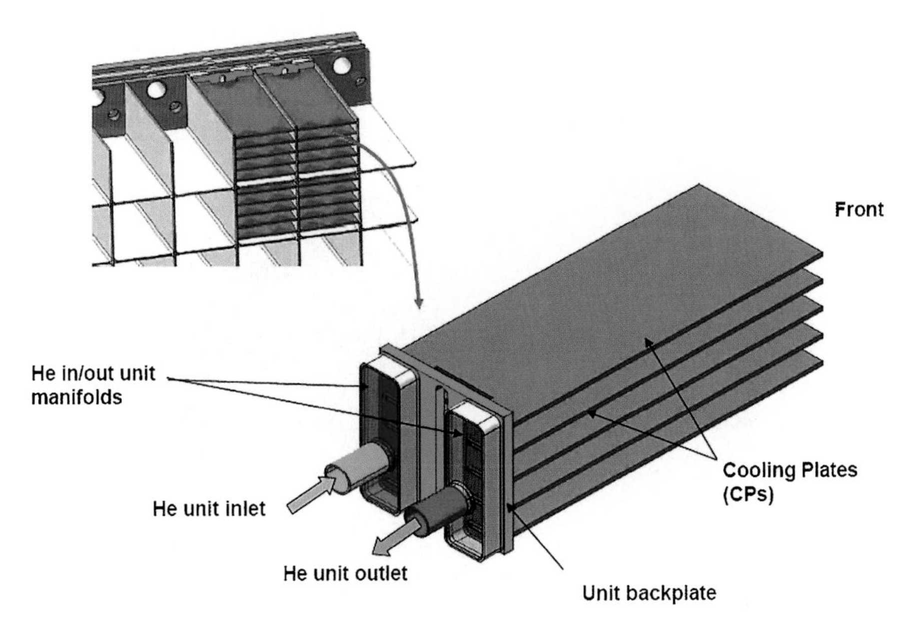

Fig. 9.15 Helium cooling arrangement in an HCLL blanket submodule [3]

in Fig. 9.14 will have circulating paths for the Pb-Li interspersed with channels for the helium coolant. The helium part is shown in Fig. 9.15, and the Pb-Li will go between the cooling plates. The tritium generated in the Pb-Li can be recovered by one of the two methods: permeation or bubbling. Hydrogen has a tendency to diffuse through walls, and tritium is just another form of hydrogen. Inside the blanket, tritium permeation into the helium coolant or other places where it does not belong is to be avoided. Outside the blanket, however, permeation windows can be made to allow hydrogen to go through and mix with a helium flow headed for a tritium separation facility. Alternatively, the Pb-Li can be formed into bubble columns where bubbles of helium capture the tritium in the liquid Pb-Li and carry it to the processing plant.

In earlier work, a molten salt called FLiBe, containing beryllium fluoride (BeF_2) and one or two parts of lithium fluoride (LiF) was proposed as a breeder fluid, but now Pb-Li is preferred. The work on FLiBe uncovered the problem of magnetohydrodynamic flow [9], which also applies to Pb-Li [10]. Both are electrically conducting fluids, and when these move inside a magnetic field, electric currents are generated in the fluid; and these currents react back on the magnetic field to produce a drag on the fluid motion. Considering how strong the magnetic fields are in a tokamak, this drag is a serious problem that increases the required pumping power. The drag is less if the flow goes along the magnetic field lines, but eventually the fluid has to cross the field lines to get out of the breeding region.

A *dual-cooled lithium lead* (DCLL) blanket uses both helium and the Pb-Li itself as coolants. This concept is shown in Fig. 9.16. Since Pb-Li is a liquid, it can be sent to its own heat exchanger and act as its own coolant. Helium is used to cool

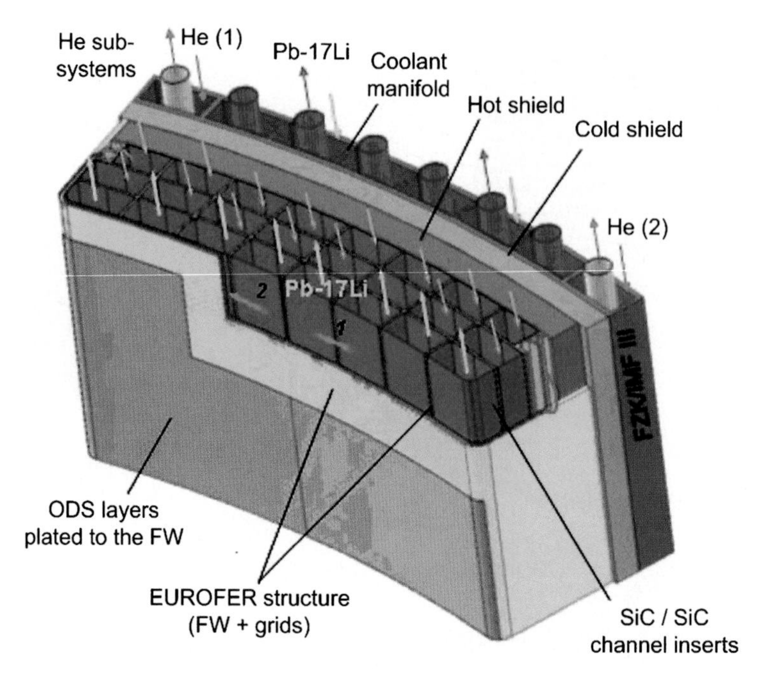

Fig. 9.16 Schematic of a dual-cooled lithium lead blanket module [34]. ODS, EUROFER, and SiC/SiC refer to high-temperature materials described under The First Wall and Other Materials

the first wall separately. The flow in the Pb-Li channels is shown in Fig. 9.17 for a case in which the magnetic field direction is into the paper. Computer models have been developed to describe the flow of the conducting liquid, including the buoyancy effect when the temperatures at the top and bottom are different. The eddies in the flow, as calculated, are shown in the inset. Since each module in a tokamak will be oriented at a different angle to the magnetic field, the structure of the flow, and hence the pressure drop, will be different at each location in the machine.

In advanced designs, the helium is eliminated, resulting in a self-cooled lithium lead breeding blanket, in which Pb-Li does all the cooling. It may take a lot of power to pump Pb-Li fast against the drag by the magnetic field. The possibility also depends on the development of the wonder-material SiC/SiC, which can operate at 1,000°C and contain a higher temperature fluid than other materials.

These blanket designs do not show all the auxiliary equipment necessary to operate the blanket. The roomful of pipes, heat exchangers, shields, and instruments for a single TBM in ITER is shown in Fig. 9.18. The blanket module itself is only the curved unit at the left, which forms part of the first wall.

Blankets for a full-scale reactor would have to satisfy many other requirements besides cooling and breeding. *Maintenance and operation* presents serious problems for a reactor designed to operate for over 25 years. The blanket material will have to be replaced many times during the life of the reactor. Solid breeders such as the

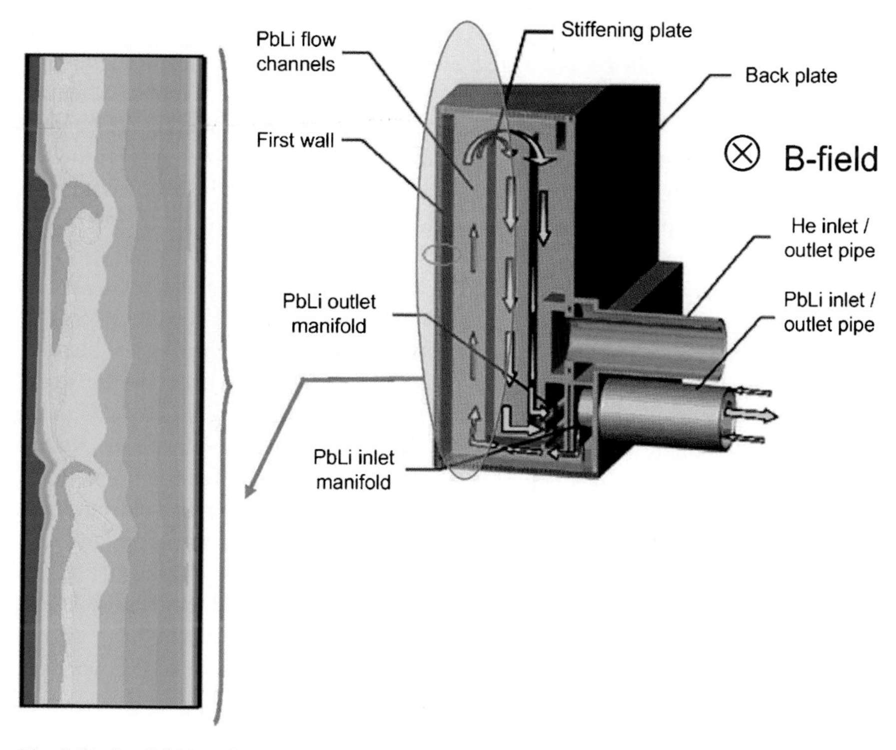

Fig. 9.17 Lead-lithium flow paths in a DCLL blanket submodule. The inset shows computer results for the eddy currents in one of the columns when the flow is perpendicular to the magnetic field [32]

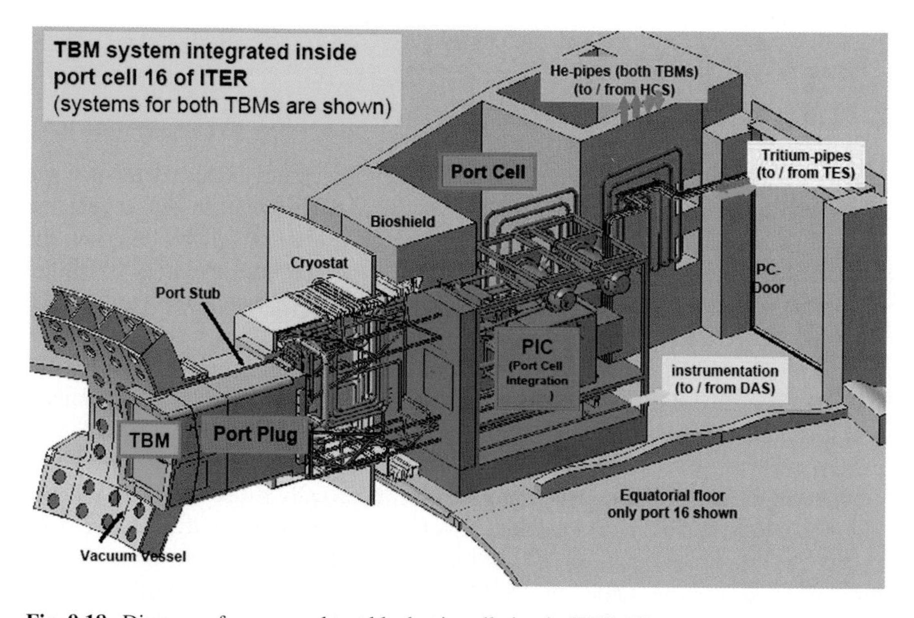

Fig. 9.18 Diagram of a proposed test blanket installation in ITER [6]

pebble-bed HCCB have to be physically removed to change the pebbles. In liquid blankets, the Pb-Li can be circulated outside the blanket and renewed without removing the blanket. Eventually, however, blankets will have to be replaced, requiring a shutdown. For easier replacement, banana-shaped blankets fitting the contour of the D-shaped plasma have been proposed. These would be lowered from the top of the tokamak during a shutdown, and all the connections to the blanket would have to come from the top. All this has to be done with remote handling, since there will be too much radioactivity for humans to work on the reactor.

Since the blankets are located inside the vacuum, they must be leak proof. Welds must be secure. Inside the blanket there are many interfaces between breeders and coolants, and a leak there would be impossible to fix without removing the blanket. There are also numerous joints in the pipes connecting the blanket to the world outside the vacuum tank. In 2008–2009, the Large Hadron Collider in Geneva suffered from a single leak in the liquid helium system which delayed the startup of the machine for over a year. In 2003, a single piece of loose foam brought down the shuttle Columbia, killing seven astronauts. Accidents happen, and extreme care must be taken in a tokamak reactor, where there are a million places where a leak can occur.

There are also safety issues in the case of an accident, including decay heat and radiotoxicity after shutdown [11]. Recycling and treatment of waste have also to be considered. However, these are not specific to blankets and will be covered in another section.

Tritium Management

Tritium Self-Sufficiency

The blanket designs shown above can barely breed enough tritium to keep a D–T reactor going. The tritium breeding ratio (TBR) is a measure of this. Each time a T fuses with a D in the plasma, one neutron is created. This neutron has to generate more than one T to re-inject into the plasma because there will be losses in the process. In addition, extra T's have to be stored to build up the inventory of tritium to run the reactor at a higher power or to fuel another fusion reactor. Only fusion can produce the enough tritium to build up its own industry.

The number of T's created in the blanket for each incoming neutron is the TBR. It has not been possible to design a blanket with a TBR larger than 1.15. That means that less than a 15% margin is available. The consequence is that tritium self-sufficiency can be achieved only after many years. The time is long because only a small percentage of the tritium injected into the plasma actually fuses with a D; most of it goes out the divertor and is recycled. This *fractional burnup* is only a few percent. Figure 9.19 shows calculations of how long it will take to double the tritium inventory. On the vertical scale, the TBR is plotted. The bottom portion, below TBR = 1.15, is what is possible. The horizontal axis shows the fractional burnup in percent. The curve labeled 1 year shows that it is not possible to double the tritium

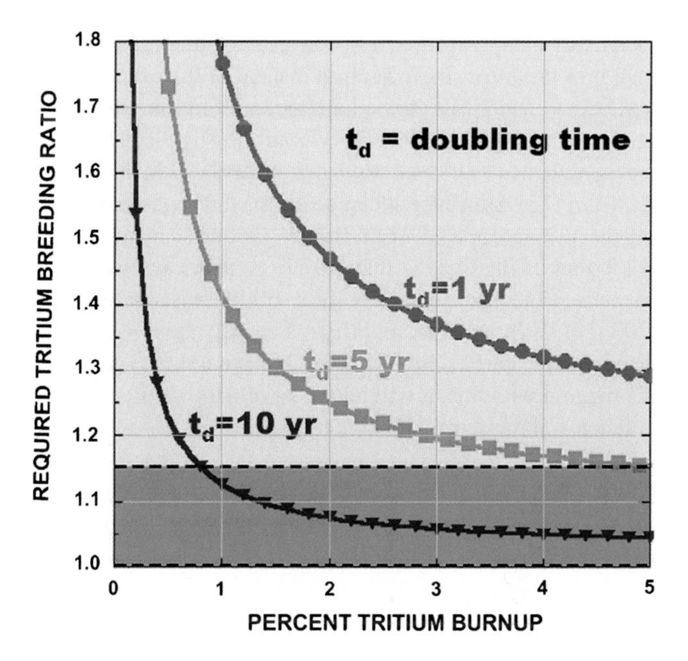

Fig. 9.19 Curves of the doubling time of the tritium inventory plotted against the TBR and fractional burnup of the tritium [32]

inventory in 1 year, since the curve never goes low enough to reach the feasible range of TBRs. The 5-year curve barely makes it if 5% burnup can be achieved. More likely, it will take almost ten years to double, and self-sufficiency can be achieved only after decades.[2]

In early tokamaks, before good divertors were developed, the fractional burnup was much larger, perhaps 30%, because of *recycling*. Ions of the plasma would hit the vacuum wall and recombine into neutral gas. This gas would go back into the plasma and be re-ionized and re-heated, thus being available again without having left the chamber. If modern divertors work well, however, ions are prevented from hitting the wall, thus preventing recycling. The ions are instead led to the divertor, where they recombine into gas and are pumped out before they can re-enter the plasma. In ITER, the fractional burnup is expected to be only 0.3%, which would be unacceptable for reactors [32]. Since burnup depends on the triple product $Tn\tau$ discussed in Chap. 8, this is another indication of the large step between ITER and a working reactor.

A fission reactor can produce only 2–3 kg of tritium a year, and tritium decays by 5.5% per year, so it is continually being lost. It will take 10 kg of tritium just to get DEMO started. ITER itself will use up most of the tritium available in the world [32]. There is therefore some urgency to develop breeding blankets with higher TBRs.

Tritium Basics

As doubly heavy hydrogen, tritium has two extra neutrons, which do not sit well with a single proton. So tritium decays by emitting an electron, a process known as beta-decay. This loss of a negative charge changes one of the neutrons into a positively charged proton and converts tritium into helium-3, a helium isotope with two protons and a single neutron instead of the usual two. This decay makes tritium radioactive, and it has to be handled carefully in a fusion plant.

Fortunately, the radioactivity is mild. The electron that is emitted has very low energy, about 19 keV. It cannot penetrate the skin, and even in air can go only 6 mm (1/4 in.) [12]. However, it can be harmful if ingested and must be carefully kept out of the water supply. Unlike fission products, tritium has a short half-life of only 12.3 years. This means that 5.47% of it decays into harmless helium each year. Because of its short life, very little tritium exists naturally. Cosmic rays make about 200 g of tritium a year, and there are only about 4 kg of natural tritium at any one time in the earth's atmosphere. Man-made tritium raises this to about 40 kg. Compared with this, it will take 1 kg of tritium just to get ITER running on DT, and a reactor may use up 100 kg per year.

The Tritium Fuel Cycle

One of the most complex technological tasks is to manage the supply of tritium. Tritium is injected into the plasma as fuel. It leaves the plasma through the vacuum pumps, most of it going through the divertor. It is generated in the breeding blankets and has to be captured and purified. It is also a contaminant in the liquids and other materials that leave the reactor and has to be removed from them. Excess tritium has to be stored safely for future use in raising the power of the reactor or starting up other reactors. Figure 9.20 shows a simplified diagram of these paths.

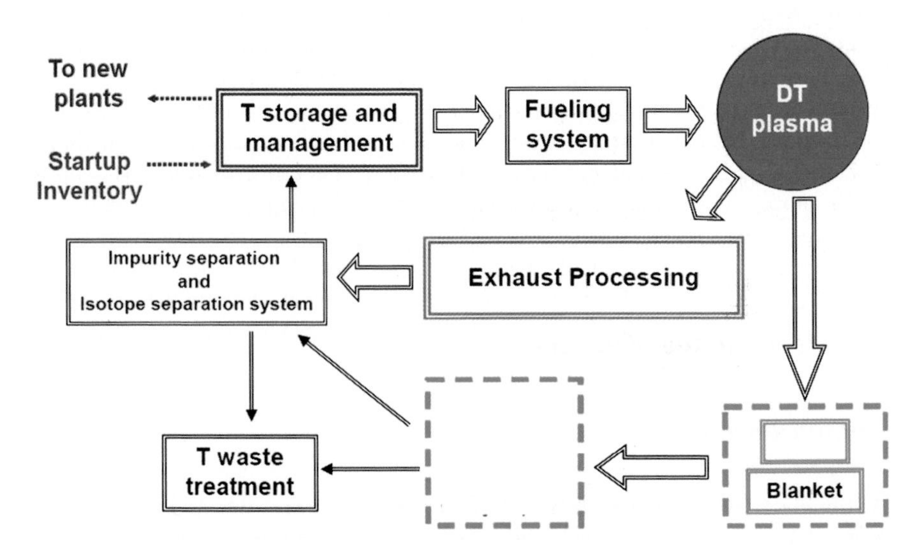

Fig. 9.20 Simplified diagram of tritium fuel cycle [32]

Tritium leaves the tokamak in two paths – either through the vacuum pumps, including those pumping the divertor, or through the first wall (FW) and the blanket. The vacuum exhaust goes directly to an isotope separation system which saves the T_2, D_2, and He and removes the impurities. Pure T_2 is sent directly to Tritium Storage and Management. The tritium generated in the blanket goes first to a tritium processing plant to remove it from the breeder materials, and then to isotope separation. Material contaminated with irremovable tritium from both streams then goes the Tritium Waste Management. The fueling system receives recovered tritium from the two paths as well as from storage or from external sources. The fueling system then injects tritium and deuterium into the plasma. Deuterium is cheap and safe and does not have to be parsimoniously recovered.

The vacuum in the torus is maintained by cryo-pumps [13]. These are porous carbon surfaces cooled by liquid helium to 5 K; that is, 5° above absolute zero, the latter being −273°C or −459°F. At that temperature, all gases except helium are condensed and stuck to the cryogenic surfaces. To release hydrogen, including tritium, the cryo-pumps are periodically heated to about 90°K, and this gas is sent to the isotope separation system. To release all the captured gases, the pumps are raised to room temperature.

Fueling is done by injecting frozen pellets of tritium and deuterium at sufficient velocity to reach the center of the plasma. This is much more efficient than injecting DT gas at the boundary, since the gas will be ionized at the surface and will not reach the interior. There is some loss of tritium in the process, and this will appear in the pumping system. The plasma is heated mainly by neutral beam injection (NBI), the beams consisting of deuterium and tritium. This system will have its own system of tritium management.

Isotope separation is done by freezing the gases to liquid helium temperatures and selective warming in four interlinked distillation columns [13]. The tritium processing plant in ITER is a large seven-story building [12]. In addition, all water in the ITER installation and all air from buildings have to pass through a detritiation plant to remove the tritium. Water released back into the environment is pure H_2O, and hydrogen released into the air is pure protium (H_2). Tritium has to be stored until it is used. This is done in metal-hydride getter beds, each capable of holding 100 g of tritium [13]. Zirconium–cobalt (ZrCo) absorbs T_2 to form $ZrCoT_3$. The reaction is reversible upon heating to release the T_2. Although techniques for tritium containment are well established in the fission industry, the amount of tritium in fusion is orders of magnitude larger. There has been no experience so far on such a large scale.

Superconducting Magnets

Introduction

The dominant features of a tokamak or any other magnetic bottle are the heavy coils that generate the large magnetic field used to confine the plasma. Until recently, all tokamaks had magnet coils made of copper, which conducts electricity better than any other metal except silver. Even so, it takes a lot of energy to drive megamperes of current through copper coils, and fusion reactors will have to use superconducting coils. Superconductors have zero resistivity, and once the current has been started in them, it will keep going almost forever. The hitch is that superconductors have to be cooled below 4.2 K with liquid helium. A cryogenic plant has to be built to supply the liquid helium, and the magnet coils (and hence the whole machine) have to be enclosed in a cryostat to insulate them from room temperature. The good news is that this technology is well developed and is *not* one of the serious obstacles to fusion power. In 1986, the world's largest superconducting magnet, the MFTF (mirror fusion test facility), was completed at the Lawrence Livermore Laboratory in California. It was a different type of magnetic bottle that we will describe in Chap. 10. However, the program was almost immediately canceled by the Reagan administration in favor of the tokamak because the USA could not afford to follow two expensive paths to fusion. The MFTF was so large that for a while it became a museum that one could walk through. Currently, three superconducting tokamaks are in operation: the Tore Supra in France, the EAST (Experimental Advanced Superconducting Tokamak) in Hefei, China, and K-STAR, in Daejon, Korea. Soon to join them is an upgrade to Japan's JT-60U (Fig. 8.6) called JT-60SA. In addition, the Large Helical Device, a superconducting stellarator-type machine, has been operating for two decades in Japan. ITER will, of course, have superconducting magnets.

Two superconducting materials are available on a large scale: niobium–titanium (NbTi) and niobium–tin (Nb_3Sn). NbTi is cheaper and easier to make, but it loses its superconductivity above 8 T. A tesla is a large unit of magnetic field equal to 10,000 G, the old unit. Common magnets rarely go above 0.1 T, but some magnetic resonance imaging (MRI) machines in medicine can go up to 1.5 T. The earth's magnetic field is only about 0.5 G or 0.00005 T. In ITER, fields up to 13.5 T are needed, so some coils are made of Nb_3Sn, and others (for lower fields) are made of NbTi. The dividing line is around 5 T [14]. Superconducting cables are complicated to make because they have to be made of a thousand thin strands. This is because the current in superconductors flows only on the surface, and thin strands have large surface areas compared to their volumes. Also, the cables have to be bendable.

ITER's Magnet Coils

Figure 9.21 shows what a niobium-tin cable looks like inside. There are over 1,000 strands in six bundles. At the center is a helix making room for the pipe that carries the liquid helium. The outer casing is a stainless steel jacket 37.5 mm (1.5 in.) in diameter. This cable, designed for the toroidal field coils of ITER, can carry 80 kA at 9.7 T. Each strand is about 0.8 mm in diameter and consists of a Nb_3Sn filament sheathed with chromium and covered with about as much copper as Nb_3Sn. The copper is necessary to mitigate quenches. A quench occurs when part of the superconductor goes normal, losing its superconductivity because of over-heating or over-current. Huge voltages would build up as the current tries to force

Fig. 9.21 Construction of a niobium-tin cable. One of the bundles has been exploded to show the strands [14]

its way through a normal conductor with resistance, and there could be an explosion. Copper can make this a gentler accident. The complexity of superconducting cables is bad enough, but to wind them into magnetic coils means that each cable has to be over 1.5 km (a mile) long.

A tokamak has many different kinds of magnet coils, and each requires a different design. Some of these can be seen in Fig. 9.22. The toroidal field (TF) coils are the large D-shaped coils. They operate up to 6 T and are the heaviest ones. Transporting them to the ITER site requires special barges, trucks, and roads. The large, horizontal ones encircling the machine are the poloidal field (PF) coils, which give the field lines their twist and shape the plasma. Because of their size, they cannot be transported and must be wound on site. The coil winding building at the ITER will be 253 m long, 46 m wide, and 19 m high.[3] A critical component is the central solenoid (CS), seen inside the hole in the torus. There is very little space there, and most of it is taken up by the interior blanket modules. This coil is the other half of the PF system that shapes the plasma and drives the tokamak current. The CS is 13 m tall and 4.3 m in diameter, weighing 1,000 tons. It also produces the highest field of 13.5 T. Figure 9.23 shows a test section of it that has been made.

There are smaller coils besides these main coils, but the difficult part is to join the superconductors to their feeds. Current is fed into the coils from normal-conducting cables, and then a superconducting switch is turned on so that the

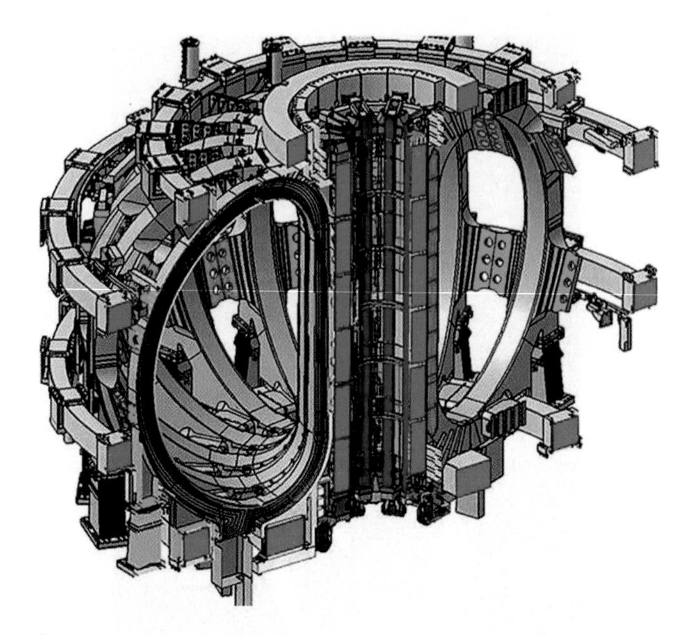

Fig. 9.22 Drawing of the magnetic coils in ITER (ITER Newsline Nos. 114 and 122 (2010). http://www.iter.org/newsline/)

Fig. 9.23 A test section of the Central Solenoid for ITER [14]

current flows only in the superconductors and the feed cables can be disconnected. These junctions are very complicated, especially since the current has to go through the wall of the cryostat from room temperature to 4 K. Almost all the nations supporting ITER participate in designing and producing the magnet system. Some make the NbTi and Nb_3Sn materials. Some make it into strands and cables. Some wind the cables into coils. And some make the feed cables and the junctions. The technology has already been developed for smaller tokamaks, and the steps to ITER, DEMO, and reactor are only matters of scale.

The Supply of Helium [4]

Helium is not a rare gas if we can afford to fill the world's balloons with it. Actually, balloons account for only 16% of helium use. Cooling of semiconductors accounts for 33%, and the rest is used for industrial and scientific purposes. The atmosphere contains four billion tonnes of He, but it is not economical to extract it by cryo-distillation. Most of our helium comes from natural gas as a by-product. Thus, helium comes from fossil fuels and will be depleted in several decades along with natural gas, as discussed in Chap. 2. In this chapter, we have seen how critically fusion reactors, as envisioned today, will depend on helium in both extremely hot and extremely cold places. In the first wall and blankets, gaseous helium is used as a high-temperature coolant. The vacuum system uses liquid helium to cool the cryo-pumps. In the magnet system, liquid helium is what produces superconductivity. It is a closed system, but there are leaks. It is estimated that ITER will lose 48 tonnes of helium a year, about 0.15% of the world's current consumption. But if eventually fusion produces a third of the world's power, those reactors would need the world's supply of helium for a whole year just to start up [4]. At some point the helium losses, say, 10% of the inventory, would exceed what comes from natural gas. You will remember that helium is one of the products of the D–T reaction. At only a few percent burnup, however, this "ash" is a negligible contribution to the total demand. Helium is needed in other industries as well; for instance, in medical equipment. The shortage is so acute that a rationing system was proposed in the USA in 2010.

High-Temperature Superconductors

In 1986, compounds were discovered that became superconducting at a critical temperature as high as 30 K. Since then, research to find better materials has been intense. The goal was to get the critical temperature above 77 K, the point at which nitrogen becomes liquid. Liquid nitrogen is much, much cheaper and easier to produce than liquid helium, which is liquid below 4 K. The 73°C difference between 77 and 4 K does not seem much. We encounter such a change every time we boil a cup of coffee. However, since one can never go below absolute zero, it is the distance from absolute zero that is important. Seventy-seven kelvin is 19 times farther from 0 K than is 4 K; and, of course, there is no shortage of nitrogen. The goal has already been achieved; three superconductors have been found that work at liquid nitrogen temperatures. The record as of 2009 is 135 K, well above 77 K. Typically, the compound is complicated: $HgBa_2Ca_2Cu_3O_x$. Until searches can be made by computer, finding new compounds will be slow; but it is a reasonable expectation that large-scale production of a high-temperature superconductor will be possible by the time DEMO is built. Maybe a room-temperature superconductor will have been found by that time. The machine would be much simpler and cheaper.

Plasma Heating and Current Drive

Introduction

Bringing the plasma up to fusion temperatures is done with the injection of neutral atoms and the excitation of different types of plasma waves. In addition, waves are also used to drive the plasma current without using transformers – so-called noninductive current drive. There are many physics problems involved in these processes. Neutral beams also fuel the plasma and give it rotational velocity. Waves not only heat the plasma and drive its current but are also used to change local conditions inside the plasma and shape the current profile. In this chapter, we are concerned with technology and therefore concentrate on the hardware and discuss only the main types of waves that can be used.

Neutral Beam Injection (NBI)

One of the aims of ITER is to reach ignition, when the alpha particles generated by the D–T reaction are able to keep the plasma hot. To get to this point, however, immense power has to be injected to raise the temperature to the order of 50 keV (500,000,000°). This is done mainly with NBI. ITER will have 33 MW of NBI. The injectors, three or four of them, are usually the largest appendages sticking out of the tokamak. In the first stage, deuterium atoms are given an extra electron to produce negative ions. Once charged, the ions can be accelerated electrostatically. Before entering the tokamak, the negative ions go through a little gas, which strips off the extra electron, restoring the atom to a neutral state. Being neutral, the atom is not affected by the magnetic field and can go well into the plasma until it is ionized by the electrons in the plasma. How far it goes depends on its energy. All large tokamaks use NBI, which is a well-established technology; but since ITER is so large, neutral beams of 1 MeV energy are needed to get to the center. NBI technology for 1 MeV has not yet been developed [15].

Ion Cyclotron Resonance Heating (ICRH)

This method heats ions by pushing them with a rotating electric field whose direction follows the ions' cyclotron motion as they revolve in their nearly circular Larmor orbits. It is sometimes more efficient to heat a minority species, such as helium-3 rather than deuterium or tritium, because of the way the energy is coupled into the plasma. The cyclotron frequency depends on the magnetic field strength, so the applied electric field has to be of a specific frequency, depending on magnetic field at the location where the ions are to be heated. In ITER, this frequency is in the range around 50 MHz. This is too low a frequency to be transmitted through a pipe, so an antenna has to be placed inside the vacuum chamber. The antenna is outside the field lines leading to the divertor (see Fig. 9.4), but it is so close to the plasma that it will be bombarded by ions. These ions will sputter antenna material into the plasma, and such contamination usually cools the plasma. ITER is to have 20 MW of ion cyclotron heating. The power is not the main problem; the problem here is to design antennas which will not affect the plasma deleteriously.

Electron Cyclotron Resonance Heating (ECRH)

In principle, what is done to the ions can also be done to the electrons, but the technology is entirely different. The electrons' cyclotron frequency is in the gigahertz range, and huge microwave generators are required. The power or current input can be deposited accurately at specific places inside the torus by adjusting the microwave frequency to match the magnetic field at those places. Since microwaves are carried through waveguides, which are specially sized and shaped pipes, they can be injected through holes in the first wall and do not require an antenna inside the vacuum chamber. The bad news is that electron cyclotron waves cannot penetrate into the plasma from the outside of the torus. A property of these waves is that they must be injected from a high magnetic field into a lower magnetic field. Since the magnetic field is highest in the hole of the torus, the launching waveguide must be located in the cramped space also occupied by the central solenoid and the inside blankets. Waves at twice the cyclotron frequency, which also resonate with the electrons' gyrations, can get in from the outer, weak-field side; but the higher frequency is more difficult to generate.

The electron cyclotron heating system in ITER calls for 20 MW of power at 170 GHz. This frequency corresponds to the cyclotron frequency at 6.0 T (60,000 G), high enough to cover ITER's magnetic field of 5.5 T at the inside radius. Although we use microwaves in everyday life, 20 MW at 170 GHz is an entirely different matter. A microwave oven puts out 1 kW at 2.45 GHz using a magnetron so small that we are not aware of its presence. Powerful microwaves are generated by gyrotrons, which work by running ECRH in reverse. In a gyrotron, an energetic electron beam is first produced. It is then injected into a magnetic field, so that the electrons undergo cyclotron gyrations. In doing so, they emit microwaves at harmonics of the cyclotron frequency which are then channeled into a waveguide leading to the tokamak. The microwaves get their energies from the electron beam, which loses part of its energy. In experiment, the remaining energy is captured in a beam dump as heat. In advanced gyrotrons, the beam can, in principle, be re-injected so that its remaining energy can be re-used. Note that the electron beam in a gyrotron cannot be injected directly into a tokamak to heat it because the electrons cannot get through the magnetic field. In a gyrotron, the electrons are injected into the magnetic field from the *ends* of the field lines. A tokamak, of course, *has* no such ends; hence the need to convert kinetic energy into microwave radiation and then injecting the radiation instead of kinetic energy directly.

High-power gyrotron research began in St. Petersburg, Russia, decades ago. Those that can operate continuously for ITER are being developed in Japan, Germany, and the USA. So far, 1 MW at 170 GHz in a long pulse has been shown to be possible. Figure 9.24 shows the size of such a gyrotron. ITER will need 24 of these to produce the required ECRH power. Figure 9.25 shows a design of a 2-MW gyrotron with superconducting magnets.

Since the gyrotron has to be under vacuum and the waveguide is at atmospheric pressure, windows have to be used to isolate the waveguide from the vacuums at both ends. At present, the only material that can transmit the wave power at that frequency is synthetic diamond. Windows 10 cm (4 in.) in diameter have been made and tested for proper cooling. In a reactor, gyrotrons and their windows have to run continuously without failure for months or years between maintenance shutdowns. This constitutes a large step in engineering that has yet to be done.

Fig. 9.24 The gyrotron room at JAERI [35]. A 1-MW gyrotron is shown at the *left*. It is 3 m (10 feet) high and covered with magnetic coils

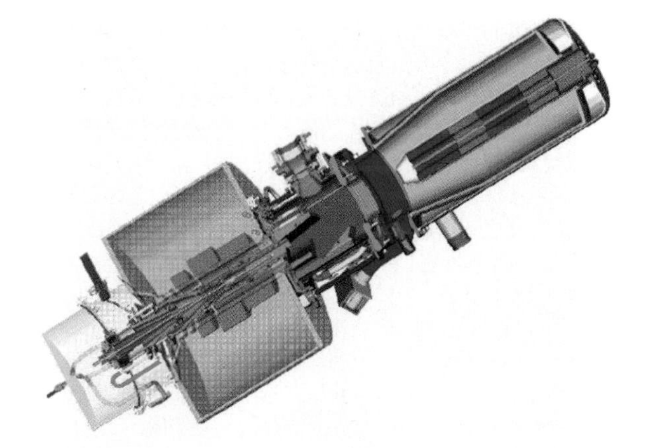

Fig. 9.25 Design of a 2-MW, 170-GHz superconducting gyrotron being developed in Germany [6]

Lower-Hybrid Heating (LHH)

A third type of wave that can be used for heating and current drive is the so-called lower-hybrid wave. This wave is particularly useful for current drive because it can control the current profile near the outside of the plasma. The lower-hybrid frequency lies between the cyclotron frequencies of the ions and electrons, or about 5 GHz in ITER. Klystrons are used to generate frequencies in this range. The wave has a long wavelength in the direction of the magnetic field, so to launch it requires a large "grill," meters in size, as shown in Fig. 9.26. Each of the openings is a waveguide fed by one or several klystrons, each with its own vacuum window. The phase of the wave emanating from each waveguide is set so that the total grill, including some dummy waveguides, forms the wave that deposits its energy in the right place. Since the launcher lies close to the plasma surface, its materials must sustain the heat and neutron damage that that implies.

In summary, the physics of auxiliary heating and current drive is well understood, but the engineering of the power supplies and the wave launchers present some difficult problems.

Remaining Physics Problems

The ITER machine is an experiment large enough to require an international consortium. Its mission is to achieve a burning plasma, one in which the alpha particles produced by the D–T reaction can maintain the plasma's temperature without external heating. At this stage of construction, not all physics problems have been solved, though they may be solved by the time construction is finished. We hope that these problems will be solved in time for DEMO. However, the physics does not have to be completely understood for something to work. Books have been written on the physics of tennis, baseball, sailing, and even pizza. Sometimes, it is easier just to get on with it.

Fig. 9.26 A lower-hybrid wave launcher of the type designed for ITER but one-fourth the size [36]

Edge-Localized Modes

Edge-localized modes (ELMs) were described in Chap. 8. They are instabilities of the H-mode pedestal which can release plasma suddenly to the wall. Although most of these particles should flow to the divertor, the sudden burst of heat can erode and damage the divertor's surfaces. The H-mode pedestal constrains one-third of the plasma's energy, and 20% of this or as much as 20 MJ can be dumped into the divertor in a fraction of a second [16]. The preferred method to suppress ELMs is to impose a rippled magnetic field at the surface of the plasma, near the pedestal. The idea is to break up instabilities that tend to be aligned with the magnetic field. The pattern of currents in the ELM coils can be varied slowly to follow changes in the magnetic field lines. This method has been tested in the DIII-D tokamak in San Diego, California, and thorough calculations have been made to design the sizes and spacings of the coils for ITER [17]. A panel of ELM coils is shown in Fig. 9.27. Figure 9.28 shows what the surface of ITER will look like with these coils installed. It will take 2.6 MW of power to drive these coils. Being in-vessel components, the coils have to withstand intense heat and neutron bombardment. In ITER, the coils are protected from the plasma by a 50-cm thick, water-cooled, nonbreeding blanket whose only function is to attenuate the neutrons.[4]

In DEMO, there would be no place for ELM coils, since breeding blankets have to cover the machine to capture as many neutrons as possible. Locating the coils behind the blanket would probably be too far. ELM coils are *ad hoc*, temporary

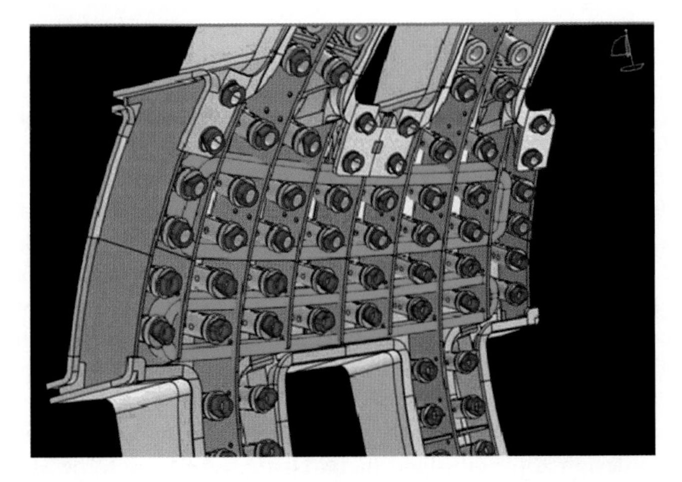

Fig. 9.27 A panel of ELM-suppression coils for ITER [6]

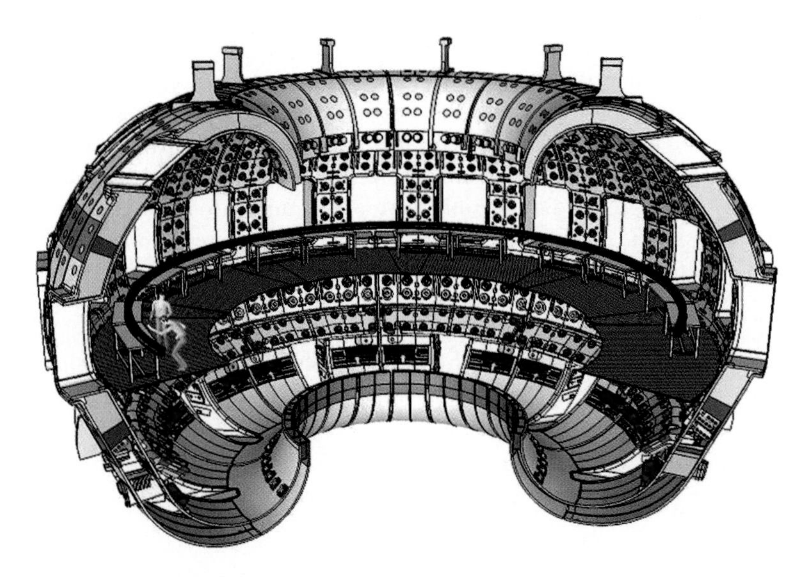

Fig. 9.28 Drawing of ELM coils installed in ITER [29]. The scale is shown by the human figures at the *left*

solutions not included in the original design of ITER since the problem had not yet arisen. The physics of ELMs has to be understood better to find passive methods for their control, but there is time to do this.

Once the ELM coils have been installed, they can also be used for other purposes. By applying a small current at a low frequency like 50 Hz, a weak instability called the RWM can be controlled. A differently spaced DC current can also be added to help prevent disruptions (described in detail in the next section).

Disruptions

As shown in Chap. 8, disruptions are disasters. Magnetic containment is suddenly lost, and the plasma drifts vertically into the walls, depositing all its thermal energy. The tokamak current tries to keep itself going as the plasma goes away, so very high voltages are generated. Runaway electrons of MeV energies are created by the high voltages, and these electrons crash into the walls, generating high-energy X-rays. The plasma current is used to generate the poloidal magnetic field, and as this field decays with the current, large forces are applied to the magnetic coils and conducting parts of the tokamak structure. The entire energy in the plasma, magnetic field, and tokamak current is something like 500 MJ, and in a disruption this is all dumped into the structure of ITER in 1/30th of a second [18]. This is like an explosion of 120 kg (260 lbs.) of TNT. Disruptions are expected in ITER, and its parts are designed to withstand them. Disruptions have to be eliminated in reactors, which would be so heavily damaged as to require lengthy shutdowns for repair.

There is a possible scenario of how a change in the magnetic structure of the tokamak discharge, such as a coalescing of magnetic islands, can cause a disruption. It has been confirmed in experiment that staying well below the known stability limits, such as the density limit, can avoid disruptions. A reactor, however, needs to operate at the highest level to lower the cost of electricity (COE). Since a disruption is now known to be a vertical displacement of the plasma, there are ideas on stopping these displacements with a coil or coils inside the chamber. Such a coil is included in Fig. 9.27. Though it is not possible to stop a disruption once it starts, there are ways to mitigate the damage. Disruptions have magnetic precursors which can be detected, and fast action can be taken. Injection of liquid jets or solid pellets of a frozen gas have been tried, but these have led to creation of too many runaway electrons. A large puff of a gas like argon can be driven well into the plasma, be ionized into high-Z ions, and increase the resistivity so that the current dies more gently. Fast gas valves have been developed for this purpose. There is then a smaller tendency to induce currents elsewhere, lower forces on the structure, and fewer runaway electrons. After a disruption, there is only gas left in the vacuum chamber. This has to be pumped out and the discharge started all over again.

Alfvén Wave Instabilities

In a burning plasma, 3.5-MeV alpha particles are generated, and as they cool down they transfer their energy to the plasma, keeping it hot. Before they become thermalized, however, the alphas are in the form of beams streaming along the magnetic field lines, and beams can excite instabilities. To do this, the velocity of the beam has to coincide with the velocity of a wave in the plasma; and the synchronism causes the beam energy to be transferred to the wave. The wave can become so strong that it disrupts the plasma. There is a plasma wave called the Alfvén wave that travels along the B-field and can have just the right velocity to match that of the alpha-particle beam. The danger that this can happen can be predicted precisely by theory [19], but whether it will actually happen or not depends on the details. ITER will be the first machine that can test for Alfvén wave instabilities in a D–T plasma. If these turn out to be important, their avoidance is a physics problem that needs to be solved.

Operating a Fusion Reactor

Startup, Ramp-Down, and Steady-State Operation

Turning on the power in a large tokamak is not an easy task. The vacuum system, the cryogenic system, discharge-cleaning of the walls, the magnetic field system, the tokamak current drive, and the various plasma heating systems, and various auxiliary systems have to be started up in sequence. Operators have learned by experience how to do this in large tokamaks. The plasma has to be maintained stably while it is being heated and while the current is being increased in synchronism with the toroidal magnetic field. Each power supply has to be ramped up at a certain time at a certain rate. Turning the discharge off also requires careful ramp-down of each system. Only after a good routine has been found can automatic controls take over.

All present tokamaks run in pulses, not continuously. Even if the pulses last for minutes or an hour, they will not uncover problems that will arise with truly steady-state operation. In the 1980s, a machine called the ELMO Bumpy Torus was run at the Oak Ridge National Laboratory. Though the magnetic configuration never caught on, the machine was run in steady-state and revealed problems that are not seen in pulsed machines. The Tore Supra tokamak in Cadarache, France, near the ITER site, has been gathering information on long-pulse operation for 20 years [20]. It is a large tokamak with high magnetic field, large current, and powerful heating. The first wall is water-cooled boronized carbon. In a deuterium plasma, the retention of deuterium by the carbon was found to be significant. This is one reason for rejecting carbon as a wall material. Damage to the ICRH antennas was noted. Electrical faults in the magnet system were found to limit the length of discharges. It was found that turning the lower-hybrid power on slowly greatly alleviated this problem. Water leaks were found to occur 1.7 times per year. The frequency of disruptions was also recorded. These were found to be caused mainly by the flaking of carbon off the walls after many days of operation. Pulses lasting 1 or 2 seconds were possible with transformer-driven currents, but with the addition of lower-hybrid current drive, 6-min pulses with 3 MW of lower-hybrid heating (LHH) were achieved in 2007. At the 2-MW level, 150 consecutive 2-min discharges could be routinely produced [21]. These are the types of problems that will be encountered when ITER is operated in continuous mode.

Maintaining the Current Profile

Advanced tokamaks utilize reversed shear and internal transport barriers for enhanced plasma confinement. These require precise shaping of the safety factor q (see Chap. 8), which determines how the twist of the magnetic field lines changes across the radius. The shape of the $q(r)$ curve controls the stability and loss rate of the plasma. Since the twist is determined by the poloidal field created by the plasma current, this current has to be shaped in a particular way. Some of the current is naturally produced by the bootstrap effect (Chap. 9); the rest has to be driven by lower-hybrid and electron cyclotron current drive. The blue curve in Fig. 9.29 shows an example of a $q(r)$ curve which stays above $q=2$ and gives reverse shear. The red curve shows the auxiliary current needed to produce this $q(r)$. Only precise control of the localized heating can produce this current profile. As the plasma starts up, the currents will be changing, and the power supplies will have to be programmed to keep the current in a stable shape.

Remote Handling

Anytime tritium or deuterium is introduced into a magnetic bottle, the wall materials will become radioactive due to neutron bombardment. It will be impossible for humans to go inside the machine or even come close to it. Robots will be used to replace parts such as blanket modules, to fix leaks and make other repairs, and to examine the interior of the chamber during shutdowns. The robotic equipment itself will be exposed to neutrons. Such remote handling has been used successfully in

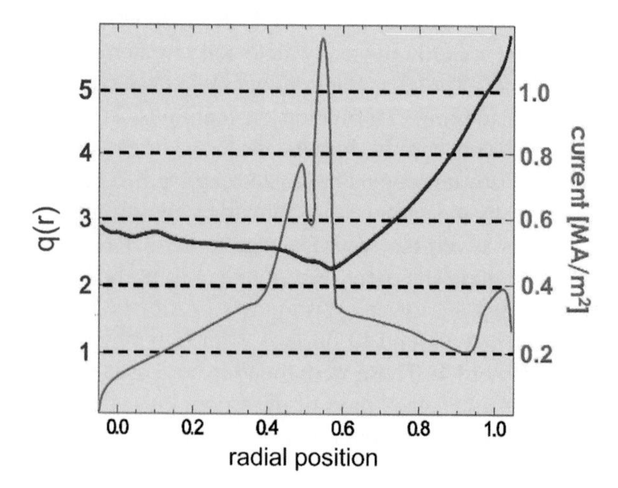

Fig. 9.29 Example of the variation of the safety factor $q(r)$ across the minor diameter of an advanced tokamak plasma (*blue*), and the plasma current distribution required to produce it (*red*) [37]

the TFTR machine at Princeton (Fig. 8.3) and the JET in England (Fig. 8.4), both of which have used DT fuel. Robots can weld joints by remote control. The first experiments in ITER will use hydrogen or helium, which produce no radioactivity. Later, deuterium experiments will give a small amount of radioactivity. In the next stage, tritium will be used; and the machine will become very "hot." ITER is much larger than TFTR or JET, and the components to be moved will be large and heavy. Remote handling is expensive and inconvenient, but it does not seem to be a technological barrier.

Fusion Development Facilities

The engineering of a fusion reactor will require solution of a number of serious technological problems, as we have seen above. ITER will take decades to build and operate, and it is not designed to solve many of these problems. It is therefore prudent to build smaller machines specially designed for technology development so that this work can proceed in parallel with ITER. Many proposals have been made for a fusion development facility (FDF). A few of these will be described here.

IFMIF: International Fusion Materials Irradiation Facility

A favorite proposal of the European Union, together with Japan, is the IFMIF, a large linear accelerator that has been in the planning stage for 16 years. A diagram of it is shown in Fig. 9.30. As you can see, this is a large installation. The accelerator occupies a building of several hundred meters in length. It is designed to produce neutrons with energies matching those that would enter a tokamak blanket. This is done by accelerating to 40 MeV a beam of deuterons onto a target of liquid lithium. Reactions like the reverse of that in Fig. 9.10 would occur: a deuteron on lithium-6 would produce beryllium and a neutron, and a deuteron on lithium-7 would produce beryllium and two neutrons. The neutrons would then be used to bombard different materials to see how they stand up.

The key parameters for assessing radiation damage are neutron flux, neutron fluence, and dpa. Flux is how many neutrons per second go through each square meter. Fluence is how many have gone through the area during the whole life of the material. Dpa measures the damage, either per year or for the whole life. The flux produced by IFMIF is comparable to that expected in ITER, and about four times less than that in DEMO. The dpa per year in IFMIF is comparable to that in DEMO (about 30) and much larger than at in ITER.[5] The fluence cannot compare with that in DEMO, but could duplicate that in the limited life of ITER.

The IFMIF will cost about $700M [22]. It has been severely criticized because only small samples, a few square centimeters in size, can be tested. This is entirely inadequate to test the large components of ITER and DEMO, especially the blanket modules.

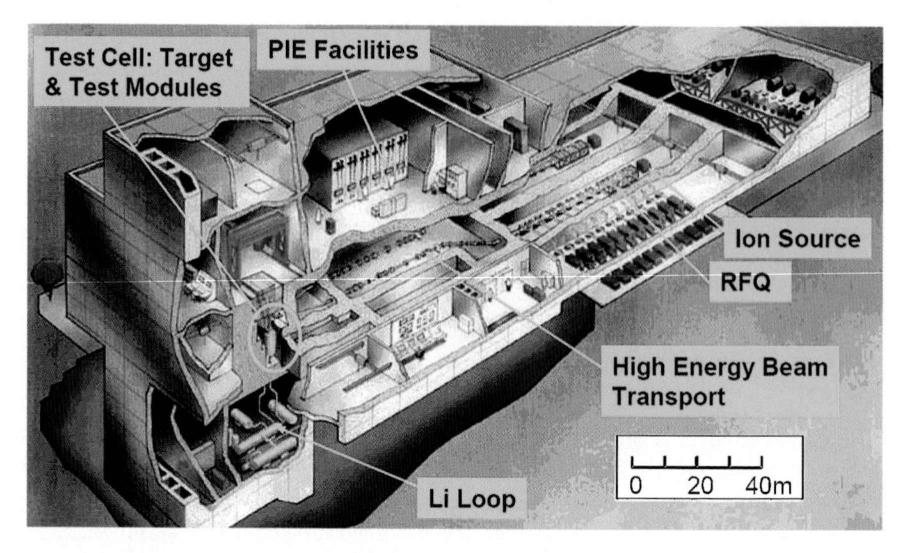

Fig. 9.30 Diagram of the International Fusion Materials Irradiation Facility [A. Möslang (Karlsruhe), *Strategy of Fusion Materials Development and the Intense Neutron Source IFMIF*]

Fusion Ignition Tokamaks

Proposals to build small but powerful tokamaks to test burning plasmas were made well before ITER. In the late 1980s, a Compact Ignition Tokamak was initiated in the USA, but was soon canceled. In 1999, Dale Meade at Princeton designed a 10-T, 2-m diameter tokamak call Fusion Ignition Research Experiment (FIRE), but this was never funded. These early ideas were based on the hope that very high magnetic fields produced without superconductivity could be used to achieve ignition on a small scale. This philosophy, promulgated by Bruno Coppi at the Massachusetts Institute of Technology, resulted in the Alcator tokamaks at M.I.T. and the Ignitor in Italy. In 2010, Italy and Russia signed an agreement to build a 13-T Ignitor-type tokamak to study burning plasma physics before ITER is finished. These small, pulsed machines cannot expose the steady-state problems that ITER will face. Engineering problems such as tritium breeding and plasma exhaust can be studied only with sufficient neutron flux. There are several proposals for large machines designed specifically for problems not tackled by ITER which will run simultaneously with ITER. None of these has been funded so far.

High-Volume Neutron Source

In 1995, noting the inadequacy of the IFMIF for blanket development, an international team headed by Abdou [23] proposed a high-volume plasma-based neutron source. A tokamak, naturally, was the best choice for a neutron source that

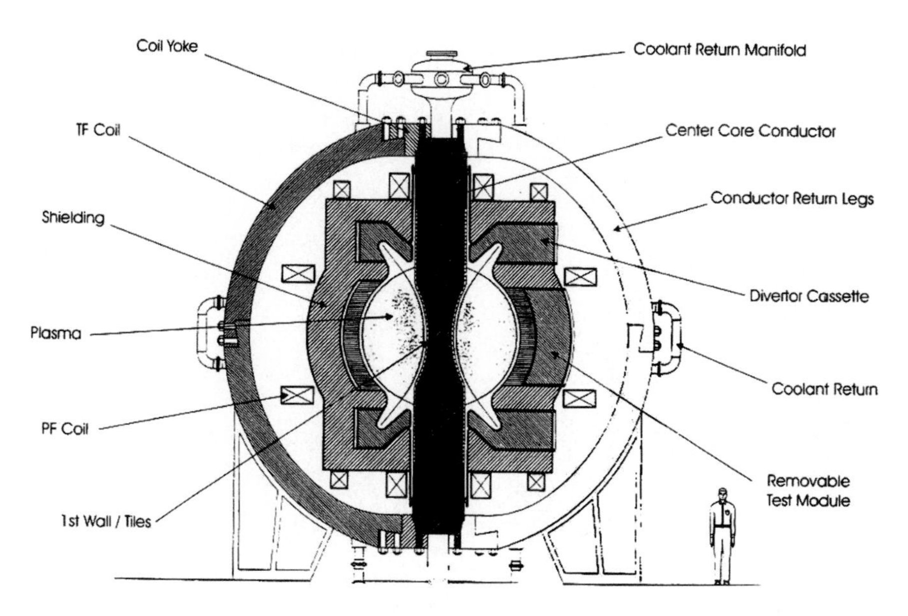

Fig. 9.31 A tokamak neutron source with single-turn normal-conducting toroidal field coils [23]

could cover large areas for blanket development. The group considered both superconducting and normal-conducting toroidal field coils, and it was found that coils made of a single turn rather than multiple windings of copper resulted in a smaller device. This is shown in Fig. 9.31. The major radius is only 80 cm and the toroidal field only 2.4 T; yet the plasma current is 10 MA and the neutron wall loading can be as large as 2 MW/m^2. The last number is indicative of how well the device can duplicate the damage to materials in a reactor like DEMO. This is done well even though the volume neutron source (VNS) is only 0.5% of ITER in volume, 2% in wall area, and 4% in fusion power produced. Significantly, the group did a risk–benefit analysis comparing the ways to obtain an 80% confidence level for DEMO to have, say, 50% availability, taking into account the mean time between failures and the time for repairs. Needless to say, operating ITER with VNS wins hands down over ITER alone. VNS also uses much less tritium in the process. The incremental cost is small: the total of capital cost and operating cost over the life of the machine is $19.6B for ITER and $24.4B for ITER plus VNS.

Fusion Development Facility

A more ambitious tokamak for technology tests has been proposed by a team at General Atomics in San Diego, California [24]. This machine is shown in Fig. 9.32. Note that this depicts only one side of the torus; the major axis is at

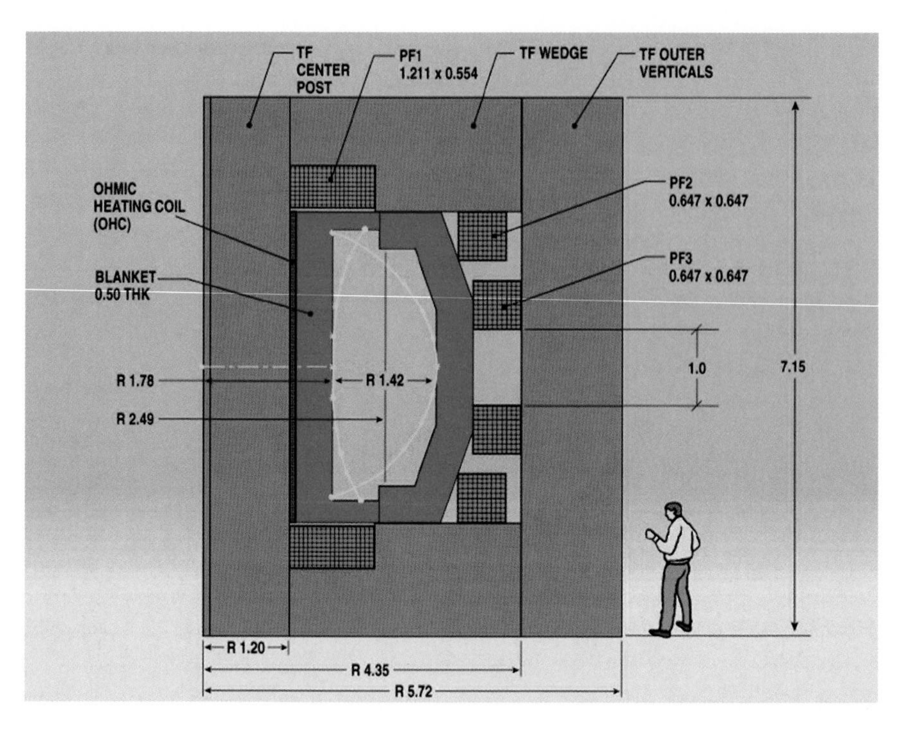

Fig. 9.32 Diagram of the cross section of the FDF tokamak [24]. The centerline of the torus is at the *left* edge of the diagram. TF is toroidal field (coil) and PF is poloidal field (coil). Dimensions are in meters

the left edge of the diagram. The dominant feature is the huge copper toroidal field coil surrounding machine. It will produce a field of 6 T (60,000 G). As seen by the size of the human figure compared to that in Fig. 8.23, FDF is actually smaller than JET. Yet the machine produces 250 MW of fusion power and can run continuously for two weeks at a time. The neutron flux is the required 1–2 MW/m^2, and the fluence is 3–6 MW-years/m^2 over a life of ten years.

Though FDF is much smaller than ITER, it can produce the neutrons for technological testing because it does not reach ignition. It runs steadily at $Q=5$, where Q is the fusion power divided by the power input to the plasma. For ignition $Q>10$ is necessary, and that is much more difficult. Nonetheless, FDF needs all the features of advanced tokamaks: high bootstrap current, internal transport barriers, radiofrequency current drive, and so forth. Remote handling will be developed, with replacement components lowered from the top, where the upper part of the toroidal field coil can be removed. Initially, blanket modules will be tested. Then, after a 2-year shutdown, a full solid ceramic blanket will be installed and tested. In the third stage, after another 2-year shutdown, a Pb-Li blanket will be installed. Only a machine with a full blanket can test such quantities as thermal stress, nuclear waste and disposal, radiation damage, and material lifetimes.

With full blankets, FDF as currently designed can demonstrate a closed fuel cycle, breeding as much tritium as it uses, reaching a TBR of 1.2. In fact, if operated at 400 MW of fusion power, it could actually breed tritium at the rate of 1 kg per year to be stored for use in DEMO. This is a very ambitious goal. In this sense, FDF is comparable to ITER in what it will accomplish. ITER will push superconducting technology, test alpha particle effects, and aim for ignition, but FDF will tackle the harder problems of technology with a smaller machine. FDF will not be cheap at perhaps one-third the cost of ITER; but since it will be a direct replacement for DIII-D, much of the expertise is already in place; and, importantly, the politics of an international project can be avoided. After the cancelation of TFTR, the USA needs to regain its position at the forefront of fusion research.

A Spherical Tokamak FDF

Spherical tokamaks are tokamaks with very small aspect ratio, which is the ratio of major radius to minor radius. They are fat doughnuts with a very small hole in the middle. These are hard to make, but they have advantages in stability. They are described in Chap. 10. Peng et al. [25] have designed a fusion development facility using a spherical tokamak (FDF-ST) with an aspect ratio of 1.5. This is shown in Fig. 9.33. The magnetic coils are normal-conducting copper, even the narrow center leg going through the small central hole. With major radius only 1.2 m, the machine is much smaller than other designs and yet can generate a neutron wall loading of 1.0 or even 2.0 MW/m^2. The toroidal field is 1.2 T, and the plasma current is 8.2 MA. The fusion power is only 7.5 MW or 2.5 times the input power. The machine can accommodate 66 m^2 of blanket area. If this can be engineered, it would be the least costly nuclear test facility to prepare for DEMO.

Fusion Power Plants

Commercial Feasibility

Industry is not interested in these technical details; it is concerned with the bottom line. RAMI is the acronym for four important criteria: reliability, availability, maintainability, and inspectability. The Electric Power Research Institute puts it in even more basic terms: economics, public acceptance, and regulatory simplicity. It is of course too soon to know how these will turn out; but designers of fusion power plants as well as fusion technology researchers are well aware of these criteria, which are always kept in mind. The fusion core is only a part of a whole power plant, a cartoon of which is shown in Fig. 9.34. The remote handling system is essential for maintainability and inspectability. The heating, current drive, and fueling systems affect reliability. The complicated fuel cycle system has to be completely

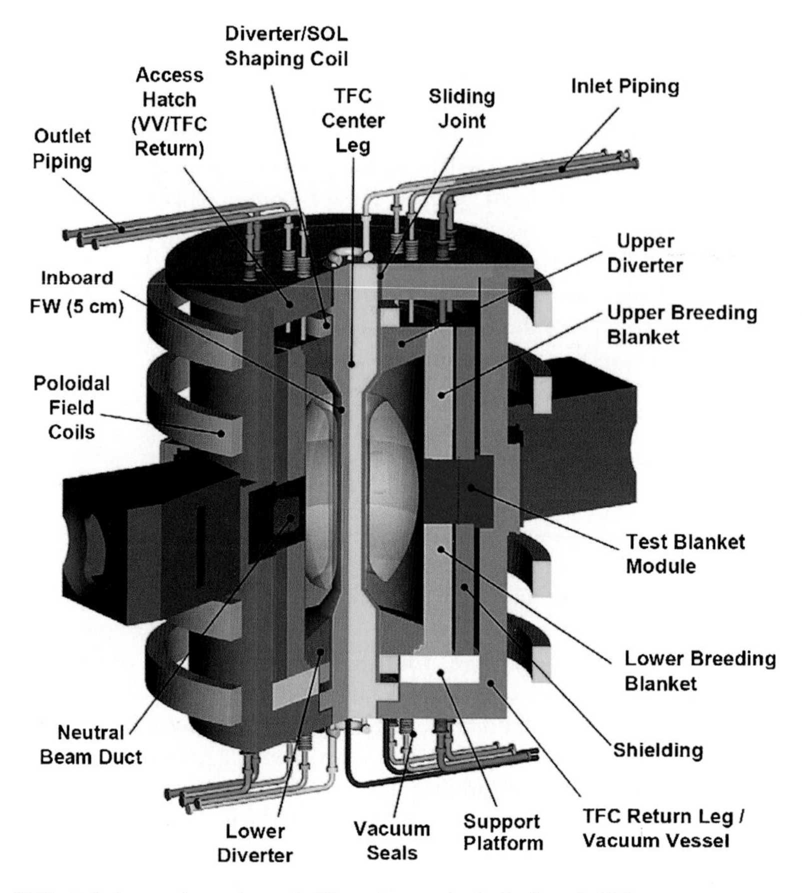

Fig. 9.33 A fusion nuclear science facility using a spherical tokamak [25]

safe in regard to tritium release. The balance of plant, the equipment that generates and transmits the electricity, is a larger part of the power plant than the power core, though it is shown deceptively as a small addition in Fig. 9.34. These are the steam turbines that drive the electric generators and the transformers and capacitors that condition the output for delivery to the transmission lines. All power stations that convert heat into electricity have this equipment, whether the fuel be coal, oil, gas, or uranium. Hydroelectric plants do not need steam; water drives the generators. Wind and solar plants produce electricity directly. Fusion plants can use the same generators and transmission lines that already exist in fossil or nuclear plants; only the power core has to be replaced. However, tokamaks are so complicated and include such temperature extremes that they will require a higher portion of the capital cost than other power cores.

Availability is an important aspect of a fusion reactor that is hard to assess. How often will leaks occur, and how long will it take to do the re-welding? How often do blankets have to be replaced, and how long will the shutdowns be?

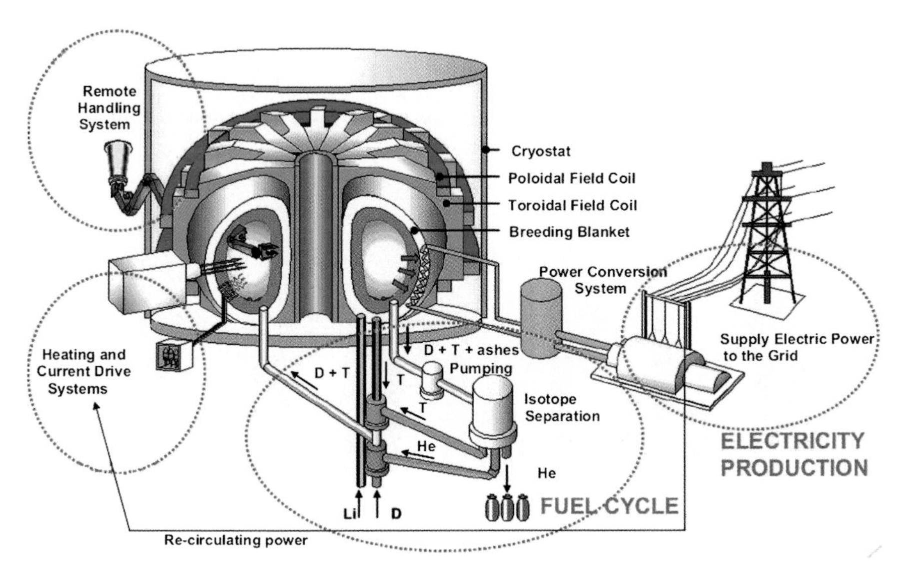

Fig. 9.34 Main parts of a fusion power plant [37]

How often will disruptions occur, and how long will it take to reassemble the machine? What percentage of the time will the machine be running during a year? During a shutdown, where will the power come from? Will we need a backup tokamak or new transmission lines from other power plants? Educated guesses are made by those who design fusion power plants based on available knowledge.

Power Plant Designs

The ARIES program in the USA is the leading group in designing fusion reactors. Originally started by Robert W. Conn in the 1980s at the University of Wisconsin and the University of California (UC) Los Angeles, it is now headed by Farrokh Najmabadi at UC San Diego. Throughout the years, new ARIES designs have been made as new physics has been discovered. The designs are not only for tokamaks; stellarators and laser-fusion reactors have also been covered. The latest designs, ARIES-AT for advanced tokamaks and ARIES-ST for spherical tokamaks, inspired the FDF proposals described above. Practical considerations such as public acceptance, reliability as a power source, and economic competitiveness pervade the studies. The designs are very detailed. They optimize the physics parameters, such as the shape of the plasma and the neutron wall loading. They also optimize the engineering details, such as how to replace blankets and how to join conductors to make the joints more radiation resistant. As new physics and new technology became available, the reactors ARIES I, II, … to ARIES-RS (reversed shear) and

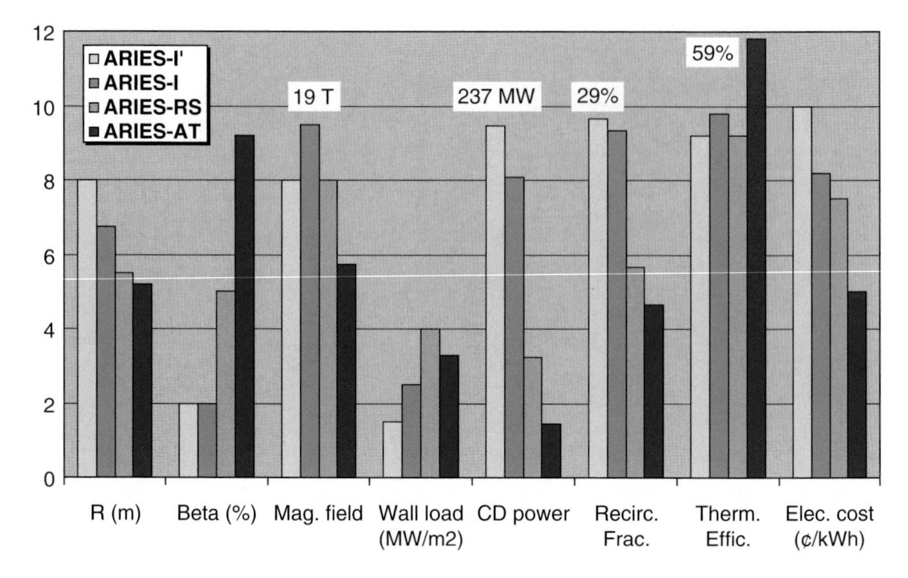

Fig. 9.35 Evolution of ARIES reactor designs. Some bars have been rescaled to fit the chart; these have their maximum values shown. *R* is the major radius. Beta is the ratio of plasma energy to magnetic field energy. The peak magnetic field is given in Tesla. The neutron wall loading is in MW/m^2. CD is current drive. Recirc. Frac. is the recirculating power fraction. Therm. Effic. is the thermal efficiency of the plant. Elec. cost is the cost of electricity in cents per kilowatt-hour

ARIES-AT (advanced tokamak) evolved to become smaller and cheaper. This is shown in Fig. 9.35. We see that as fusion physics advanced from left to right in each group of bars, the size of the tokamak, the magnetic field, and the current-drive power could be decreased while increasing the neutron production. This is due to the great increase in plasma beta that the designers thought would be possible. The recirculating power fraction is the power used to run the power plant; the rest can be sold. It dropped from 29 to 14%. The thermal efficiency in the latest design breaks the 40% Carnot-cycle barrier by the use of a Brayton cycle. Finally, we see that the COE is expected to be halved from 10¢ to 5¢ per kWh with advanced tokamaks.

ARIES-AT is shown in Fig. 9.36. Unlike existing tokamaks, this reactor design has space at the center for remote maintenance and replacement of parts. The philosophy in reactor design is to assume that the physics and technology advancements that are in sight will actually be developed and, on that basis, optimize a reactor that will be acceptable to industry and the public. It is not known whether high-temperature superconductors will be available on a large scale, but this would simplify the reactor. The blankets will be of the DCLL variety, and it is predicted that the Pb-Li can reach 1,100°C without heating the SiC walls above 1,000°C. This high temperature is the key to the high thermal efficiency. For easier maintenance and better availability, the blankets are made in three layers, two of which will last the life

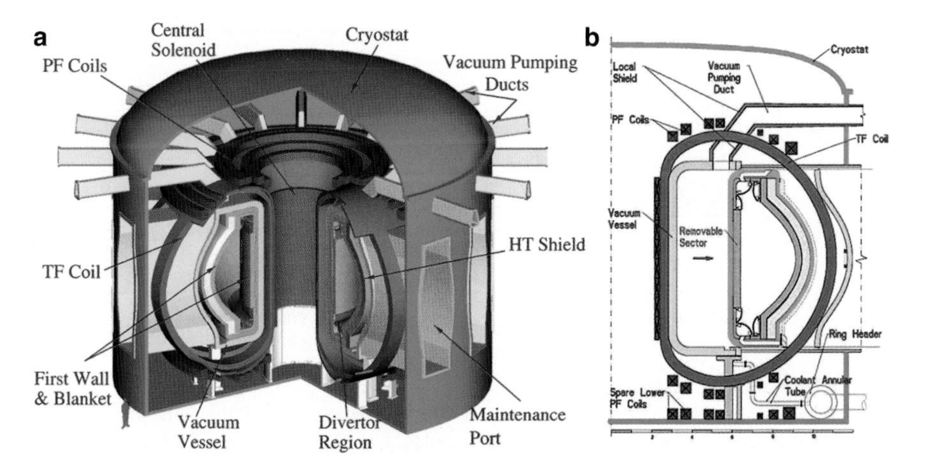

Fig. 9.36 Drawing of the ARIES-AT reactor design and its cutaway view at the *right* [38]

of the reactor. Only the first layer, along with the divertor, has to be changed out every five years. Sectors are removed horizontally and transported by rail in a hot corridor to a hot cell for processing. Shutdowns are estimated to take four weeks.

Turbocharging and supercharging in automobiles are terms that are well known to the public. Airplanes engines are turbocharged. Modern power plants use thermodynamic cycles that have higher efficiency than the classic Carnot cycle. The ARIES-AT reactor will use one of these called a Brayton cycle. The hot helium from the tokamak blanket is passed through a heat exchanger to heat helium that goes to electricity-generating turbines. The two helium loops are isolated from each other because the tokamak helium can contain contaminants like tritium. The turbine also runs with cooler helium at a different flow rate. The Brayton cycle precompresses the helium three times before it goes into helium turbines. The heat of the helium coming out of the turbines is recovered in coolers that cool the helium before it is compressed. It is this system that achieves the 59% thermal efficiency of the ARIES-AT design.

ARIES-AT will produce 1,755 MW of fusion power, 1,897 MW of thermal power, and 1,136 MW of electricity. The radioactive waste generated will be only 30 m^3 per year or 1,270 m^3 after 50 years. The plant will run for 40 of those years if availability is 80%. Ninety percent of this waste is of low-grade radioactivity; the rest needs to be stored for only 100 years. No provisions for public evacuation are necessary, and workers are not exposed to risks higher than in other power plants. The COE from ARIES-AT is compared with other sources in Fig. 9.37. We see that electricity from fusion is not expected to be extravagant.

Europeans have also made reactor models in their Power Plant Conceptual Studies (PPCS) [26]. Figure 9.38 is a diagram of the tokamak in those designs. As with the ARIES studies, Models A, B, C, and D in PPCS (Fig. 9.39) trace the evolution of the design with advances in fusion physics and technology, with Model D using the most speculative assumptions. All these models produce about 1.5 GW of

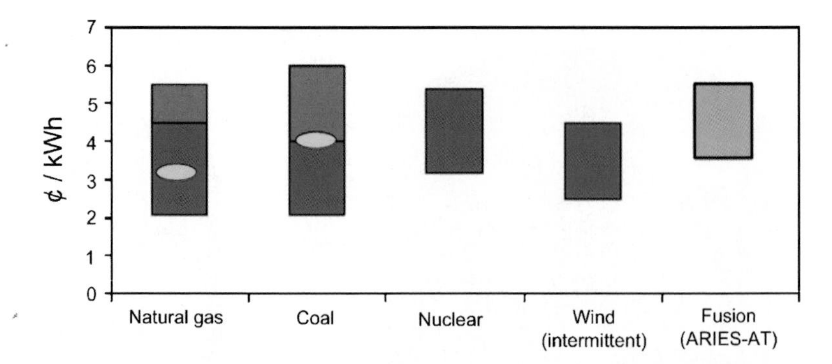

Fig. 9.37 Estimated year 2020 cost of electricity in US cents per kilowatt-hour from different power sources [graph adapted from [25], but original data are from the Snowmass Energy Working group and the US Energy Information Agency (*yellow ellipses*)]. The *red* range is the cost if a $100/ton carbon tax is imposed. The fusion range is for different size reactors; larger ones have lower cost

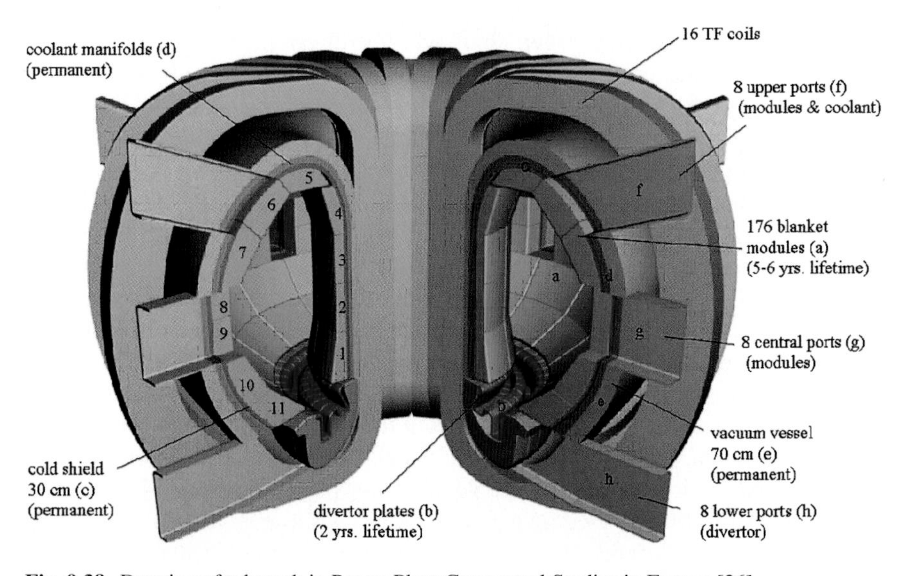

Fig. 9.38 Drawing of tokamak in Power Plant Conceptual Studies in Europe [26]

electricity, but they are smaller and use less power with gains in knowledge. The recirculating fraction and thermal efficiency of Model D matches that of ARIES-AT. Safety and environmental issues were carefully considered. The cost estimates are given in Fig. 9.40, also in US cents per kWh. The difference between the wholesale price of electricity and that available to consumers is clearly shown. It is seen that fusion compares favorably with the most economical sources, wind and hydro.

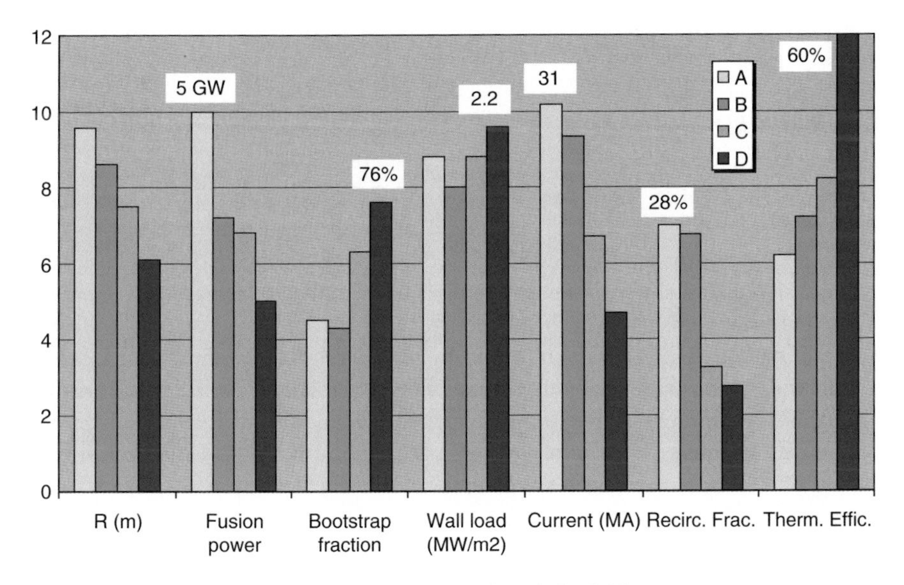

Fig. 9.39 Evolution of PPCS designs [26]. See caption of Fig. 9.35

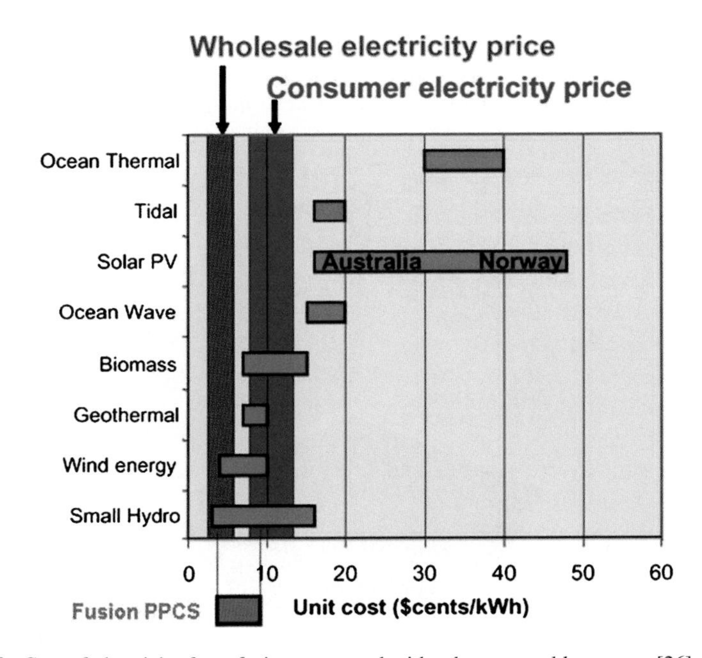

Fig. 9.40 Cost of electricity from fusion compared with other renewable sources [26]

The Cost of Electricity

Methodology

In spite of the fact that we do not yet know how a fusion reactor will be constructed, or even if it is at all possible, detailed calculations have been made on the COE based on the reactor models described in the previous section. The work of Ward et al. [27], which we will summarize here, is based on the European PPCS designs. Their calculated costs for each component of a power plant compare well with those from the ARIES studies in the USA. Being a renewable power source, fusion shares with wind, solar, and hydro the benefit of essentially zero fuel cost. However, the capital cost is large. A breakdown is given by Ward [28] in Fig. 9.41. The capital cost of the tokamak power core is almost as large as that of the balance-of-plant, which is the power conversion system and electrical generators shown in Fig. 9.34. Compared with fossil fuel plants, the capital cost and replacement of blankets and divertors take the place of fuel costs. These fusion-specific costs depend on the reactor model. The models A, B, C, and D in Fig. 9.39 range from ITER-like primitive designs with steel chambers and water cooling to speculative advanced designs with Pb-Li liquid cooling and SiC/SiC first walls. Computer programs are used to calculate the costs of each component under different assumptions.

Important Dependences

The COE depends on some factors that are independent of the power core and others that are specific to fusion. These factors appear in the following formula for COE,

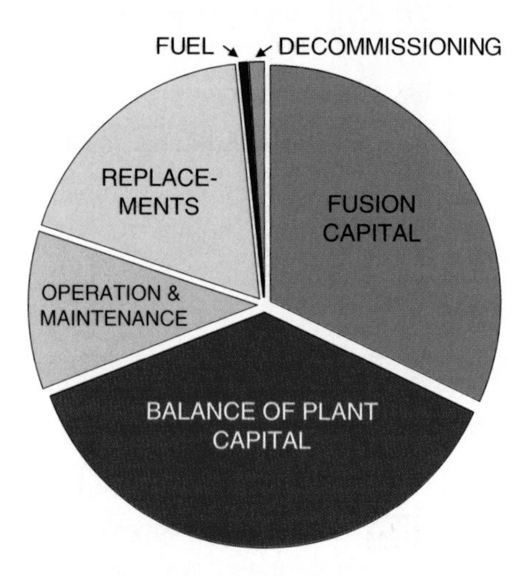

Fig. 9.41 Cost breakdown of a fusion power plant [28]

which at first seems rather daunting. However, it is not necessary to know what the formula means in detail; it is used here just as a convenient way to show what affects the cost. The COE is proportional to the quantities in the parenthesis times those in the

$$\mathrm{COE} \propto \left(\frac{rL}{A}\right)^{0.6} \frac{1}{\eta_{\mathrm{th}}^{0.5} P_{\mathrm{e}}^{0.4} \beta_{\mathrm{N}}^{0.4} N^{0.3}}$$

denominator of the fraction following it. Inside the parenthesis, r is the discount rate, a financial factor similar to interest rate that will be explained later. L is a learning factor which takes into account that the first of a kind is always more expensive than the tenth one made. L starts at 1 and gets smaller with learning, so COE drops. A is the availability, which is the fraction of time the plant is running rather than shut down for repairs. Larger A means lower costs. The fusion reactor designs have A's ranging from 60 to 80%.

The first two quantities in the denominator at the right have to do with the whole plant, and the last two concern the quality of the plasma in the tokamak. Eta-thermal (η_{th}) is the efficiency of converting heat into electricity. P_{e} is the size of the plant in terms of electrical power produced. The larger the better because of economy of scale. Beta-normalized (β_{N}) expresses the efficiency with which the plasma current can confine a large amount of hot plasma by creating the right amount of twist in the magnetic field. Finally, N is the ratio of the plasma density to that predicted by Greenwald limit (Chap. 8) for a stable plasma. In the different reactor models, r varies from 5 to 10%, L from 0.5 to 0.7, A from 0.6 to 0.8, η_{th} from 35 to 60%, P_{e} from 1 to 2.5 GW, and N from 0.7 (safe) to 1.4 (speculative). Most importantly, β_{N} varies from 2.5 to 5.5, representing the progression from well-established data to hopefully achievable advanced tokamak operation. Figure 9.42 shows the COE predicted from the PPCS models A–D as a function of the learning factor L.

As an example of how sensitive the COE is to assumptions made in the models, Fig. 9.43 shows how the availability factor A changes with the lifetime of the materials

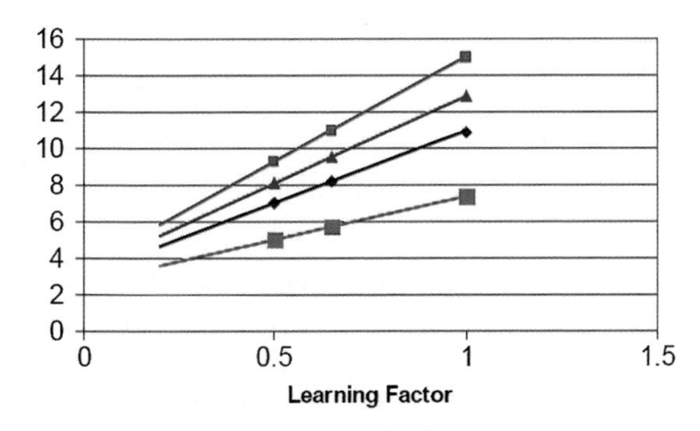

Fig. 9.42 The cost of electricity, in euro cents per kilowatt-hour, calculated for various reactor models as a function of the learning factor L [28]. Model A is an ITER-like machine, and D is the most advanced reactor envisioned at present. Power plants start at $L = 1$ and progress leftward to lower costs as more are built

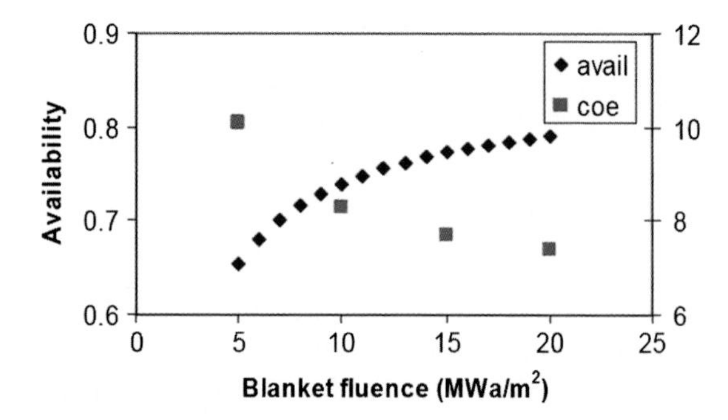

Fig. 9.43 Dependence of the power plant availability and cost of electricity on the degree to which materials in a tokamak reactor can withstand neutron damage [28]

in the first wall and blankets. The lifetime is expressed as the neutron fluence that the materials stand before they have to be replaced. The fluence is measured in years at an equivalent neutron energy flux of 1 MW/m^2. The shorter the lifetime, the more often the blankets will have to be replaced, and hence the lower the availability. This then increases the cost (the higher blue points at the left).

Cost Levelization/Discounting

Expenses and income are both functions of time. Costs start accruing when a power plant is proposed and initial studies are made, for instance, on environmental impact. Land is purchased, the plant is designed, equipment is ordered, and construction begins. This takes many years. The plant is finished and begins producing power. Profits begin to be made, year by year. At the same time, there are expenses for operating the plant, and for repairing and replacing equipment. To get a reasonable number for the COE, one has to adjust all the expenses and income forward or backward to the same date. Time is money. This is called discounting. It is done with another formula:

$$\mathrm{COE}=\frac{\Sigma_t (C+\mathrm{OM}+F+R+D)_t/(1+r)^t}{\Sigma_t E_t/(1+r)^t}$$

This is a formula unfamiliar to physicists but may be more familiar to readers involved with business or finance. Here C is the capital cost, OM is operation and maintenance, F is the fuel cost, R is the cost of replacements, D is the cost of decommissioning at the end of life, and r is the *discount rate*. In the denominator, E is for earnings. The sum is over time t. To derive a value at time zero for an expense or income occurring at another time, a discount has to be applied. The discount rate is like an interest rate but includes also expectations of what the market will be like, how much inflation there will be, and factors like those. Financiers normally assign a discount rate between 5 and 10%.

Suppose we want to calculate the COE as of the start of planning. We set that as $t = 0$. For simplicity, let us do the accounting annually, not monthly or daily. Suppose it takes five years to get ready, five years to build the plant, and it has been operating for another five years. For years $t = 1$–5, we have the money $C_1 - C_5$ spent in those years, which is only interest on money borrowed, salaries, and rental for office space. For years $t = 6$–10, C will be much larger, as the plant is built. For years 11–15, we have $C + \mathrm{OM} + F$ for those years, and also E earned in those years. Each year's amounts are divided by $(1 + r)$ raised to the power t in order to get the value as of $t = 0$. Both the numerator and the denominator are summed over the years, and the ratio is the COE. In later years, there will also be values for R and D.

To get a better idea of what discounting means, let us consider a simple example. Suppose you borrow \$1M to build a machine, taking five years to do so. At the end of the five years you sell the machine for \$1M. However, you could not have sold that machine for \$1M at Year 0, since that machine did not exist yet and you could not make any money with it. It has a smaller discounted value at Year 0, given by $C/(1 + r)^5$, according to the formula above. If $C = \$1\mathrm{M}$ and the discount rate is $r = 5\%$, we have a value at $t = 0$ of $C/(1.05)^5$, which works out to be only \$0.784M. The reason is that you had to pay compound interest during the five years. One million dollars compounded annually at 5% is \$1M times $(1.05)^5$, which is \$1.276M. You had to pay \$0.276M in interest, so you made only \$0.724M, and that is closer to the value of the machine at $t = 0$. Actually, you did not have to borrow all the money at once, so the discounted, or levelized, value is \$0.784M, which is exactly the reciprocal of \$1.276M.

This exercise points out that a large part of the cost of any power plant, regardless of its power source, is interest during construction. If the discount rate is 7.5% (halfway between 5 and 10%), and the plant takes five years to construct, summing over the discounted value of one-fifth of the capital cost for each of five years shows that 20% of the cost is interest and other financial factors. The levelized COEs of all different kinds of power plants (except fusion) in many different countries have been analyzed in exhausting detail by the International Energy Agency.[6]

The Cost of Fusion Energy

Figure 9.44 shows how the COE from fusion compares with that from other energy sources [28]. Each entry has two bars showing a minimum and a maximum value, the difference depending partly on location and partly on technology. For fossil fuels, the maximum is the cost including the expense of carbon sequestration. For fusion, the maximum and minimum represent the range of the reactor models ABCD described above. These data for other energy sources are from the IEA report of 1998. Fuel prices and interest rates have fluctuated so violently in recent years that the comparison has not been updated. However, the levelized COEs of nonfusion sources are available for 2005[6] and 2010.[7] The data for 2010 are shown in Fig. 9.44. For comparison, the fusion COE given in Fig. 9.44 is reproduced in Fig. 9.45. That graph shows also the breakdown between capital costs and

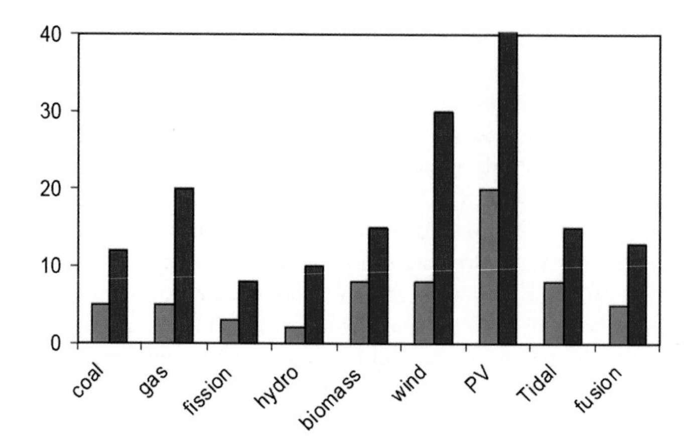

Fig. 9.44 Comparison of the cost of electricity from conventional and renewable energy sources [28]

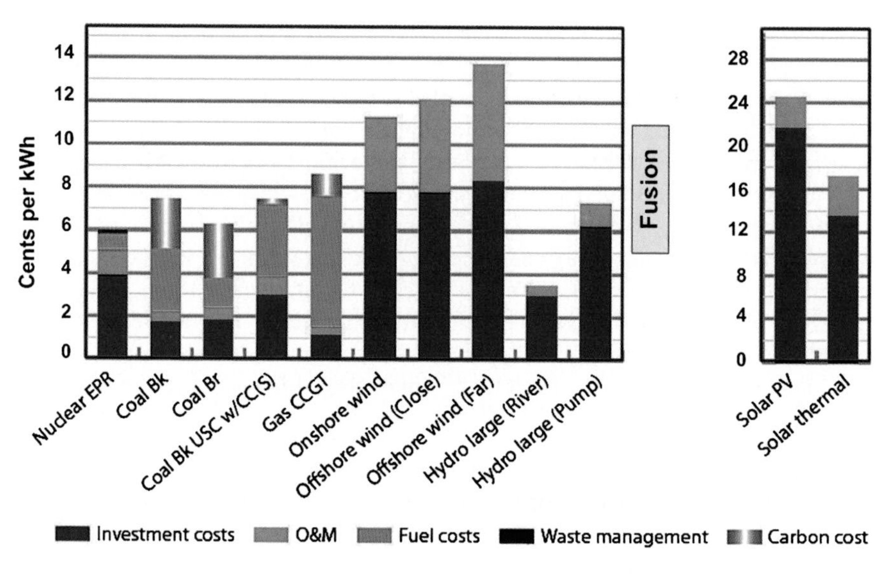

Fig. 9.45 Estimated cost of electricity in Europe from nuclear, fossil-fuel, and renewable sources assuming a 5% discount rate[7]. The color code gives the breakdown among capital costs, operation and maintenance (O&M), and fuel costs. For nuclear plants, there is charge for nuclear waste management. For fossil-fuel plants, there is a cost for carbon management under certain assumptions. The estimated cost range for fusion plants has been added. The solar photovoltaic (PV) and solar thermal costs have to be plotted on a different scale

operation and maintenance costs, as well as the estimated cost of carbon capture and sequestration for fossil-fuel plants. The data are from different time periods, but the difference is insignificant in view of the uncertainties involved. It is seen that *the COE from fusion plants will be competitive with that from other renewal sources and from fossil-fuel plants with carbon management.*

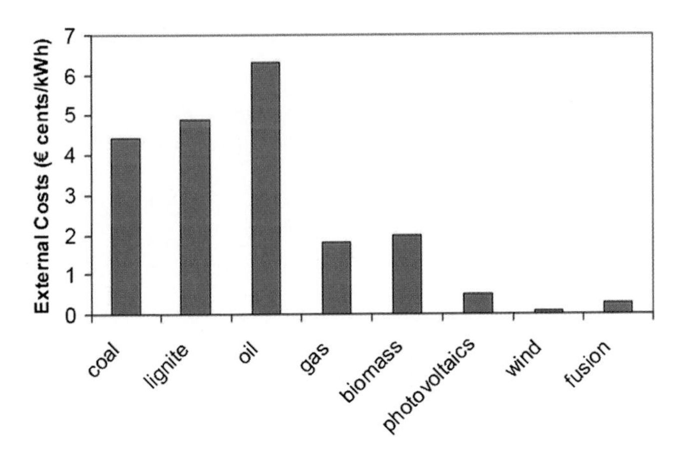

Fig. 9.46 External costs of fusion compared with other energy sources [27]

It is interesting to note that the large variability of the COE is reflected in the IEA's 2010 report[7]. The figures for each energy source vary greatly from country to country. In addition, the sensitivity of the estimates to such factors as corporate taxes, discount rate, and fuel cost is emphasized.

Not included in the above analyses are external costs, which include damage to the environment, general health, and human life. Such costs have been evaluated by site to eliminate location biases. For instance, one considers the difference when a fusion plant is put in place of a coal plant in the same location. It turns out that the external costs of fusion are extremely low, ranging from 0.07 to 0.09 euro cents per kWh. Comparison with other energy sources is shown in Fig. 9.46.

The net present value of fusion takes into account the probability of success or failure. Though this obviously has a high degree of uncertainty, there is a large margin for error, since the annual world energy expenditures exceed the annual cost of fusion development by 1,000 times. It has been estimated that if fusion captures 10–20% of the electricity market in 50 years, the discounted future benefit of fusion is $400–800B; or, if the probability of failure is counted, it is still $100–400B. This means that development of fusion is worthwhile even if fusion captures only 1% of the world electricity market [27].

Notes

1. However, a vertical Allure Ignition Stellarator with a liquid Li wall was proposed in 2010 to be built in Spain.
2. It has been pointed out that tritium is also generated in the beryllium multiplier, an effect usually neglected in estimates of breeding ratio [3].
3. ITER Newsline Nos. 114 and 122 (2010). http://www.iter.org/newsline/.
4. M. J. Schaffer (General Atomic), private communication.
5. A. Möslang (Karlsruhe), *Strategy of Fusion Materials Development and the Intense Neutron Source IFMIF*.

6. *Projected costs of generating electricity, 2005 update*, published by the Nuclear Energy Agency and the International Energy Agency of the Organization for Economic Cooperation and Development (OECD).
7. *Projected costs of generating electricity, 2010 Edition*, published by the Nuclear Energy Agency and the International Energy Agency of the Organization for Economic Cooperation and Development (OECD).

References

1. L.L. Snead, *Ceramic Structural Composites, the Most Advanced Structural Material, 9th Course on Technology of Fusion Tokamak Reactors* (International School on Fusion Reactor Technology, Erice, Italy, 2004)
2. L. Giancarli et al., Fusion Eng. Des. **61–62**, 307 (2002)
3. L. Giancarli, *The PPCS In-Vessel Component Concepts* (International School on Fusion Reactor Technology, Erice, Italy, 2004)
4. D. Stork, *DEMO and the route to fusion power*. 3rd Karlsruhe International School on Fusion Technology, Sept 2009
5. A. Ying et al., *Current status of the test blanket program in ITER and implications for blanket component testing requirements and goals*. Symposium on Fusion Energy, San Diego, CA, June 2009
6. G. Janeschitz, *The development of commercial fusion energy in the EU*. Seminar, University of California, Los Angeles, Jan 2008
7. S. Casadio, *Ceramic Breeder Technology* (International School on Fusion Reactor Technology, Erice, Italy, 2004)
8. A. Abou-Sena, A. Ying, M. Abdou, Fusion Sci. Technol. **56**, 206 (2009)
9. J. Takeuchi et al., Fusion Eng. Des. **83**, 1082 (2008)
10. S. Smolentsev, R. Moreau, M. Abdou, Fusion Eng. Des. **83**, 771 (2008)
11. A. Pizzuto, *Comparison of Breeder Blanket Designs* (International School on Fusion Reactor Technology, Erice, Italy, 2004)
12. M. Glugla et al., *Review of the ITER fuel cycle systems and recent progress*. Symposium on Fusion Energy, San Diego, CA, June 2009
13. I.R. Cristescu, *The Fuel Cycle System* (International School on Fusion Reactor Technology, Erice, Italy, 2004)
14. E. Salpietro, *Magnet Design and Technology* (International School on Fusion Reactor Technology, Erice, Italy, 2004)
15. Report of the Research Needs Workshop (ReNeW), Bethesda, MD, June 8–12, Office of Fusion Energy Sciences, US Department of Energy, 2009
16. P.J. Heitzenroeder et al., *An overview of the ITER in-vessel coil systems*. Princeton Plasma Physics Laboratory Report PPPL-4465, Sept 2009
17. M.J. Schaffer et al., Nucl. Fusion **48**, 024004 (2008)
18. L.R. Baylor et al., *Disruption mitigation technology concepts and implications for ITER*. Symposium on Fusion Energy, San Diego, CA, June 2009
19. H.L. Berk et al., Phys. Rev. Lett. **68**, 3563 (1992)
20. G. Giruzzi et al., Nucl. Fusion **49**, 104010 (2009)
21. J. Bucalossi et al., *Performance issues for actuators and internal components during long pulse operation*. Symposium on Fusion Energy, San Diego, CA, June 2009
22. J. Rathke, *Engineering overview: International fusion materials irradiation facility*. FESAC Subcommittee Review (FESAC is the Fusion Energy Sciences Advisory Committee of the US Department of Energy), San Diego, CA, Jan 2003
23. M.A. Abdou et al., Fusion Technol. **29**, 47 (1996)
24. A.M. Garofalo et al., IEEE Trans. Plasma Sci. **38**, 461 (2010)

25. Y.K.M. Peng et al., Fusion Sci. Technol. **56**, 957 (2009)
26. D. Maisonnier, *PPCS Reactor Models* (International School on Fusion Reactor Technology, Erice, Italy, 2004)
27. D.J. Ward et al., Fusion Eng. Des. **74–79**, 1221 (2005)
28. D.J. Ward (EURATOM/IKAEA Fusion), *Impact of Physics on Power Plant Design and Economics* (International School on Fusion Reactor Technology, Erice, Italy, 2004)
29. N. Holtkamp, *Status of ITER*. Fusion Power Associates meeting, Washington, DC, Dec 2009
30. G. Federici, *Plasma Wall Interactions in ITER and Implications for Fusion Reactors* (International School on Fusion Reactor Technology, Erice, Italy, 2004)
31. G. Janeschitz, *Divertor Physics and Technology* (International School on Fusion Reactor Technology, Erice, Italy, 2004)
32. M. Abdou, *Challenges and development pathways for fusion nuclear science and technology*. Seminar, Seoul National University, S. Korea, Nov 2009
33. N.B. Morley, M. Abdou, in *Fusion Power Associates Annual Meeting*, Washington, DC, Oct 2005
34. D. Ward (UKAEA, Culham, UK), *Impact of Physics on Power Plant Design and Economics* (International School on Fusion Reactor Technology, Erice, Italy, 2004)
35. K. Sakamoto et al., Nucl. Fusion **43**, 729 (2003)
36. G.T. Hoang et al., Nucl. Fusion **49**, 075001 (2009)
37. J. Pamela et al., *Key R&D issues for DEMO*. Symposium on Fusion Technology, Rostock, Germany, Sept 2008
38. F. Najmabadi, in 18th KAIF/KNS Workshop, Seoul, Korea, 21 April 2006

Chapter 10
Fusion Concepts for the Future*

Advanced Fuel Cycles

Some day the inhabitants of this planet will look back at the clumsy magnetic bottle, the D–T tokamak, which is described in the previous chapters. The tokamak will seem like an old IBM Selectric typewriter with font balls compared to Microsoft Word on a 2-GHz notebook computer. The deuterium–tritium reaction is a terrible fusion reaction, but we have to start with it because it is easy to ignite. It generates power in neutrons, which make everything radioactive so you cannot go near the reactor. The neutrons are hard to capture and also damage the whole structure of the machine. And you have to breed the tritium and keep it out of the environment. There are much cleaner fusion fuels that we can use in next-generation magnetic bottles.

These future magnetic bottles will hold denser, hotter plasmas for a longer time. Then we can use reactions that do not produce the intense flux of energetic neutrons that plagues D–T reactors. Here is a list of the main possibilities.

$$\mathbf{D} + \mathbf{D} \rightarrow \mathbf{T} + \mathbf{p} \quad \text{(half the time)}$$
$$\mathbf{D} + \mathbf{D} \rightarrow \mathbf{He}^3 + \mathbf{n} \quad \text{(half the time)}$$
$$\mathbf{D} + \mathbf{He}^3 \rightarrow \alpha + \mathbf{p}$$
$$\mathbf{p} + \mathbf{B}^{11} \rightarrow 3\alpha$$
$$\mathbf{p} + \mathbf{Li}^6 \rightarrow \mathbf{He}^3 + \alpha$$
$$\mathbf{He}^3 + \mathbf{Li}^6 \rightarrow 2\alpha + \mathbf{p}$$
$$\mathbf{p} + \mathbf{Li}^7 \rightarrow 2\alpha$$
$$\mathbf{He}^3 + \mathbf{He}^3 \rightarrow \alpha + 2\mathbf{p}$$
$$\mathbf{D} + \mathbf{Li}^6 \rightarrow 2\alpha$$

*Numbers in superscripts indicate Notes and square brackets [] indicate References at the end of this chapter.

F.F. Chen, *An Indispensable Truth: How Fusion Power Can Save the Planet*,
DOI 10.1007/978-1-4419-7820-2_10, © Springer Science+Business Media, LLC 2011

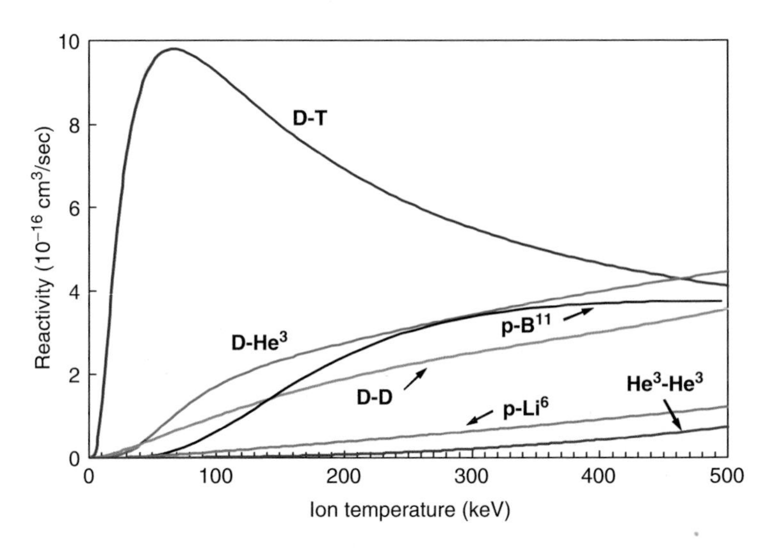

Fig. 10.1 Reactivities of several fusion reactions versus ion temperature in keV [1, 2]

In this list, **D** stands for a deuteron, **T** for a triton, **p** for a proton, and **α** for an alpha particle (He^4 nucleus). $\mathbf{He^3}$ is a rare isotope of helium with only one neutron instead of two. Figure 10.1 compares some of these reactions with D–T. What is shown are their reactivities, which show how fast fusion occurs for each fuel mix at each ion temperature.

The special role of D–T is immediately apparent. Not only does it fuse much faster than anything else, but the peak occurs at a much lower temperature. The 50-keV temperature of the peak can already be achieved. We next describe the advanced fuels in groups.

The first group involves only deuterium, which is plentiful in water. It can fuse with itself two ways, either producing T and p, or He^3 and n. When it goes the first way, the proton is harmless, but the T will quickly react with D in a DT reaction and produce a 14-MeV neutron. When D+D goes the second way, it produces harmless He^3 and a weaker neutron. So the DD reaction is not completely clean; there are neutrons, but much fewer of the dangerous ones. Forty percent of the energy comes out as charged particles (p, T, He^3, and α), which keep the plasma hot and can give up their energy electrically instead of through a thermal cycle. The neutron damage to materials is greatly reduced. These two DD reactions will occur simultaneously, but their reactivities are very low even when summed. However, there is a gain of a factor of 2 because both reactants are the same. That is, each deuteron can react with all the other ions instead of with only the half that are tritons, as in a DT reactor. However, this still leaves the DD reaction with a much lower rate than DT.

The reactions in the second group have the next highest reactivities and are the most promising ones. D-He^3 has sizeable reactivity at low temperature and produces no neutrons. Unfortunately, you cannot keep deuterium from fusing with itself, so there are DD reactions going on at the same time. But the energy in neutrons is reduced by a factor of 20 relative to DT, and this is an *almost* clean reaction. The problem is that He^3 does not occur naturally. It can, however, be mined on the moon. It is estimated that there are a billion tons of He^3 just under the surface of the moon, enough to supply the world for 1,000 years if it could be brought down here [3]. Mining machines have been designed which could dig 1 km^2 of the moon's soil, down to 3 m depth, to get 33 kg of He^3 a year [4]. If the moon is ever colonized, this would be the fuel used. Finding deuterium there may not be as easy, and the much harder He^3–He^3 reaction (Fig. 10.1) would have to be used. Burning D–He^3 on earth will have to wait until space shuttles can reach the moon. Nonetheless, the simplicity of the engineering is so attractive that a D–He^3 reactor has been designed [5].

The p-B^{11} reaction is the most attractive one at present. The reactants are not radioactive, and only helium is produced. Without neutrons, all the shielding and blankets of DT reactors are unnecessary. Fusion power plants can dispense with the tritium recovery and processing plant, as well as with remote handling equipment. Only hydrogen and boron are used. Boron is plentiful on earth, and B^{11} is its main isotope. We commonly use 20 Mule Team Borax, a cheap cleanser. All the energy comes out as fast alpha particles. Since these are charged particles, there may be a possibility of direct conversion of the energy into electricity without going through boilers and turbines. This can be done by leading the alphas into a channel where they can be slowed down with electric fields, thus producing electricity directly, or by capturing the synchrotron radiation emitted by the alphas spiraling in a magnetic field. However, boron is not a light element; it has a charge of 5 ($Z=5$) when it is fully ionized. When electrons collide with ions, they produce X-rays at a rate that increases with Z^2. Though it is not hard to shield against these X-rays, they represent energy that is lost to the plasma, and it is harder to raise the plasma temperature. Special methods being developed to overcome this is described in a later section.

All the other reactions on our list have very low reaction rates, as exemplified by p-Li^6 and He^3–He^3 in Fig. 10.1. Reactants with atomic number Z above 2 have two other problems besides low reactivity. First is the synchrotron radiation loss mentioned above. Second, there are competing reactions when there is a large number of proton and neutrons, and they can combine in different ways. For instance, p-Li^7 looks like a great reaction, producing two alphas. However, $\mathbf{p}+\mathbf{Li}^7\rightarrow\mathbf{Be}^7+\mathbf{n}$ (a neutron) is also possible [6], and this happens 80% of the time. The two reactions in the third group above form a chain reaction in which the He^3 generated by p-Li^6 can react with Li^6 to regenerate the proton, and only alphas are the result. However, the reaction rate is low, and there are competing reactions.

Speaking of chain reactions, it was Hans Bethe who invented the famous carbon cycle that allows hydrogen fusion to occur in the sun at a comparatively low temperature. Carbon is used as a catalyst that regenerates itself. Other chain reactions for the sun have been devised since then. No one so far has found a chain reaction for advanced fuels that will allow them to burn at lower plasma temperatures on earth. However, there has never been a large-scale effort to find such a chain reaction.

The last reaction listed above, $\mathbf{D + Li^6 \rightarrow 2\alpha}$, looks very attractive, but there are five competing reactions that produce nasty products. It has an interesting story. Lithium is the lightest solid element. Lithium hydride (LiH) is a glassy, solid material. It is one of the hydrides mentioned in Chap. 3 for carrying hydrogen in hydrogen cars. If we replace natural Li with Li^6 and H with D, we get lithium-6-deuteride, a similar solid that is easy to transport and store. There is no information on this reaction on the Web because apparently it is useful for making hydrogen bombs, being easy to carry and producing a large amount of energy, 22 MeV. In a bomb, the reaction is set off by neutrons, and the nasty by-products would be just fine for the purpose. Mention of tests involving this reaction can be found in public histories of atomic bomb development in the USA. In a fusion reactor, however, a deuterium–lithium plasma would be hard to ignite, and the neutrons and gamma rays emitted would be hard to manage. The reaction rate [7] is about 28% of that of He^3–He^3, the lowest curve in Fig. 10.1. Furthermore, the competing reaction $\mathbf{D + Li^6 \rightarrow Be^7 + n}$ occurs 3.5 times more often, producing neutrons. This is the reason many clean-looking reactions are not actually viable for a reactor.

Stellarators

Research on closed magnetic bottles started with stellarators such as the figure-8 stellarator shown in Fig. 4.18. In 1969, the Model C Stellarator in Princeton, the largest at the time, was converted to a tokamak because of the good results coming from Russia with their configuration. Soon almost all new machines were tokamaks. This was because of the self-healing properties of tokamaks, as described in Chap. 7. When the temperature profile in the plasma became too peaked, sawtooth oscillations would arise and smooth it out to maintain stability. All this is now changed, and stellarators have come back as a hope for the future.

The difference between these two toroidal devices, tokamaks and stellarators, is the way the poloidal magnetic field (the component that gives the field lines their helical twists) is generated. In tokamaks, a large current in the plasma generates that field. In stellarators, external coils generate that field, and no large plasma current is necessary. But these external poloidal-field coils are hard to make. Present-day Advanced Tokamaks no longer rely on the self-healing features that were initially useful. We have learned how to shape the plasma current with radiofrequency and microwave power to keep the plasma stable. In fact, sawtooth oscillations are now deliberately eliminated. Since self-organization is no longer necessary, we can reconsider stellarators. Stellarators are less subject to effects such as disruptions that are connected with the large plasma current. In effect, they eliminate a source of energy that allows a plasma to self-organize destructively in its efforts to escape from confinement. Furthermore, since transformer action to drive the plasma current is not necessary, stellarators are more suitable for steady-state, continuous operation.

Wendelstein

The largest programs that kept stellarators alive during the tokamak era were the Wendelstein program in Germany and the large helical device (LHD) in Japan. How stellarators look nowadays is a far cry from the first primitive devices. A schematic of the magnet coil structure in a classical stellarator is shown in Fig. 10.2. The circular coils generate the toroidal field, and the helical coils add the poloidal field. The number of conductors on a minor circumference determines the periodicity of the helical field. The magnetic island structure is fixed externally and not by the internal plasma current. Note that the plasma is no longer circular; it follows the helical structure of the coils.

Now imagine that the circular and helical coils are combined into individual coils that produce the same magnetic fields. We then have the structure shown in Fig. 10.3, which is a diagram of Wendelstein 7-X, the newest stellarator being constructed in Greifswald, Germany (formerly East Germany). These coils are easier

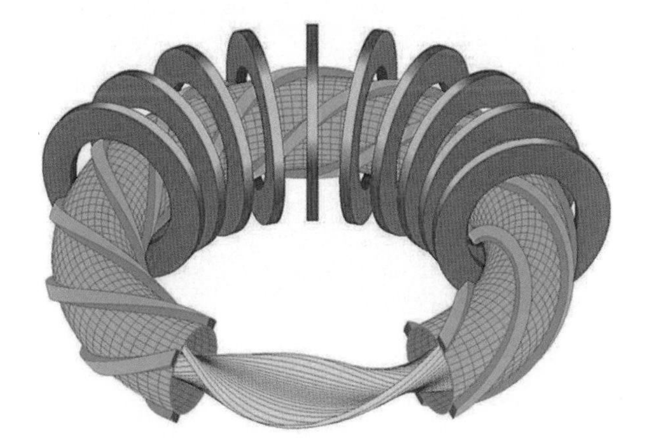

Fig. 10.2 Schematic of a stellarator with separate toroidal-field coils and helical windings [8]

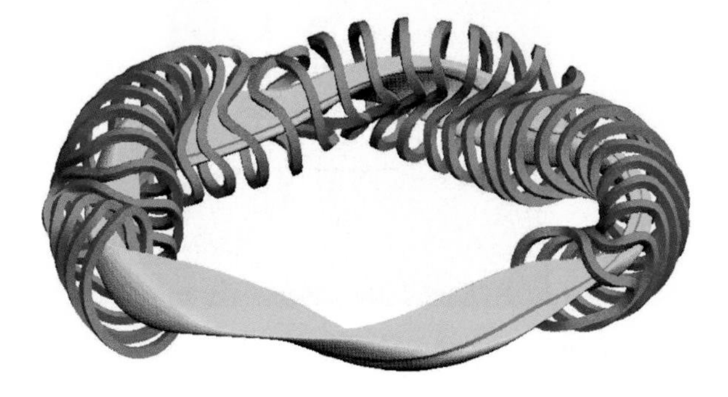

Fig. 10.3 Magnet coil structure of Wendelstein 7-X [10]

to assemble, since the poloidal-field coils do not have to be threaded through the toroidal-field coils. To conserve magnetic-field volume (which is costly), the coils are also shaped to conform to the shape of the plasma.

Design of the coils can be done with computers, but these unusual coils have actually been constructed in special jigs. The coils are of superconducting NbTi cooled with liquid helium. Not all the coils are different, of course. There will be seven different types, 10 of each, and 70 coils overall. Figure 10.4 shows two of these being lifted. The vacuum chamber will also conform to the plasma shape; a section of it is shown in Fig. 10.5. This *will* link the coils together and have to be assembled with them.

Wendelstein 7-X is a very large and complicated machine to be finished in 2014. Plans are to reach 30-min pulses with 40 MW of heating power, reaching conditions approaching those in ITER but only with aneutronic fuels.

Fig. 10.4 Two Wendelstein 7-X coils [8]

Fig. 10.5 A section of the Wendelstein 7-X vacuum chamber [8]

Large Helical Device

The first noncircular stellarator may have been the Heliac in Canberra, Australia, but the granddaddy of them all is the LHD in Toki, Japan, shown in Fig. 10.6. Envisioned by Koji Uo while on sabbatical at Princeton in the 1960s and completed in 1998, this machine showed that large superconducting coils producing 30-T magnetic fields could be manufactured and operated reliably for years. The most amazing accomplishment, however, was the demonstration that the weird, twisting vacuum chamber and similarly complicated magnet coils could actually be manufactured to the required tolerances. Figure 10.7 shows an artistic photograph of the vacuum chamber.

Fig. 10.6 Overall view of the large helical device [10]

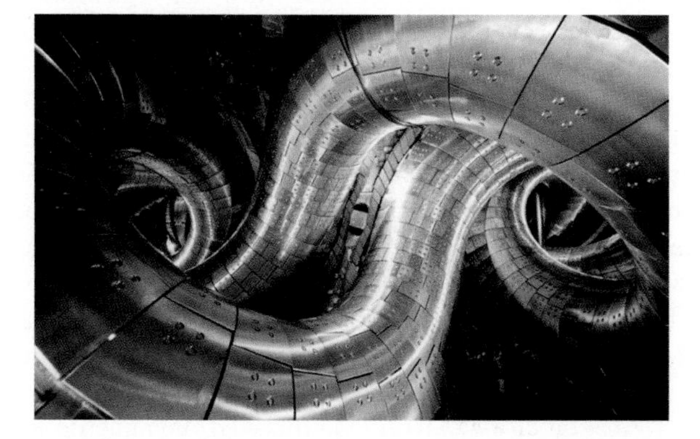

Fig. 10.7 The vacuum chamber of the LHD (www.nifs.ac.jp/en/introduction_e.html.)

In operation, the LHD has outperformed tokamaks in several aspects. The plasma density has reached 10^{21} m^{-3} (10^{15} cm^{-3}), which is many times larger than the Greenwald limit (Chap. 8). This shows that this unexplained, empirical limit may apply only to tokamaks and can be exceeded in stellarators. The maximum ion and electron temperatures achieved were 13.5 and 10 keV, respectively, though not at the same time. Nonetheless, T_i exceeds T_e in normal operation, as is desirable since it is T_i that causes fusion. Beta, the ratio of plasma pressure to magnetic-field pressure (Chap. 8), is an important measure of the quality of a fusion plasma. The beta value of 5% achieved in LHD is higher than is normal for tokamaks. Not all these record-breaking numbers can be obtained at the same time, of course. What counts is the triple product $Tn\tau$, the simultaneous ion temperature, density, and confinement time plotted in Fig. 8.1. On that scale, the LHD would be at 0.44, at about the middle of the plot. With fueling by pellet injection, discharges an hour long can be produced in LHD when the power is lowered so that $Tn\tau$ is at 80% of its maximum value.

Benefits of Nonaxisymmetry

Tokamak plasmas are basically symmetric around the major axis. They may have D-shaped rather than circular cross sections, but they still look the same from any direction. The figures here show that stellarators are far from symmetric. Instead of using the plasma current to shape the plasma, external coils are used, and these can produce shapes that cannot be formed by plasma current alone in a self-organized tokamak. It is precisely the *lack* of self-organization that gives stellarators their advantage [11]. Nonaxisymmetric shaping can be used to improve plasma stability, control ELMs, and eliminate disruptions. Indeed, the ELM coils being added to ITER to suppress ELM instabilities do so by spoiling the axisymmetry. In DEMO, the bootstrap current is relied on to supply at least 80% of the plasma current. This is extremely difficult to produce and control when self-organization is strong. In stellarators, a large plasma current is not at all necessary, since the rotational transform is generated by external coils.

In addition to their suitability for steady-state operation, stellarators have some unexpected advantages as reactors. Very small errors in the magnetic field (0.01%) have been found to cause problems with plasma confinement. Originally, stellarators' problems were believed to be due to magnetic errors, but it has been found that once axisymmetry is broken, the wild shapes shown above are actually less sensitive to magnetic errors. Data from all stellarators have been found to follow a scaling law and fall on the same curve, as shown in Fig. 8.21 for tokamaks, so that extrapolation can be used to design larger machines. In addition, higher density and beta values have been achieved in stellarators. A purported benefit [11] of higher density is the formation of a MARFE (Multifaceted Asymmetric Radiation From the Edge), a "detached" layer which forms when plasma recombines before reaching the divertor. The energy is then radiated away before it reaches the divertor, sparing the divertor of the large heat load. The energy, however, has then to be taken up by the first wall. The advantages of stellarators come at a price: the difficulty of making and assembling the weirdly shaped coils and vacuum chambers; but this technology has already been demonstrated.

Compact Stellarators

Stellarators like the Wendelsteins are large machines with large aspect ratios R/a, where R is the major radius of the ring and a is the radius of the cross section. There is a movement to build smaller, more economical machines by shrinking R to get aspect ratios of 3–5 instead of 10 or more. Proposed compact stellarators have been designed with different magnetic-field configurations to see which would work better. This freedom of design is not available for tokamaks, but it also means that it is harder to converge on the optimal design. The National Compact Stellarator Experiment (NCSX) was funded and under construction at the Princeton Plasma Physics Laboratory, but the project was canceled during the 2009 worldwide economic depression. Figure 10.8 shows the NCSX and its coil structure. There are only 18 coils of three different shapes. Although this machine was well designed and would have complemented the Wendelstein 7-X nicely, its discontinuation was reasonable. Tokamaks are far ahead in development, and to get a fusion reactor working the fastest way is to give them the highest priority.

Stellarators are second-generation confinement devices. They are probably better suited for reactors than tokamaks, but we need much more experience with how they run. An obvious question is: Where do you put the blankets in a DT stellarator

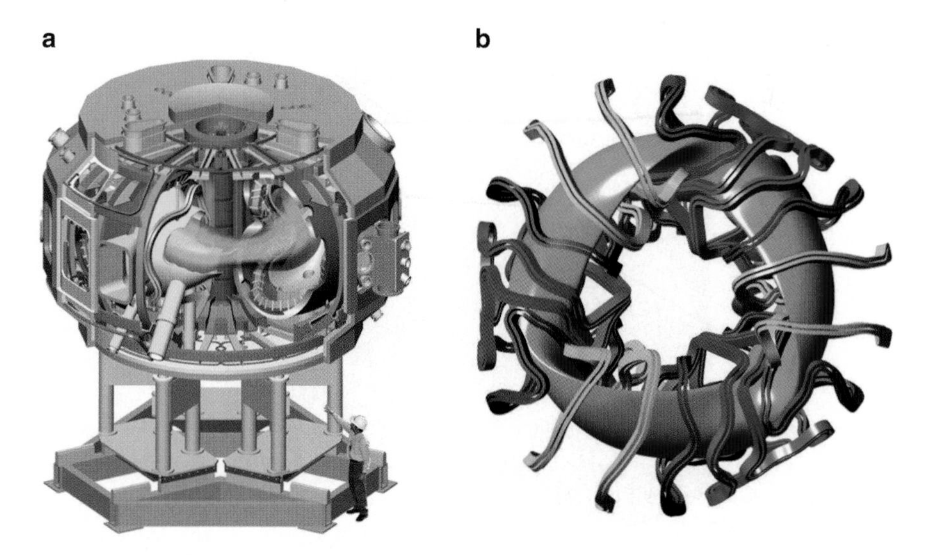

Fig. 10.8 (**a**) Diagram of the National Compact Stellarator Experiment and (**b**) its coil structure [12]

reactor? The problem is that the magnet coils are not circular but have small twists and bends. The coils have to be close to the plasma for these fine features to be felt; too far, and the details will be smeared out. That's why the vacuum chamber has to be shaped to fit the coils. In a reactor, one still has to leave room for the tritium-breeding blanket, and the only way to do this is to scale the whole machine larger. There have been several reactor studies from Germany, Japan, and the USA. The ARIES-CS design is shown in Fig. 10.9 and the overall view in Fig. 10.10. It was found possible to place the blanket modules between the plasma and the vacuum wall and superconducting coils.

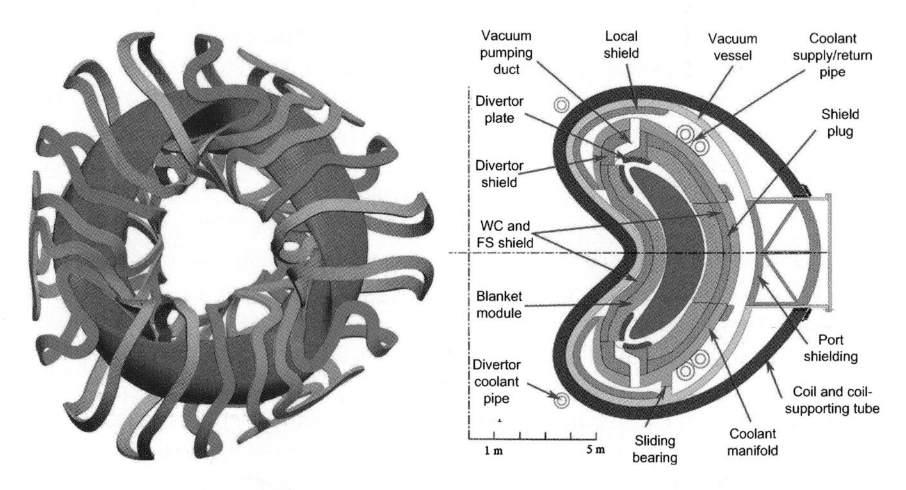

Fig. 10.9 Design of the ARIES-CS compact stellarator reactor [13]

Fig. 10.10 Overall of the ARIES-CS reactor conceptual design [13]

Spherical Toruses

Spherical Tokamaks

In Chap. 3, we carefully showed that a magnetic bottle has to be doubly connected and not a sphere; hence tokamaks are toruses.[1] How, then, can a tokamak be spherical? No, *spherical tokamak* is not an oxymoron. A tokamak can be spherical as long as there is still a hole in the middle. This is shown in Fig. 10.11. These small, fat tokamaks have typical aspect ratios A between 1 and 2. There are many advantages to having small A, but the problem is how to fit all the necessary equipment into the small hole. Spherical tokamaks (STs) are so attractive that many clever ideas have been proposed for treating the small hole, and there are over two dozen STs all over the world testing these ideas.[2] In fact, one can eliminate the hole in the vacuum chamber altogether as long as the magnetic field is still toroidal.

Aside from the small size and the consequent cost savings, STs have a large advantage in plasma stability. This is explained in Fig. 10.12, which shows the magnetic-field structure in an ST. The field lines behave very differently from those in a normal tokamak (Fig. 6.1). A particle following a field line spirals around the central column before returning to the outside of the plasma. Good and bad curvatures are shown in Fig. 7.10. In good curvature, the bend is toward the plasma, and in bad curvature, it is away from the plasma. We see that there is a lot of good curvature around the central column, and a region of weaker bad curvature when the field line returns to the top. Since particles spend more time in good curvature than in bad, there are strong forces pushing the plasma inwards. Much smaller magnetic fields are needed in STs because of the good confinement.

In a 1986 paper [16], Martin Peng and D.J. Strickler noted that the vertical field needed in tokamaks (Fig. 6.19) had a natural tendency to elongate the plasma, and they laid out the basics for the design of STs. Elongation is the vertical length of the plasma compared with its minor diameter, and it has good consequences for STs. As the aspect ratio goes down from 2.5 to 1.2, the elongation increases from 1.1 to 2, and the magnetic field that gives the needed quality factor q for a given

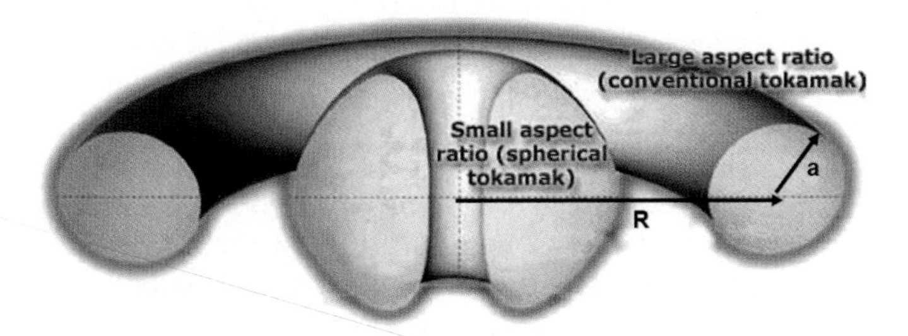

Fig. 10.11 A spherical tokamak has an aspect ratio much smaller than a normal tokamak [15]

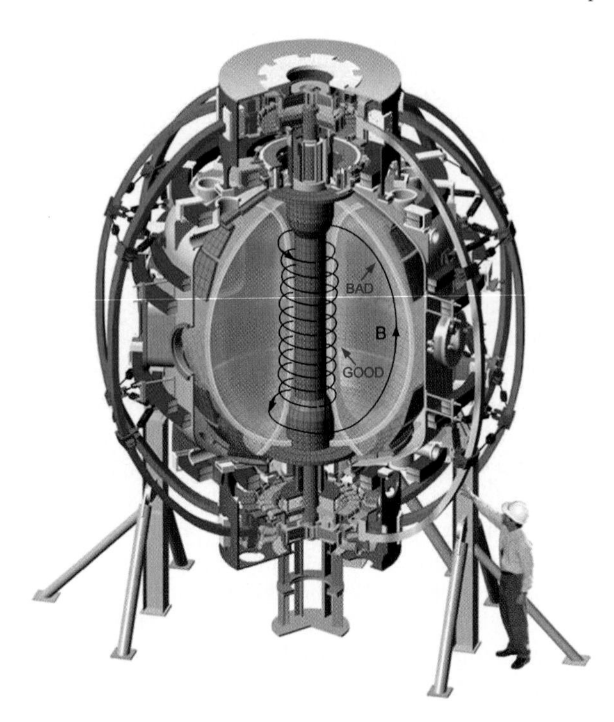

Fig. 10.12 Sketch of one magnetic field line in a spherical tokamak with a current-carrying central column. The regions of good and bad curvature are marked (Adapted from S. Prager (University of Wisconsin), *Magnetic Confinement Fusion Science Status and Challenges*, February 2005)

plasma current falls by a factor of 20! [15] The value of beta (ratio of plasma pressure to magnetic-field pressure) is therefore very high in STs.

The British machines START (Small Tight Aspect Ratio Tokamak) and its successor MAST (MegAmpere Spherical Tokamak) have given the most information on STs. A photograph of the spherical plasma in START is shown in Fig. 10.13. The graph of beta values obtained in START (Fig. 10.14) shows the great improvement over normal tokamaks. In that graph, β_T is the toroidal beta (that calculated with the toroidal magnetic field), and β_N is the normalized beta, as defined in Chap. 8 under Troyon Limit. The recent data (red dots) show that the density limit can be exceeded in a spherical torus.

In spite of their physical appearance, STs exhibit the same phenomena observed in large-A tokamaks; the H-mode and ELMs, for instance. MAST is suitable for studies of ELMs and was used for the design of ELM-suppression coils. The shape of the field lines also gives STs a natural divertor.

Now we tackle the question of how to minimize the width of the central column. The toroidal *magnetic field* in a tokamak is generated by coils that thread through the hole, as shown in Fig. 6.1. All the coil legs that go through the hole can be combined into a single copper bar carrying all the current, as shown in Fig. 10.13.

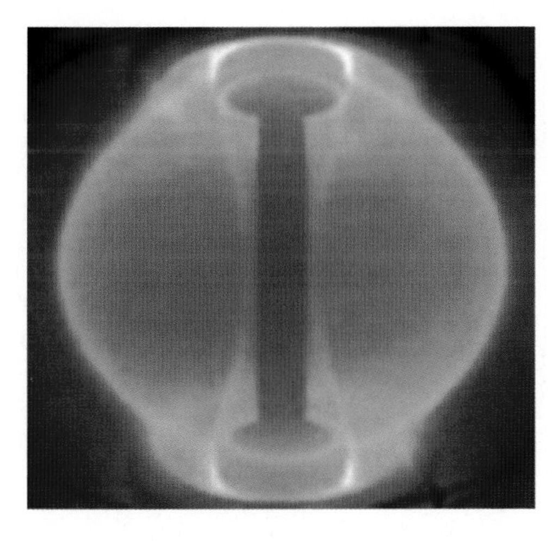

Fig. 10.13 The spherical plasma in START [15]

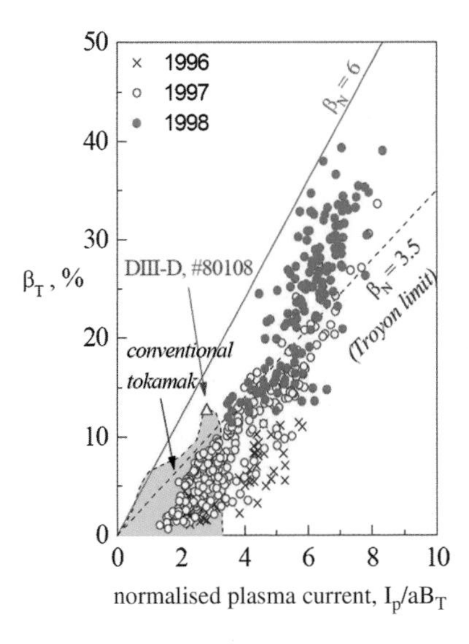

Fig. 10.14 Plot of toroidal beta (β_T) in START and normal tokamaks [15]

This is possible because the B-field is small in an ST, so the coil currents are reduced. To drive the toroidal *plasma current*, the brute force way is to put an iron core through the hole and drive the current by transformer action, as shown in Fig. 7.14. Most tokamaks use air-core transformers that have no iron. These consist of toroidal coils around the plasma, including some inside the hole. This is shown in Fig. 7.15. These methods are called *inductive drive*. The disadvantage is that the current has to be increasing to excite the current; and since it cannot increase forever, the tokamak has to be pulsed. Modern tokamaks use *noninductive* drive,

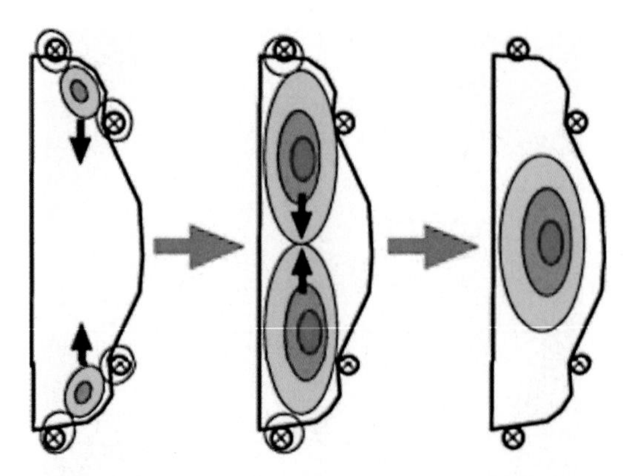

Fig. 10.15 Creation of a toroidal plasma in a spherical tokamak with no central column by the merging of two plasmas [15]

which consists of bootstrap current and wave-driven currents (Chap. 9). This would eliminate the need for toroidal coils inside the hole.

The problem is that you can't launch a wave unless there's a plasma, and you can't confine a plasma unless there is already a rotational transform. So it seems that at least some small toroidal coils have to be crammed into the hole, but there may be a solution. Neutral-beam injection is the usual way to heat a large tokamak. Currently, there has been some success (in MAST [15]) in ramping up the NBI in such a way that it drives a current also. It is also possible to create plasmas in corners of the chamber where poloidal coils can be inserted, and to have these plasmas drift and merge into the center. This is illustrated in Fig. 10.15.

While experimentation on STs is being conducted intensely worldwide, reactor studies have been made both in Europe and the USA. The ARIES-ST design of 1999 is shown in Fig. 10.16. The central column is made to be slid out and replaced easily. All blanket modules are on the outside. Note the natural divertors at the top and bottom.

Spheromaks

A spheromak (Fig. 10.17) is a toroidal plasma in a chamber with no hole in the middle. There can be toroidal coils to generate a poloidal B-field, but there cannot be any coils to generate a toroidal B-field since that would require a conductor going down the middle. Plasma with imbedded fields is injected into the chamber from external sources, and then the plasma self-organizes into a toroidal shape with both toroidal and poloidal fields (Fig. 10.18). Contrary to stellarators, which do away with the self-organization of tokamaks, spheromaks depend *completely* on self-organization. The classic method of injection with "plasma guns" is shown in Fig. 10.19.

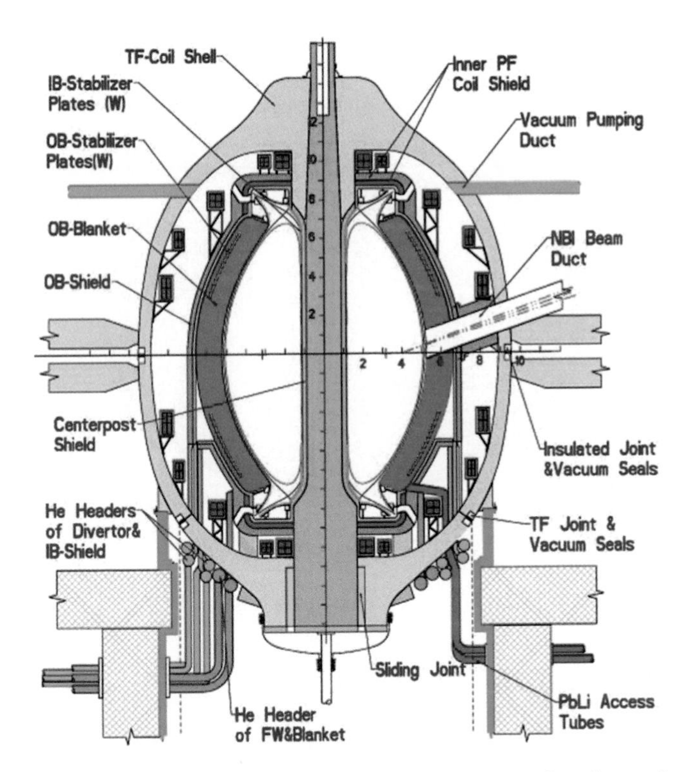

Fig. 10.16 Design of the ARIES-ST spherical tokamak reactor (http://www-ferp.ucsd.edu/ARIES/Docs/ARIES-ST/.)

Fig. 10.17 Artist rendition of a spheromak [18]

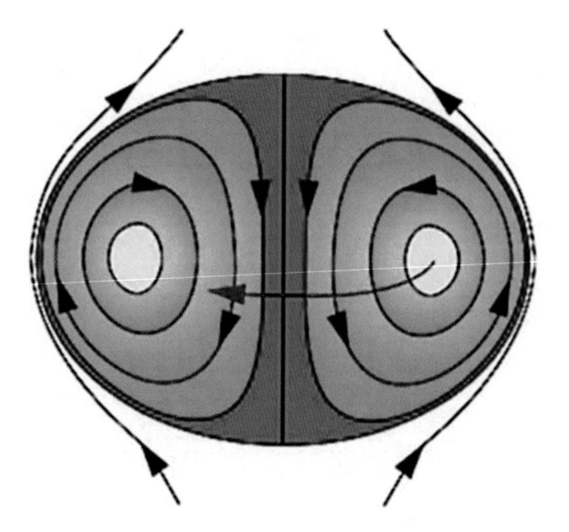

Fig. 10.18 Toroidal and poloidal magnetic fields in a spheromak [19]

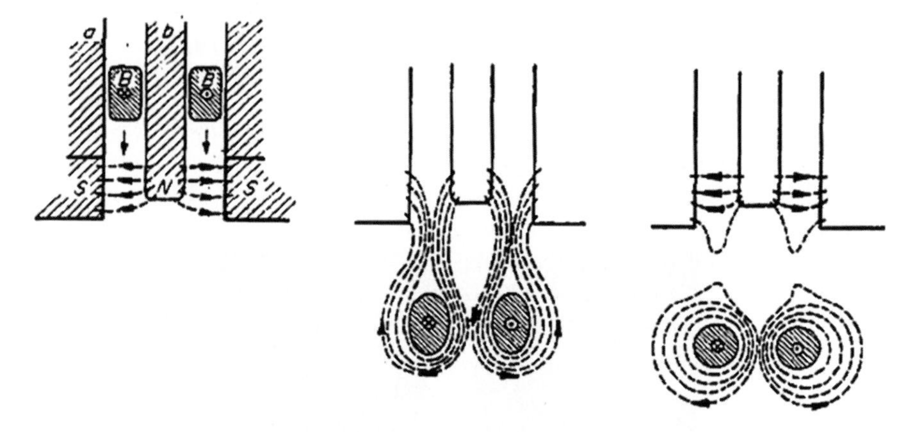

Fig. 10.19 Injection of a toroidal plasma into a spheromak with a plasma gun [20]

The stable configuration that results from a period of instability and rearrangement of fields and currents is predicted by the classic theory of J.B. Taylor [9]. The main point is that the force exerted by a current in a B-field is perpendicular to **B**. These forces will move the plasma around until there are no more forces. That happens when the current **J** is parallel to **B** everywhere, so that there is no perpendicular force. Then *the lines of* **B** *are the same as the lines of* **J**. The B-field created by each element of **J** is just that which the neighboring elements of **J** follow. This means that the field is purely poloidal on the outside and purely toroidal on the minor axis, but the fields do not have to be neatly arranged as in Fig. 10.18. There can be a jumble of field lines that satisfy the minimum-force condition. The plasma will organize itself. Also needed is a conducting shell that keeps the whole plasma from expanding. When it tries to expand, image currents in the shell will push the plasma back.

These force-free configurations are interesting to physicists because they occur in many places, including outer space. However, spheromaks are unlikely candidates for fusion reactors. So far the confinement times have been short, and the plasmas have to be pulsed. Experiments have been aimed mostly at the problem of magnetic reconnection, which is important in space physics.

Magnetic Mirrors

How Mirrors Work

Mirror machines, together with stellarators, were the strongest proposals for plasma confinement devices when fusion research started in the early 1950s. The effort was led by R.F. (Dick) Post, who is still actively pursuing the mirror concept today. Unfortunately, the mirror program at Livermore was canceled in 1986 by the USA in order to concentrate on the tokamak. Research on mirror confinement has continued in Russia, under the guidance of Dmitri Ryutov, and in Japan, with the Gamma 10 machine. Reactors using the mirror principle would have the great advantage of direct conversion of energy to electricity without a thermal cycle, the same advantage that hydro, wind, and solar power have over other power sources.

A mirror machine is a leaky magnetic bottle. In Fig. 10.20, a pair of coils generates a magnetic field that bulges out between them. An ion or electron will gyrate in a Larmor orbit around the field lines, as shown in Fig. 4.10. As the orbit approaches the strong-field region at the ends, the Larmor orbit becomes smaller and smaller. To conserve angular momentum, the particle has to gyrate faster and faster, just as a skater does when she pulls her arms in during a spin. But energy is conserved, and to get this extra rotational energy, the particle has to take it out of its translational motion. It slows down in its efforts to escape out the end. Finally, all its translational energy is lost, and the particle has to turn around and go back.

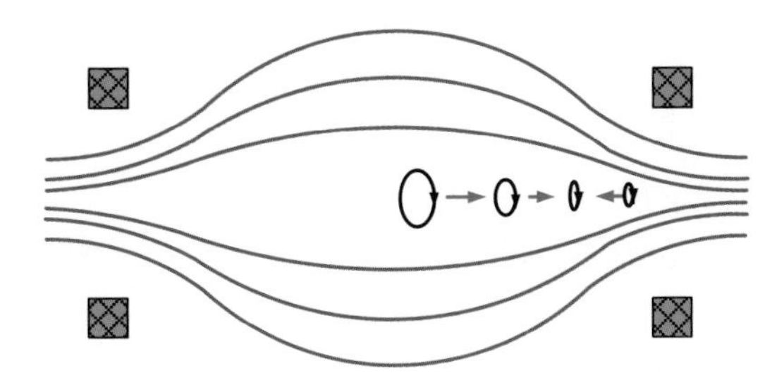

Fig. 10.20 Illustration of magnetic mirroring

It is reflected by the magnetic mirror, which is just the strong-field region. The particle bounces back and forth between the mirrors at each end.

When a plasma is created, ions (and electrons) have both translational energy and gyrational energy. The ones with lots of translational energy and little gyrational energy are lost out the ends. If the mirrors are strong, meaning that the *mirror ratio* between the fields at the throat and at the midplane is large, only a few particles are lost; and the rest of the plasma is confined. However, this plasma is not in thermal equilibrium; some of the velocities are missing. These velocities are in the *loss cone*. The plasma will then devise an instability to regenerate those missing velocities and fill the loss cone. There is then a continuous loss, giving magnetic mirrors a short confinement time. Such microinstabilities, however, are not the main problem in mirrors.

Ioffe Bars and Baseball Coils

Note that a simple mirror has unfavorable curvature (Fig. 7.10) and is unstable to the basic Rayleigh–Taylor interchange instability (Fig. 5.5). This problem was solved by M.S. Ioffe [21] by adding what are now known as Ioffe bars, shown in Fig. 10.21. These are four conductors parallel to the axis adding a poloidal field to the mirror field. The plasma is squeezed into a peppermint-candy shape. The strength of the magnetic field now increases outward in every perpendicular direction, so it is energetically impossible for a Rayleigh–Taylor instability to develop and push the plasma out. This is called a *minimum-B* configuration, since the plasma sits in a minimum of the B-field. Of course, the plasma can still leak out the ends.

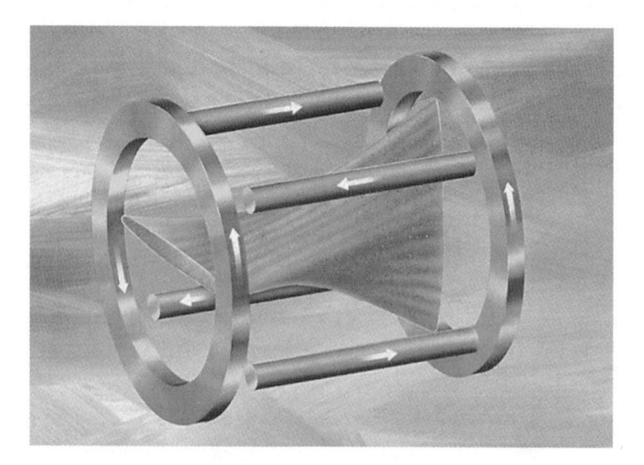

Fig. 10.21 A magnetic mirror with Ioffe bars (An old diagram or picture originally from Lawrence Livermore National Laboratory.)

Now imagine how to combine the Ioffe bars with the circular coils into a single coil. This proceeded in two steps. First, one can combine them into two identical coils, called *yin–yang* coils, shown in Fig. 10.22. This was such an attractive shape that an artist made a sculpture of it (Fig. 10.23). Finally, all the necessary currents can be combined into a single coil called a *baseball coil* because it resembles the seam on a baseball. This is shown in Fig. 10.24.

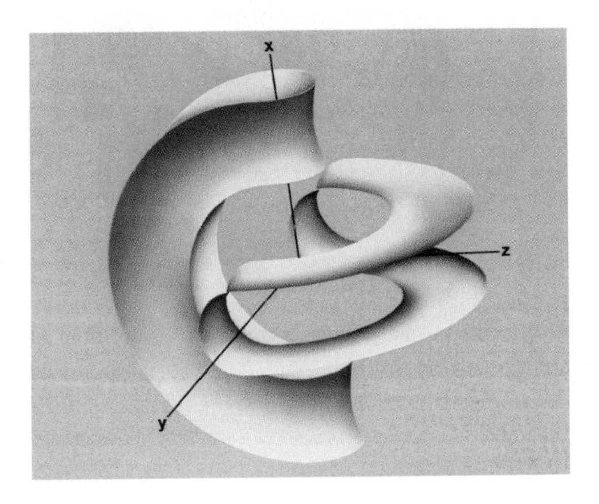

Fig. 10.22 Yin–yang coils (An old diagram or picture originally from Lawrence Livermore National Laboratory.)

Fig. 10.23 A yin–yang coil sculpture (Photo by the author at the 1977 meeting of the Plasma Physics Division of the American Physical Society in Atlanta, GA.)

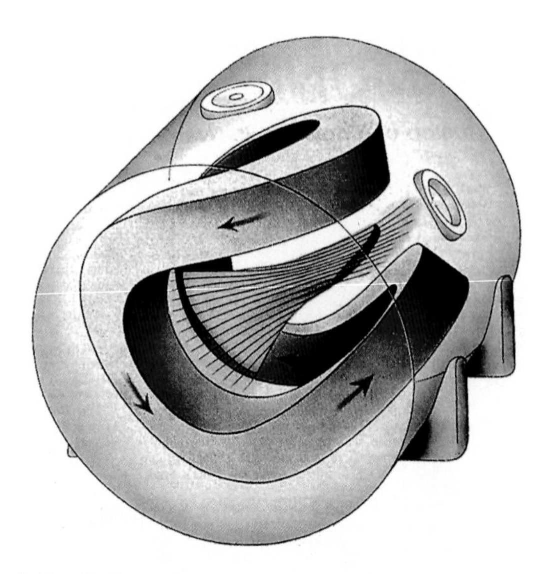

Fig. 10.24 A baseball coil (An old diagram or picture originally from Lawrence Livermore National Laboratory.)

Mirror Machines

Although these coils provide good stability, they do not enclose a large volume of plasma. They can, however, be used to stabilize a large volume of plasma attached to them. A series of large machines called tandem mirrors was built at Livermore with a long region of uniform B-field, which has neutral stability and is stabilized at the ends with yin–yang or baseball coils. One of these, the TMX, is shown in Fig. 10.25. The end coils of these machines became more and more complex as each difficulty was overcome. Intense heating produced enough density in the baseball coil to stabilize the main plasma in the weaker central region. Thermal barriers used electrostatic potentials to keep the plasma hot in the baseball coils. Sloshing ions were used to shape these potentials. Circularizer coils matched the flattened plasma in the baseball coil to a round one on either side. Anchor coils with higher field were the final plugs at the ends.

The successor to TMX was to be the MFTF-B installation, whose size can be appreciated in Fig. 10.26, which shows one of the yin–yang magnets being moved using the old Roman method of rolling logs. In spite of an earthquake occurring while the coil was being lifted into place, the machine was finished just in time for the entire mirror project to be canceled, much to the dismay of its leader, Keith Thomassen, and, of course, Dick Post.

The 27-m long Gamma 10 machine at Tsukuba, Japan, however, continued to operate and has improved confinement by increasing the potential barrier confining the ions [24]. Instabilities have also been eliminated by producing electric field shear [25]. Results from tandem mirrors, however, pale compared with those from toruses. Densities peak at 4×10^{18} m^{-3} (4×10^{12} cm^{-3}), ion temperatures at a keV or

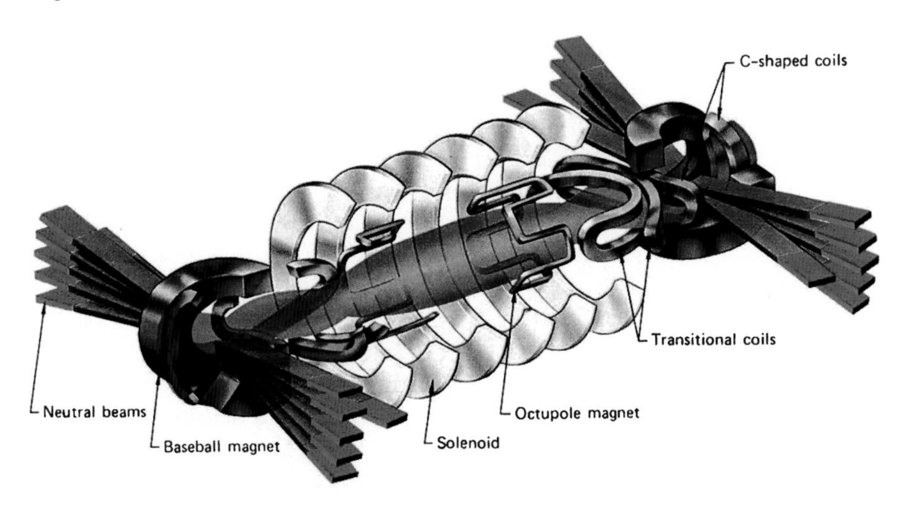

Fig. 10.25 Diagram of the tandem mirror experiment [23]. The flat bars represent neutral beams heating the plasma in the stabilizing coils

Fig. 10.26 Moving the MFTF-B yin–yang magnet (An old diagram or picture originally from Lawrence Livermore National Laboratory.)

two, electron temperatures around 250 eV, and energy confinement times of order 10 ms. In addition, control of electric potentials sometimes requires plasma contact with conducting walls. Though the present state of the art on magnetic mirrors does not suggest their reactor relevance, they may be useful for other tasks that do not require net energy output. These include creating plasmas for transmutation of nuclear wastes or energy production in fission–fusion hybrids. First and foremost, however, is the proposed use of mirror machines as economical neutron sources for materials testing, as described in Chap. 9. Such a machine burning D–T fuel would produce 2 MW/m^2 fluxes of 14-MeV neutrons over a sizable area using only 200 g of tritium per year [26].

Axisymmetric Mirrors

During the mirror hiatus of the last two decades, however, new ideas have emerged that revive the possibility of mirror reactors. The yin–yang and other end coils of tandem mirrors have large magnetic stresses because of their twisted shapes. The new idea is to make mirror machines completely axisymmetric, using only simple circular coils. The feasibility of this was proved by Gas Dynamic Trap experiments in Novosibirsk, Russia [14]. The mirror field can be made extremely strong, creating a large mirror ratio (as large as 2,500), thus reducing the size of the loss cone. A schematic of this is shown in Fig. 10.27. It looks like a large plasma with a pinhole leak at each end, but the pinhole is not in real space but in velocity space.

The Gas Dynamic Trap produced 10-keV ions with a peak density of 4×10^{19} m^{-3} (4×10^{13} cm^{-3}) and electron temperatures of 200 eV. The beta value was 60%, compared with only a few percent in tokamaks, since only a weak central field is needed to contain large-orbit ions with mainly perpendicular energy.[3] In mirrors, neutral beams are used to inject ions, and no energy is wasted in heating the electrons. The machine is pulsed for only 5 ms, and the confinement time is only a millisecond or so. Electric fields are produced by applying voltages to different parts of the walls where the field lines end.

In an axisymmetric tandem mirror, the complexity of the stabilizing coils is gone; the circular coils are easy to make. How, then, can the plasma be stable? It turns out that the outside plasma *beyond* the mirrors can play an essential role. There the field lines have *favorable* curvature, bulging inwards toward the plasma. The stability there can overcome the instability driven by the bad curvature at the ends of the central region. It turns out that the density in the outside region does not have to be very high for this to happen as long as the plasma diameter there is large. One can shape the outside field with large coils, of which one is shown, so as to optimize the stabilizing effect [27]. A "kinetically stabilized tandem mirror" machine has been proposed [26, 27] to test this principle. That machine, shown in Fig. 10.28, uses multiple axisymmetric mirrors and injection of ions into the diverging region to improve stability.

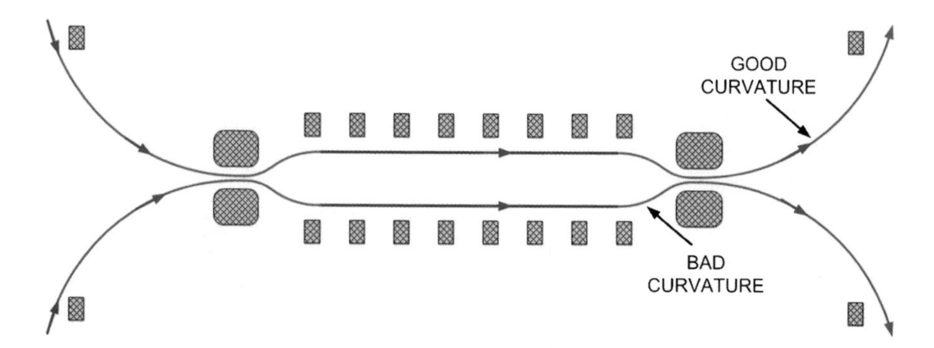

Fig. 10.27 Magnet system of a totally axisymmetric mirror machine

Direct Conversion

If these theoretical ideas prove to be feasible, the escaping plasma can be used to generate electricity directly, as shown in Fig. 10.29. The ions pass through electrodes, inducing electric currents in them. Note that this exhaust is a natural divertor spreading the heat over a large area. The product alpha particles are well contained in the main plasma and keep it hot; their release rate through the mirror can be controlled by design.

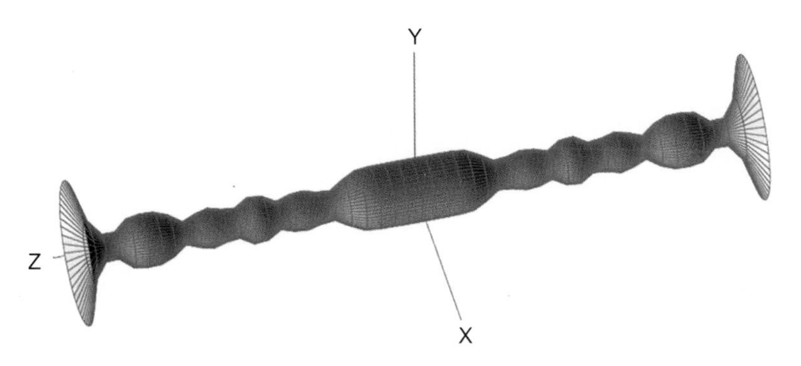

Fig. 10.28 Proposed kinetically stabilized tandem mirror machine [27]

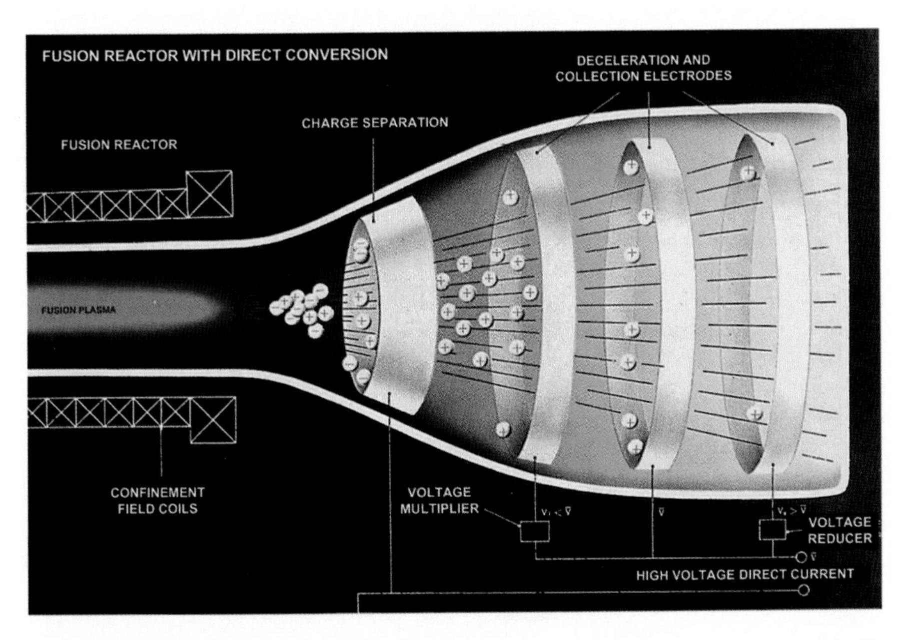

Fig. 10.29 Cartoon of a magnetic mirror direct converter (An old diagram or picture originally from Lawrence Livermore National Laboratory.)

Magnetic Pinches

Reversed-Field Pinch

A pinch is a plasma carrying a current so large that the surrounding magnetic field that it generates confines and compresses it. It is basically unstable to the kink instability (Fig. 6.2). A toroidal pinch has the current running around in a torus, so that it is also subject to the gravitational interchange instability (Fig. 5.7). A reversed-field pinch (RFP) adds a toroidal field imposed by external coils, as in a tokamak, and has special properties. The Zeta machine at Harwell, England, one of the first fusion experiments revealed to the world at the 1958 Geneva Atoms-for-Peace Conference, was an RFP (*cf.* Chap. 8). That machine suffered from a misinterpretation of the neutrons it generated and was abandoned, but research on RFPs has continued since that time.

The Zeta experiment showed that, after an initial period, the plasma settled into a quiescent, stable state. This was explained by Bryan Taylor's theory [9], which predicted that the plasma would self-organize into a minimum-force, maximum inductance state. In an RFP, this state has a current distribution that reverses the direction of the helical field lines, as shown in Fig. 10.30. This looks like the tokamak field of Fig. 5.9, but notice that the outermost field lines are going backwards in the toroidal direction compared to those near the center. Hence the name reversed-field pinch.

In spite of the seeming quiescence of the Taylor state, the RFP suffers from magnetic fluctuations caused by the basic hydromagnetic instabilities, especially the tearing mode (Chap. 6). A conducting shell close to the plasma is needed to

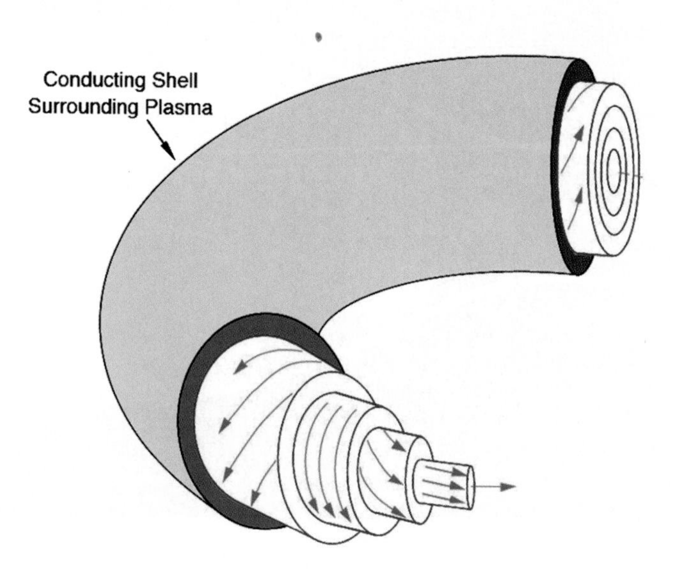

Fig. 10.30 Field lines on the magnetic surfaces of an RFP [30]

control the resistive wall mode, and active feedback stabilization is needed also. If it can be made to work, the RFP has the great advantage of self-generated magnetic field, requiring the addition of only a small toroidal field from external coils. These coils need not be superconducting, since they consume little power. The relatively weak magnetic field means that very high beta values can be achieved. However, for a reactor, the conducting shell makes the design of the blanket and first wall problematical. The bootstrap current is small, so the large toroidal current has to be driven inductively with a transformer. That means that the plasma has to be pulsed. There is some evidence that a DC current can be created with a oscillating drive [30], but this is at a primitive stage.

In spite of doubts about its reactor relevance, considerable progress has been made in understanding the physics of RFPs. This research is also of interest to space scientists, since processes like reconnection also occur in space. New results come mainly from the RFX machine in Padua, Italy [31], and the MST in the University of Wisconsin [30]. At low power, the RFP does not self-organize sufficiently, and magnetic surfaces of many helicities are all tangled up. When the dominant mode exceeds 4% at a current of 1.5 MA, however, the plasma snaps into a single helix, whose cross section is shown in Fig. 10.31. The plasma moves off-center into a helical shape, the magnetic surfaces are no longer jumbled, and confinement is much improved. The electron temperature is seen to increase a factor of 2 to about 850 eV.

In the MST, magnetic chaos is suppressed actively by shaping the current profile with pulsed poloidal current drive. Figure 10.32 is a computer simulation of the magnetic surfaces before and after the current drive is imposed. Plasma confinement time is increased ten times by this, up to 12 ms, which is comparable to that in small tokamaks. Electron temperature reaches 2 keV, and ion temperature 1.3 keV. Beta values of order 26% have been achieved. The plasma density surpasses the Greenwald limit (Chap. 8) by 20% [30]. In spite of the weak magnetic

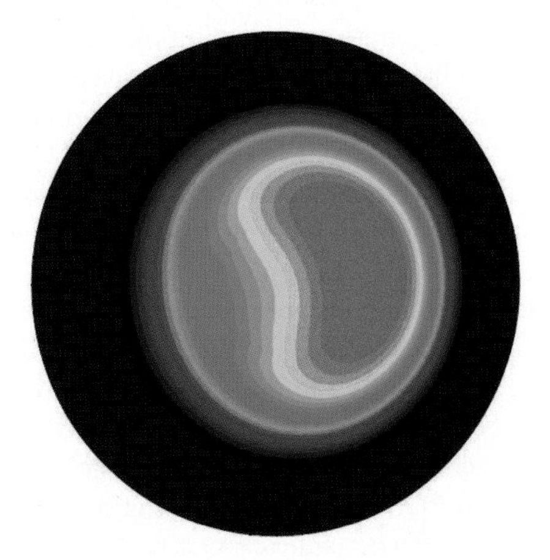

Fig. 10.31 Temperature distribution in an RFP in the single-helicity mode [31]

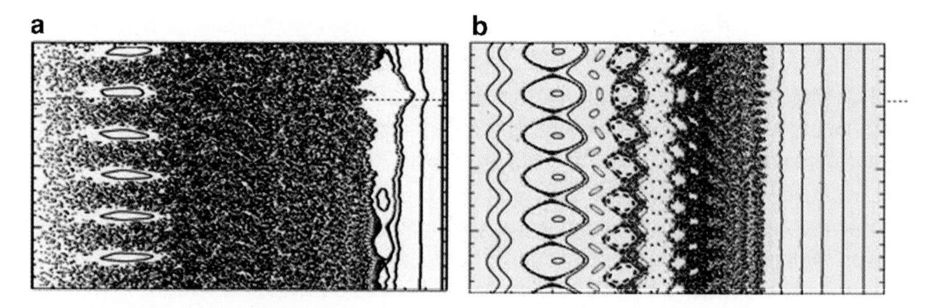

Fig. 10.32 Simulations of magnetic field lines in an RFP cross section (**a**) before and (**b**) after the application of pulsed poloidal current drive [30]

field, energetic ions have been found to be well contained. This was found by injecting 20 keV neutral beams, which turn into 20 keV deuterons. However, ions in RFPs are not heated by neutral beams because they are naturally heated by reconnection. This is a process in which magnetic-field lines merge, destroying some B-field and converting its magnetic energy into plasma energy. This phenomenon also occurs in the earth's magnetic field, so that RFP research, as well as spheromak research, has relevance to other fields of science.

Field-Reversed Configuration (FRC)

This interesting device is not really a pinch; it has characteristics of spheromaks, pinches, inertial confinement, and even mirrors. A simple diagram is shown in Fig. 10.33. If you rotate the diagram 90°, it looks like a spheromak (Fig. 10.18), but it has one essential difference. *There is no toroidal magnetic field.* The toroidal direction is indicated by the ellipses for the current and an ion Larmor orbit. Toroidal coils on the outside create a magnetic field (B-field) going from right to left in the diagram. A toroidal current driven in the plasma creates a B-field opposite to the external field. The current is in the same direction as the electron diamagnetic current (Chap. 6), which adds to it. When the current is large enough to cancel the external field, there is a radius R at which the B-field is zero. This is the center of the tubular plasma. It is confined by a purely poloidal B-field. Inside of R, the B-field is opposite to that which was applied. Outside of R, the B-fields from the internal current and the external coils are squeezed up to the vacuum wall, which, being conducting, is a *flux conserver*. The field lines are divided into two types divided by a *separatrix*, shown by the dashed line, which represents a field line which leads to a $B=0$ point on the axis. The field there has to be zero because it cannot point in two directions at the same time. Plasma inside the separatrix is confined in closed magnetic surfaces; those diffusing outside the separatrix are lost out the ends of the machine. There is therefore a natural divertor; and mirror coils, of which one is shown, can be designed to treat the escaping plasma the same way as in a mirror machine (Fig. 10.27), including the possibility of direct conversion.

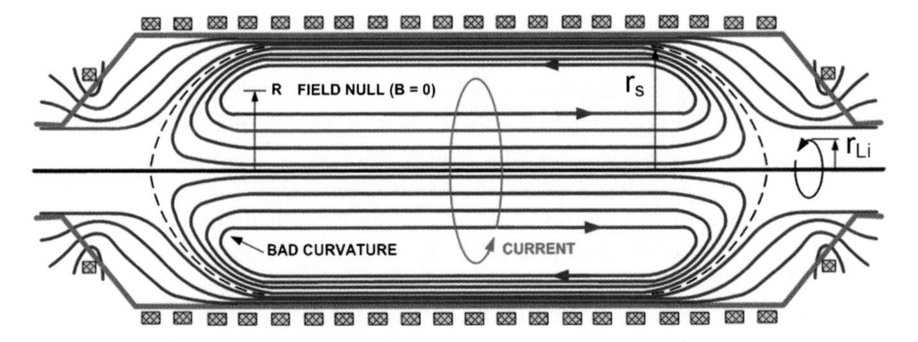

Fig. 10.33 Schematic of an FRC showing the *poloidal field lines* and the toroidal current that shapes them. The *dashed line* is the separatrix, with maximum radius r_s. An ion orbit is shown to define the Larmor radius r_{Li}. R is the major radius of the center of the plasma, located at the field null. The thick *gray line* represents the vacuum wall and flux conserver. The regions of bad (convex) curvature are shown

Although the field lines on which the plasma lies are closed, this should be better than a mirror machine, but the FRC configuration is highly unstable. There are no helical field lines to link regions of good and bad curvatures, as there are in a spherical tokamak (Fig. 10.12). In fact, there is no good curvature anywhere. The curvature is especially bad at the ends of the machine, as shown in Fig. 10.33. How can an FRC plasma be stable against the main hydromagnetic instabilities? The FRC depends on finite-Larmor-radius effects (Chap. 6). The Larmor radius r_{Li} of the ions at the bad-curvature regions is not negligible compared with the size of, say, the Rayleigh–Taylor instability (Chap. 5). That means that ions can travel across field lines far enough to short-circuit the voltages that the instability generates, keeping it from growing. Electrons, with much smaller Larmor radii, cannot do that; they are tied tightly to the field lines.

How large does r_{Li} have to be? It has to be a sizable fraction of the plasma width, which can be measured by the distance between the center of the plasma and the last closed surface at r_s. This is $R-r_s$. The number of Larmor radii in that width is then $s=(R-r_s)/r_{Li}$. The parameter s has to be *small* to keep the plasma stable. In early FRC experiments, s was only 2 or less. However, plasma diffuses at a rate proportional to r_{Li} via electron–ion collisions (Chap. 6), even if it is stable. So s has to be *large* to get long confinement times, and there is always this struggle to get stability at as large an s as possible.

If instabilities can be controlled, FRCs could have advantages as reactors [17]. They are small and do not require large B-fields. They naturally have high beta, since beta actually goes to infinity at the field null. Longer machines are predicted to be more stable, giving an easy way to get more plasma volume. FRCs have natural divertors and the possibility of direct conversion. Once created, an FRC plasma can be moved into a compression chambers, where pulsed coils can pinch them to higher density and temperature. Research on FRCs has always been on the back burner, so they have not had the support of large computing efforts that tokamaks and laser fusion have had. Expensive equipment like neutral-beam heating has also not been available. There is precious little information on how the early plasmas were created, but recent success in using rotating magnetic field (RMF) current drive has given the program new impetus. Invented more than 30 years ago by Ieuan Jones and Lance McCarthy at the University of Adelaide in Australia, this method applies a transverse magnetic field that rotates at radiofrequencies in the toroidal direction. Fig. 10.34 shows an end-on view of the RMF lines as they are affected by the plasma. Electrons are entrained by these field lines and rotate with them to the best of their ability, but they are slowed down by collisions with the ions. The rotating field has to have enough power to overcome this drag. There is also a radiofrequency skin depth so that the field does not penetrate all the way into the plasma. In the original Rotomak, the field lines were not closed, so confinement could not be good; but RMF works well in an FRC.

Experiments on the science of FRCs have been carried out in a series of machines in the Redmond Plasma Physics Laboratory of the University of Washington. The most troublesome instability has been the tilt mode, shown in Fig. 10.35. By 1995, stability had been obtained up to $s=5$ [17]. It was found that energy was lost mainly by radiation due to impurities coming off the walls. Conditions were greatly improved with a new vacuum system in the TCSU machine.

The reduced drag on RMF current drive allowed it to produce denser and hotter plasmas. The total temperature (T_i+T_e) increased about a factor of 4 to ≈200 eV, the plasma

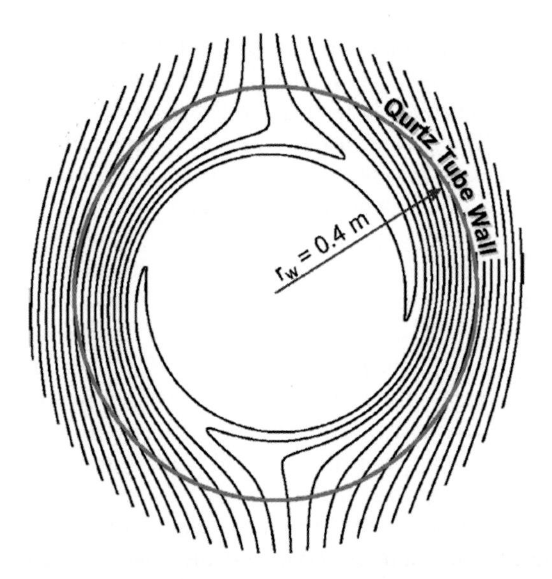

Fig. 10.34 Field lines of RMF current drive. The pattern rotates clockwise at about 150 kHs [32]

Fig. 10.35 Simulation of the tilt mode in an FRC using the NIMROD code [33]

density about a factor of 3, and the plasma pressure about 50%. Diagnostics are still rudimentary. The plasma pressure can be expressed as a magnetic field B_e which has the same pressure. Compared with the RMF amplitude B_ω, B_e is 4.9 times larger. RMF current drive in principle allows steady-state operation. These are encouraging results, but the plasma parameters are still very modest. It may be a long time before the conditions of an old reactor study [34] can be realized.

The high betas in FRCs make them suitable for advanced fuels, which require hotter and denser plasmas to ignite. This is being pursued in private industry with a FRC-type machine in which hydrogen ions are injected into a boron plasma for the p-B^{11} reaction. Tri-Alpha Energy in Irvine, California, was named for the three alpha particles which result from that reaction. The innovation involves adding rotation to the plasma in an FRC and is based on a theory by renowned plasma theorist Norman Rostoker [35].

Z-Pinches

A Z-pinch, called a Zed-pinch in England, is the simplest of all fusion approaches. It involves no more than pulsing a large current between two electrodes immersed in deuterium or DT. A column of plasma is ionized by the current and is confined by the magnetic field generated by the current itself, with no need for external coils. The B-field is very strong, since it surrounds the plasma closely, and it compresses the plasma until its density and temperature are high enough for fusion. It is very unstable, of course, because of the kink instability (Fig. 6.2). Since Z-pinches are so easy to make, much effort had been spent in trying to stabilize the pinch or to make it so fast that fusion occurs before the instability breaks it apart. These unsuccessful efforts became outmoded with the invention of wire arrays.

Starting with a current through a tungsten wire was found to improve Z-pinches because of the initial straight path and the slower motion of the heavy ions. A ring of tungsten wires, as shown in Fig. 10.36, was found to be qualitatively better because of the blending of their magnetic fields. If the wires are close enough together, the B-fields outside the circle form an overall azimuthal field that compresses all the plasma into the center without kink instabilities because the current through each wire is comparatively small.

Figure 10.37a is a photograph of a 4-cm diameter array of 240 tungsten wires, each 7.5 μm (0.0075 mm) in diameter, at Sandia National Laboratories in New Mexico. In the Z-machine (described later), some 20 MA of current was pushed through the wires, forming a dense Z-pinch in the center. The aim was to generate X-rays, and about 200 TW (200 trillion watts) of these were generated [36]. This is a spectacular result, since the total electrical generating capacity of the USA is only about 1 TW. Of course, the X-ray pulse lasts only a nanosecond or so. Figure 10.37b is the same pinch fitted with an inner array of 120 wires. The plasma from the inner wires creates a plasma that smoothes out the instabilities that start to develop in the outer plasma, and the X-ray power is increased to 280 TW [37]. However, *these pinches cannot produce continuous energy for primary power.*

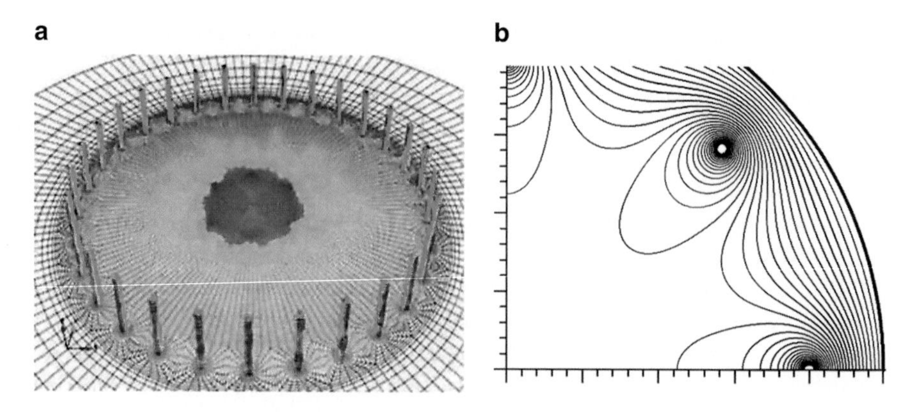

Fig. 10.36 (**a**) Diagram of a circular array of wires for a Z-pinch (Sandia National Laboratories, Albuquerque, NM. www.sandia.gov). (**b**) The magnetic field around each wire

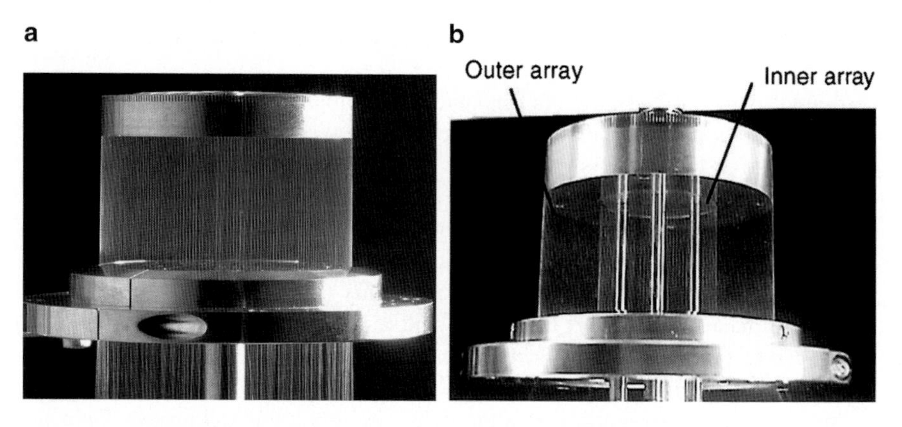

Fig. 10.37 Single (**a**) and double (**b**) wire arrays for Z-pinches at Sandia [36, 37]. The wires are about 2 cm long

The Magpie Project of Imperial College, London, conducts innovative research on wire-array Z-pinches in which the wires point radially outward from the center [38]. External magnetic fields can also be imbedded in the pinch. Though the work is interesting, it is done for other purposes than energy production.

Plasma Focus

Also called the dense plasma focus (DPF), this is one of the oldest devices invented to create fusion. Because of its simplicity, it is used in small laboratories over the world for instructional research. A diagram is shown in Fig. 10.38. A plasma is formed by discharging a large capacitor between the center electrode and the outer cylinder. An ionization front, shown by the white curve, travels rapidly to the end

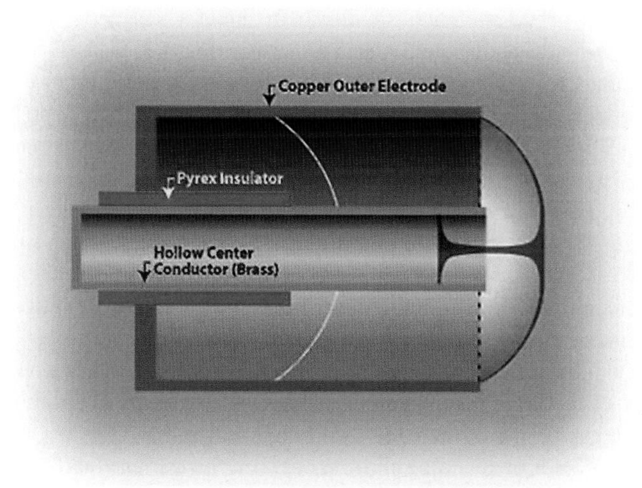

Fig. 10.38 Diagram of a dense plasma focus (http://www.plasma-universe.com.)

at the right. There, the current flows between the electrodes in the crown-shaped plasma consisting of streamers. In the center of the crown is a dense Z-pinch which can reach fusion conditions for a brief instant.

Intense X-rays are generated, and with deuterium for DT, neutrons are produced for 10–20 ns [39]. Both diagnostics and theory are difficult for the DPF, and it is not well understood. Nonetheless, some groups are proposing the DPF for p-B^{11} fusion. There is interesting physics to be studied in the DPF; but, as with all single-pulse machines, it is not suitable as an energy source.

Inertial Confinement Fusion

Introduction

When high-intensity lasers became available around 1970, people like Ray Kidder at Livermore and Keith Brueckner at the University of California, San Diego, began to think about inertial fusion. If it's so hard to hold a plasma with a magnetic field, what about heating a plasma so fast with a laser that it fuses before it can fly apart? The idea was to fill a very small glass sphere with deuterium or DT fuel and zap it with lasers from all sides. The glass would evaporate and expand outwards, and the reaction would push the fuel inwards into a small hotspot where it would fuse before it could turn around and blow out. They worked out the numbers and made a proposal to the Atomic Energy Commission to start a laser-fusion program at Livermore. This proposal was reviewed by a committee, chaired by Lawrence Hafstad, which included the author. The proposal was accepted, and the rest is history.

Starting with a budget much smaller than that for magnetic fusion, the laser program was very successful, and a series of larger and larger lasers was built, with names like Janus, Argus, Shiva, Nova, and now NIF, the National Ignition Facility. The success depended in large part on an intricate computer program by John Nuckolls, the first of its kind, which could predict what would happen in the implosions. At $458M, the budget for inertial confinement has overtaken that for magnetic confinement ($426M) [29]. However, inertial fusion is not primarily funded for energy. Although some scientific support comes from Fusion Energy Sciences, the main support is from the National Nuclear Security Administration. That's because the miniexplosions that lasers can create are powerful enough to mimic the effects of hydrogen bombs on materials. Data needed to maintain the nuclear stockpile and develop new weapons can be obtained without underground testing with real explosions. In addition, the study of the behavior of matter under extreme pressure and temperature conditions is vital to our understanding of astrophysical objects in our universe.

One might object to spending more money on the military part of fusion rather than on the energy part, but that expenditure is essential. National security must come first. Without freedom, we can't do anything. Laser fusion is sold to the public as an energy source because of its glamorous achievements. It will reach ignition decades before ITER can. However, it is a pulsed system like the pinches in the previous section and is difficult to make into a steady power source.

The main problem is the lack of a suitable driver. The term *inertial confinement fusion* was coined to include drivers that are not lasers. To have a steady power output, a laser-fusion power plant has to implode a pellet at least ten times a second. A car runs smoothly with 3,200 explosions a second (four cylinders at 800 rpm), but 10 explosions per second would be enough for a power plant. However, lasers can't pulse that fast. The most powerful ones use neodymium-doped glass disks a couple of feet in diameter. As much light as possible is passed through the glass for amplification. This heats up the glass almost to the point of cracking. It takes hours for the glass to cool. With earlier lasers, two shots a day were all that could be expected. There are thousands of these disks in a megajoule laser. If one of them cracks, the whole system shuts down.

The main task, then, is to find a better driver. Ion beams have been tried, but they are hard to focus down to a small target and also have to be pulsed. Krypton-fluoride (KrF) lasers use no glass and can be pulsed more rapidly. They have some promise, but pulsing at five times a second has been proved possible only at low power. Systems based on pulsed power (discussed later) are also pulsed infrequently. Laser fusion should be considered as the fantastic technological achievement that it is, but not as a promising base-load energy source.

General Principles

Laser fusion would require a separate book to describe. Here it is treated briefly as an alternative to tokamak fusion. The idea is to put a deuterium–tritium mixture

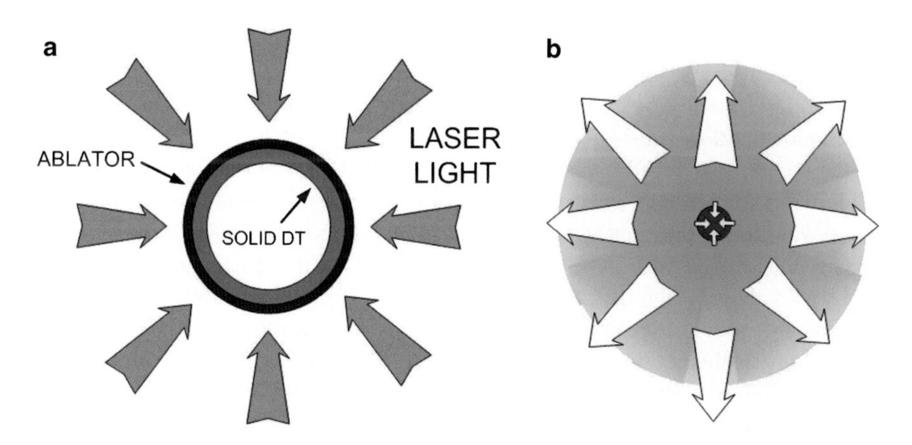

Fig. 10.39 Mechanism of inertial fusion. (**a**) Laser light impinging on a spherical capsule. Actual targets are much more complicated. (**b**) The capsule being compressed by the expanding plasma blown off from the ablator

into a capsule about 2 mm (~1/16 in.) in diameter and hit it from all sides with a huge amount of laser power for several billionths of a second. The power from NIF in one short pulse is 500 times the total electric power capacity of the USA. This is what is supposed to happen.

In Fig. 10.39a, the laser or ion beam energy impinges on a capsule uniformly from all sides. The capsule contains fuel in the form of solid DT, covered by a sacrificial layer called an *ablator*. The ablator is immediately ionized into a dense plasma, which expands violently away from center. The capsule is compressed as if jet engines had taken off on all sides. Figure 10.39b shows the expanding plasma compressing the capsule. With sufficient laser power, the DT fuel is compressed to a density of 1,000 g/cm^3, approximately 100 times the density of lead; and the temperature reaches 10 keV. The breakeven condition equivalent to the Lawson criterion (Chap. 5) is $\rho R > 1$ g/cm^2, where ρ is the density in g/cm^3 and R is the final radius in centimeters. Fusion occurs, and there is a miniature explosion releasing the helium and neutron products. The energy of an NIF pulse is about 1.8 MJ, and the energy generated could be as much as 100 MJ, equivalent to 24 kg of TNT. To produce 1,000 MW of thermal power would require ten explosions per second. Glass lasers, however, can pulse only once every few hours.

Instabilities

The beauty of inertial confinement was supposed to be its freedom from the instabilities in magnetic confinement. No such luck: there are new instabilities! First there is an old one, the Rayleigh–Taylor instability (Chap. 5), which occurs whenever a light fluid pushes against a heavy one. The expanding plasma pushes against the capsule with a huge force. If there is any deviation from smoothness in either the capsule or the laser light, small ripples will grow and destroy the compression before it gets very far. Figure 10.40 shows what can happen.

Parametric instabilities are a new class of instabilities caused by laser radiation [41]. In Fig. 10.41, a laser ray enters the blown-off plasma from the right and generates a wave in the plasma shown by the curly line. This wave has maxima in plasma density at the vertical bars. The laser ray reflects off these density stripes coherently, as if they were a diffraction grating. The reflected ray goes off to the right. The incoming and reflected waves interfere constructively to strengthen the plasma waves, which then reflect more strongly yet. The net result is that much of the incoming light is reflected back toward the laser. Less light reaches the capsule, but that is not the worst part.

Extreme care must be taken to prevent the reflected beam from being amplified as it goes back through the laser. Otherwise, it will fry the laser. There are two kinds of plasma waves that can be generated in a parametric instability. One is an

Fig. 10.40 Computer simulation of a Rayleigh–Taylor instability [40]

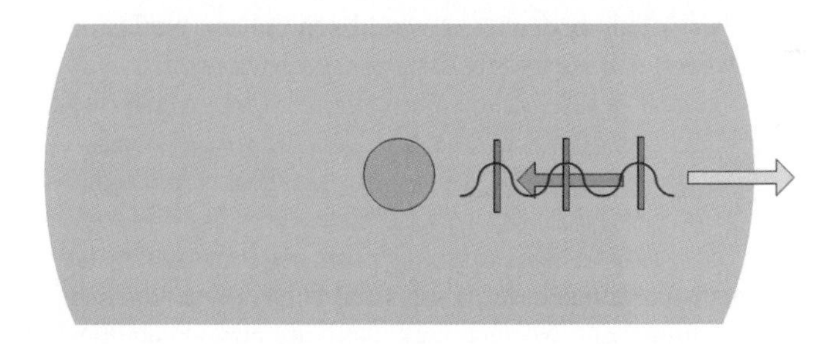

Fig. 10.41 Schematic of a parametric instability; explanation in text

ion acoustic wave, in which case the instability is called stimulated Brillouin scattering (SBS). The other is an electron plasma wave, in which case the instability is called stimulated Raman scattering (SRS). The worst part about SRS is that the plasma wave accelerates a beam of electrons. This can preheat the DT fuel so that it cannot be compressed to the required size. All this happens in a very small space, so the beam of electrons is very narrow, forming a pinch. The magnetic field generated by this beam is measured in megagauss (100s of tesla). It is not true that inertial fusion avoids magnetic fields! The instabilities, however, are not those in magnetic fusion.

The higher the frequency of the laser light, the higher are the densities where SBS and SRS occur. Higher frequencies will penetrate more deeply into the plasma corona and minimize these instabilities. This is the reason the NIF laser will use the third harmonic ("3ω") of its fundamental frequency even though almost half the light intensity will be lost in the conversion.

Glass Lasers

Major laser facilities are so large that they cannot be shown in a single picture. A simplified schematic is shown in Fig. 10.42. A weak pulse of the right spatial and temporal profiles is generated in an oscillator at the left. The same pulse is sent into each laser chain. Each identical chain consists of increasingly large amplifiers to raise the power of the beams. At the right-hand side, the beams enter a *switch yard* consisting of mirrors to bring the beams into the target chamber (the white sphere) at the desired angles. The beams have a finite length, since light travels at 1 feet (30 cm) per second, so a 1-ns pulse is only a foot long. Each beam path has to have the same length for the beams arrive at the same time. In between the amplifiers are

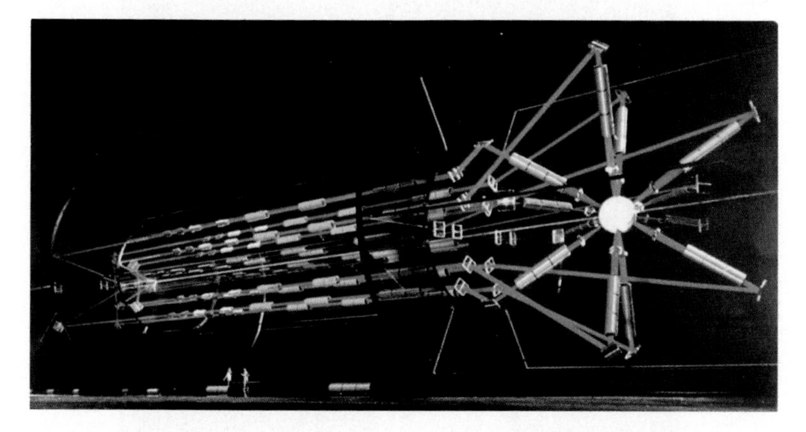

Fig. 10.42 Simplified schematic of a glass laser installation (Photo from the author's archives; original from a national laboratory: Livermore, Los Alamos, or Sandia.)

optical units to reject the reflected light and to maintain the same time variation, spatial profile, and smoothness that the beams started with. The light is divided into multiple beams not only to illuminate the target uniformly, but also to avoid overheating the glass in each beam.

The NIF laser has 192 beams divided into 48 groups of four each. The neodymium in the doped glass is driven into an excited state by a pulse of light from flash lamps. Originally, the lamps were like the electronic flashes in cameras. Recent conversion to solid-state units like LED flashlights have greatly reduced the complexity and cost. It takes 400 MJ of capacitors to store the energy for the lamps. An excited amplifier lases when it is tickled by the light from the previous stage. The total length of each light path is 300 m, the length of three football fields for either kind of football. Nd-glass lasers produce infrared light with 1.06 μm wavelength. This is upshifted to 3ω with single crystals of potassium dihydrogen phosphate. The 3ω light has a wavelength of 351 nm, which is in the ultraviolent, so different optical materials have to be used. Figure 10.43 shows individual beam tubes in the earlier Nova laser. Figure 10.44 shows the NIF laser bay before it was all covered up. The optical elements in each beam tube have to be held in exact alignment and kept completely dust free. In NIF, the optical equipment between each amplifier stage is preassembled in refrigerator-size boxes so that spares can be slipped into place from below if one element fails.

Nd-glass lasers were developed also at the Institute for Laser Engineering in Osaka, Japan, under the leadership of Prof. Chiyoe Yamanaka. Other important participants in the development of glass lasers were Academicians N.G. Basov and A.M. Prokhorov in competing groups at the Lebedev Institute in Moscow; Kip Siegel, who founded KMS Fusion in Michigan; Moshe Lubin, who founded the Laboratory for Laser Energetics in Rochester, New York; the group at the Rutherford-Appleton Laboratory in England; and Edouard Fabre's laboratory

Fig. 10.43 View of the Nova laser bay (https://lasers.llnl.gov/multimedia/photo_gallery/.)

Fig. 10.44 View of the NIF laser bay (https://lasers.llnl.gov/multimedia/photo_gallery/.)

at the Ecole Polytechnique in Palaiseau, France, which led to the Laser Mégajoule being constructed in Bordeaux.

Other Lasers

Carbon dioxide (CO_2) lasers were the earliest high-power lasers and were pursued vigorously at the Los Alamos National Laboratory. First the MARS, then the Helios, then the Antares lasers were built. The Antares has the size of a large locomotive, but it was never finished. CO_2 lasers have a wavelength of 10.6 μm, in the far-infrared, and it was soon found that parametric instabilities could not be controlled at such long wavelengths. A large CO_2 laser was developed at the Naval Research Laboratory in Washington, DC. A large volume of a He–N_2–CO_2 mixture was ionized by an electron beam in the form of a large sheet. This technique was used in the 3-kJ Nike krypton-fluoride laser, which has a wavelength of 248 nm, shorter and better than the 351-nm 3ω light from NIF. Being a gas, KrF does not need to be cooled between pulses, and the 700-J Electra laser at NRL has pulsed at 5 Hz for days or weeks at a time [22]. The KrF laser could be a suitable driver for inertial fusion if the unwieldy and expensive electron beam could be replaced by a simpler ionizer.

Target Designs

Originally, glass microballoons containing DT gas were used as targets. They were like the glass beads used to coat projector screens but had to be perfectly round and smooth.

One is shown in Fig. 10.45a , and a number of them are shown on a coin in Fig. 10.45b. When hit with lasers, the glass exploded, half going out and half going in, compressing the gas. This is how the first fusion neutrons were observed.

Later targets used low-Z ablators to have a more controlled compression (Z is the atomic number). Examples of target designs are shown in Fig. 10.46. All of them have a shell of frozen DT as the fuel. In panel (a), there is also a bit of DT at the center, confined by a heavy pusher. This is supposed to ignite first, giving energy to help ignite the main fuel. In panel (b), the ablator is polystyrene foam, which allows DT gas to be permeated into the capsule without using a fine tube, as in Fig. 10.45a. The DT is frozen at cryogenic temperatures, and is melted and smoothed by the little bit of heat from the decay of the tritium. In panel (c), a beryllium

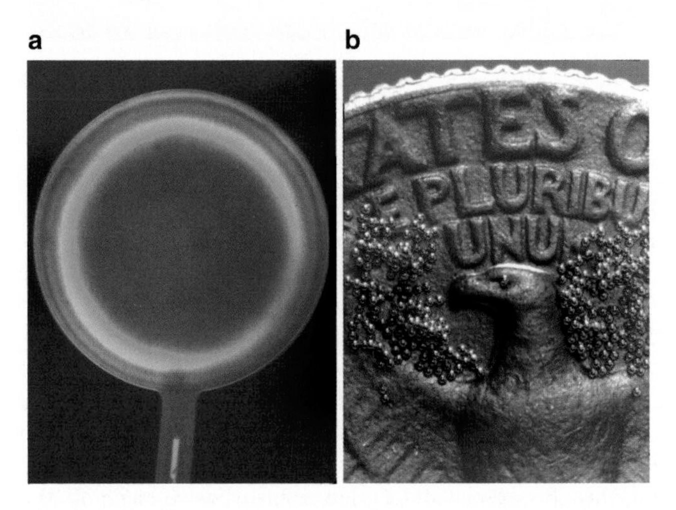

Fig. 10.45 Glass microballoons used as laser fusion targets, (**a**) magnified, and (**b**) in real size. (Photo from the author's archives; original from a national laboratory: Livermore, Los Alamos, or Sandia.)

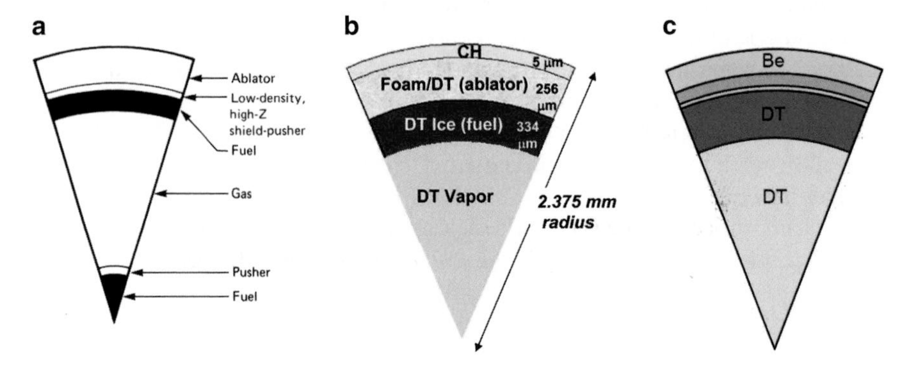

Fig. 10.46 Examples of capsule designs: (**a**) with central ignition [43]; (**b**) with plastic foam [42]; (**c**) with beryllium ablator [44]

ablator is used in a design to optimize shock heating. To improve compression, multiple shocks can be created by shaping the laser pulse into increasingly strong steps. Since strong shocks travel faster than weak ones, multiple shocks can be timed to catch up with one another just when they reach the center.

Target design is very computation-intensive, since the progression of the implosion has to be predicted. Designs differ depending on their purpose and the driver. Making just one of these targets takes great skill and cost. In a reactor, each pellet can cost no more than $0.50. Surprisingly, it is predicted that in mass production, these targets can be made for only $0.16 each [45]. Tens of thousands can be made at once in fluidized beds, and the infusion of DT into the spheres and the freezing of a layer at 18 K can be done to a whole batch at once since injection of DT through individual micron-size tubes is no longer necessary.

Direct and Indirect Drive

Direct drive is what we have pictured so far: a spherical target is compressed by laser light impinging on it uniformly from all directions. The main problem is that the laser beams must have no hotspots that can cause Rayleigh–Taylor instabilities to develop. Research on direct drive is the main mission of the Omega laser in the Laboratory for Laser Energetics in Rochester, New York. After years of trials, optical tricks have been devised to produce beams of the required uniformity.

Indirect drive is considerably more complicated. The laser is first fired into a cylindrical cavity called a *hohlraum*, German for "hollow space." Upon striking the inside wall of the hohlraum, which is usually made of gold, the laser light generates intense X-rays. The capsule in the center is bathed in a sea of X-rays, which compress it uniformly. Because of their high frequencies, X-rays are not subject to parametric instabilities. However, the laser beams must enter the hohlraum through a small hole in either end. Any stray light that hits the side of the holes will generate plasma and excite parametric instabilities there. Figure 10.47 is a view of a gold hohlraum, and Fig. 10.48 is an artist's rendering of laser beams entering a hohlraum with a capsule in the center.

Indirect drive, the main emphasis of the programs at Livermore, is known to work well in bombs, but it is much more complicated for fusion than direct drive is. The hohlraums are hard to make, and the capsules have to be suspended at the center. (For this, there has been talk of using spider-web strands, for which there is no man-made replacement.) The hohlraums have to be shot to the center of the target chamber because the DT would melt if they were dropped slowly. Even then, cooled holders, shown in Fig. 10.49, have to be used to keep the hohlraum at cryogenic temperature during its transit through the chamber. The holders also help protect the hohlraum from the force applied to accelerate them. *Fast ignition* is a new complication. To achieve better efficiency, this new method uses a very short prepulse focused with a cone (Fig. 10.50) to ignite the DT gas at the center of the pellet. The fusion energy from that burn helps to ignite the main fuel.

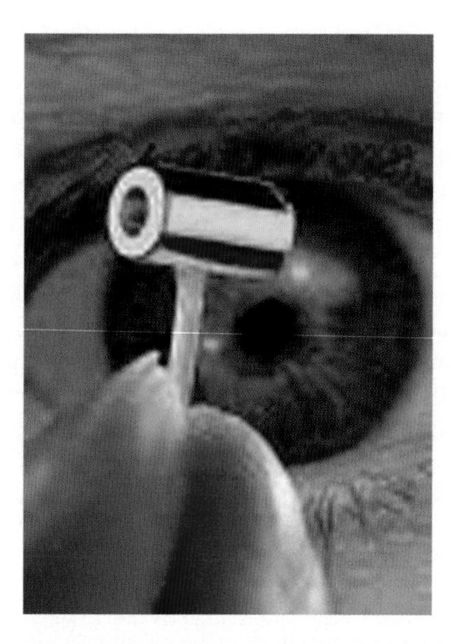

Fig. 10.47 A hohlraum (https://lasers.llnl.gov/programs/nic/.)

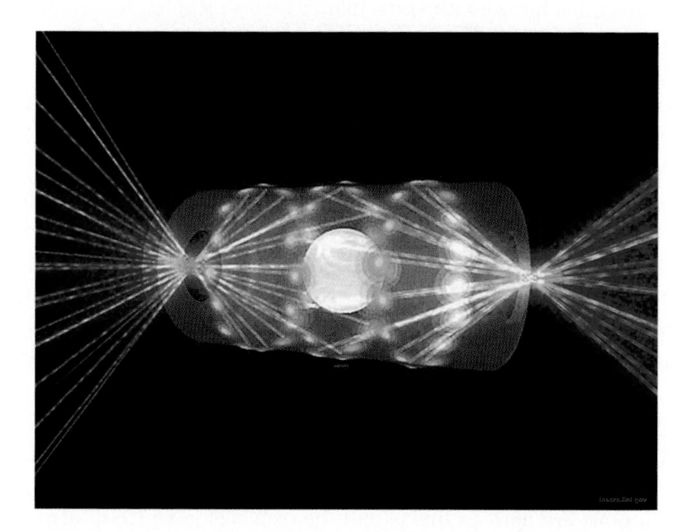

Fig. 10.48 Mechanism of indirect drive (www.flickr.com/photos/llnl/2843501990/.)

Imagine the sequence of each shot. A laser pulse is generated in an oscillator and is divided into 196 beamlets, each of which is passed through numerous amplifiers and optical switches in a 300-m path until the total energy exceeds 1 MJ. These beamlets form 48 beams, which the switch yard sends into the target chamber, shown in Fig. 10.51. Each beam is focused onto the target with micron accuracy in space and nanosecond accuracy in time. For indirect drive, the beams are divided

Fig. 10.49 A hohlraum held between cooling fingers (https://publicaffairs.llnl.gov/news/news_releases/2010/nnsa/NR-NNSA-10-01-02.html.)

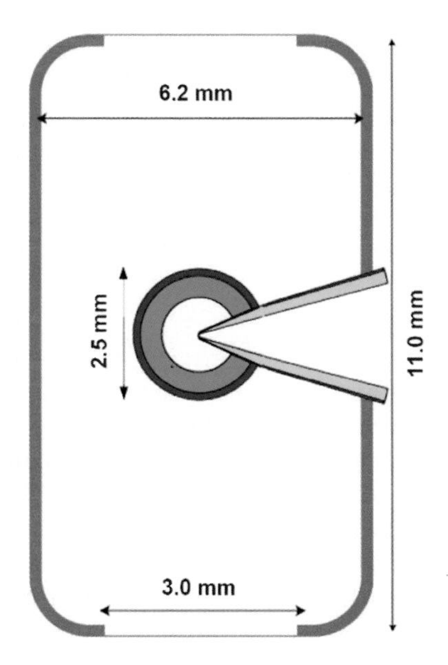

Fig. 10.50 Diagram of a fast-ignition hohlraum [44]

into two bunches, each entering the hohlraum at one end. The beams must not spill over onto the edges of the holes, or else they would make plasma and block the entrance. The cylinder must be aligned perfectly with the beams. In fast ignition, the hohlraum must also be in the right rotational position for the cones to be

Fig. 10.51 The NIF target chamber being lifted into the building (https://www.llnl.gov/str/Atkinson.html.)

aligned. After the shot, everything is vaporized, and the chamber has to be cleared for the next shot. In experiments, the targets are rigidly held by an arm, and successful implosions of the pellet have been achieved.

Reactor Technology

Even though a suitable driver has not yet been found, reactor studies can still be made, especially for the simpler direct drive case [42]. The energy released from each capsule is equivalent to tens of kilograms of TNT, but the blast waves of TNT cannot be produced because there is so little material involved [43]. The only ions are from the tiny capsule and the DT fuel, plus the helium that is produced. Much of the energy comes out as radiation, and the first wall has to withstand that. The neutrons, as usual, go through the first wall into breeding blankets. The first wall has to withstand the radiation, mostly in the form of X-rays. Inertial fusion has the advantage over tokamaks in that there is a much larger distance between the energy source and the wall. The main candidates for the first wall are (1) a solid material like the SiC/SiC compounds proposed for tokamaks, (2) a wall thinly wetted with a liquid, and (3) a waterfall of liquid FliBe (Chap. 9) covering a solid wall. In laser fusion, solid walls would suffer from repeated thermal cycling, which greatly shortens their life.

In direct drive, 71% of the fusion energy comes out as neutrons, 27% as ions, and only 1.4% as X-rays. In indirect drive, 69% comes out as neutrons, 5.8% as ions, and a whopping 25% as X-rays since the hohlraums are designed to produce X-rays [47]. The ions and X-rays deposit their energy in a very thin layer on a *dry wall* [48], which must be well cooled to take the heat. A more serious problem is the deposition of the fast alpha particles into the wall, forming helium bubbles that cause the wall to exfoliate. A method to avoid this is to impose a cusp magnetic field (Fig. 7.8) to protect the wall and lead the ions into divertors. However, this requires strong superconducting coils as in magnetic fusion.

A *wetted wall* can be a thin layer of FliBe injected through small holes in the first wall and protecting the wall from ions and X-rays. The liquid is collected at the bottom, re-processed, and re-injected at the top of the chamber. A *thick liquid wall* [43] is a cylindrical waterfall of FliBe or PbLi between the target and the solid wall. The waterfall intercepts the fusion products, goes into a tank below the chamber, and is re-processed and re-injected at the top. In this case, the target has to enter from the top or bottom. Sethian et al. [42] have compared direct-drive reactors based on diode-pumped glass lasers and KrF lasers. Both kinds have been shown to withstand repeated pulsing at 5–10 Hz at low power. They have similar wall-plug efficiencies: 10% and 7%, respectively; they are compared in case high-power pulsing can be developed.

In inertial fusion, there is the problem of restoring the vacuum in the 100 ms between shots. The remaining gas must not be ionized by the laser. The laser beams have to strike the target with 20-μm accuracy from 10 to 20 m away, and a small amount of gas will deflect the beams. A "glint" system has been tested to overcome this [42]. As the capsule nears the center of the chamber, a small laser is fired to illuminate it. The direction of the reflection is detected, and mirrors are moved to keep the beam on target. To do this with 48 beams, however, is a daunting task, and only spherical targets in direct drive can be used. *There is no clear path to fusion energy with lasers.*

Pulsed Power

This term describes systems which can deposit huge amounts of energy in a short time, but directly, without lasers. Alan Kolb, one of the earliest fusion researchers, left that program to start the field of pulsed power by founding Maxwell Laboratories in San Diego, California, to develop large, fast capacitors for storing energy. They were the first to put "a megajoule in a can." A megajoule is not an incomprehensible amount of energy. It is the heat energy of a pot (3 L) of water at boiling temperature. A 50 ampere-hour car battery contains 2 MJ. What matters is how fast the energy can be released to get power. Power is the rate of energy delivery. While a car battery can be drained in an hours, capacitors can release their energy in nanoseconds. Capacitors can store over 2 J/cm^3. A megajoule can be crammed into 500,000 cm^3, which is the size of a cube 80 cm (30 in.) on a side. A pulsed power installation has hundreds of these.

To get high voltage, the capacitors are hooked up in a *Marx bank*, shown in Fig. 10.52. In this arrangement, a DC power supply is connected to each capacitor as shown in the top half of the figure. After charging, the switches between the

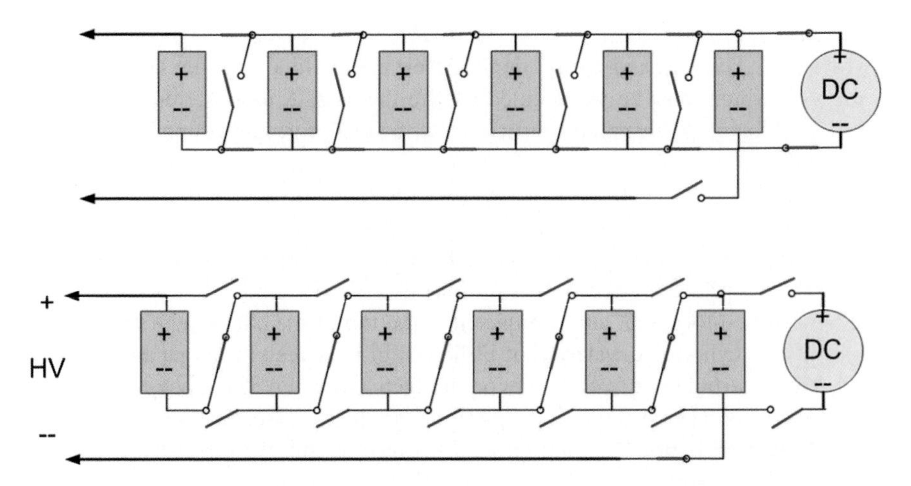

Fig. 10.52 Schematic of a Marx bank. At the *top*, the capacitors are charging in parallel; at the *bottom*, they are discharging in series

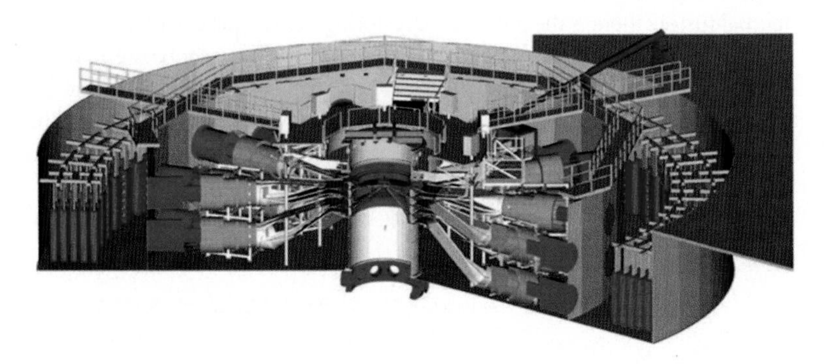

Fig. 10.53 Diagram of the Z-machine, the world's largest pulsed power machine [46]

capacitors are opened, as shown in the bottom half, and the diagonal switches are closed, connecting the capacitors in series. A much higher voltage is then produced than a single power supply can generate.

The current is then carried to the machine in a *Blumlein*. This is a big, specially designed transmission line that can handle the huge currents and voltages that the Marx bank can provide. The Blumlein uses water as the insulator, and also has magnetic insulation by the B-field generated by the current. The pulses can also be made shorter in the process. The spark-gap switches are perhaps the most important high-tech elements in the system.

Figure 10.53 is a diagram of the Z-machine at Sandia National Laboratories in Albuquerque, New Mexico, and Fig. 10.54 shows the actual machine. The capacitors surround the machine, and the cylinders are the Blumleins carrying the energy pulses into the vacuum chamber at the center. The capacitors in Z store 11.4 MJ,

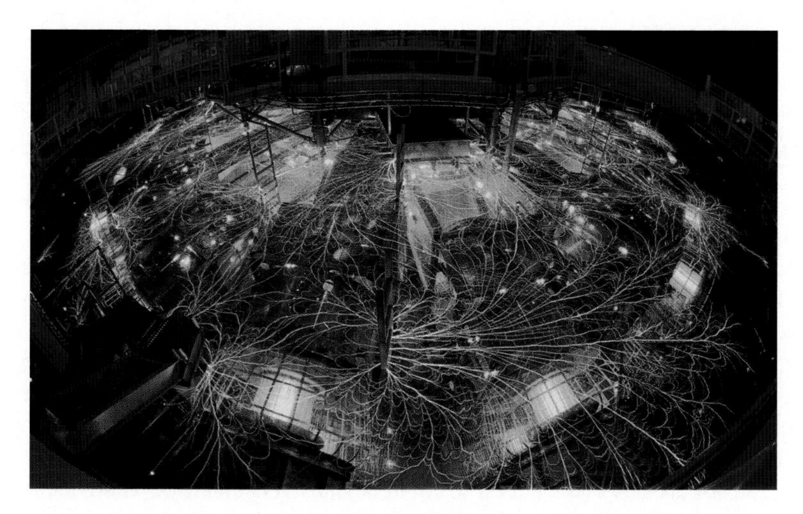

Fig. 10.54 The Z-machine during a discharge (http://www.sandia.gov/media/). This publicity photo shows arcs which occur only during abnormal operation

of which 5 MJ is delivered by the Blumleins to the load. A 100-ns pulse can carry 20 MA of current and 60 TW of power. For military applications, the machine can produce 2 MJ of X-rays per pulse at a power of 200 TW.

For fusion applications, the Z-machine can produce heavy- or light-ion beams to transport to a capsule larger than those in laser fusion because of the higher energy here. The problem is in the transport. Ion beams are hard to keep in focus across the large distance to the pellet. When the beam becomes narrow near the target, its space charge will tend to expand it unless the charge is neutralized. The best way to do that is to send the ions through a preformed plasma, whose electrons can neutralize the space charge. This is a perfect setup for a beam-plasma instability. Ion-beam drivers have not been successfully developed. The plans now are to use the intense X-rays from pulsed power to fill a hohlraum. Even if this works, it cannot work at 10 Hz. Pulsed power is not a promising source for an inertial fusion driver.

Hoaxes and Dead Ends

Cold Fusion

There has been much ado about almost nothing following the 1989 announcement by Fleischmann and Pons that they had produced energy in a flask of heavy water. The experiment consisted of electrolyzing the D_2O into gaseous products by applying a DC voltage between an anode and a cathode, the latter made of palladium. The energy input and output from the apparatus had to be carefully measured.

There was energy balance for several weeks, but then they found that the output was a few watts larger. Since then, the experiment has been repeated hundreds of times by reputed scientists without similar success. There have also been many believers in cold fusion who accuse the scientific community of snobbish exclusivity, and who occasionally report observations of excess energy generation. The American Physical Society has held conference sessions and panel discussions on cold fusion with the conclusion that it is impossible.

Electrochemical potentials are sometimes surprisingly strong. The hydrogen car fuel cell (Fig. 3.51), for instance, uses a platinum or palladium catalyst to dissociate and ionize hydrogen magically before it has been heated. There is, however, a huge difference between the 10 eV in ionization and the 10 keV in overcoming the Coulomb barrier in fusion. In cold fusion perhaps the deuterium seeps into the palladium after some time, and eventually two D's get very close together and somehow the applied voltage can cause them to fuse. Sometimes a few neutrons are observed, but these could be due to cosmic rays. There is interesting physics in these infrequent events, and the International Conference on Cold Fusion has been meeting annually since 1990. Institutes for cold fusion have been established in some countries. However, cold fusion power is so miniscule that it would not pay for the palladium, much less a whole power plant. And it is only thermal power, not direct electrical power. Cold fusion may have interesting scientific aspects, but it has no relation to power production.

The uproar over cold fusion has had one benefit, however. It shows that the public is not disinterested in controlled fusion power, as long as it is cheap. It simply does not understand why it is so hard to achieve, and why there are no shortcuts leading to the gold at the end of the rainbow. This book attempts to explain why.

Bubble Fusion

Sonoluminescence is a phenomenon in which megahertz sound waves in a liquid can cause a bubble to collapse into a very small dot, creating a high temperature there. Using deuterated acetone as the liquid, some researchers reported detecting fusion neutrons created by the collapsing bubble. However, experts on sonoluminescence, including Seth Putterman of UCLA, were not able to reproduce these results and have categorically stated that this is *not* a way to produce fusion. It appears that this is an even more extreme farce than cold fusion.

Muon Fusion

This is the original idea on cold fusion, having been disclosed by Luis Alvarez in his Nobel Prize speech in 1968 [49]. Muons are fundamental particles like electrons but 207 times heavier. They are produced in accelerators and live for 2 μs (an eternity!)

before decaying. As you know, elementary particles and photons have a dual nature, sometimes behaving like particles and sometimes like waves. As waves, they have a wavelength, called the deBroglie wavelength, which is inversely proportional to their masses. Being some 200 times heavier, muons have wavelengths 200 times shorter. A negative muon can take the place of an electron in an atom, and the "cloud" of negative charge is then 200 times smaller, bringing the nuclei of molecules closer together. The muon-fusion process for DT molecules is shown in Fig. 10.55.

In the first line of that figure, normal D and T atoms with their large electron clouds can combine into a DT molecule, just as two H atoms can form H_2. In the second line, a μ-meson (muon) replaces the electron in the tritium atom, and the resulting muonic tritium atom has a smaller size. Next, a deuterium nucleus joins the triton inside the muon cloud, forming muonic DT with the two nuclei close together. Normally, the D and the T repel each other with their positive charges and cannot fuse into helium at room temperature. However, in quantum mechanics, particles can tunnel through the Coulomb barrier if it is thin enough. In a muonic DT molecule, this can happen very fast, and the entire process can happen several hundred times during the 2-μs lifetime of the muon. In the last line of Fig. 10.55, DT fusion has occurred, creating the usual products of a neutron and an alpha

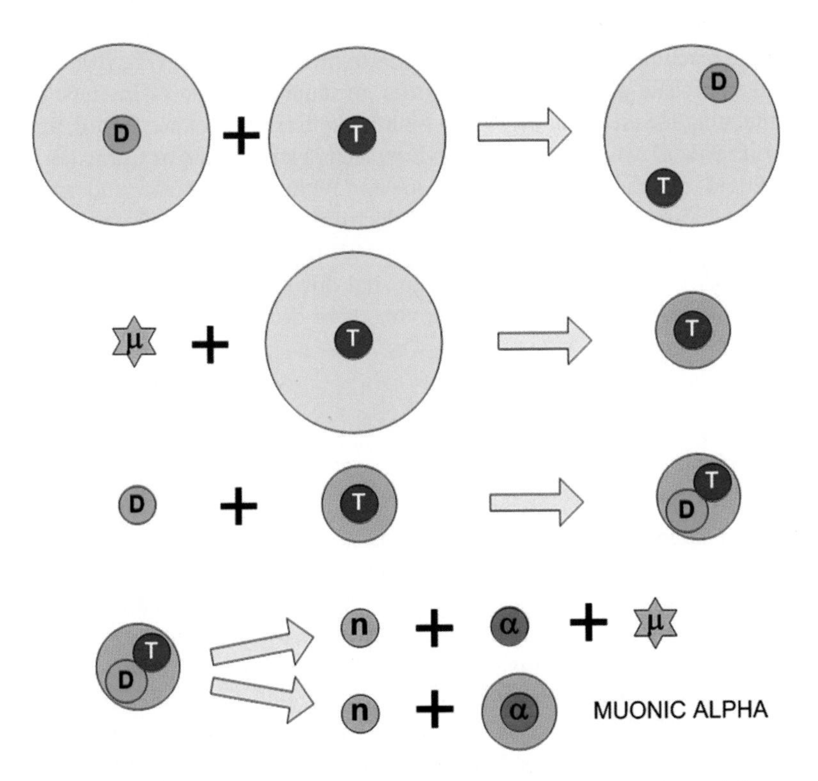

Fig. 10.55 Steps in muon fusion, ending with some muons "sticking" to alpha particles

particle. What the muon does then is essential. If it flies off, it can catalyze another fusion again and again. However, if it "sticks" to the alpha particle, it is carried off and is lost. The sticking fraction is between 0.4% and 0.8%, and this limits the number of reactions that one expensive muon can catalyze.

Experiments are being done in accelerator laboratories like RIKEN-RAL[4] in England and TRIUMPH in Vancouver, Canada. About 120 DT fusions per muon have been observed [28]. At 17.6 MeV per event, this amounts to over 2 GeV of energy. However, it takes 5 GeV to make each muon. There are ways to improve this ratio, by using polarized deuterons, by working at high temperatures, or by making cheaper accelerators. At this stage, the physics of muon fusion is still in its infancy.

Astron

The story of Astron is more about a person than about a fusion concept [50]. Nick Christofilos was a self-made Greek physicist who independently co-invented the alternating-gradient focusing principle for accelerators and later the Astron machine. Since he was a Greek citizen working on US-classified material, he was not allowed to access his own work once it had been filed away. The Astron was a very large machine at Livermore which was to produce an FRC (Fig. 10.33) with a ring of relativistic electrons injected from an induction linac of his own design. Accumulating the electron layer from multiple pulses was not successful, and only 6% field reversal was attained. Meanwhile, Hans Fleischmann at Cornell achieved 100% field reversal using pulsed power. Without sufficient understanding, Christofilos also did not realize that the electrons would lose their energy by synchrotron radiation. But his persuasiveness finally gave way to reason, and the Atomic Energy Commission prepared to shut down the project. Before this happened, however, Christofilos, a hard-driving, hard-drinking smoker, died of a heart attack at 55 in 1972.

Electrostatic Confinement

An electric field pushes ions and electrons in opposite directions, and it makes sense that a steady electric field cannot confine a plasma. However, Bob Hirsch, who later headed the AEC's fusion division, proposed a machine which has become popular with amateur fusioneers because of its simplicity. The device has two spherical grids one inside the other, the outer one grounded and the inner one at a large negative potential [51]. Gas is ionized between the grids, and ions are accelerated toward the center, where they accumulate and create a large positive potential. Subsequent ions are repelled by this "virtual anode" and bounce away back to the grids. They then oscillate inside the sphere and can collide with one another for an occasional fusion. This suffers from the original reason for a thermal plasma, as explained in Chap. 4. Streaming ions fuse only once in 10,000 collisions. The other collisions degrade their energies so that they no longer can fuse and eventually diffuse out of the system. Grids are OK for small experiments, but they will melt at fusion densities. Furthermore, Debye shielding at these densities will prevent the applied voltage from reaching the center of the holes in the mesh.

Migma

Early in the game, colliding accelerator beams were proposed, and several migmatrons were built. With accelerators, it is easy to get ions up to the energy of the peak of the DT reaction, nearly 80 keV, or even the p-B^{11} reaction, nearly 300 keV. The beams are of low density, but they can be put into storage rings to circulate past the collision point many times. Elastic scattering, however, generates bremsstrahlung radiation, and there are always instabilities with streaming particles. A comprehensive stability theory has never been worked out for migmas.

Ultimate Fusion

One hundred years from now, what will a fusion reactor look like? Most of the ideas described in this chapter will have been discarded, and a few will have been combined. Once the period of patched-up Rube-Goldberg-like experiments is over, private industry will develop a simpler and cheaper system that self-organizes into a stable configuration and keeps itself hot without much external power. The reactor will probably have a roundish shape, like that of a compact tokamak. The fuel will probably be p-B^{11}, which does not require tritium breeding and generates few neutrons. Or it could be d-He^3, though He^3 would have to be made in an auxiliary fission reactor. The magnetic surfaces will be closed and have the interior good curvature of a spherical tokamak (Fig. 10.12). They may look like those of the Chandrasekhar–Kendall–Furth force-free configuration shown in Fig. 10.56.

There will be natural divertors at the top and bottom. The exterior regions above and below the divertor necks can be expanded like those of an axisymmetric mirror (Fig. 10.27) to create more good curvature for stabilization. High-energy alpha particles leaving the divertors can be channeled into direct converters to generate high-voltage DC directly. The central core can be slid up or down continuously to be refreshed without a shutdown.

This is a dream, but we can hope.

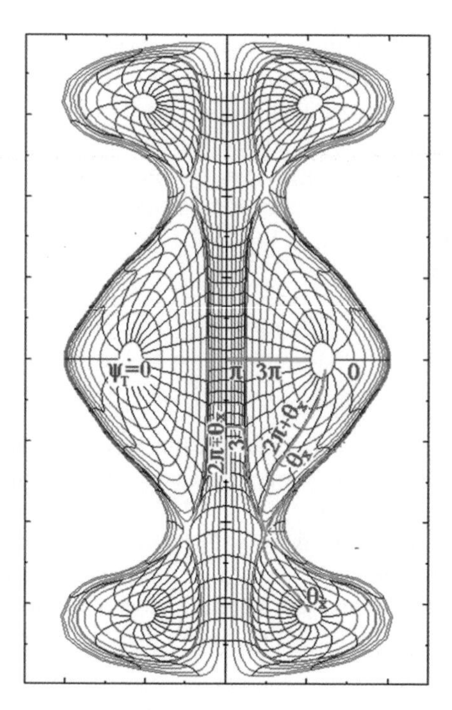

Fig. 10.56 Conceptual magnetic configuration of a third-generation fusion reactor (http://www.frascati.enea.it/ProtoSphera/ProtoSphera%202001/3.%20Chandrasekhar-Kendall-Furth.htm.)

Notes

1. The Latin plural is, of course, *tori*; but we use *toruses* here so as not to confuse *tori* with *torii*.
2. http://www.toodlepip.com/tokamak/spherical-tokamaks.htm
3. Energy, of course, has no direction; it is velocity that is perpendicular to the B-field. In this section, we use this loose term, which might be easier for a nonscientist to understand.
4. RIKEN standes for Rikagaku Kenkyusho, a private research foundation in Japan. RAL stands for the Rutherford-Appleton Laboratory in England.

References

1. H.-S. Bosch, G.M. Hale, Nucl. Fusion **32**, 611 (1992). for deuterium fusion
2. W.M. Nevins, R. Swain, Nucl. Fusion **40**, 865 (2000). for boron fusion
3. L.J. Wittenberg, J.F. Santarius, G.L. Kulcinski, *Lunar source of ^{3}He for commercial fusion power*. Fusion Technol. **10**, 167 (1986)
4. J.F. Santarius, *Role of Advanced-Fuel and Innovative Concept Fusion in the Nuclear Renaissance*, 48th Annual Meeting of the Division of Plasma Physics, Philadelphia, PA, Oct. 31, 2006 (Bull. Amer. Phys. Soc. Abstract No. BAPS.2006.DPP.JM2.4)
5. F. Najmabadi, R.W. Conn, et al. *The ARIES-III D-He^3 Tokamak Reactor Study*, 14th Symposium on Fusion Engineering, San Diego, CA, 1991 (IEEE No. 91CH3035-3, p. 213) (IEEE, Piscataway, 1992)

6. E. Fermi, *Nuclear Physics*, Notes by J. Orear, A.H. Rosenfeld, R.A. Schluter (University of Chicago Press, 1950), p. 152
7. R. Feldbacher, *The AEP Barnbook* (Alternate Energy Physics Program, Institute for Theory of Physics, Graz, Austria, 1987). Published by the International Atomic Energy Agency, Nuclear Data Section
8. H.-S. Bosch (Max-Planck Institute), *Construction of Wendelstein 7-X: Engineering a Steady State Stellarator*. 23rd Symposium on Fusion Engineering, San Diego, CA 2009
9. J.B. Taylor, Phys. Rev. Lett. **33**, 1139 (1974)
10. J.F. Lyon (Oak Ridge National Laboratory), *The World Stellarator Program*, Fusion Power Associates Symposium, Washington, DC, 2006
11. A.H. Boozer, Plasma Phys. Control Fusion **50**, 124005 (2008)
12. H. Neilson (Princeton Plasma Physics Laboratory), *The Promise and Status of Compact Stellarators*, Fusion Power Associates Symposium, Gaithersburg, MD, 2004
13. F. Najmabadi and the ARIES Team (University of California, San Diego), *ARIES-CS Compact Stellarator Study*, Report UCSD-CER-06-05 (2006).
14. A.A. Ivanov, Paper EX/P5-43, 22th IAEA fusion energy conference, Geneva, Switzerland, 2008
15. A. Sykes (Culham), *The Development of the Spherical Tokamak*. International Conference on Plasma Physics, Fukuoka, Japan, 2008
16. Y.-K.M. Peng, D.J. Strickler, Nucl. Fusion **26**, 769 (1986)
17. J.T. Slough et al., Phys. Plasmas **2**, 2286 (1995)
18. P.M. Bellan, *Spheromaks* (Imperial College Press, London, UK, 2000)
19. S. Woodruff (University of California, Berkeley), *Alternative Pathways to Fusion Energy*, Fusion Power Associates Meeting, Washington, DC, 2006
20. H. Alfvén, L. Lindberg, P. Mitlid, J. Nucl. Energy Part C Plasma Phys. **1**, 116 (1959)
21. M.S. Ioffe, J. Nucl. Energy, Part C Plasma Phys. **7**, 501 (1965)
22. M.C. Myers et al., Nucl. Fusion **44**, S247–S253 (2004)
23. A Livermore drawing. See, for instance, Richard F. Post, *Thoughts on Fusion Energy Development*, Fusion Power Associates Meeting, Livermore, CA, December 2008.
24. K. Yatsu et al., Nucl. Fusion **43**, 358–361 (2003)
25. T. Cho et al., Paper EX/9-6Rd, 20th IAEA Fusion Energy Conference, Vilamoura, Portugal, 2004
26. W. Horton et al., *Axisymmetric tandem mirror D-T neutron source* (2008), http://burning-plasma.org/web/ReNeW/whitepapers/5-24%20horton_renew_whitepaper.pdf
27. T. Simonen et al., *The Status of Research Regarding Magnetic Mirrors as a Fusion Neutron Source or Power Plant*, Summary of workshop held in Berkeley, CA, September 8–9, 2008.
28. K. Ishida (RIKEN), *Muon catalyzed fusion, recent progress and future plan*, International Workshop on Neutrino Factories and Superbeams, Irvine, CA, 2006.
29. FY 2010 estimates: Phys. Today, April 2010
30. J. Sarff (University of Wisconsin), *Physics Progress of Reversed Field Pinch Magnetic Confinement*, American Physics Society Division of Plasma Physics Meeting, Atlanta, GA, 2009
31. R. Lorenzini et al. (Padua), Nat. Phys. Lett. (online), June 14, 2009
32. H.Y. Guo et al., Phys. Plasmas **14**, 112502 (2007)
33. R.D. Milroy et al. (Redmond), *FRC Formation and Sustainment with RMF Current Drive*, American Physics Society Division of Plasma Physics Meeting, Atlanta, GA, 2009
34. J.F. Santarius et al., *Field-Reversed Configuration Power Plant Critical Issues*, University of Wisconsin Report UWFDM-1084 (1998).
35. N. Rostoker, A. Qerushi, Phys. Plasmas **9**, 3057 (2002). 3068
36. R.B. Spielman et al., Phys. Plasmas **5**, 2105 (1998)
37. C. Deeney et al., Phys. Rev. Lett. **81**, 4883 (1998)
38. F. Suzuki-Vidal et al., IEEE Trans. Plasma Sci. **38**(Part 1), 581 (2010)
39. V.A. Gribkov et al., Physica Scripta **81**, 035502 (2010)
40. D.S. Clark et al., Phys. Plasmas **17**, 952703 (2010)

41. F.F. Chen, *Introduction to Plasma Physics and Controlled Fusion*, 2nd ed., vol. 1: "Plasma Physics" (Plenum, New York, 1984), p. 309ff.
42. J.D. Sethian et al., IEEE Trans. Plasma Sci. **38**, 690 (2010)
43. J.J. Duderstadt, G.A. Moses, *Inertial Confinement Fusion* (Wiley, New York, 1982)
44. D. Clark et al. (LLNL), *Indirect Drive Fast Ignition Target Designs for the National Ignition Facility*, FESAC Subpanel Workshop, Washington, DC, August 2008.
45. D.T. Goodin et al., Nucl. Fusion **44**, S254 (2004)
46. R.B. Spielman et al., Plasma Phys. Control Fusion **42**, B157 (2000)
47. F. Najmabadi et al., Fusion Sci. Technol. **46**, 401 (2004)
48. A.R. Raffray et al., Fusion Sci. Technol. **46**, 417 (2004)
49. S.E. Jones, Nature **321**, 127 (1986)
50. E. Coleman, *Greek Fire: Nicholas Christofilos and the Astron Project in America's Fusion Program*, w3.pppl.gov/post/docs/coleman.pdf
51. R.L. Hirsch, J. Appl. Phys. **38**, 4522 (1067)

Chapter 11
Conclusions

Ah, but a man's reach should exceed his grasp, or what's a heaven for?

Never have these oft-quoted words by Robert Browning been more pertinent. The very existence of man depends on his ability to get energy when nature's bounty runs out. We may not succeed in creating our own Promethean fire, but it's within reach.

Fusion is a solution to both climate change and energy shortage. Fusion energy is inexhaustible and nonpolluting.

Fusion will cure our dependence on oil. There will be no need to wage wars in the Middle East. With unlimited energy, there will be electricity or hydrogen to run cars.

With unlimited energy, desalination can provide fresh water in all coastal regions.

Fusion cannot explode or be proliferated.

Fusion does not need to disturb the environment or wildlife habitats. Reactors can be located on the sites of aging coal or nuclear plants. In particular, they can be located near population centers. No new cross-country transmission lines need to be built urgently.

Fusion is the only energy source that can sustain mankind for future centuries and millennia. The sooner we get it, the less we need to spend on temporary solutions.

Scientific Summary

In Chap. 1, we summarized the scientific evidence for global warming caused by carbon dioxide emitted by human activity, especially the burning of fossil fuels. In Chap. 2, we summarized the known facts on fossil-fuel reserves, especially the critical shortage of oil. We showed the difficulty of and dangers in extracting the last reserves as well as the expense in sequestering the greenhouse gases emitted in their use. In Chap. 3, we surveyed alternative energy sources and found that none of them, except nuclear energy, can provide dependable backbone power, although many are suitable as supplementary power sources.

F.F. Chen, *An Indispensable Truth: How Fusion Power Can Save the Planet*,
DOI 10.1007/978-1-4419-7820-2_11,

In Chaps. 3–5, we introduced the concept of fusion power and explained why a magnetic bottle holding a hot plasma is needed to fuse hydrogen into helium to get energy from water. In Chaps. 7 and 8, we explained the physics of plasma containment in a device called a tokamak and summarized all the difficult problems that have already been solved. In Chap. 9, we gave details on all the extremely difficult engineering problems that have yet to be solved. Finally, in Chap. 10, we showed other ways to achieve fusion power which have not yet been explored extensively but which may make better reactors than the tokamak.

Cost of Developing Fusion

Financial Data

The benefits of fusion will not come cheaply, but the cost is smaller than that of other projects that the USA has undertaken with success. Figure 11.1 compares the costs of the Manhattan Project, the Apollo Program, and the Iraq and Afghanistan wars (up to 2010) with the projected cost of developing a fusion reactor. In constant 2010 US dollars, the Manhattan Project cost $22.6B, and the longer Apollo program cost $100.8B.[1] Other estimates are twice as high.[2] The two current wars have

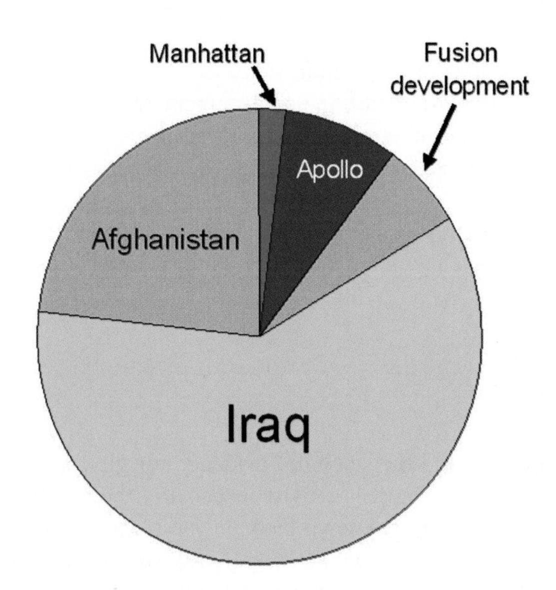

Fig. 11.1 Comparative costs of the Manhattan Project, the Apollo Program, the Afghanistan and Iraq wars, and the conjectural cost of development of fusion reactors. All costs are normalized to 2010 US dollars

cost \$732B and \$282B, respectively, so far.[3] The cost of developing fusion is a highly conjectural estimate. The cost of ITER, originally set at €5B (\$6.3B), has risen to €16B (\$21B).[4] Engineering research will require fusion development facilities (FDFs). These have not been costed out, but one design is 45% the linear size of ITER, and the cost rises as the size squared. With the higher projected cost of ITER, this would make an FDF cost about \$4.2B. Perhaps three of them would be required for a total of \$12.6B. The DEMO would cost at least twice as much as ITER or \$42B. The total is \$75B, less than that of the Apollo program, which did not solve any pressing problems. After DEMO has been run successfully, further development would be turned over to private industry, and federal support would no longer be needed. The fusion cost given here is a guess, but it is clear that the USA has the resources to develop fusion without outside help. It is only a matter of priorities. Jack Kennedy showed that it can be done.

Figure 11.2 gives a breakdown of the \$5.1B FY 2011 budget request of the Department of Energy's Office of Science.[5] Fusion Energy Sciences is the item that supports magnetic fusion research. It is the smallest item there. Basic Energy Sciences is deservedly the largest item because it supports current renewable energies like wind and solar. High-energy physics traditionally has a large budget because it is the community that gave us the hydrogen bomb to win WWII. It still has a large budget for accelerators and experiments that can improve our knowledge of the structure of matter. This is the forefront of science, but mankind may or may not need to know this to survive.

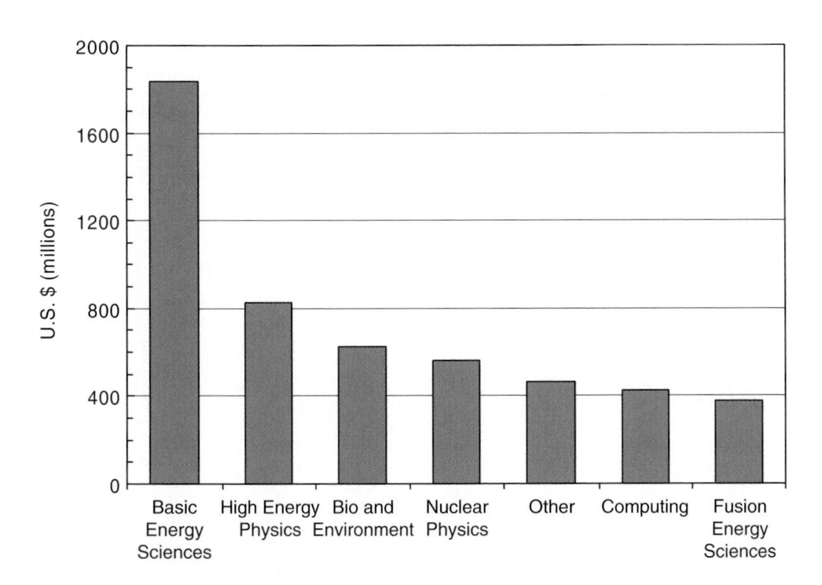

Fig. 11.2 Support for different divisions in the US Office of Science

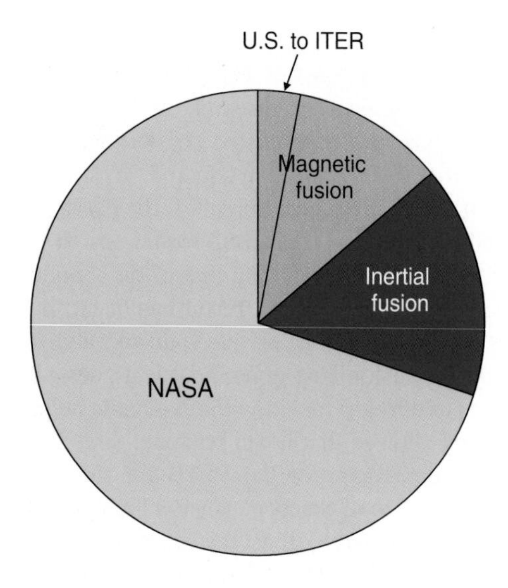

Fig. 11.3 Comparison of the annual budgets of the space and fusion programs in the US

Figure 11.3 compares the annual budgets in the USA for magnetic and inertial confinement fusion research with the \$1.9B for the NASA space program. The magnetic fusion budget includes a paltry \$80M contribution to ITER, equivalent to *four hours* of expenditures in the Iraq war. Exploration of the solar system (NASA) and study of the behavior of matter under extreme conditions (ICF) are exciting extensions of modern knowledge which scientists are happy to have funded because of their importance to national security. These programs, however, contribute little to the solution of environmental and energy issues. We are spending more money looking for the Higgs boson than for a solution to global warming and oil shortage. Re-examination of priorities is in order.

Figure 11.4 shows the cost of the ITER experiment, including construction but not operation. The expense is shared by seven nations. It is the first giant step toward fusion power. Compared with this is the amount spent by the USA alone to wage the war in Iraq for *one month*.[6] The graph speaks for itself. The USA could easily have taken this step alone had it not been so dependent on Mid-Eastern oil.

Conclusion

- Developing fusion power will cost less than putting a man on the moon. The Manhattan and Apollo programs have shown that the scientific and engineering communities have the ingenuity to achieve almost unimaginable goals once it is driven by national priorities, a sense of urgency, personal challenge, and a sense of national pride. With seven nations having banded together to push forward on fusion, the USA has lost its chance to do this alone. However, we are still far

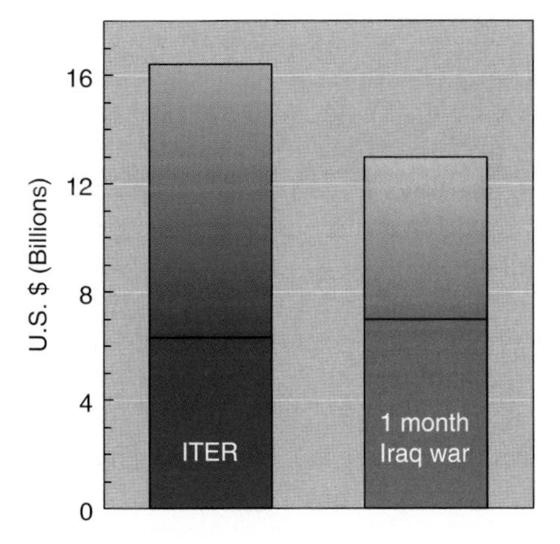

Fig. 11.4 Comparison of the cost of ITER with that spent on the war in Iraq in one month. For ITER, the *lower* part is the original budget, and the *upper* part the added cost in the revised estimate. For Iraq, the line divides the minimum and maximum estimates of the cost.[6] The maximum includes the costs of occupation and repatriation of troops

from the goal because the most difficult problems of materials engineering have yet to be solved. The USA can regain its former leadership in fusion research by building one or more large FDFs to solve these problems simultaneously with ITER to shorten the time to a working reactor.

- The development of wind and solar power in private industry has stimulated the economy. Fusion machines are big and must be funded by the government, but the economic stimulus can also be generated by the subcontracts awarded to small companies. For instance, such components as superconducting strands, silicon carbide tiles, blanket modules, RF antennas, and even 3D computations can all be parceled out to start-up companies. New jobs will be created, and new financing will be secured.
- *A high-priority Apollo-like program to put fusion on a fast track will cost less than Apollo did and will solve the* CO_2 *problem, the fossil-fuel shortage problem, and the oil dependence problem all at once.*

Epilogue

Research on space science, astronomy, and high-energy particles has produced incredibly detailed knowledge of our environment on both macroscopic and microscopic scales. It has been a long journey for *Homo Sapiens* to have evolved from simple food gathering to these intellectual heights. This knowledge, however, will be of little comfort if we cannot find the means to assure the preservation of our species.

We have benefited from the discoveries made by adventurers driven by the urge to explore the unknown and to reach the inaccessible, even at great risk or expense. Magellan, Columbus, Roald Amundsen, Edmund Hillary, Roger Bannister,

Neil Armstrong... One climbs Mt. Everest because it is there. To shrink from pursuing the goal of unlimited energy borders on cowardice.

We close on a philosophical note. We have been incredibly lucky. Our planet settled at just the right distance from the sun so that H_2O, a very stable molecule, is in liquid form most of the time, forming the basis for life. As plant life lived and died, its fossils lay buried for millennia as human life developed. This legacy of fossil energy allowed humans to form a civilization and develop brains that could think abstractly and explore our surroundings and the whole universe. Our intellectual capacity grew to such an extent that we could design and make computers that someday can do the thinking for us. The energy source that allowed all this to happen will soon be depleted; but, luckily, we now have the smarts to create our own energy source. But do we have the wisdom to actually do it?

Notes

1. D.D. Stine, *The Manhattan project, the Apollo program, and federal energy technology R&D programs: a comparative analysis*, Congressional Research Service RL34645 (2009).
2. Physics World, July 2007.
3. http://www.costofwar.com.
4. IEEE Spectrum, September 2010. Some say that it might be as high as $20–25B, but this still puts the cost of fusion below that of the Apollo program.
5. Request to Congress as of February 1, 2010.
6. Congressional Budget Office per http://www.usgovinfo.com/library/weekly/aairaqwarcost.htm.

Index

This index serves also as a Glossary. Pages shown in **boldface** contain also a definition of the item.

F.F. Chen, *An Indispensable Truth: How Fusion Power Can Save the Planet*, DOI 10.1007/978-1-4419-7820-2,

Printed in the United States of America